PRINCIPLES OF
Statistics

PRINCIPLES OF
Statistics

Paul A. Herzberg

York University
Toronto, Ontario

ROBERT E. KRIEGER PUBLISHING COMPANY
MALABAR, FLORIDA
1989

Original Edition 1983
Reprint Edition 1989 with corrections

Printed and Published by
ROBERT E. KRIEGER PUBLISHING CO., INC.
KRIEGER DRIVE
MALABAR, FLORIDA 32950

Library of Congress Cataloging-in-Publication Data

Herzberg, Paul A., 1936-
    Principles of statistics/Paul A. Herzberg.
       p.   cm.
    Reprint. Originally published: New York: Wiley, 1983.
    Includes bibliographical references and index.
    ISBN 0-89464-374-6 (alk. paper)
    1. Mathematical statistics. I. Title.
  [QA276.H485 1989]
519.5--dc19
                                          89-2542
                                          CIP

10   9   8   7   6   5   4   3   2

To the memory of
my mother, Luise Herzberg
and my cousin, Peter G. Thurnauer.

Preface to the Reprint Edition

I am very pleased that Krieger is publishing my book. Students and teachers have appreciated the features summarized in the orginal preface. In particular, *Principles of Statistics* emphasizes a number of themes, including:

- Sketches and diagrams are an aid to problem-solving. (E.g., tree diagrams in Chapters A and 5; sketches of the normal distribution in Chapter 7; sketches of the normal and binomial distributions to solve hypothesis-testing problems in Chapter 9 and later.)
- The mean is the balance point and the standard deviation is approximately half the 68% range of all distributions, whether empirical or theoretical. (E.g., Chapters 4, 6, 7, 9.)
- Elementary statistics is, to a large extent, the study of the relationship of two variables. (E.g., Chapters 8, 11, 14, 15, 16, 17, D.)
- The distinction between a score (or interval of scores) on the one hand and an area (probability, relative frequency) on the other is essential for understanding distributions and using statistical tables. (Almost every chapter.)
- The distinction between measures of the size of the relationship and tests of the existence of a relationship helps resolve some of the paradoxes of hypothesis testing and also helps summarize dozens of formulas. (E.g., Chapters 8, 11, 13, C, 16, 17, D.)

I have taken the opportunity, in this Krieger edition, to correct errors which I, my students, and others have discovered. The *Study Guide*, described in the following "Preface" and "Teaching Package", is available from Krieger.

Paul A. Herzberg
Department of Psychology
York University
North York, Ontario   M3J 1P3
Canada

Preface

This book is an introduction to statistics. It describes the basic theoretical ideas of statistics and shows how statistics is applied to research studies. The mathematical skills required are minimal. Lengthy written explanations of the basic ideas are given instead of using mathematics. (The actual mathematical skills required are described and reviewed in Chapter 1 of the *Study Guide*. The practice quiz found there can be used to determine if a student's skills are adequate for a course based on this book.)

This is, therefore, not a cookbook. It is intended to convey a solid understanding of basic statistical theory and to help students carry out a variety of statistical procedures. Enough topics are included to fit a variety of course outlines but the book is organized so that many chapters may be omitted. (See below.)

In addition to covering basic theory and applications, the book has three main features:

- It emphasizes principles rather than rules.
- It provides pedagogical approaches and aids.
- Together with the associated study guide and the regular quizzes, it forms an integrated and thoroughly tested set of teaching materials.

The book is called *Principles of Statistics* because it emphasizes principles rather than rules. Many students find statistics to be a very difficult subject and look to a set of rules (and formulas) that will help them. Teachers, on the other hand, usually believe that the best long-term strategy for learning statistics is to learn a set of general principles. Hence, in their lectures, they emphasize understanding and comprehension of particular topics rather than simply following various rules and formulas. Most of the standard formulas and rules of statistics are given, however, and I have tried in several ways to help readers understand them: for example, 14 general ideas are labeled *principles*, in the hope that students and their instructors will focus on them.

Other pedagogical approaches and aids are provided:

1 The use of sketches and diagrams for problem solving.
2 Techniques to estimate the approximate value of a statistic or the result of a calculation.
3 Emphasis on the study of relationships between variables by means of the regression function.
4 Four review chapters that serve to integrate preceding material.
5 A comprehensive and detailed treatment of hypothesis testing.
6 Statistical tables designed to facilitate student use.

Shortly after I began to write the manuscript, I persuaded Elke Weber to write the study guide and quiz questions for it. The writing of such supplementary materials is often left until the manuscript is completed, but we both felt that the supplements were too important to be written in such haste. Instead Ms. Weber's study guide and quiz questions have been written in parallel with the writing of the manuscript. The full set (text, study guide, and quiz questions) has been class tested and revised over several years and is an unmatched teaching package for instructors of statistics.

Notation. I have generally adhered to standard notation but instructors should be aware of a few choices I made on pedagogic grounds.

(a) I begin with the *unadjusted* sample standard deviation (s') that has divisor n; later I introduce the *adjusted* sample standard deviation (s) that has divisor $(n - 1)$. s' is used in all correlation and regression formulas in order that, for example, the Pearson correlation formula has n and not $(n - 1)$ in the denominator. s is used for all tests and estimates of means.

(b) The basic variable, beginning in Chapter 2, is called y (not x), and all formulas for means, standard deviations, and so on are expressed in terms of y. Many elementary texts use x when only one variable is discussed but have to switch awkwardly to y for the dependent variable when two variables are discussed (using x for the independent variable). Since I emphasize that statistics is a study of relationships between variables and since many procedures require the calculations of means and standard deviations of the *dependent* variable, I believe it is better to work with y, rather than x, from the beginning.

(c) χ^2_{obs} is used for the test statistic in chi-square tests in parallel to $z_{obs}, t_{obs}, F_{obs}$. I would have preferred to use X^2 for this statistic since most modern, advanced treatments of the analysis of categorical data use X^2. But on the advice of several reviewers, I have used χ^2_{obs}.

Terminology. As with notation, standard terminology is kept unless there were pedagogic reasons to do otherwise.

(a) The term "univariate scatterplot" refers to a graph which statisticians often use but for which there does not seem to be a name. (See, for example, Figure 2.7.)

(b) Although I strongly stress in Section 10.2 the distinction between failing to reject the null hypothesis (H_0) and accepting it, I first introduce acceptance and rejection regions in Section 9.1 by the slightly unusual names of "acceptance region for H_0" and "acceptance region for H_1".

(c) Types of variables (discrete, continuous, numerical, categorical, etc.) are discussed since these distinctions are very helpful for students in learning how to choose the appropriate procedure. However, a discussion of levels of measurement (ordinal, interval, ratio, etc.) is omitted since I do not believe that the level of measurement determines the statistical procedure.

(d) I prefer the phrase "distribution of the statistic" to "sampling distribution" because students have difficulty discriminating between the latter term and "sample distribution." For particular statistics I use phrases such as "distribution of the sample means" and "distribution of number of successes".

Organization. This text provides enough material to meet many teaching needs. It is organized so that certain chapters and sections can be omitted according to the goals of the course. The basic *core* of the book consists of the following chapters: 1, 2, 3, 4, 6, 7 (except Section 7.5, Example 3), 8, 9 (except Sections 9.2 and 9.4), and 10. Sections 6.4, 7.6, and B.2 provide a brief introduction to statistical inference, principally to motivate the discussion of distributions in these chapters. These sections could be omitted, since Chapters 9, 10, and 12 are self-contained discussions of hypothesis testing and of estimation. Several *continuations* can then be selected more or less independently. Tests and estimates of one and two means are covered in Chapters 11, 12, and 13. Procedures for dichotomous, categorical, and rank variables are covered in Chapters 14 and 15. (A selection from Chapters 14 and 15 would allow only chi-square tests to be covered.) Correlation and regression are introduced in Chapter 8 (included in the core) and are more extensively covered in Chapter 16. Analysis of variance is covered in Chapter 17, which depends on Section 16.3 and some of the earlier material on tests and estimates of one and two means. Probability theory, beyond the bare minimum introduced in the core, is covered in Chapters 5 and 18. Finally, the *review chapters* (A, B, C, D) integrate and synthesize the preceding material.

Acknowledgments. I am grateful to two former students of mine, Margo Cronyn and Elke Weber, who were helpful and cogent critics of the entire manuscript. Elke also worked closely with me while she wrote the study guide and the quiz questions. Grant Austin helped me greatly during the last stages of manuscript preparation. I have been assisted by comments from many teaching assistants and students in my courses as well as the following professional reviewers: Nancy S. Anderson, University of Maryland; David Alan Bozak, SUNY–Oswego; Curtis W. Carlson, Marian College; Donald A. Cook, Competence Assistance Systems; Virginia Falkenberg, Eastern Kentucky University; Marvin J. Homzie, University of Virginia; Irwin J. Jankovic, Loyola Marymount University. Their contributions have been invaluable. Some of the ideas in this book were first tried out in a statistics course that Ron Sheese and I shared for several years. My dean, Harold Kaplan, has been extraordinarily supportive of my work and of Elke Weber's work on the study guide and quizzes. I am also grateful to Doris Rippington and Nadia La Penna, who typed and retyped many versions of the manuscript. My greatest debt is to my wife, Louise, who has supported me unfailingly.

Paul A. Herzberg

Teaching Package

The supplements for *Principles of Statistics* were written at the same time as the text and are uniquely integrated with it. They include features not available in the supplements of other statistics textbooks.

Each chapter of the *Study Guide*, written by Elke U. Weber, has three parts:

1 Objectives. Each of the 20 to 30 behavioral objectives is of the form: "after reading the text you should be able to . . . ." The objectives are categorized by cognitive level (knowledge, comprehension, and application) and by major content area in the chapter.

2 Tips and Reminders. This section addresses topics that students find problematic or difficult as shown by Ms. Weber's analysis of recurrent mistakes made by 300 students.

3 Practice Quiz. Each of the 10 multiple-choice questions is based on one or two of the chapter's objectives. The questions usually test comprehension or application objectives in contrast to most multiple-choice questions, which usually test knowledge objectives. The wrong answers for each question were chosen on a rational basis and the reason for each wrong answer is included in the study guide.

The *Instructor's Manual* includes several regular quizzes for each text chapter (except review chapters). These quizzes are identical in format and parallel in content to the practice quiz in the study quide. The manual also analyzes these quizzes in detail: the objective(s) on which each question is based, the parallel question in the practice quiz, the reason for each wrong answer, and the proportions of students choosing each wrong answer in the last year of class testing.

Contents

xiii

XVI Contents

PRINCIPLES OF
Statistics

1 THE STUDY OF STATISTICS

Statistics is a set of techniques for analyzing numbers. This chapter offers four examples that will help you to understand where such numbers come from, why it is important to analyze these numbers, and some things you need to do in order to learn the techniques of statistics.

Statistical methods are widely used in industry, business, and government, as well as in the physical sciences, the agricultural sciences, and, perhaps most extensively, the social sciences. Because the author is a psychologist, many of the examples in this book are taken from the social sciences in general, and from psychology in particular. However, the techniques and the principles are general and may be applied to the other areas in which statistics is used.

1.1 Four Statistical Examples

A common way to study a set of numbers is to represent the numbers pictorially: that is, in a graph. After reading this book, you should be able to interpret a wide variety of graphs, some of which are included in the following examples. Do not be surprised if you do not fully understand these graphs now; they are included here only to give you a sample of the data and graphs with which statistics deals. These four examples are discussed again in Section D.3, at the end of the book.

Example 1— The Visual Cliff

A classic study of an infant's perceptual abilities is the visual cliff experiment. An infant is placed near an edge, beyond which apparently is a drop of several feet. In fact, a horizontal piece of transparent plastic extends beyond the edge to prevent the child from falling. Some infants refuse to crawl beyond the edge. Others crawl beyond the edge (onto the plastic), apparently oblivious of the cliff.

Consider a study in which each infant is tested twice, with the mother nearby. In one condition, the mother is standing near the deep end: that is, the infant has to cross over the edge and crawl along the plastic to get to the mother. In the other condition, the mother is standing near the shallow end: that is, the infant has to crawl away from the edge and does not have to crawl over the plastic. Is the infant's perception of depth sufficiently developed so that he refuses to cross the edge?

Suppose that of 29 infants tested, only four cross over the edge to the mother standing near the deep end, whereas 25 crawl to the mother when she is standing near the shallow end. These numbers are represented in the graph in Figure 1.1. Note that the graph vividly shows the great difference in the number of infants who crawl toward their mothers in the two conditions.

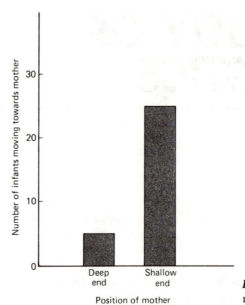

Figure 1.1 The visual cliff experiment.

**Example 2—
Religiosity and
Church
Attendance**

Suppose that you want to see if attendance at church is related to strength of religious belief (religiosity). You interview 200 persons and rate each of them on a four-point scale ranging from 1 (no religious belief) to 4 (very strong religious belief). You also inquire about their frequency of attendance at church. Each person is classified as never attending, sometimes attending, or attending at least once per week. The numbers of persons falling into each of the 12 combined categories of religiosity and attendance are shown in Table 1.1. For example, there are 31 persons who are of moderate religiosity and who attend church rarely.

It is much more difficult to comprehend the data in Table 1.1 than it is to understand the few numbers reported for the visual cliff experiment in Example 1. It is also more difficult to draw a graph of these data, and such a graph is not shown here. Nevertheless, later chapters will discuss ways to interpret data such as these. (In fact, these data show *no* evidence for a relationship between religiosity and church attendance.)

**Example 3—
A Learning
Study**

Many studies in psychology focus on the rate at which animals or humans learn. Usually, such experiments focus on several conditions that are believed to affect the rate of learning. Consider, for example, a study with two conditions: in one, the subjects are given a reward for every correct response; in the other, the subjects are rewarded for only 50% of the trials in which they make a correct response. The first condition may be

TABLE 1.1 Religiosity and Church Attendance

| | | RELIGIOSITY | | | |
		NONE	WEAK	MODERATE	STRONG
	NEVER	11	7	10	3
ATTENDANCE	RARELY	32	20	31	9
	REGULAR	27	16	26	8

TABLE 1.2 Number of Correct Responses

100% REWARD CONDITION	50% REWARD CONDITION
8	21
18	15
12	20
10	11
12	13
9	14
11	11
11	16
15	10
8	19
Average = 11.4	Average = 15.0

called the 100% reward condition and the second condition the 50% reward condition. Each subject is studied for 30 trials. The number of trials on which the subject makes the correct response is a measure of the rate of learning.

Suppose that there are 10 subjects in each of the two conditions and that their scores are those shown in Table 1.2. The first step in analyzing such data is to compute the averages of the scores in the two conditions. These averages are shown at the bottom of each column of numbers in the table, and also in Figure 1.2. We see that subjects in the 100% reward condition made an average of 11.4 correct responses in 30 trials. whereas subjects in the 50% reward condition made an average of 15.0 correct responses in 30 trials. It appears, therefore, that subjects did better in the 50% reward condition than in the 100% reward condition. As later chapters will make clear, however, such a conclusion can properly be made only after further analysis of the data.

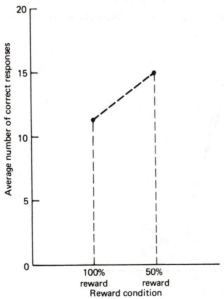

Figure 1.2 A learning study.

**Example 4—
Blood Groups**

The four major blood groups have the following relative frequencies in the general U.S. population.

O: 45%; A: 41%; B: 10%; AB: 4%

Suppose that a blood distribution clinic is located in an area that contains a large immigrant population. It is decided to determine whether the relative frequencies of the four blood groups in that special area differ from those in the general population. In a sample of 500 persons from that area, the relative frequencies are

O: 39%; A: 47%; B: 3%; AB: 11%

These two sets of relative frequencies (or percentages) are shown in Figure 1.3.

A typical question that statistics attempts to answer is, Do these data indicate that there is an actual difference between persons in the special area and the general U.S. population? It would appear from the figure that there is a difference, but statistical techniques, including some discussed later, are required to answer this question conclusively.

1.2 Learning Statistics

This text provides information about many statistical techniques. But you need more than this information in order to learn statistics. You must become familiar with numbers and graphs and be able to use them with ease. One way to improve your skills is to try to understand the numbers and graphs in books and magazines. As you progress through this text, your ability to understand them should increase. You will also find that most graphs are summaries of a great deal of information. If you understand a printed graph, you may be able to skip over several paragraphs in a magazine article or in a book.

Another reason to look for numbers and graphs in your reading is that you will see the relevance of statistics to subjects in which you are interested. Most technical and scientific writing includes many numbers, graphs, and statistical techniques. Your interest in these subjects should help increase your motivation to learn statistics.

A difficulty in learning statistics is that you may get lost in details and fail to appreciate the important ideas underlying those details. This book has a

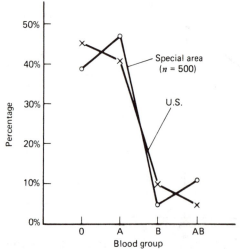

Figure 1.3 Blood groups.

feature that may help you in this regard. Some of the most important ideas in statistics are described in this book as *principles* and labeled P1, P2, etc., for emphasis. These principles, which on the whole are quite general and non-numeric, comprise the foundation of statistical practice. It is much more important to remember these principles than to remember the details of many statistical procedures.

Finally, statistics is a mathematical subject. If you are in doubt about your basic mathematical skills, you should test and, if necessary, improve those skills soon. Chapter 1 of the Study Guide for this book contains a review of basic mathematics and also a practice quiz.

2 DISTRIBUTIONS

This chapter describes a number of basic techniques for summarizing data. A set of data is difficult to study unless its essentials are made obvious. One basic technique of clarification is to construct a distribution of the data. Several kinds of distributions are described below.

2.1 Frequency Distribution Tables

You are teaching arithmetic to an elementary school class, and you want to study how fast the students can do a particular addition problem. The students come to your desk in turn, and you time them as they do the problem. Their times, to the nearest second, are given in Table 2.1.

The table lists 39 numbers. What can we make of these numbers? The list is useful only if we want to know a particular student's time; for that purpose, an alphabetical list would even be better. But we really do not want the individual times. We want to know, in general, how fast the students can do the problem. Perhaps we want to compare the performance of the class on this problem with their performance on other problems. In short, we are looking for ways to summarize the data in the table.

What is the shortest (fastest) time? Several students completed the problem in 9 seconds. What is the longest (slowest) time? One student took 16 seconds. These questions give us one useful description of the data: "The times range from 9 to 16 seconds".

These are the extreme times, but what about the students in the middle? What can we say about their times to give us knowledge of the class without having to look at each person's time? We want to see how the students' times are spread, or *distributed*, over the values ranging from 9 to 16 seconds. How many students took 9 seconds? 10 seconds? 11 seconds? and so on.

We therefore want to construct a table that lists the possible times (9, 10, . . . 16 seconds) together with the number of students associated with each time. Such a table is called a *frequency distribution table*. (See Table 2.2.) The frequencies in the last column are determined from the column of tallies, which is compiled by recording a tally (or mark) for each number in the original list of data (Table 2.1). (The tallies are usually omitted from frequency distribution tables, since they are no longer needed once the table has been constructed.) The table is called a distribution table because it shows how the students' times are distributed over the range of values from 9 to 16 seconds.

We see from the table that three students took 9 seconds, nine students took 10 seconds, etc. Once the distribution table has been constructed, we probably need not refer again to the original list of times in Table 2.1. The distribution table shows us that most students took 10, 11, or 12 seconds; that no student took 14 seconds; that two students took 15 seconds, etc. The fact that you can see these things at a glance, instead of searching for them, is the advantage of constructing a frequency distribution table such as Table 2.2.

TABLE 2.1 Time Taken to Complete an Addition Problem

STUDENT	TIME (SECONDS)	STUDENT	TIME (SECONDS)
Gillian	11	Kathy	10
Deborah	11	Michael	13
Susan	15	Heather	11
Raffaela	10	Debbie	11
Marc	10	Marvin	12
Jerry	15	Bruce	12
Eugene	12	Rena	9
Laury	12	Nyla	13
Claire	9	Charmaine	11
Paul	11	Clare	13
Linda	12	John	9
Sandi	10	Michelle	11
Ingrid	16	Karen	11
Robin	11	Alex	12
Cindy	11	Lois	11
Fern	10	Margo	11
Kathryn	12	Agnita	10
Judy	10	Ross	12
David	11	Elke	10
Jim	10		

So far, a frequency distribution has been displayed only in the form of a table. Other ways to display a frequency distribution are described in the rest of the chapter.

2.2 Frequency Distribution Graphs

Frequency distributions show all the scores (different numbers) in a set of data, together with the frequency of each score. Since pictures and drawings usually have more impact than do tables, it is useful to convert a frequency distribution table into a *frequency distribution graph*. Such a graph allows us to compare the frequencies and to make general statements about the frequency distribution.

A frequency distribution graph is easy to draw once the frequency distribution table has been constructed. Figure 2.1 shows the graph of the addition times introduced in the previous section. Note that the possible scores (times) are given on the horizontal axis of the graph; this axis is known as the *abscissa*. The frequencies are given on the vertical axis, known as the *ordinate*. This

TABLE 2.2 Frequency Distribution Table of Addition Times

TIME (SECONDS)	TALLIES	FREQUENCY
16	|	1
15	| |	2
14		0
13	| | |	3
12	| | | | | | | |	8
11	| | | | | | | | | | | | |	13
10	| | | | | | | | |	9
9	| | |	3
		Total = 39

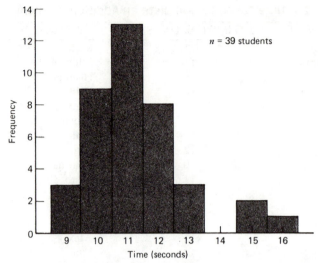

Figure 2.1 Frequency distribution graph of addition times (histogram).

particular type of graph is called a *histogram*. Each frequency is represented by a bar, or column, whose height equals the frequency. Notice that the bars in the histogram are drawn so that each score (time) is at the center of the base of its bar.

Another type of graph, called a *polygon,* is shown in Figure 2.2. Instead of drawing a bar whose height equals the frequency, you mark a point at that height above the score value and then connect the points by straight lines. Note that times 8 seconds and 17 seconds have been added to the abscissa and their frequencies recorded as zero, so that the lines in the polygon return to zero at each end of the graph.

The histogram and polygon forms of a frequency distribution include the same information. The histogram form is preferable if it is important to empha-

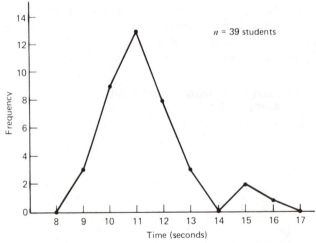

Figure 2.2 Frequency distribution graph of addition times (polygon).

size that only certain scores have frequencies. (In Figure 2.1 it is clear that a time of, for example, 9.4 seconds did not occur—only integer times were recorded—whereas in Figure 2.2 it is not so clear that 9.4 seconds did not occur, since a frequency can be read for that time.) On the other hand, the polygon form is preferable if two or more graphs are to be superimposed, because there are fewer lines in the polygon than in the histogram. (See Figure 2.5 for an example.)

Frequency and Area

In both types of frequency distribution graph the vertical distance represents frequency. The height of each bar (in a histogram) or dot (in a polygon) represents the frequency of the corresponding score on the abscissa. There is, however, another way to look at frequency distribution graphs. Consider the histogram of Figure 2.1 as redrawn in Figure 2.3. Each bar has been divided into small blocks, each of which represents one student. Think of each block as having one unit of area. Then the number of units of area in each bar, or in several bars, represents the frequency of a score, or of an interval of scores.

For example, there are 25 units of area in the bars for 9, 10, and 11 seconds, and there are 14 units of area in the remaining bars. Note that the number of students who take 11 seconds or less is almost twice the number of students who take more than 11 seconds. This observation can be made, in fact, from the original Figure 2.1, without drawing in the individual unit blocks. The graph shows us visually, without making any reference to numeric frequencies, that most times are 11 seconds or below. A powerful feature of frequency graphs, then, is that their area represents frequency. (This feature is discussed further later in the chapter.)

The polygon form of frequency distribution graph cannot be divided into units of area so readily as can the histogram form. In a rough manner, however, we can also treat the area under the polygon over any interval of scores as representing the frequency of scores in that interval. By comparing Figures 2.1

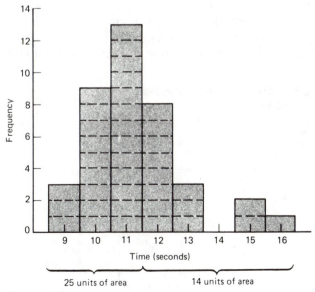

Figure 2.3 Frequency as area.

and 2.2, you should be able to see that both figures show that more students took 11 seconds or less than took more than 11 seconds.

2.3 Grouped Distributions

Data sets similar to the preceding example are uncommon, because in that example the range of observed times was small and the scores were recorded to the nearest second. Since there were only eight possible scores within the observed range (9, 10, . . . , 16), it was natural to construct the frequency table with these eight possibilities and their frequencies (Table 2.2). In most situations, however, there will be dozens or even hundreds of possible scores. We must modify our procedure in order to make a useful frequency table for such data.

Consider the data in Table 2.3, taken from a study of two classes of patients in a mental hospital. The patients are classified as either neurotic or psychotic. The scores in the table are on a *depression scale*, which is derived from a test for depression consisting of 100 questions. High scores on the scale indicate high depression, and low scores indicate low depression.

Obviously, the scores cover a large range: the neurotic scores range from 30 to 81; the psychotic scores from 43 to 83. In order to summarize the neurotic scores in the manner of Table 2.2, we would have to list all the possible scores from 30 to 81. The frequency of most of these scores would be zero; a few would have a frequency of one, and very few would have frequencies greater than one. Such a lengthy table would not be very useful, and therefore it would not be a good summary. We need to condense these scores.

It is natural to group the scores into *intervals* and then to calculate the frequency of scores falling in each interval. This procedure is illustrated in Table 2.4, in which the frequency distributions for neurotics and psychotics are shown separately. Such distributions are usually called *grouped-frequency distributions*. In the case of the neurotics, two scores (30 and 38) fall between 30 and 39, inclusive; hence, the frequency of this interval is recorded as two. There is one score (40) in the interval 40–49, seven scores in the interval 50–59, etc. The sum of the frequencies is 15—the number of neurotics—and it is customary to record this sum at the bottom of the table. The frequency distribution of the psychotics is constructed in the same way.

You may be wondering how to decide what intervals to use. Why were

TABLE 2.3 Scores on the Depression Scale

NEUROTICS		PSYCHOTICS	
53	81	60	78
67	50	78	57
68	59	68	83
30	53	69	64
40	63	43	73
61	38	75	51
54	55	64	62
58		71	69
		51	78
		69	83
n = 15 subjects		*n* = 20 subjects	

TABLE 2.4 Grouped-frequency Tables

NEUROTICS		PSYCHOTICS	
SCALE SCORE	FREQUENCY	SCALE SCORE	FREQUENCY
80–89	1	80–89	2
70–79	0	70–79	6
60–69	4	60–69	8
50–59	7	50–59	3
40–49	1	40–49	1
30–39	2	30–39	0
	15		20

intervals such as 30–39, 40–49, etc., used? Why not 30–36, 37–43, etc.? Or, 30–36, 37–46, 47–49, etc.? It is hard to make firm rules about the choice of intervals for a grouped table, but you should follow some general principles.

First, the intervals should be "nice". The intervals 30–39 and 40–49 are nice because each starts with a multiple of 10, so that all scores in the thirties are in one interval, all scores in the forties are in the next interval, and so on. It is much easier to construct the table using such nice intervals than it would be using intervals such as 30–36, 37–43, etc.

Second, the original data should be represented fairly and usefully, and the summary table should be easy to interpret. This very important general principle applies to most work in statistics. As mentioned previously, it is not very useful to record the frequencies of each possible score: 30, 31, 32, etc. A summary in this form is not easy to interpret and is scarcely an advance on the original data. The set of intervals 30–36, 37–46, 47–49, etc. suffers from a different problem: the intervals are not all the same size. If we were to calculate the frequencies of these unequal intervals, we would find it difficult to compare frequencies among the neurotics and to compare those frequencies with the frequencies of psychotics.

This observation seems to indicate a rule: intervals should all be the same size (have the same length). You should, however, learn the general principle from which the rule derives: namely, that the data should be represented fairly. The general principle is always valid but the rule may have exceptions. Consider, for example, data such as those in Table 2.1, which lists the times, in seconds, to complete a task. Suppose that five of the students took much longer than the others: 20, 79, 81, 240, and 500 seconds, respectively. How could you construct a frequency distribution with equal intervals for such data? If you try, you will either end up with a very large number of intervals (not very useful, since most intervals will have zero frequency) or you will be forced to use unequal intervals. For example, the intervals you might use are 9–10, 11–12, 13–14, 15–16, and 17–500. The last interval differs from the others but probably is the most useful way to summarize the exceptional scores. In using such intervals, you would be following the principle of making a useful summary but are violating the rule of equal intervals. Hence the advice to remember the principle so that you will be able to modify it. If you remember only a rule, you will not be able to modify it if the data demand change. Principles are adaptable; rules are not.

The intervals used in Table 2.4 supposedly are nice. But other nice intervals could be chosen, such as 30–34, 35–39, 40–44, etc. These intervals are equal (each contains five possible scores), but they are shorter than the previous intervals. Note that the intervals start on nice numbers: namely, multiples of five. But why is this choice of intervals not as good as the choice used in Table 2.4? The reason is that we would now have 12 intervals (instead of the six in Table 2.4)—too many for 15 scores, because there would be many intervals with a frequency of either zero or one. Hence, the choice made in Table 2.4 is better. In order for a frequency distribution to be a useful summary of a set of data, the number of intervals should be neither too many nor too few. In any particular case the choice of the number of intervals depends on the total number of scores. If you have 1,000 depression scores instead of only 15, then intervals such as 30–34, 35–39, 40–44, etc. should be used. But even with a thousand scores it would not be wise to use too many intervals. A frequency table is a summary, and if there are more than 10 to 20 intervals the summary loses its usefulness.

The above discussion may seem unsatisfactory, since we have not provided definite rules for constructing grouped frequency distributions. We have tried instead to illustrate how to apply a more important general principle rather than a particular rule. The general principle involves vague words such as "nice", "useful", "fairly", and "easy". Statistics is not an exact science. To summarize a set of data is partly an art. The aims of this art may be stated as a principle.

P1: Summarization Principle
Statistical summaries should
(a) Be as useful as possible.
(b) Represent the data fairly.
(c) Be easy to interpret.
When the summary is a frequency distribution, the intervals should be "nice".

The frequency distribution is the only summarization technique we have considered thus far. The summarization principle can be used to make decisions on the number of intervals to use, the width of the intervals, etc., for a given set of data that you wish to summarize. However, almost all of the statistical techniques presented in this book are techniques to summarize data, and the summarization principle can be applied to any of them.

2.4 Variables

Section 2.3 described how to make a summary of a set of data in the form of a grouped-frequency table. The natural next step is to draw a graph of such a distribution. As pointed out in Section 2.2, such graphs can be made in either of two forms, histogram or polygon. Histograms of the neurotic and psychotic groups are drawn separately in Figure 2.4. In order to compare the two distributions, it is useful to superimpose the two graphs. If we try to do this with the histogram, however, the graph will be confusing because of the many overlapping vertical and horizontal lines. A polygon is much more satisfactory. The two groups are graphed as polygons in Figure 2.5.

The graph in Figure 2.5 shows that the scores in the psychotic group tend to be greater than the scores in the neurotic group, even though there is

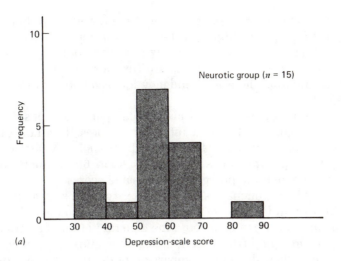

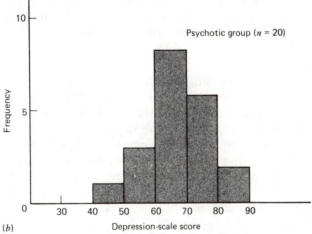

Figure 2.4 Grouped-frequency histograms. (*a*) Neurotic group (*n* = 15). (*b*) Psychotic group (*n* = 20).

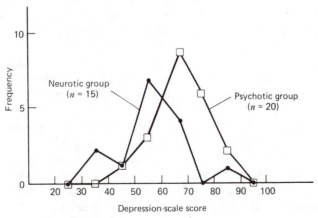

Figure 2.5 Grouped-frequency polygons.

considerable overlap of the scores. Note that the psychotic graph is above the neurotic graph for high (greater than 60) depression scores, but is generally below the neurotic graph for low (less than 60) depression scores. The combined graph clearly shows this difference in the scores of the two groups, indicating that there is a tendency for psychotics to have higher scores than neurotics.

The two polygons can also be compared by considering the area under each polygon over the same intervals of scores. Recall the discussion in Section 2.2 of the fact that areas represent frequencies. Visually, the area under the psychotic polygon for scores greater than 60 is much greater than the area under the neurotic polygon for the same scores.

But how, you may ask, should these graphs be drawn? For example, which values should be shown in the abscissa (horizontal axis)? In order to answer this question we must discuss the general notion of a variable and its values—the main topic of this section. Another question you may have is whether Figure 2.5 is really a fair comparison of the two groups, since the groups have different sizes: the neurotic group has 15 people, whereas the psychotic group has 20. This question will be discussed in Section 2.6, which presents techniques to make the groups more comparable.

Definition of a Variable

Students often have difficulty with the term "variable". You may think of a variable as something that varies, but this is not a very satisfactory definition. Although the sun seems to vary in its brightness during the day, it is the intensity of the light, and not the sun itself, that varies. Brightness is a variable because it has possible values: namely, "high", "medium", or "low". This is the important aspect of a variable—that it can have a variety of values. (Note that the word "value", as used here, means "amount" or "quantity"—*not* "moral value"!) In scientific work these values often will be numeric. So the brightness of the sun might be 150 or 43.6 or 0.63 in some appropriate units of measurement.

The best way to think of a variable is to think of it as a quality that can have different values under different circumstances. A variable can always be denoted by a name ("brightness", "intelligence") or by a letter or abbreviation (B, IQ). Since the variable can take values, we can write, for a particular case, "Brightness = 150", or "B = high", or "IQ = 112", or "IQ = 97". These equations are all of the form "Variable = specific value", illustrating the definition of a variable as something that has values. As a final example, consider the variable "sex". The possible values of this variable are "male" and "female". So we can write "sex = male" or "sex = female". In some situations it is convenient to use numbers even for this variable. So we may declare a convention that male is coded "0" and female is coded "1". Then we could write "sex = 0" or "sex = 1".

Outside of mathematics, variables generally are measurements or characteristics of objects, animals, or persons. So brightness is a measurement of the sun at different times, IQ is a measurement of persons, and sex is a characteristic of persons. In the example given in Section 2.3, the depression-scale score represented a measurement on certain persons in two groups. It has possible values (43, 76, etc.) and therefore is a variable.

Graphs and Intervals

Now let us return to the graphs in Figures 2.4 and 2.5 and compare them with Figures 2.1 and 2.2. In Figures 2.1 and 2.2, which are frequency distributions of addition times, the variable (time) is shown on the abscissa, or horizontal axis, and its possible values (9, 10, . . . , 16 seconds) are explicitly shown on this axis. The frequency of each of these possible values is shown on the ordinate.

Figures 2.4 and 2.5 are similar to Figures 2.1 and 2.2. Now the variable on the abscissa is the depression-scale score and the ordinate, as before, shows the frequency. In these graphs, however, *intervals* of values of the variable are shown with their frequencies. For example, the neurotic histogram seems to show that the frequency of scores in the interval from 30 to 40 is two. But Tables 2.3 and 2.4 show that the interval extends only from 30 to 39. If 40 were included in this interval, there would actually be three scores in the interval, since one of the neurotic scores is exactly 40. If you think about it, you will conclude that the histograms have not been drawn consistently with Table 2.4.

There are several ways to change the graphs to make them consistent with the table. In explaining this, we will introduce some important new concepts. However, the graphs in Figure 2.4 are still the "nicest" and the most "useful" graphs.

The first way to change a grouped-frequency histogram is shown in Figure 2.6a, which illustrates explicitly that the first interval extends only from 30 to 39, the second from 40 to 49, etc. The problem with this method is that on a complete graph of the size shown in Figure 2.4, this distinction is too fine and subtle to be shown. You should keep in mind, however, that the intervals in Figure 2.4 are not quite correct.

Intervals, Limits, and Midpoints

Let us reconsider the histogram of addition times in Figure 2.1. What does a time of 9 seconds really represent? Recall that the times are recorded to the nearest second, so that a recorded time of 9 seconds does not mean exactly 9 seconds, but only approximately 9 seconds. The real time that the task took could range from 8.5 seconds to 9.5 seconds. Similarly, if the task had taken between 9.5 and 10.5 seconds, the time would be recorded as 10 seconds. So the score or time of 9 seconds really represents an interval of scores. The limits of this interval (8.5 and 9.5) are called the *real limits* of the interval. These real limits are shown in Figure 2.6b for part of the histogram of addition times.

How does this notion of real limits help us with the distribution of depression-scale scores? The measurement of depression by the particular test consisting of 100 questions is surely not a perfect measure of depression. Is there any reason to think that "depression" can only have the values 31, 32, 33, etc., and never a decimal value such as 32.37? Certainly, our scale cannot have such decimal values, since the score is defined as the number of questions answered affirmatively in the direction indicating depression. Nevertheless, it is useful to think of a score of 32 as representing scores between 31.5 and 32.5. Then the interval of scores from 30 to 39 really represents scores from 29.5 to 39.5, since 30 includes scores from 29.5 to 30.5 and 39 includes scores from 38.5 to 39.5. In order to keep a clear distinction between the two kinds of intervals we will call an interval such as 30–39 an *apparent interval* and an interval such as 29.5–39.5 a *real interval*. This terminology reflects the fact that whereas the interval 30–39 seems to include scores from 30 to 39, it really contains scores

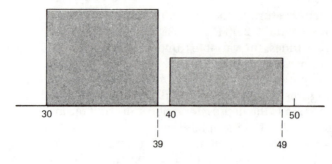

Depression-scale score

(a)

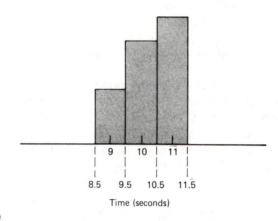

Time (seconds)

(b)

Figure 2.6 Graphing intervals. (a) Using apparent limits. (b) Using real limits—continuous variable measured imprecisely. (c) Using real limits—discrete variable treated as continuous. (d) Using midpoints.

from 29.5 to 39.5. We can refer to 29.5 as the *lower real limit* of the interval, to 39.5 as the *upper real limit*, to 30 as the *lower apparent limit*, and to 39 as the *upper apparent limit*.

If we accept this argument, we can draw histograms for real intervals (Figure 2.6c). But just as the solution shown in Figure 2.6a was not practical, this subtlety is usually impossible to show on a full graph such as the one in Figure 2.4. One advantage of drawing the intervals as in Figure 2.6c is that it is then clear that a score of 40 belongs, as it should, in the interval 40–50 and not in the interval 30–40.

The distinction between apparent and real intervals helps us to determine how long the interval is. In Figure 2.6a, in which apparent limits are used, the actual width of each bar is in fact 9 points. However, when real limits are used for the intervals, as in Figure 2.6c, the width of each bar becomes 10 points. We are going to treat as synonymous the terms "length" and "width" of intervals. We will define the *width* (or *length*) of an interval as the width of the *real* interval, so that the width of the depression score intervals is 10 points. Similarly, the addition-time intervals have a width of 1 second, since each of

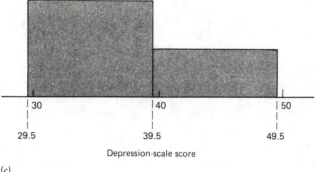

(c)

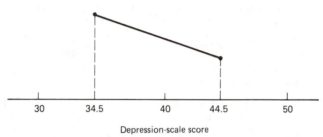

(d)

Figure 2.6 (continued)

the bars in Figure 2.6b has this width. The width is the difference between the upper and lower real limits of the interval.

This completes our discussion of the grouped-frequency histogram (Figure 2.4). Let us now turn to Figure 2.5, which shows two grouped-frequency polygons. The depression-scale score is shown in the abscissa, as usual. The points should be drawn over the *midpoints* of each interval. Since the real interval for scores from 30 to 39 extends from 29.5 to 39.5, the midpoint of this interval must be 34.5, since 34.5 is halfway between 29.5 and 39.5. A part of a polygon using these precise midpoints is shown in Figure 2.6d. When an actual polygon is drawn, however, the distinction between 34.5 and 35 cannot be made, since the numbers are so close together. Polygons should be drawn as demonstrated in Figure 2.5, in which only nice numbers such as 30, 40, etc. are shown and the points are plotted over midpoints that are not explicitly shown.

Continuous and Discrete Variables

Underlying the specific ideas in this section is the distinction between discrete and continuous variables. Remember that variables are qualities that can have values. The values of a *continuous variable* are numbers, and these numbers can take any possible value within a certain range. A good example of a continuous variable is *time*. Consider the different times it can take to complete a 100-m race—say, from 10 to 20 seconds. Since *any* value between 10 and 20 is possible, time is continuous. For example, 12.3 is possible, as is 12.4—but so too are 12.3478 and 12.3479. And between the last two times there are infinitely many other possible times, such as 12.347816 and 12.347837.

Discrete variables can have values that are either names or numbers. (In the case of number values, not all numbers are possible.) Consider the discrete variable "psychiatric diagnosis". This variable can have values: neurotic, schizophrenic, paranoid, etc. You cannot make infinitely fine divisions between these labels. Hence the variable is discrete. Or consider a discrete variable with numeric values: for example, the number of errors before learning a task. This variable can have values: 0, 1, 2, etc. A value such as 1.24 is not a possible number of errors. This variable is also discrete.

It would seem that the definitions of continuous and discrete variables would be easy to use. But this is not true. Consider again the time to complete the arithmetic task. Is this variable discrete or continuous? Surely, time is continuous. But the teacher measured time to the nearest second. The measured variable was discrete, since only values such as 9, 10, 11, etc. were possible. From this point of view, any measured variable must be discrete, because our measuring instruments can never be infinitely precise. So in reality continuous variables are only a theoretical possibility; measured variables are always discrete. This does not mean, however, that continuous variables are not important. Continuous variables are often easier to handle mathematically. In many cases we proceed as if a variable were continuous even though we know it is actually discrete.

When studying a particular variable, the relevant question is not "What type of variable *is* this?" but rather "How is this variable being *treated*?" In some cases a truly continuous variable is treated as if it were discrete; an example is "time taken to complete a task", just discussed. In other cases, a truly discrete variable is treated as if it were continuous; an example is "number of errors before task completed successfully". The number of errors is 0, 1, 2, etc., a discrete variable. However, some methods of analysis treat this variable as if it were continuous.

2.5 Univariate Scatterplots

The graphs of frequency distributions that we have been studying are very useful, since they show us at a glance several characteristics of the original scores. We can see the values of the largest and smallest scores and the range of the scores. We can see if the scores tend to cluster by looking for peaks in the graph. And we can see whether the distribution is reasonably smooth or whether the frequency of scores varies up and down.

In the examples we have seen so far, the number of scores in the distributions ranged from 15 to 39. Grouped-frequency graphs are often drawn for sets of data ranging from 15 or so up to hundreds and thousands of scores.

Suppose, on the other hand, that we have a very small set of scores. When a set is very small, there are not many data to summarize. We can look at the numbers themselves. But sometimes it is still convenient to use a graphic representation of the set of scores, particularly if we want to compare several sets. Grouped distributions are unsatisfactory for very small sets since, if the intervals are very broad, there will be very few intervals. Conversely, if the intervals are narrow, most intervals will have frequencies of zero or one.

An alternative to grouped distributions is called a *univariate scatterplot*. Look at Figure 2.7, which shows the depression data once more. For each group, a single line is drawn, representing the possible values of the variable—here, the depression-scale score. Each score is indicated on the line by a cross (or other suitable mark). These scatterplots can provide the same information as a histogram or polygon. Compare Figure 2.7 with Figures 2.4 and 2.5. Fre-

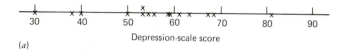

(a)

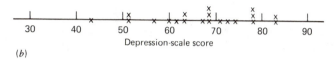

(b)

Figure 2.7 Univariate scatterplots. (a) Neurotic group (n = 15). (b) Psychotic group (n = 20).

quency is indicated in scatterplots by the clustering or density of the crosses. In the neurotic group, most scores fall between 50 and 70, whereas in the psychotic group the scores fall mainly between 60 and 80. This is easily seen on the scatterplot.

You will notice one difficulty in drawing univariate scatterplots. There are two neurotic scores of 53, both of which must be indicated. Since both cannot be placed right on the line, the cross for the second score is placed above the line. This, however, tends to spoil the visual appearance of the line of crosses of varying density. In the psychotic group this problem occurs more frequently. There are even two scores, 69 and 78, that occur three times each.

The depression-score data are really too numerous to be plotted in this way; frequency graphs are better in this case. But for smaller sets of data the univariate scatterplot works well. Consider the data shown in Figure 2.8. The numbers are normal maximum temperatures, by month, for two cities. Each set of data has 12 numbers, or scores. These scores have been shown in two scatterplots. In this example, the lines have been drawn vertically, instead of horizontally. These two scatterplots show clearly the difference in weather between New York and San Francisco. New York maximum temperatures range widely (from the 30s to the 80s), whereas San Francisco maximum temperatures cover a much narrower range.

Univariate scatterplots are useful for very small sets of data, and particularly for measurements of a continuous variable where duplicate values are unlikely. Such scatterplots are also important in that they form the basis of more-complicated scatterplots that will be described in later chapters.

2.6 Relative-frequency Distributions

It is common in ordinary conversation to refer to the percentage of people with a certain characteristic rather than to the absolute number or frequency of such persons. If 3 million wage earners have incomes over $25,000 and there are 50 million wage earners in all, then we can say that 6% of wage earners earn over $25,000. For most purposes the 6% figure is more useful than the 3 million figure. The percentage figure allows us to compare easily the number of high-income wage earners in different groups. Knowing that 2 million males earn over $25,000 and 1 million females earn over $25,000 provides no basis for comparison unless we know the total number of male and female wage earners. However, if we are told that 10% of male wage earners but only 2% of female

Normal Daily Maximum

Temperature (°F)

Month	New York	San Francisco
Jan	38.5	55.3
Feb	40.2	58.6
Mar	48.4	61.0
Apr	60.7	63.5
May	71.4	66.6
June	80.5	70.2
July	85.2	70.9
Aug	83.4	71.6
Sept	76.8	73.6
Oct	66.8	70.3
Nov	54.0	63.3
Dec	41.4	56.5

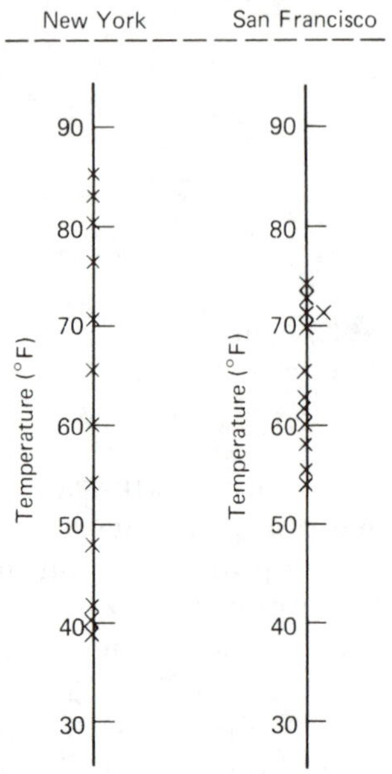

Figure 2.8 Univariate scatterplots of temperature. [From Table no. 367 of U.S. Bureau of the Census, *Statistical Abstract of the United States: 1978* (99th edition). Washington, D.C. (1978).]

wage earners have incomes over $25,000, then we have some basis of comparison.

Percentages are common in statistical work. However, the terms "proportion" and "relative frequency" also are often used. These three terms are defined as follows.

$$\text{Proportion} = \text{Relative frequency} = \frac{\text{Frequency}}{\text{Total frequency}}$$

$$\text{Percentage} = \text{Proportion (or relative frequency)} \times 100$$

"Proportion" and "relative frequency" are synonymous: they each mean the ratio of the frequency in the category of concern to the total frequency of persons or things in the reference set. In the above example the category of concern is persons with income over $25,000. The frequency of this category is 3 million. The reference set is all wage earners. The total frequency of persons in this set is 50 million. Hence, the proportion of persons who earn over $25,000 is 3 million/50 million = 0.06. So the relative frequency (or proportion) of such persons is 0.06. In order to use percentages, simply multiply 0.06 by 100 (that is, move the decimal place two places to the right), which yields the result that 6% of wage earners earn over $25,000.

Review once again the frequency distribution tables and graphs for the neurotic and psychotic groups (Table 2.4, Figures 2.4 and 2.5). If you wish to compare the groups with each other, you first should note that the groups are

TABLE 2.5 Relative Frequencies

DEPRESSION-SCALE SCORE	NEUROTIC GROUP PROPORTION	PSYCHOTIC GROUP PROPORTION
80–89	.07	.10
70–79	.00	.30
60–69	.27	.40
50–59	.47	.15
40–49	.07	.05
30–39	.13	.00
	1.01	1.00

not of equal size: the neurotic group has 15 persons; the psychotic group, 20 persons. Suppose you want to compare the number of persons in the two groups who have depression scores between 60 and 69. Four neurotics and eight psychotics have such scores. To make a fair comparison, you must convert these numbers to relative frequencies, as is done in Table 2.5. The relative frequency (or proportion) of neurotics scoring between 60 and 69 is $4/15 = 0.27$. Similarly, the proportion of psychotics in this interval is $8/20 = 0.40$. The psychotic figure is still larger than the neurotic figure (0.40 versus 0.27), but the difference is not quite so great as the original figures would indicate (eight versus four).

Table 2.5 shows the proportion for each interval of depression score. These proportions should add to 1.00, since all scores are included in some interval. (Sometimes, because of rounding, the sum of the proportions will not equal 1.00. For example, the neurotic proportions add to 1.01. Despite this problem, it is recommended that proportions be reported to only two decimal places, since additional digits do not provide useful information and make the table hard to read.[1])

Proportions or relative frequencies can be plotted on a graph. Figure 2.9 shows relative-frequency polygons for the two groups. Histograms could also be constructed, but the two distributions are shown more clearly by polygons. This figure is very similar in appearance to Figure 2.5. The general shape of the

[1] Rounding is discussed in detail in Section 3.6.

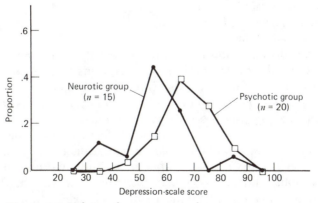

Figure 2.9 Relative frequency graphs.

distributions is the same, since the horizontal scale is identical and the vertical scale has simply been changed from frequency to relative frequency. Note, however, that the two curves have a different relative vertical position in the relative-frequency graph than they had in the original figure. This happens because the total frequency of the two groups is different. We can say that the relative-frequency graph represents the data more fairly than the original frequency graph does. Recall that fairness was one of the aspects of the summarization principle of Section 2.3.

The main purpose of converting frequencies to relative frequencies is to allow for comparison of two or more distributions. We have seen, however, that the general shape of a distribution is not changed by such conversion. Section 2.7 will show you how to describe the shape of distributions in words. These descriptions apply equally to frequency and to relative-frequency distributions.

Relative Frequency and Relative Area

Section 2.2 pointed out that the area of a frequency distribution represents frequency. This interpretation requires that the area under the graph be divided into units, with each unit of area representing one observation or score. Then the number of units of area over an interval represents the number of scores in that interval.

Dividing the area into units is not very convenient. It is easier to take the *ratio* of the area above the interval to the area under the whole distribution. This ratio represents the relative frequency of the interval of scores. This interpretation of relative area is exact for histograms but only approximate for polygons; however, the approximation is good enough to use when "eyeballing" a distribution graph. (The interpretation can be used with either frequency or relative-frequency graphs.)

Look again at any of the histogram or polygon graphs in this chapter. Consider an interval of scores. The ratio of area above the interval to the total area of the graph is the relative area; the relative area is (approximately) the same as the relative frequency of the interval.

Example

Look at Figure 2.9 (or Figure 2.5). Suppose you want to compare the proportion of scores above 70 for neurotics and psychotics. The area under the neurotic graph above the interval from 70 to 100 is a very small proportion of the total area of the graph. On the other hand, the area under the psychotic graph above the same interval is almost one half the total area of that graph. Hence, scores above 70 are more frequent among psychotics than among neurotics.

2.7 Shapes of Distributions

Now that you have learned how to construct distributions, you must learn ways to describe these distributions. Descriptions can be formulated in terms of numbers or in terms of words. The next two chapters will teach you various ways to give numeric summaries of distributions. In this section you will learn a few useful words to describe distributions.

All the distributions presented so far may be called *empirical* distributions, since they are obtained from actual data. Such distributions typically have a jagged or uneven shape. In a study with a very large number of observations and with very many narrow intervals, the distribution is much smoother than it is in the typical empirical distribution. In the limit of infinite

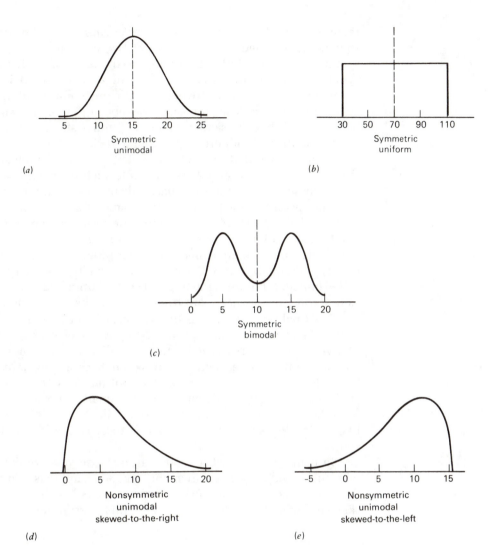

Figure 2.10 Some idealized distributions.

sample size and infinitesimally small interval size, such distributions would be perfectly smooth. These distributions may be called *idealized* distributions, some examples of which are given in Figure 2.10. (These distributions may be thought of either as smooth polygons or as smooth histograms with only the tops of the bars drawn.)

A vertical scale is not shown on these graphs, because the magnitude of the frequencies or relative frequencies is irrelevant to the shape of these distributions. (The horizontal scale also is unnecessary—it is included so that parts of the distribution may be referred to.) Remember too that the shape of a distribution is the same whether it is drawn as a frequency or a relative-frequency distribution.

The first characteristic of the shape of a distribution is *symmetry*. A distribution is symmetric if it can be folded about a vertical axis (the line of symmetry) so that the right half falls exactly on the left half. The point on the horizontal axis at which the fold is made is called the center of symmetry.

Figures 2.10*a*, 2.10*b*, and 2.10*c* are all symmetric. The line of symmetry is shown by a dashed line. In Figure 2.10*a* we would say that the scores are symmetrically distributed about a score of 15. The fact that the distribution is symmetric means that the frequency of a score of 20 is the same as the frequency of a score of 10 (since both are equidistant from the center of symmetry, 15). Similarly, for each score above 15 there is a corresponding score below 15 that has the same frequency. Similar statements can be made about the distributions in Figures 2.10*b* and 2.10*c*.

Distributions that are not symmetric are called *nonsymmetric* or *asymmetric*. Figures 2.10*d* and 2.10*e* illustrate such distributions, neither of which can be folded so that the right part falls exactly on top of the left part.

The second characteristic of the shape of a distribution is the number of peaks it displays. Distributions with one peak, or maximum, are called *unimodal* distributions. (The statistical word for "peak" is "mode", so distributions with one peak are called unimodal.) Figures 2.10*a*, 2.10*d*, and 2.10*e* are clearly unimodal. Figure 2.10*c* has two equal peaks, so this distribution is called *bimodal*. Figure 2.10*b* might be called unimodal but usually is described as *uniform*, since every score from 30 to 110 has the same frequency.

The final characteristic is the *skewness* of a distribution. Distributions are said to be *skewed* or *not skewed*. Although skewness can be defined quantitatively, we will reserve the term for distributions in which skewness is obviously present or absent. Skewness is determined visually, by looking at the two *tails* of the distribution. Most distributions have the property that, for large and small scores, the frequencies become smaller and smaller as one goes away from the center of the distribution, so that the two ends of the distribution look like "tails". In a symmetric distribution these two tails have the same shape, of course.

In many nonsymmetric distributions these tails are obviously different—one stubby and the other drawn out. A good example is Figure 2.10*d*, in which the left tail is stubby or short and the right tail is drawn out. We call this a skewed distribution. More specifically, this distribution is *skewed-to-the-right*,

Figure 2.11 Some empirical distributions. (a) Distribution of IQs on the Wechsler Adult Intelligence Scale of a national sample of 2,052 U.S. adults aged 16 and over. (b) Percentage (i.e., relative-frequency) distributions of scores on the Army Alpha intelligence test of white draftees in groups I, II, and III who took Alpha only (upper panel) and of scores on the Army Beta intelligence test of white draftees in these groups who took Beta only or Beta and Alpha (lower panel). (c) Change in the distribution of reaction time to an auditory stimulus under the influence of incentives. In the "incentive series", the subject was informed of his last reaction time; in the "punishment series", he received a shock in the finger when the reaction was at all slow. Each curve shows the distribution of 3,600 single reactions obtained from three subjects whose times were nearly the same. [Sources: (a) From David Wechsler, *The Measurement and Appraisal of Adult Intelligence*, copyright © 1958 Williams & Wilkins Co. Reprinted by permission. (b) Reproduced from "Psychological Examining in the United States Army", in *Memoirs of the National Academy of Sciences*, National Academy Press, Washington, D.C., 1921. (c) From *Experimental Psychology*, revised edition, by Robert S. Woodworth and Harold Schlosberg. Copyright 1938, 1954 by Henry Holt and Company, Inc. Reprinted by permission of Holt, Rinehart and Winston, CBS College Publishing.]

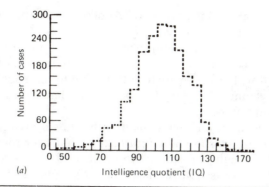

(a)   Intelligence quotient (IQ)

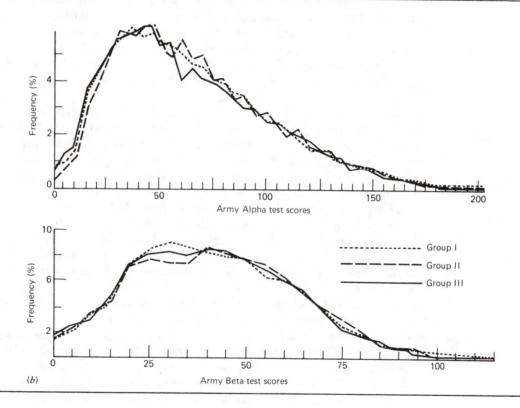

(b)

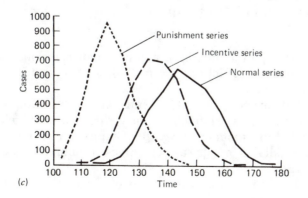

(c)

since the right tail is drawn out. Another way to describe the same distribution is to say that it is *positively skewed*.

Conversely, a distribution such as that in Figure 2.10e is called *skewed-to-the-left*, since it is the left tail that is drawn out. This distribution is also known as a *negatively skewed* distribution. The terms "positively" and "negatively" reflect whether the long tail occurs for large (positive) values or for small (negative) values.

Let us now consider some distributions drawn from real data. Three such distributions are shown in Figure 2.11. First, note that these empirical distributions are "rough" in contrast to the smooth, idealized distributions of Figure 2.10. This means that terms such as "symmetric" and "bimodal" cannot be precisely applied to empirical distributions. It is almost impossible for an empirical distribution to be perfectly symmetric. Nevertheless, we could describe the distribution of IQ in Figure 2.11a as being approximately symmetric. If you look at this figure more closely, though, you will see that it is very slightly skewed-to-the-left, since the left tail trails off a little bit more than the right. This judgment is difficult to make, however; the predominant characteristic of the figure is symmetry.

The distributions of Figure 2.11b are definitely not symmetric. Those in the top part of the figure are certainly skewed-to-the-right. Those in the bottom probably are also skewed in that direction, but this judgment is difficult to make. It can be stated, however, that the distributions are unimodal and nonsymmetric. Note that we ignore the many small peaks in an empirical distribution and call it unimodal if there is one main peak.

The three distributions of reaction time shown in Figure 2.11c certainly are unimodal. Although they are not perfectly symmetric, it is difficult to judge whether they are appreciably skewed in either direction. The main differences among the three distributions in Figure 2.11c arise in their general location and spread. These characteristics will be discussed at length in the next two chapters.

The shape of a distribution can also be judged from a univariate scatterplot. In the examples shown in Figure 2.12, note that the peaks of the distributions are associated with a concentration of points in the scatterplot. The distributions in Figures 2.12a, 2.12d, and 2.12e are all unimodal, since there is one concentration. The distribution in Figure 2.12c has two concentrations and is, therefore, bimodal. There are no concentrations in Figure 2.12b and, further, the points are equally spaced; hence, the distribution is uniform. You can also see that the symmetry or skewness of a distribution can be judged from the univariate scatterplot.

2.8 Summary

The first step in summarizing any set of data is to construct a distribution of the data. General guidelines for making such a distribution are given by the *summarization principle*.

Moderate to large sets of numbers can be summarized in a *frequency distribution* or a *relative-frequency distribution*. Such distributions can be presented in a *table* or in a *graph*. Graphs can be of the *histogram* or *polygon* form. Small sets of data can sometimes be summarized usefully in a *univariate scatterplot*.

Distributions are distributions of a *variable*, which is a measurement or

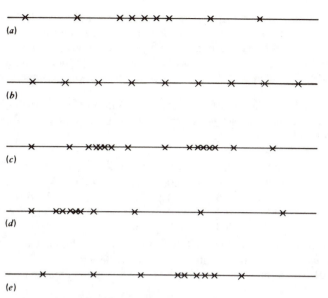

Figure 2.12 Shapes of some univariate scatterplots. (a) Symmetric unimodal. (b) Symmetric uniform. (c) Symmetric bimodal. (d) Nonsymmetric unimodal, skewed-to-the-right. (e) Nonsymmetric unimodal, skewed-to-the-left.

characteristic of an object, animal, or person. It is more important to consider whether a variable should be *treated* as a *discrete* or as a *continuous* variable than to decide whether the variable is *really* discrete or continuous.

Distributions can be *grouped* or *ungrouped*. When the data is grouped into *intervals*, one must be clear about the distinction between *real* and *apparent intervals* and *limits*. Each interval also has a *midpoint* and a *width* (or *length*).

Relative frequencies are the same as *proportions*. *Percentages* are simply relative frequencies multiplied by 100. A relative-frequency distribution is usually preferable to a frequency distribution if two or more sets of data are to be compared. Relative frequencies can be determined from relative areas under a frequency graph or a relative-frequency graph.

The shapes of distributions can be described in many ways: *symmetric, nonsymmetric, unimodal, bimodal, uniform, skewed-to-the-right (-left)* or *positively (negatively) skewed*. The skewness of a distribution is determined from the *tails* of the distribution. All these terms can be more precisely applied to *idealized* distributions than to *empirical* ones.

Exercises

Note: The graphs you construct for Questions 1, 2, and 4 also will be required for Question 5, and for the Exercises for Chapter 3. Therefore, you should retain these graphs for later use.

1 (Sections 2.1 and 2.2) Fifty-two children are tested for the ability to hit the bulls-eye in a dart board. They are each given additional turns until they succeed in hitting the bulls-eye. The number of turns they took is recorded below. Make a frequency distribution of these data in three forms: table, histogram, and polygon.

   The data are 3, 6, 3, 3, 4, 4, 4, 2, 5, 3, 4, 3, 1, 4, 4, 3, 4, 6, 4, 3, 4, 3, 3, 2, 2, 3, 3, 3, 4, 3, 1, 4, 4, 3, 5, 3, 5, 4, 4, 3, 2, 4, 4, 3, 2, 4, 4, 3, 3, 3, 4, 5.

2 (Section 2.3) Forty-nine persons show up for the first fitness class. The time each person takes to run 100 meters is measured. (Some can only walk the distance.) The times, in seconds and fraction of a second, are given below.

(a) Make a grouped-frequency table and graph (either histogram or polygon) for these data. Choose appropriate intervals and justify your choice using the summarization principle.

(b) State another set of intervals that would not satisfy the summarization principle. Explain why this set is not satisfactory. Do *not* make a table or graph using these intervals.

The data:

31.3	26.2	45.0	34.2	30.1	35.1	25.5	29.7	37.2	16.5
42.0	16.1	31.0	33.9	20.0	26.0	29.8	37.7	37.9	32.7
20.6	41.3	34.9	27.3	31.3	30.7	30.0	35.1	27.3	37.9
34.7	34.6	20.7	23.0	37.2	31.6	41.2	57.6	41.6	28.8
27.6	31.0	28.3	46.8	33.5	38.8	26.6	23.6	29.8	

3 (Section 2.4) Consider the following three sets of data. Only a few numbers are given in each case, as these data are not to be analyzed.

(i) The number of pins knocked down in one game by each bowler: 187, 206, 113, 153, . . .

(ii) The distance in meters jumped by each long jumper: 8.73, 10.07, 7.22, 9.35, . . .

(iii) The score on a test of mathematics achievement by each child: 78, 35, 93, 86, 57, . . .

For *each* set of data, do the following.

(a) Name the variable that is measured.

(b) Give an example of a value of the variable.

(c) State whether the variable, *as measured*, is discrete or continuous.

(d) Suppose that each variable is *treated* as a continuous variable. State one possible interval of values of the variable that might be used in a grouped-frequency distribution. State the apparent and real limits of this interval, and state its midpoint.

4 (Sections 2.5 and 2.6) A set of data can be summarized by a univariate scatterplot or by one of several types of graph. Choose an appropriate method of representation for each of the two following sets of data. Justify your choice of method in terms of the summarization principle. Make the summary scatterplot or graph you have chosen for each set.

(i) The times, in seconds, that each of 10 students takes to complete an arithmetical problem: 13.9, 11.8, 10.7, 16.7, 14.8, 17.8, 12.6, 13.1, 15.6, 18.7.

(ii) Fourth- and fifth-grade students take a test of reading ability that gives their reading level in terms of grade level. The reading levels are summarized in the following frequency table:

READING GRADE LEVEL	GRADE-FOUR STUDENTS	GRADE-FIVE STUDENTS
2.5–2.9	2	1
3.0–3.4	5	5
3.5–3.9	10	7
4.0–4.4	5	10
4.5–4.9	12	14
5.0–5.4	8	24
5.5–5.9	1	18
6.0–6.4	0	7
6.5–6.9	0	2
	43	88

5 (Section 2.7) In answering the preceding questions you have drawn five distributions.

Question 1 Distribution of number of turns to hit bulls-eye.
Question 2 Distribution of times to run 100 meters.
Question 4(i) Distribution of times to complete arithmetical problem.
Question 4(ii) Distribution of reading levels in Grade 4.
Question 4(ii) Distribution of reading levels in Grade 5.
(a) Are these empirical or idealized distributions?
(b) Describe the shape of each distribution.

3 CENTRAL TENDENCY

Chapter 2 described how to summarize data by means of a frequency distribution. In this chapter you will learn how to make summaries that are even more concise. These new summaries are measures of the center of the distribution. There are three commonly used measures of the central tendency of a distribution: the mean, the median, and the mode. These three measures are defined and the relationships among them discussed.

A calculator is very useful in statistical work. Some general principles governing the use of any calculator are described. These principles help to ensure that calculated answers are correct and that they are reported to a suitable number of digits.

3.1 The Average
The notion of the average of a set of numbers or scores is familiar to everyone. In order to calculate the average of a set of scores, you simply add up the scores and divide the sum by the total number of scores. As this section will make clear, however, this average is not the only useful measure you can calculate. Since there are several possible measures of an average, the broader meaning of the term "average" will not be used after these new measures have been defined. What is generally called the average of a set of numbers will be called the *mean* of these numbers.

The last chapter demonstrated that a frequency distribution is a good way to summarize data. Since such a distribution is still fairly complex, it is useful to reduce it further. It is natural to ask where most of the scores tend to fall, or where the center of the distribution is. These questions are described in statistical work as questions about the *central tendency* of the distribution of scores. Statisticians prefer to use the term "central tendency" rather than the term "average" because the latter term is often thought of as denoting the mean of a set of numbers.

There are various ways to measure the average or central tendency of a set of numbers. Table 3.1 gives the frequency distribution of the salaries of 26 employees of the fictitious XYZ Company. Note that many salaries are close together (from $8,000 to $11,700), whereas the remaining salaries are widely dispersed, extending all the way up to $50,000. It is not very useful to group these numbers into intervals.

The graph of the data in Figure 3.1 does not follow exactly the form of either a histogram or a polygon. A vertical line is used instead of a bar, as in an ordinary histogram. Since the data have not been grouped, it is not possible to give a width to the bars, as is done in a histogram. One purpose of this example is to show that you must be flexible and creative in tabulating and graphing a set of data. Every set of data has some unique peculiarities, so it is impossible to give rigid rules for making tables and graphs. As stressed in the summarization principle given in the last chapter, any summary of a set of data must be useful and easy to interpret. Do you agree that Table 3.1 and Figure 3.1 satisfy this general principle?

TABLE 3.1 Salaries of Employees of XYZ Company

SALARY (S)	NUMBER OF EMPLOYEES (f)	COMPUTATIONS (S × f)
$50,000	2	$100,000
20,000	1	20,000
15,000	2	30,000
11,700	1	11,700
11,000	3	33,000
9,700	4	38,800
9,000	1	9,000
8,000	12	96,000
	26	Sum = 338,500
		Mean = 338,500/26 = $13,019.23

What is the central tendency or average of these salaries? First, consider the familiar average, which we now call the *mean*, defined as the sum of all salaries divided by 26 (the number of employees). This mean can be computed in either of two ways.

(a) By adding the salaries: 50,000 + 50,000 + 20,000 + 15,000 + 15,000 + ... + 8,000 + 8,000 + 8,000 = 338,500. The mean is then 338,500/26 = $13,019.23. (Note that the 26 individual salaries are included in the sum.)

(b) By taking advantage of the duplication of many of the salaries—an easier method. Instead of adding 50,000 + 50,000 we can compute 50,000 × 2 = 100,000. Similarly, instead of adding 8,000 twelve times, we can compute 8,000 × 12 = 96,000. These computations are shown in the last column of

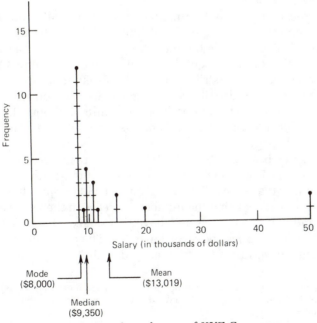

Figure 3.1 Salaries of employees of XYZ Company.

Table 3.1. The products of salary times frequency ($S \times f$) are then added, giving the same sum (338,500) as before. The mean is then 338,500/26 = $13,019.23. [Note carefully that the sum is still divided by the number of employees (26), not by the number of products (8).]

To the nearest dollar, the mean is $13,019. Now we must determine whether this figure of $13,019 is a good representation of the central tendency, or average value, of the salaries of employees in XYZ Company. Look again at the table of salaries. You will see that only five employees earn more than $13,019, whereas 21 earn less. If so few employees earn more than $13,019, perhaps $13,019 is not a good measure of central tendency.

Is there a salary with the property that half of the employees earn more than it and half earn less? Table 3.1 reveals that 13 employees earn $9,700 or more and that 13 employees earn $9,000 or less. (Check this by adding the frequencies in the table.) So a figure around $9,700 or $9,000 should be a good measure of the central tendency of these salaries. In order to get a single number, find the number halfway between the two salary levels: $\frac{1}{2}(9,000 + 9,700) = 9,350$. This measure of central tendency is called the *median*. The median is $9,350 because half the employees earn more than $9,350 and half earn less than $9,350.

You will probably agree that $9,350 is a better measure of the average salary than $13,019. However, you might propose a still better measure for these data, noting from the table of salaries that no employee earns exactly $9,350. How can this number be used as a measure of the average salary when no one earns exactly $9,350? (The same criticism applies to the mean salary of $13,019.) This argument would lead you to conclude that the best measure of average salary is $8,000, since more people earn that salary than any other salary. As Table 3.1 shows, 12 people (almost half the employees) earn $8,000. This measure of central tendency is called the *mode*. We say that the *modal* salary is $8,000.

Recall the terms "unimodal" and "bimodal" from the previous chapter. They describe the shape of distributions. These terms are of course related to "mode". The mode is the most frequent score in a distribution, and therefore the location of the peak of the distribution of the scores.[1] When a distribution is unimodal (as this distribution of salaries is), then the location of the peak is at the mode of the distribution. When a distribution has two modes (bimodal), however, the location of both peaks would be reported. (In general, a distribution is considered to have two modes even when the two peaks are not exactly the same height.)

Let us briefly review the three measures of central tendency as shown in Figure 3.1. The mode is $8,000 because the greatest frequency (longest line) occurs for that salary. The median is $9,350 because half the salaries are below that salary (check that the number of pieces in the vertical lines below $9,350 is 13) and half the salaries are above this salary (check that the number of pieces in the vertical lines above $9,350 is 13). The mode and median, then, are easy to describe in terms of the frequency distribution.

---

[1] The word "mode" has two distinct meanings. It can mean "peak" or it can mean "location of the peak". In the above paragraph, the word "peak" is used for the first meaning and the word "mode" for the second.

It is more difficult to see that the mean ($13,019) is properly placed in the frequency distribution. The basic property of the mean is that it is the *balance point* of the distribution. In order to understand this concept, imagine that each of the 26 persons is sitting on a seesaw (or balance bar) represented by the horizontal axis. The 12 persons at $8,000 are piled on top of one another, as shown in the figure. Similarly, the one person who earns $9,000 is located at the indicated position on the abscissa, the four persons at $9,700 are piled on top of each other at the $9,700 position, etc. Assume that each person has the same weight. Now, where should the fulcrum (axis) of the balance bar be placed so that the bar balances? The answer is that the bar will balance if the fulcrum is placed at the mean, which is $13,019. In other words, the five persons with salaries over $13,019 balance the 21 persons below $13,019! This result occurs because the five high-salaried persons are seated very far from the balance point (particularly the two persons at $50,000). As children on seesaws quickly discover, one person who moves far enough away from the fulcrum can balance two persons sitting closer to the fulcrum on the other side.

The purpose of this section has been to give you an introduction to the three important measures of central tendency. The next three sections describe these measures in more detail.

3.2 The Mean

Formal definitions of the mean, median, and mode are presented in this section and the following two sections. We will return to a more intuitive discussion of the relationships among these measures in Sections 3.5 and 3.6. Section 3.6 also shows you how to ensure that your calculations are correct.

Symbol Shock

This section introduces symbols in definitions and calculations. It is well known that mathematical symbols cause some readers considerable anxiety. Statistics instructors sometimes refer to this common reaction to symbols as symbol shock. The following suggestions may help you overcome the problem of symbol shock.

1 Keep in mind that each mathematical formula and procedure in this book is based on simple ideas that you should try to understand intuitively. When you get lost in computations, make yourself return to these intuitive ideas. In the present chapter, return to the underlying ideas discussed in the preceding section. These ideas provide the basis of every technique discussed later in the chapter.

2 Each symbol can be said aloud. *Be sure and learn how to pronounce each symbol.* One of the reasons many students have problems with mathematical symbols is that they do not say them aloud. You cannot learn to *discriminate* one symbol from another unless you say them aloud. We will be using many symbols, including some Greek letters which are probably unfamiliar to you. It is important that you learn how to tell one symbol from another.

3 Each symbol has a certain *meaning*. So in addition to saying a symbol out loud, be sure and concentrate on its meaning. Similarly, each formula you will use can be described and explained in words. Constantly remind yourself of these descriptions and explanations. If you do so, symbols and formulas will not cause you so much difficulty.

Definition of the Mean

The mean of a set of scores is the sum of the scores divided by the number of scores. Suppose there are five scores: 8, 12, 9, 6, 10. The mean is $(8 + 12 + 9 + 6 + 10)/5 = 45/5 = 9.0$. This is simple!

In order to define the mean in mathematical notation, we must first define some symbols. In the above example there are five scores: 8, 12, 9, 6, 10. We could give each of these scores a different symbol—for example, y, z, a, b, c—but we would run out of letters very quickly with larger sets of scores. So instead, we use the more complex symbol y_1 to represent the first score, y_2 to represent the second score, and so on. The numbers written below y are known as *subscripts*. The symbol y_1 is pronounced "y sub 1", y_2 is "y sub 2", and so on. The five scores are represented by y_1, y_2, y_3, y_4, and y_5, respectively. Similarly, a set of 100 scores could be represented as y_1, y_2, . . . , y_{100}. It is important to understand that each symbol stands for an actual score, so that, for the set of five scores, y_1 stands for the score 8, y_2 stands for the score 12, etc. In brief,

$$y_1 = 8, y_2 = 12, y_3 = 9, y_4 = 6, y_5 = 10$$

We use the symbol n to represent the *number of scores*. In the example, $n = 5$. The fifth score in our example is represented by y with subscript 5. Since it is the last score, it can also be represented as y with subscript n: that is, as y_n. We can now give, in symbols, the scores of any set of size n.

$$y_1, y_2, . . . , y_n$$

Notice that the last score has the symbol y_n.

Let us now write a formula to express the mean of a set of n scores. Recall that the mean is the sum of the scores divided by the number of these scores (n). So the mean is $(y_1 + y_2 + . . . + y_n)/n$. This formula clearly instructs us to take the first score and to add to it the second score, and to keep adding scores until we get to the last score (y_n). We then divide this sum by n to get the mean.

This formula is still lengthy to say out loud. You would say, "The mean is y sub 1 plus y sub 2 plus and so on plus y sub n, all divided by n". There is an easier way to say this: "The mean is the sum of the ys divided by n". Since mathematicians like to abbreviate formulas as much as possible, a symbol is used to represent the sum. The symbol is Σ, or capital sigma (the Greek letter S). The phrase "the sum of the ys" is written "Σy" and pronounced "sum of the ys" or "sigma y". Thus we can simply write

$$\text{Mean} = \frac{\Sigma y}{n}$$

Remember that this should be read either as "the mean is the sum of the ys divided by n", or, if we wish to read the Greek letter as a Greek letter, as "the mean is sigma y divided by n".

The usual symbol for the mean itself is "$\bar{y}$", which is read as either "mean" or as "y-bar", referring to the bar on top of the y. So, at long last, we have the most condensed formula for the mean.

$$\bar{y} = \frac{\Sigma y}{n} \quad \text{(definition of mean)}$$

Example

The formula for the mean applies to any set of data, large or small. Let us apply the formula to the depression scores of the psychotic group in Table 2.3. Here the number of scores, n, is 20.

The scores are $y_1 = 60$, $y_2 = 78$, etc. The mean is

$$\bar{y} = \frac{\Sigma y}{n} = \frac{y_1 + y_2 + \ldots + y_{20}}{20} = \frac{60 + 78 + \ldots + 83}{20}$$

$$= \frac{1,346}{20} = 67.3$$

Mean as Balance Point

Remember that the mean is a measure of the central tendency of a set of scores. The mean represents the center of the scores, in a clearly specified sense. One way to understand this sense is to use the notion of balancing a seesaw or balance bar, as suggested in the last section. Look at Figure 3.2 for an illustration of the 20 scores of the psychotic group and the balancing of these

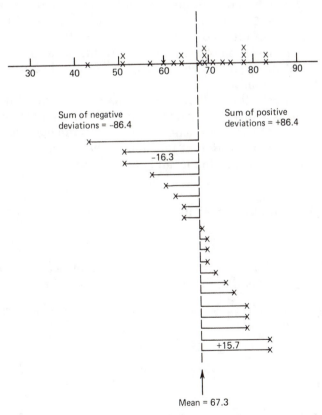

Figure 3.2 The mean as the balance point.

scores about the mean. The scores have been represented in a univariate scatterplot (redrawn from Figure 2.7).

The mean, 67.3, is marked. Notice that there are eight scores below (to the left of) the mean and 12 scores above (to the right of) the mean. This discrepancy would appear to produce an imbalance if the mean is used as the position of a fulcrum for the horizontal bar. How can eight persons on the left balance 12 persons on the right? Balance occurs at the mean of 67.3 because the eight persons below the mean tend to be further from 67.3 than are the 12 persons above the mean. This is another way of saying that the distribution is skewed-to-the-left.

The technical term for the distance of a score from the mean is *deviation*. Each score's deviation from the mean is shown in the lower part of Figure 3.2. The deviations of scores to the left generally are longer than the deviations of scores to the right. If we measure the lengths of these deviations and add them up, we find that both the eight deviations on the left and the 12 deviations on the right add to 86.4. Since it is a person's deviation or distance from the fulcrum that determines the effect of that person on the balance of the bar, and since these deviations total to the same amount on each side of the mean, the bar will balance at the mean of 67.3. (*Note:* this argument assumes that the "persons", represented by the crosses, have equal weight.)

In mathematical terms, the deviation of a score, y, from the mean, $\bar{y}$, is given by $(y - \bar{y})$. For example, a score of 83 has a deviation from the mean of $(83 - 67.3) = 15.7$. Notice that this deviation is a positive number, which is consistent with this score (83) being greater than (or above) the mean. Conversely, the deviation of 51, a score below the mean, will have a negative value of $(51 - 67.3) = -16.3$. The deviation is negative because the score is less than (or below) the mean.

The sum of the negative deviations is -86.4 and the sum of the positive deviations is $+86.4$, so when we sum *all* the deviations, positive and negative, we get *zero*. This is the essence of the idea of the mean as a balance point. *The mean has the property that the sum of the deviations of all scores from the mean is zero.*[2] Hence, the mean is the balance point.

3.3 The Median

In order to find the median of a set of scores it is first necessary to *rank* the scores.

Ranking Scores

Ranking a set of scores consists of putting the scores in order from lowest to highest and assigning to the ordered scores the numbers 1, 2, 3, . . . , n.

[2] An algebraic proof of this fact is easily given but is not required for understanding the above. The sum of the deviations is:

$$
\begin{aligned}
\Sigma(y - \bar{y}) &= \Sigma y - \Sigma \bar{y} & (1)\\
&= \Sigma y - n\bar{y} & (2)\\
&= \Sigma y - n(\Sigma y/n) & (3)\\
&= \Sigma y - \Sigma y & (4)\\
&= 0 & (5)
\end{aligned}
$$

Step (1) follows from the fact that the sum of differences is the difference of two sums. Step (2) follows from the fact that the sum of a constant ($\bar{y}$) taken n times is just n times the constant. In Step (3), the mean has been replaced by its formula ($\Sigma y/n$). In Step (4), the ns have been canceled. Step (5) then follows from the difference of two equal quantities.

For example, the six numbers 8, 20, 14, 3, 9, 6 would be ranked by putting them in ascending order 3, 6, 8, 9, 14, 20 and assigning the ranks 1, 2, 3, 4, 5, and 6 to these scores, respectively. In other words, the score of 3 is ranked one or first, the score of 6 is ranked two or second, etc.[3]

When there are several equal scores or tied scores, the procedure remains the same. For example, the scores 115, 122, 115, 130, 120, 122, 127 would be placed in the following order: 115, 115, 120, 122, 122, 127, 130. Then the ranks 1, 2, 3, 4, 5, 6, and 7 are applied in the usual way. Note that the highest rank is always equal to the number of scores, n, even when there are tied scores. In this example we would say that a score of 115 has ranks one *and* two; a score of 120 has rank three; a score of 122 has ranks four *and* five; a score of 127 has rank six; and a score of 130 has the highest rank, seven.[4]

Definition of the Median

The median of a set of scores has the property that half the scores are less than (or equal to) the median and half the scores are greater than (or equal to) the median. This definition is quite satisfactory for understanding the median. In particular cases, however, there are some minor complications in actually calculating the median. Let us start with the two examples given in Section 3.2.

Example 1

There are five scores: 8, 12, 9, 6, 10. In order to calculate the median, we must first rank the scores. The ranked scores are 6, 8, 9, 10, 12. The median can be thought of as the *middle score* in this ranking. Since there are five scores, the middle score is the third score, 9. So the median is 9. Notice that there are two scores below the median and two above it. If we take the score of 9 itself and consider it split in half, we can say that $2\frac{1}{2}$ scores are less than or equal to the median and $2\frac{1}{2}$ scores are greater than or equal to the median. We have therefore satisfied the condition of the definition given above.

This procedure works whenever the number of scores is an *odd number*.[5] The middle score is then always unambiguously defined. Suppose we had 109 scores and had ranked the scores from smallest to largest. Which score would be the median? The easiest way to find the middle score is to take 109 and divide by 2: $109/2 = 54\frac{1}{2}$. Then the next higher number, 55, is the rank of the score at the median. Note that the fifty-fifth score has 54 scores below it and 54 scores above it. Or, if we split that score in half, there are $54\frac{1}{2}$ scores below the median and $54\frac{1}{2}$ scores above it.

Example 2

The other example given in Section 3.2 was the set of 20 depression scores of psychotics (given originally in Table 2.3). In order to get the median of these scores, we must rank them from lowest to highest. The ranked scores are 43, 51, 51, 57, 60, 62, 64, 64, 68, 69, 69, 69, 71, 73, 75, 78, 78, 78, 83, 83. Now we look for the middle score. However, for an

---

[3] Sometimes ranking is done by putting the scores in *descending* order. The highest score is ranked one, the second highest score is ranked two, etc. With the exception of one procedure described in Chapter 15, it does not make any difference in which direction the scores are ranked. Using the ascending order has the advantage that high scores have high numeric ranks and low scores have low numeric ranks.

[4] In Chapter 15 this method of assigning ranks to tied scores will be modified so that the tied scores get a rank equal to the *mean* of the ranks assigned by the above system. However, the system described above is the appropriate method of ranking as a preliminary to finding the median.

[5] An odd number has a remainder of 1 when divided by 2. For example, 1, 3, 5, 7, . . . , 21, . . . , 109, . . . , 655, etc. are odd numbers.

even *number*[6] of scores there is no middle score—there are *two* middle scores. In this example, the middle scores are the tenth and eleventh scores, which are 69 and 69. Here the two middle scores are equal, so it is clear that the median is 69.

What do we do if the two middle scores are different? In the salary data given in Section 3.1, the two middle salaries were $9,000 and $9,700. (Since there were 26 salaries, the middle scores were the thirteenth and fourteenth, which are $9,000 and $9,700.) We then define the median as the sum of these two scores divided by two. In other words, the median is the average or mean of these two scores. So the median is ($9,000 + $9,700)/2 = $9,350. To find the ranks of the middle scores when n is even, we divide n by two. So when n is 20, we get 20/2 = 10. The two middle scores are the tenth and the eleventh scores. The median is halfway between these two scores, so that there are 10 scores below it and 10 scores above it. In summary, to find the median,

> If n is *even*, n/2 will be an integer. Find the score with that rank and the next higher rank; the median is the mean of these two scores.
>
> If n is *odd*, n/2 will be an integer plus 1/2. Take the next higher integer; the median is the score with that rank.

In practice it can be very time-consuming to rank the scores before calculating the median. If a frequency distribution has already been obtained, the labor involved in calculating the median will be greatly reduced.

Example 3

Consider again the 39 addition times presented in Section 2.1. Their frequency distribution is given in Table 2.2. The median is the twentieth score (39/2 = 19½; the next higher integer is 20). In order to find the twentieth score in the distribution, we start at the bottom of the table and accumulate frequencies until we reach 20. The cumulative frequency at 9 seconds is three; at 10 seconds, 12; and at 11 seconds, 25—more than 20. Clearly, the twentieth score is 11 seconds. Hence, the median is 11 seconds.

Example 4

Finally, we once again graph the data given in Example 2 (20 depression scores) to demonstrate how the median differs from the mean. The univariate scatterplot of Figure 3.2 is repeated in Figure 3.3. The median (69) is shown, along with the mean (67.3). There are 10 scores below or equal to the median and 10 scores above or equal to the median. Three scores are exactly equal to the median: one of them has been included in the 10 scores on the left and the other two have been added to the scores on the right. This figure shows clearly that the median is the middle score. *The median divides the scores in half.* The mean does not have this property. Rather, as shown in the last section, the mean divides the scores so that the distances on the two sides balance each other.

3.4 The Mode

The mode of a set of scores is the most frequent score. This definition is clear and straightforward. As with the median, however, there are certain sets of data for which some complications arise in applying this definition.

Refer to Figure 3.3. *Two* scores, 69 and 78, have the greatest frequency;

---

[6] An even number can be exactly divided by 2. For example, 2, 4, 6, 8, . . . , 20, . . . , 120, . . . , 168, etc. are even numbers.

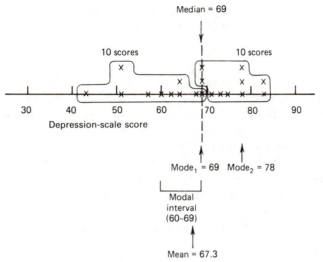

Figure 3.3 Central tendency of depression scores of psychotic group.

each occurs three times in data. We could therefore describe the data as *bimodal*; that is, there are two modes. In the figure, these modes are marked as mode$_1$ = 69 and mode$_2$ = 78.

We have defined the mode for *ungrouped* data. In this way the definition of the mode is unambiguous. You simply look for the score that is most frequent. In some cases, there may be two or more such scores; then there are two or more modes. However, it is sometimes useful to define the mode in a *grouped* distribution. If this is done, the mode will depend on the particular intervals used for the grouping.

For the 20 depression scores of the psychotic group, the frequency distribution is given in Table 2.4 and graphed in Figures 2.4 and 2.5. The graphs clearly show that the most frequent score is in the interval between 60 and 69. We will call this interval the *modal interval*. (See Figure 3.3.) The grouped distribution is clearly unimodal, and so it is useful to report the mode of the distribution. Strictly speaking, the mode is not a single score but an interval of scores, so we use the term "modal interval". You will perhaps agree that for these data reporting the modal interval is more useful than reporting the two particular scores that happen to have the highest frequency.

There is not much more that can be said about the mode. For practice, you should consider the various graphs in Figure 2.11. For each one, find the mode or the modal interval.

3.5 Relationships Among the Measures of Central Tendency

This section outlines the relationships among the three measures of central tendency: the mean, the median, and the mode. The discussion is focused on the graphs of four types of distributions (Figure 3.4). For each type, two graphs are shown. The left member of each pair is an empirical grouped-frequency distribution. The right member is an idealized distribution. (The distinction between empirical and idealized distributions was made in Section 2.7. All the empirical distributions in the figure are histograms, but they could also have been drawn as polygons.) Knowing the relationships among the measures and how these relationships are affected by the type of distribution helps us to decide which measure is most appropriate to calculate for each type of distribution.

Symmetric Unimodal Distributions

Consider first the symmetric and unimodal distributions of Figure 3.4a. In the empirical graph, the modal interval is clearly the interval 35–40, since this interval has the greatest frequency.[7] The median would be somewhere in this interval, since it is clear from the figure that half the scores are above the indicated median and half the scores are below it. The figure also shows that the mean is in this interval. Remember that the mean is the balance point. Since the distribution is symmetric, the balance point must be in the middle.

For an idealized graph, such as that displayed on the right of Figure 3.4a,

[7] We ignore here the distinction between real and apparent limits.

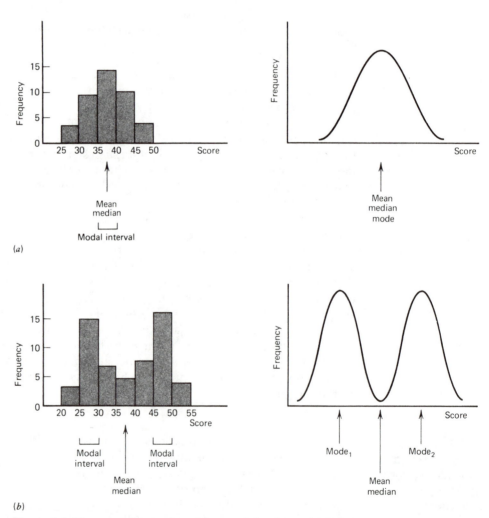

(a)

(b)

Figure 3.4 Measures of central tendency and the shapes of distributions. (a) Symmetric (and unimodal). (b) Bimodal. (c) Skewed-to-the-right (positively skewed). (d) Skewed-to-the-left (negatively skewed).

the same relationships hold. For an exactly symmetric distribution, the mean and median coincide. Furthermore, in a unimodal distribution that is exactly symmetric, the mean, median, and mode will all be identical.

Bimodal Distributions

Next, look at the bimodal distributions in Figure 3.4b. These distributions might also be symmetric, but the important characteristic for our purposes is their bimodality. For the empirical distribution, two modal intervals are shown. Notice that we still consider the distribution to be bimodal even if the frequencies of the two modal intervals are not exactly the same. Because the distribution is approximately symmetric, and because the median divides the scores in half and the mean is the balance point, both the mean and the median must be in the middle of the distribution. This is also true in the idealized distribution. Again, the mean and median will be exactly equal when the distribution is exactly symmetric. In real data the distribution won't be exactly symmetric; hence, the mean and median will be near each other but usually will not be identical.

From Figures 3.4a and 3.4b we see that the three measures of central tendency will be equal or almost equal in symmetric unimodal distributions,

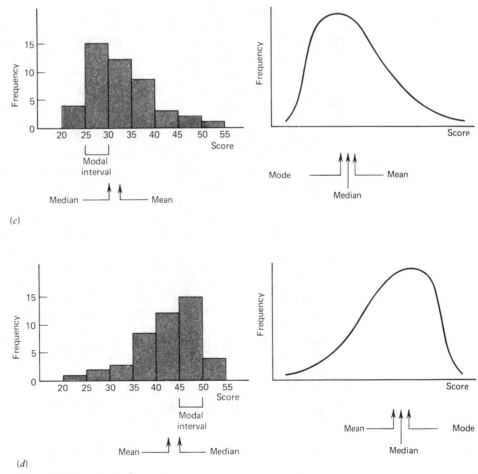

(c)

(d)

Figure 3.4 (continued)

but that the mode will be quite different from the mean and the median in bimodal distributions. Hence, if we are working with a distribution that is unimodal and symmetric, or close to symmetric, it does not really matter very much which measure we use. All three measures will give about the same value. But for a bimodal distribution it does make a difference which measure is chosen. In the bimodal distribution the mean and median do not tell us where most of the distribution falls, even though they correctly tell us where the balance point and the middle score are located. The two modal intervals, or the two modes, tell us where the bulk of the distribution lies. For a bimodal distribution, therefore, we usually report the location of the two modes.

Skewed Distributions

Finally, we turn to nonsymmetric, skewed distributions (Figures 3.4c and 3.4d). The modal interval, or mode, is easy to find, but the mean and median are more difficult to determine. You will notice from the figures that the mean is greater than the median (which is greater than the mode) in positively skewed distributions and that the reverse ordering is true for negatively skewed distributions. This is an important relationship, but you should *not* attempt to memorize it. You are just as likely to get it wrong as right! It is much easier to learn why this ordering occurs and then apply the reasoning to particular distributions.

The reasoning behind this ordering is illustrated in Figure 3.5, which shows a positively skewed distribution. In both the empirical and idealized distributions the median is to the right of the mode (or modal interval). Be sure you appreciate that this must be correct. (If, instead, the median were exactly at the mode, fewer than half the scores would be to the left and more than half to the right of the median. But the median must have half the scores on each side. Hence, the median is somewhat to the right of the mode in a distribution that is positively skewed).

Now think of the horizontal axis as representing a lever on which each of the persons (of equal weight) is sitting at the position of his or her score. An even easier way to think of the situation is to imagine that the histogram or idealized distribution was cut out of a piece of heavy plywood and placed on the axis that is the lever. Suppose now that the fulcrum is placed at the median. Will the lever balance? You may at first think that it will, arguing that there are equal numbers of persons on each side of this fulcrum. In terms of the plywood, you would argue that there is an equal area or an equal weight of plywood on each side of the fulcrum.[8] This is true, but it is not the condition for balance! Note that because of the positive skew, the persons on the right tend to be further from the fulcrum than are the persons on the left. Hence, the lever will fall to the right. The lever is not balanced. In order to make it balance, we must move the fulcrum a small distance to the right. In fact, we must move it to the mean. Therefore, the mean is to the right of the median (in a positively skewed distribution).

Since the mean is to the right of the median, less than 50% of the scores are to the right of the mean and more than 50% of the scores are to the left of the mean. (See the bottom of Figure 3.5.)

---

[8] Frequency may be represented by area. (See Sections 2.2 and 2.6.) The median divides the total area into two equal parts.

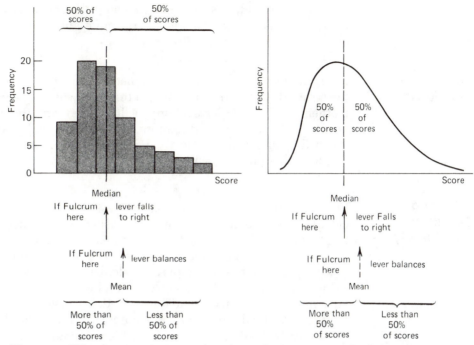

Figure 3.5 Why the mean is greater than the median in a positively skewed distribution.

Return now to Figure 3.4*c*. Note that the mode, median, and mean have been properly placed. In Figure 3.4*d* the three measures are in the opposite order because the distribution is negatively skewed. As an exercise, draw sketches of negatively skewed distributions and show that the order of the three measures is correct. Use a similar argument to that given in discussing Figure 3.5.

What do our discoveries about the relationships among the mean, median, and mode tell us about which measure to use when a distribution is skewed? Recall the salary data from the beginning of this chapter. This distribution was very skewed. It was clear that the median better represents the data than does the mean. For distributions that are only slightly skewed, both are suitable, however. What about the mode? The mode is rarely used as a measure of central tendency even in skewed distributions, unless the peak of the distribution is unusually high (as it was in the salary data).

Extreme Scores   Another way to think of the effect of skewness on the measures of central tendency is to consider the extreme scores in the distribution. In particular, consider the scores in the upper tail of the distribution in Figure 3.4*c*. Suppose that one of the scores now in the interval from 50 to 55 falls instead between 70 and 75. This might happen because of a very unusual subject in the study or simply because of an error in recording. What effect would this have on the mean and median? The mean would be affected since the balance point would move to the right. This one person would now have more influence on our imaginary balance bar, so the fulcrum would have to move to the right to counterbalance this person. But the median would not change at all. Since the median is the middle score, it is unchanged if one score at the extreme of the distribution is moved. This is a very important property of the median. *The median is insensitive to extreme scores.*

Summary of Relationships

For many purposes the median is a better measure of central tendency than is the mean. However, the mean is still the most widely used measure, because its mathematical properties are simpler. It is important, though, to remember the virtues of the median. Furthermore, in distributions that are not badly skewed the median and mean are nearly the same. We can therefore estimate one from another. This property will be exploited in Section 3.6.

3.6 Using a Calculator Intelligently

Before calculators became so readily available, students had to do statistical calculations by hand. With an inexpensive pocket calculator, students often feel that their difficulties with calculations are over. Unfortunately, however, it is still quite possible to make mistakes. The purpose of this section is to show that there are certain principles of calculation that should be followed when using a calculator or doing a calculation by hand. The principles are particularly important when a calculator is used, because a person using a calculator tends to do calculations in a more automatic and unthinking way.

Consider the data in Table 3.2, which contains 26 numbers that you could think of as being the times, in seconds, chalked up by a sample of rats running the length of a runway. Our task is to compute the mean of these numbers. The calculations seem straightforward. We simply add the numbers on our calculator and divide the result by 26. What could be easier?

Reasonableness of an Answer

Suppose we obtained an answer of 9.37 on a calculator. Should we accept this answer? No, we should not. Scanning the 26 numbers in the table, we see that the smallest number is 2.8 and the largest is 8.3. *The mean of a set of numbers cannot be greater than the largest number, nor smaller than the smallest number.* Why is this true? Think of the mean as the balance point, and you will

TABLE 3.2 Twenty-six Numbers

4.6	8.3	7.0	8.2	6.0	6.4
7.3	8.2	4.6	3.5	6.7	
2.8	5.3	4.9	7.3	7.9	
8.2	5.7	7.3	7.7	7.1	
7.6	6.0	7.2	6.8	7.3	

INTERVAL	FREQUENCY
8.0–8.9	4
7.0–7.9	10
6.0–6.9	5
5.0–5.9	2
4.0–4.9	3
3.0–3.9	1
2.0–2.9	1
	26

realize that the balance point cannot be outside the range of the numbers. A lever won't balance if all the persons (scores) are on one side of the fulcrum.

This is, of course, obvious. But the type of thinking it illustrates is so important that we will state a general principle. The simple observation made about the mean is an illustration of the principle.

P2: The Principle of Reasonableness

Check the results of all calculations for reasonableness. The way to do this depends on the problem. You may have to look at the distribution of scores or at a graph, or you may have to apply some theoretical result. You may even have to invent a new way to check the reasonableness of the answer.

The answer that we obtained for the mean was 9.37, which clearly was wrong. Instead of immediately doing the calculation over again, it would be best to estimate roughly what the mean is going to be. Then when we repeat the calculation, we can check our answer against the estimate. One way to estimate what the mean will be is to make a frequency distribution of the data. The distribution does not have to be very detailed. Just use simple, convenient intervals. For these data the distribution at the bottom of Table 3.2 is easy to construct, since each interval contains scores that start with the same digit. For example, the interval 8.0–8.9 contains only scores starting with 8.

It is quite easy to estimate the mean from the frequency distribution. First, estimate the median. By counting the frequencies from the bottom (or the top), you can see that the thirteenth and fourteenth scores are in the interval 7.0–7.9. (Remember that for 26 scores, 26/2 = 13, so the median is found from the thirteenth and fourteenth scores.) Furthermore, the median will be near the low end of the interval, since there are already 12 scores below 7.0. *You need not work out the median exactly, however.* Simply estimate the median roughly, since you want to estimate the mean roughly. And you do not want to spend too much time making this estimate. You can be sure that the median is somewhere around 7.1, 7.2, 7.3, or so.

What does this tell us about the mean? The discussion of the preceding section showed that the mean and median will usually be similar. Notice further that the distribution is skewed toward the smaller scores (that is, negatively skewed, or skewed-to-the-left). We know, therefore, that the mean will be less than the median and that it will be below 7.0—say, somewhere between 6.5 and 6.9.

Let us return to the calculator. We now compute the mean as 6.5346154. This answer is reasonable. Its value fits with our estimate. We are confident in reporting this result as the mean of the data. Notice how the principle of reasonableness gives us confidence in our result.

Usefulness of an Answer

Even though the answer (6.5346154) is a reasonable value of the mean, it is not completely satisfactory. Surely we don't want all these decimal places. How do we judge how many decimal places to report? Considerations of this kind will be called considerations of the *usefulness* of our results. They can be summarized as a principle.

P3: The Principle of Usefulness

Report a useful and appropriate number of digits in any calculated number. The useful number of digits depends principally on the number of digits in the original data and on the size of the data set. It also depends on the use to which the calculated result will be put. Digits that are essentially "noise" should not be reported.

There are really two aspects to this principle. First, you must decide what number of digits to retain in the answer. Second, you must *round* the answer to the chosen number of digits. It is not possible to give hard-and-fast rules for deciding on the number of digits to retain; a few hints on how to do this are given in the above statement of the principle. However, it *is* possible to give definite rules on rounding.

Take the numbers in Table 3.2, all of which are recorded to one decimal place. If the numbers are times in seconds, the time has been recorded to the nearest tenth of a second. Surely, it is useful to report the mean to a tenth of a second, at least. Since there are 26 numbers, the mean of these numbers will be more precise than the individual numbers. So it makes sense to report an additional decimal place in the mean. You might even report two extra decimal places in the mean of a large data set (several hundred scores). You should report the mean of these numbers as 6.53, or possibly 6.535, but definitely not to more than three decimal places. Is there any purpose for which it is *useful* to know that the mean is 6.53462?

The last sentence of the definition of the principle of usefulness refers to digits that are essentially "noise". The technical definition of this concept cannot be given here, but an easy way to understand the concept is to imagine that one of the scores in the data set was slightly changed. For example, suppose the first score in the table was 4.7 instead of 4.6. Our results should not be appreciably affected by this small change. If we now compute the mean of the data, we get 6.5384615. In comparing this number with the previous result of 6.5346154, we see that the two results differ in the third decimal place. This is a rough indication that the third and later decimal places are noise and should not be reported. Hence, we should report the mean of these data as 6.53.

This discussion highlights one problem in using a calculator. The calculator automatically produces a large number of decimal places. If the calculation had been done by hand, most of these digits would not have been calculated. An appropriate (useful) number of decimal places would have been chosen. The principle of usefulness emphasizes the importance of making such a judgment even though the calculator gives a large number of digits.

Rounding

You have learned some ways to decide on a useful number of digits to report in an answer. How do you actually reduce your answer to this number of digits? The procedure is called *rounding*. First, decide on the number of digits to keep, using the ideas given above. Then, *if the first digit to be dropped is 6, 7, 8, or 9, round up*. (For example, rounding 76.5821 to one decimal place gives 76.6, since the first dropped digit is 8.) *If the first digit to be dropped is 0, 1, 2, 3, or 4, round down*. (For example, rounding 76.5304 to one decimal place gives 76.5, since the first dropped digit is 3.) Note that rounding depends on the first digit to be dropped; rounding is *not* carried out one digit at a time, starting at the right.

The only complicated case in rounding occurs when the first digit to be dropped is 5. The rule for rounding numbers *when the first digit to be dropped is 5 is as follows.*

(a) If the digits following the 5 are *not* 000 . . . , round *up.* (For example, when rounding 76.8531 to one decimal place, you get 76.9 by rounding up.)

(b) If the digits following the 5 are exactly 000 . . . , of if there are no digits following the 5, round so that the last retained digit is an even number (0, 2, 4, 6, 8). (For example, rounding 76.85000 or 76.85 to one decimal place gives 76.8. However, rounding 76.95000 or 76.95 to one decimal place gives 77.0—in order that the last retained digit be 0, an even number.)

The effect of these rounding rules is that you will round up half the time and round down half the time.[9] Further examples of the application of these rounding rules may be found in Table 3.3.

There are two final points to be made. *When the last digit retained after rounding is zero, be sure that you report this digit.* For example, suppose you round 76.042 to one decimal place. Your answer should be 76.0, not 76. You may think that these two numbers are the same, but they are not. The first number (76.0) indicates that the answer is useful to the first decimal place, whereas the second number (76) indicates that the answer is only useful to the nearest integer, or whole number.[10]

[9] Consider numbers with just one decimal place that are to be rounded to the nearest integer: for example, 7.6, 7.2, 7.7, 7.5, 7.0, etc. Suppose that all 10 possible digits in the first decimal place are equally frequent. If we round up when this digit is 6, 7, 8, or 9, we will have rounded up four-tenths of the time. If we round down when this digit is 1, 2, 3, or 4, we will have rounded down four-tenths of the time. When this digit is 0, we drop the 0, but the number itself is unchanged. Now, if we always round up those numbers ending in 5, we will be rounding up five-tenths of the time and rounding down four-tenths of the time, given a bias toward higher numbers. Hence, a special rule must be made for numbers ending in 5, to ensure that half the time such numbers are rounded up and half the time they are rounded down. The rule given in the text ensures this, but other, equally satisfactory rules could be stated. The rule given is the most common one, however. Note that the special rule is only for those numbers that end in exactly 5 or in 5 followed by one or more zeros.

[10] The difference between 76.0 and 76 can be seen in another way. Recall the distinction between real and apparent limits. The number 76.0 indicates that the measurement falls somewhere between 75.95 and 76.05, whereas the number 76 indicates a measurement between 75.5 and 76.5. Hence, 76.0 and 76 are different.

TABLE 3.3 Examples of Rounding

ORIGINAL NUMBER	ROUNDED TO ONE DECIMAL PLACE	ROUNDED TO NEAREST INTEGER
17.374	17.4	17.
17.521	17.5	18.[a]
17.486	17.5	17.[a]
17.500	17.5	18.
16.500	16.5	16.
0.486	0.5	0.
0.513	0.5	1.
0.500	0.5	0.
1.500	1.5	2.

[a] Note that the rounding to the nearest integer starts from the original number, not from the result rounded to one decimal place.

Finally, the notions of reasonableness and usefulness apply to all statistical work. This section has illustrated the principles by calculating a mean. These principles, however, are applied throughout this book.

3.7 Summary

A basic part of summarizing any set of data is to determine an *average* or representative value. The technical term for average is *central tendency*. Three measures of central tendency are the *mean*, the *median*, and the *mode*. The first two measures are always calculated from the raw data. Although the mode also may be calculated from raw data, more often it is calculated from grouped data, in which case it is called the *modal interval*.

The mean can be thought of as the *balance point* of a distribution. This terminology reflects the fact that the sum of the *deviations* of all scores from the mean is zero. The median is the *middle score* of a distribution, whereas the mode is the most frequent score. These three measures have the same value in symmetric, unimodal distributions and have special relationships in skewed distributions. The *extreme scores* in a distribution have considerable effect on the mean, but no effect on the median.

Two general principles apply to all calculations: the principles of *reasonableness* and *usefulness*. By making a rough estimate of a value, we can check the reasonableness of the value we obtain by calculation. When we choose the appropriate number of digits to retain in an answer, we are making sure the result is useful. After deciding the useful number of digits we must *round* our answer to that many digits.

Exercises

1 (Sections 3.1 and 3.2) For the set of 24 numbers given below,
   (a) Compute the mean of these numbers in two ways.
   (b) Illustrate by a univariate scatterplot that this mean is the balance point of these numbers.
   The numbers are 43, 45, 45, 46, 41, 44, 44, 42, 44, 48, 42, 44, 45, 45, 46, 43, 43, 44, 42, 44, 45, 45, 44, 42.

2 (Sections 3.3 and 3.4) For the data set given below,
   (a) Compute the median of these data.
   (b) Compute the mode or modal interval of these data.
   The data are 26, 35, 27, 17, 29, 25, 28, 6, 30, 20, 16, 11, 22, 27, 27, 32, 27, 26, 31, 31, 26, 22.

3 (Section 3.5)
   (a) Calculate the mean of the numbers in Question 2.
   (b) Are the values of the mean, median, and mode (or modal interval), relative to one another, consistent with the theory described in Section 3.5 of the text?

4 (Section 3.6) This question is based on the five graphs constructed for Questions 1, 2, and 4 in the Exercises for Chapter 2.
   (a) For each graph, estimate, roughly, the value of the median and the mode (or modal interval).
   (b) From the values obtained in *(a)*, and the shape of the graphs, estimate the value of the mean for each distribution.

5 (Section 3.6) In the preceding question you estimated, roughly, the mean of five distributions. In this question you will check one of these answers—that of Question 2 in the Exercises for Chapter 2 (the data of the time to run 100 meters).
   (a) Compute the mean of these data, using the original 49 numbers.

(b) Compare the computed value with your estimated value. Are your satisfied with the agreement?

(c) To how many decimal places should the computed mean be reported? Why?

(d) Round your answer to that many decimal places.

6 (Section 3.6) Round the following numbers to the indicated number of decimal places.

NUMBER	NUMBER OF DECIMAL PLACES
46.987	2
46.985	2
46.995	2
46.047	1
46.05	1

4 VARIABILITY

The central tendency of a distribution indicates the location of the bulk of the distribution. Distributions with the same central tendency can differ in the degree to which they are spread about their mean, median, or mode. The spread, or variability, of a distribution can be measured in a number of ways.

The most important measure of variability is the standard deviation. However, some simpler measures of variability will be introduced first, and the standard deviation then related to other measures of variability by a general principle. The first section of the chapter describes percentiles, since the measures of variability are related to percentiles.

The chapter, therefore, is organized as follows. First, percentiles are described. Second, some measures of variability are defined in terms of percentiles. Third, the standard deviation is related to these measures. Finally, the formulas for the standard deviation are discussed.

4.1 Percentiles

The percentile is a number that describes the position of a score in a distribution of scores. The percentile of a score is the percentage of scores less than (or equal to) that score. The percentile can also be interpreted as the percentage of the area of the distribution that is below the given score. Figure 4.1 shows a distribution of heights. John's height is given as 188.3 cm. Since 90% of the persons in the distribution are shorter than John (and 90% of the area falls below 188.3), we say that John is at the 90th percentile of the distribution of heights. (See Sections 2.2 and 2.6 for a discussion of the relation between frequency and area.)

There is a slight complication in the definition of percentiles.[1] The following example shows that the percentile must be defined as follows: *the percentile of a score is the percentage of scores less than the midpoint of the score.*

---

[1] The topic of percentiles has a very confusing terminology. Some authors make a clear distinction between a percentile and a percentile rank. They define, for example, the 75th percentile as that score below which 75% of the distribution falls. So the 75th percentile might be a score of 142. On the other hand, a score of 142 has a percentile rank of 75 if 75% of the distribution falls below that score. Percentile is, therefore, a score and percentile rank a number on the scale from 0 to 100, but it is very difficult to keep this terminological distinction clear. In fact, many writers, including developers of psychological tests, do not recognize this distinction. For example, they may refer to a person's percentile as being 75. In the terms of the above example this is wrong, because the percentile *rank* is 75, whereas the 75th percentile is 142.

The author does not believe that this fine distinction is worth maintaining. As Cronbach states, "Various writers use various terms: percentile score, percentile rank, percentile, centile—all have the same meaning" [L. J. Cronbach, *Essentials of Psychological Testing*, 3rd edition, New York: Harper & Row (1970), p. 89]. Here we concentrate instead on the two inverse processes: going from a score to a percentile and going from a percentile to a score.

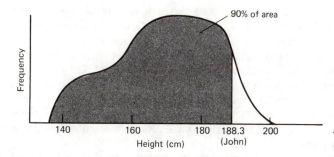

Figure 4.1 Percentile as area.

Example 1

Suppose that John is one of 100 persons in a sample. Some of the heights in this sample are shown in Table 4.1. The heights have been ranked from 1 to 100. John's height of 188.3 cm is 90th from the bottom of the list of 100 heights. Hence, it is easy to see that John is at the 90th percentile, since the percentage of heights less than or equal to John's height is 90 divided by 100, which equals 0.90 or 90%. This example has, of course, been chosen to make the calculations very simple. Usually, the number of scores will not be such a nice number as 100.

You can see, however, that there is a difficulty in calculating percentiles in this way. Consider the percentage of heights that are *greater* than (or equal to) John's height. There are 10 persons taller than John. This means that there are 11 persons whose heights are greater than or equal to John's height. (John must be included in this count, just as he was in the count of 90.) Hence, the percentage of heights greater than or equal to John's

TABLE 4.1 100 Heights Converted to Percentiles

HEIGHT (CM)	RANK	PERCENTILE = PERCENTAGE BELOW MIDPOINT	ROUNDED PERCENTILE
202.6	100	99.5	100
193.7	99	98.5	98
192.8	98	97.5	98
192.6	97	96.5	96
191.3	96	95.5	96
.	.	.	.
.	.	.	.
.	.	.	.
188.4	91	90.5	90
188.3 (John)	90	89.5	90
188.0	89	88.5	88
.	.	.	.
.	.	.	.
.	.	.	.
167.6	52	51.5	52
167.3	51	50.5	50
166.7	50	49.5	50
166.0	49	48.5	48
165.9	48	47.5	48
.	.	.	.
.	.	.	.
.	.	.	.
139.8	1	0.5	0

height is 11%. But 90% + 11% = 101%, not 100%, which would be a more satisfactory result. The difficulty is, of course, that John has been counted in both groups—those who are shorter than John and those who are taller than John.

The solution is to think of John's height or score of 188.3 as being spread out over an interval. The measurement of height is not exact. The heights are shown to the nearest 0.1 cm. This means that a height of 188.3 could actually be a height anywhere between 188.25 and 188.35. (Recall the discussion of real limits in Section 2.4.) Imagine that half the score is below 188.3 and half the score is above 188.3. Note that 188.3 is the *midpoint* of the interval of 188.25 and 188.35. So the measured heights can all be thought of as midpoints.

The number of scores less than 188.3 is 89 + 1/2 = 89.5. The 89 scores are certainly less than John's score, and the half-score comes from half of John's score. Then, since there are 100 scores in all, the percentile of John's score is 89.5. (Note that the number of scores greater than 188.3 is 10 + 1/2 = 10.5; the sum of the numbers 89.5 and 10.5 is 100, as we want.)

Our conclusion from this simple example is that John is at the 89.5th percentile of the height distribution. You are probably wondering about retaining the decimal (0.5) in this percentile. In some sense, certainly, 89.5 is the "exact" answer. But you should remember from the last chapter that an important consideration in the reporting of any number is whether all the digits are *useful*. It is hard to imagine a situation in which we would want to know that the percentile is precisely 89.5. Usually, we would be satisfied with a percentile rounded to the nearest whole number or integer. Following the rounding rule of Section 3.6, we would report the percentile as 90. (Recall that, when the digit to be dropped is 5, with no following digits, we round to the nearest even digit.)

It is quite common to round percentiles to the nearest integer. However, it is not essential to do so if one or more decimal places in the percentile are useful. Rounding to the nearest integer sometimes can produce slightly strange results. Consider the person just above John in the ranking in Table 4.1. The height of 188.4 is ranked 91st. The percentage of scores below the midpoint is 90.5. So the height of 188.4 is at the 90.5th percentile, which if rounded to the nearest integer following the rounding rules, is the 90th percentile. So John and the person just above him have the same percentile, when the percentile is rounded to the nearest integer. This is true throughout the example—all percentiles are even numbers and each even percentile occurs twice. (See the last column of Table 4.1) This unusual result only occurs, however, because we have exactly 100 scores and choose to round to the nearest integer.

Changing Scores to Percentiles

This subsection describes how to convert a given score to a percentile. (The following subsection describes the reverse process—going from a percentile to a score.) The formula to use is

$$\text{Percentile} = \frac{\text{Cumulative frequency below midpoint}}{n} \times 100$$

The *cumulative frequency* is the number of scores less than a certain score, so that the cumulative frequency below midpoint is the number of scores less than the midpoint of the score. The total number of scores is n. The formula is simply a translation of the definition of percentile of a score as the percentage of scores less than the midpoint of the score.[2]

Example 2

Look at the 19 scores given in Table 4.2. Suppose we want to know the percentile of the score of 76. Having determined that the rank of this score is 17 (the score is 17th in the distribution), we have only to compute the numbers in the outlined row for 76. (It is not necessary to calculate all the percentiles if only one percentile is desired. However, all the percentile calculations are shown in the table for purposes of illustration.)

The third column shows that 16.5 scores are below the midpoint of the score of 76. Remember that we think of the score of 76 as spread uniformly from 75.5 to 76.5, so that the midpoint is 76.0. There are, then, 16 scores completely below 76 and half of the score below the midpoint. So there are 16.5 scores below 76.0. This number, 16.5, is the cumulative frequency below midpoint. We then convert this number to a percentage. There are 19 scores in all, so 16.5 scores is $(16.5/19) \times 100 = 86.84\%$ of the scores.

We see, therefore, that the score of 76 is at the 87th percentile in the distribution of scores. We interpret this number as indicating that 87% of the scores are below 76 and 13% $(= 100\% - 87\%)$ are above 76.

Now look at Figure 4.2, in which this result is shown in a univariate scatterplot. The figure shows that 87% of the scores are below the score of 76. Notice that the dividing line is drawn exactly through the cross for the score of 76. This is done to ensure consistency with the method of calculating percentiles in which we divide the score of

---

[2] Not only can the terminology of percentiles be confusing, but the calculations themselves can be made very elaborate. Again, keep your attention focused on the two basic processes of going from a score to a percentile and going from a percentile to a score. To help you, a method of calculation has been chosen that has few steps and that avoids unnecessary complications. This method has three main features: calculations are always made from ungrouped data, the percentage of scores is calculated below the midpoint of the given score, and answers are rounded to a useful number of digits.

TABLE 4.2 Score to Percentile (*n* = 19)

SCORE	RANK	CUMULATIVE FREQUENCY BELOW MIDPOINT	PERCENTILE = PERCENTAGE BELOW MIDPOINT	ROUNDED PERCENTILE
81	19	18.5	97.37	97
80	18	17.5	92.11	92
76	17	16.5	86.84	87
75	16	15.5	81.58	82
74	15	14.5	76.32	76
69	14	13.5	71.05	71
68	13	12.5	65.79	66
65	12	11.5	60.53	61
64	11	10.5	55.26	55
62	10	9.5	50.00	50
60	9	8.5	44.74	45
59	8	7.5	39.47	39
58	7	6.5	34.21	34
57	6	5.5	28.95	29
50	5	4.5	23.68	24
48	4	3.5	18.42	18
42	3	2.5	13.16	13
40	2	1.5	7.89	8
39	1	0.5	2.63	3

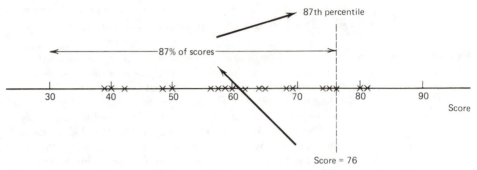

Figure 4.2 Score to percentile (n = 19).

76 in half. Notice the arrows that remind us that in this subsection we go *from* a score *to* a percentile. (The next subsection will explain how to carry out the reverse process.)

Example 3

In the previous example, all the scores were different: that is, there were no tied scores. How do we calculate the percentage of scores below a given score if several persons obtained that score? Consider the data in Table 4.3, which lists 143 scores ranging from 17 to 24. Each possible score has a certain frequency, which is given in the second column of the table.

We follow the same procedure as before. To calculate the percentile of a score of 22, we determine the number of scores below 22 and convert this number to a percentage. From the second column we see that there are 37 + 26 + 14 + 6 + 3 = 86 scores definitely below 22. But we have to follow the "midpoint procedure" as well. Here there are 32 scores of 22. Half of 32 is 16.0. Therefore, the number of scores (cumulative frequency) below the midpoint (22.0) is 86 + 16.0 = 102.0.

The next step is to convert this number to a percentage. We divide by the total number of scores and multiply by 100. The results are shown in the last two columns of the table. As usual, we round the percentile to the nearest integer. So the percentage of scores below 22 is 71%, and hence 22 is at the 71st percentile.

This result is shown graphically in Figure 4.3. The frequencies are represented by bars centered on the individual scores. The frequencies of scores below 22 are shown by the colored bars. These bars represent a total frequency of 86. Notice that half the bar at 22 is also shaded, corresponding to the method we used to calculate the percentile. Half

TABLE 4.3 Score to Percentile (*n* = 143)

SCORE	FREQUENCY	CUMULATIVE FREQUENCY	CUMULATIVE FREQUENCY BELOW MIDPOINT	PERCENTILE = PERCENTAGE BELOW MIDPOINT	ROUNDED PERCENTILE
24	8	135	135 + (8/2) = 139.0	97.203	97
23	17	118	118 + (17/2) = 126.5	88.462	88
22	32	86	86 + (32/2) = 102.0	71.329	71
21	37	49	49 + (37/2) = 67.5	47.203	47
20	26	23	23 + (26/2) = 36.0	25.175	25
19	14	9	9 + (14/2) = 16.0	11.189	11
18	6	3	3 + (6/2) = 6.0	4.196	4
17	3	0	0 + (3/2) = 1.5	1.049	1
	143				

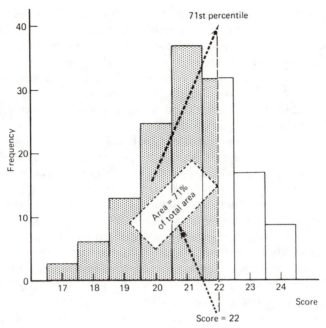

Figure 4.3 Score to percentile (n = 143).

this bar represents a frequency of 16.0. Hence, the cumulative frequency below the midpoint (22.0) is 86 + 16.0 = 102.0, as we found before. Think of this frequency of 102.0 as being represented by the colored *area* of the bars, which amounts to 71% of the total area. This gives us a graphical representation of percentile. *The score of 22 is at the 71st percentile because 71% of the area of the frequency distribution is below 22.*

Finally, note that the arrows in the figure remind us that we have proceeded *from* a *given* score *to* a *calculated* percentile.

The examples in this subsection have illustrated the calculation of a percentile from a score. In all cases, we work from the raw scores: that is, from an ungrouped frequency distribution. We calculate the percentage of scores below the midpoint of the given score. Usually, the percentage is rounded to the nearest integer. The percentage of scores below a given score can be thought of in terms of the graph of the frequency distribution of the scores. The percentage of scores below a given score is represented by the percentage of the area of the distribution below that given score.

Changing Percentiles to Scores

The preceding subsection showed how to answer questions such as What is the percentile of John's height of 188.3 cm? The answer was the 90th percentile. This subsection explains how to answer a question such as What height is at the 90th percentile? The answer, of course, is 188.3 cm. It is important to see that these two questions are related but are not the same. Each question is the reverse of the other. The first question asks for the percentile of a given score; the second asks for the score of a given percentile.

The distinction between these two kinds of questions is so important for later work that we will attach an informal label to each. The first kind of question (going from a score to a percentile) will be called a *forward problem*. The second kind of question (going from a percentile to a score) will be called a *backward problem*. Forward problems are, in a sense, simpler and more natural. You are given a score and have to calculate the percentage of scores below that score. In a backward problem, this process is reversed. You are given a percentage and must find the score that has the given percentage of scores below it. Even the description of the problem is awkward; we think of these problems as backward problems.

The calculations in a backward problem are somewhat more complicated than those in a forward problem. The method will be illustrated by several examples. You should not have any difficulty, however, if you keep in mind that the basic process in backward problems is going from a given percentage (that is, a percentile) to a calculated score. The method to be described has been kept as simple as possible so that you will not forget the basic process.[3]

The two basic processes, forward and backward, are illustrated in Figure 4.4. In a forward problem (Figure 4.4a) we start with a score of 16, for example. From the frequency distribution we find the cumulative frequency below the midpoint (CFBM) of the score of 16. The percentage of scores below this midpoint, the percentile, is then

$$\text{Percentile} = \frac{\text{CFBM}}{n} \times 100 \quad \text{(Forward problem)}$$

where n is the total number of scores. In the figure the result is assumed to be 57%—meaning that 57% of the area of the frequency distribution is below the score of 16. Hence, the score of 16 is at the 57th percentile. Notice the direction of the arrows. In forward problems the arrows go "up" from a score to a percentage.

Conversely, in a backward problem (Figure 4.4b), we start with a percentile—here, the 23rd percentile, which means that 23% of the area is below the required score. If there are n scores in all, then the cumulative frequency below the midpoint (CFBM) of the required score is

$$\boxed{\text{CFBM} = n \times \frac{\text{Percentile}}{100} \quad \text{(Backward problem)}}$$

We then determine that a score of 11 has this cumulative frequency below midpoint. Notice here that the arrows go "down" from a percentage to a score, indicating that this is a backward problem.

Let us now consider three examples of backward problems, based on the same data sets used in the previous examples.

Example 4
Refer back to the 100 heights summarized in Table 4.1. Suppose we want to know what height is at the 48th percentile. Since there are 100 scores, we want to know what height

---

[3] The method for backward problems presented in this text does not employ interpolation. Although interpolation is required for complete consistency between the results of forward and backward problems, in most practical work in statistics this refinement is unnecessary. The method used here always produces an observed score rather than a number between two observed scores. In the terms of Chapter 3, it is rarely *useful* to have a more precise result. The method also assumes that an ungrouped frquency distribution is available; hence, interpolation is not required to find a score within an interval of scores.

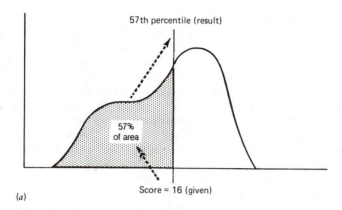

(a)

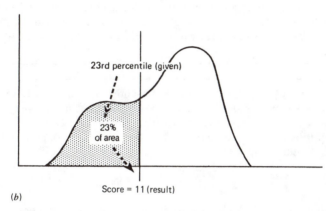

(b)

Figure 4.4 Forward and backward problems.
(a) "Forward" (changing scores to percentiles).
(b) "Backward" (changing percentiles to scores).

is ranked 48th. From the column of ranks in the table, we see that a height of 165.9 is ranked 48th. This answer, possibly rounded to 166, is good enough for our purposes. However, for consistency with the results of the next examples, note that we really should look at the column of *percentages below midpoint*, since that was the column used in forward problems. We see that a rank of 48 is midway between the percentages of 47.5 and 48.5. This would indicate that the correct answer is the mean of the two corresponding scores of 165.9 and 166.0, or 165.95, which rounded is again 166.

An alternative method is possible if the whole of Table 4.1 is available. Given any percentile—say, 48—we simply look down the last column of the table until we locate 48. Here it occurs twice, indicating that the 48th percentile is the mean of 165.9 and 166.0, as we found above. The backward nature of this problem is easily seen. The table originally was constructed to show the results of forward problems. For each score in the first column, the corresponding percentile is given in the last column. To get the result of a backward problem, we simply read the table backward: start with the last column and find the corresponding result in the first column.

Example 5

In Table 4.2 there are 19 scores. Suppose we want to know what score is at the 60th percentile. Imagine first that only the scores and their ranks have been calculated. How would we proceed? We must calculate the cumulative frequency below the midpoint of a score that has 60% of the scores below it. This cumulative frequency is

$$\frac{60}{100} \times 19 = 0.60 \times 19 = 11.4$$

We now must find what score has a cumulative frequency below the midpoint of 11.4. Of course, no score has precisely such a cumulative frequency. A score of 64 has 10.5 scores below its midpoint. A score of 65 has 11.5 scores below its midpoint. Since the required 11.4 is closer to 11.5 than to 10.5, we settle on the corresponding score of 65 as the required score. In summary, 65 is at the 60th percentile.

The above calculation was made assuming we were "starting from scratch", with only the scores and their ranks. If the entire Table 4.2 had already been calculated, the third column would give the number of scores below each midpoint and we would immediately see that 11.4 is closest to 11.5, giving the answer as the score of 65. It is even easier to ignore all columns of the table except the first and the last. We do not need to calculate the number 11.4. We simply start with the required percentage of 60% and look down the last column until we find the closest number. Here the closest number is 61%, which corresponds to the score of 65 in the first column—the required answer. Again we see that backward problems are very easy to solve if the forward problems have been worked out for all scores.

Figure 4.5 illustrates this example. Notice that the vertical dashed line representing the 60th percentile is drawn slightly to the left of the cross for the score of 65. But this line is closer to 65 than to any other score, and hence 65 is the 60th percentile in this set of data.

Example 6

Table 4.3 shows the frequencies of 143 scores ranging from 17 to 24. Suppose we would like to know what score is at the 85th percentile. As usual, we can proceed in one of two ways. If we have only the information in the first two columns, we would start by computing the cumulative frequency corresponding to a percentage of 85%. This cumulative frequency is

$$\frac{85}{100} \times 143 = 0.85 \times 143 = 121.55$$

We then have to find what score has this cumulative frequency below midpoint. We will have to calculate most or all of the third and fourth columns of the table, since we need to know the cumulative frequencies below the midpoints in order to find the cumulative frequency closest to 121.55. As Table 4.3 indicates, the closest cumulative frequency below midpoint is 126.5 for a score of 23. Hence, we conclude that 23 is at the 85th percentile. (You might think that we should compute the 85th percentile as somewhere between 22 and 23, to take account of the fact that 121.55 is not exactly the same as 126.5. However, as indicated in footnote 3, such a technique, called *interpolation* between 22 and 23, is not used here. Certainly, if we want an integer result for the percentile, then 23 is the required answer.)

The second way to proceed assumes that the last column of Table 4.3 is already calculated. We simply find the number in the last column that is closest to 85%. The closest number is 88%, indicating that the required score is 23.

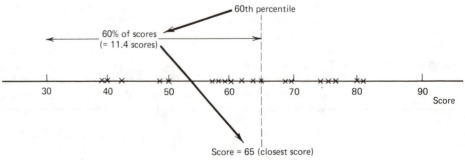

Figure 4.5 Percentile to score ($n = 19$).

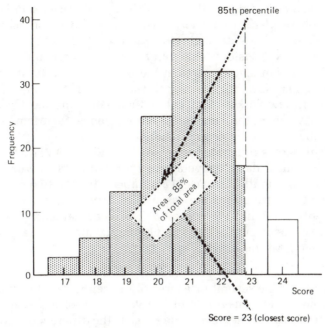

Figure 4.6 Percentile to score (n = 143).

Figure 4.6 illustrates this result. In a backward problem such as this we go *down* from a percentage (85%) to a score. Here the area of 85% extends to the vertical dashed line that is not quite at the score of 23. However, 23 is the closest score and hence is the 85th percentile.

The Median and Quartiles

The median was defined in Section 3.3 as the middle score, or that score which has half the scores below it and half the scores above it. This definition indicates that *the median is the 50th percentile*. Finding the median is a backward problem, since we start with 50% and have to find a score that has 50% of the scores below it. Let us return to Example 1 and Table 4.1. The last column of the table shows that the scores of 166.7 and 167.3 are at the 50th percentile, so that the median is $(166.7 + 167.3)/2 = 167.0$.

Just as it is often convenient to divide a set of scores into halves (giving the median or the 50th percentile), so it is often convenient to divide a set of scores into quarters. The scores that divide the set of scores in this way are called *quartiles*. The *first quartile* is the 25th percentile; hence, one-quarter of the scores are below the first quartile. The *second quartile* is the 50th percentile; therefore the second quartile is simply another name for the median. Finally, the *third quartile* is the 75th percentile; three-quarters of the scores are below the third quartile, and one-quarter of the scores are above it.

4.2 Some Measures of Variability

The ability to calculate percentiles is important in its own right, but the main reason for the extensive treatment of percentiles in the preceding section was to assist you in understanding the important concept of the *variability* of a distribution.

We often compare the temperatures of various cities. In the northern hemisphere, places in the north usually have lower temperatures than places in the south. Such a statement refers to the central tendency of the distribution of temperatures. Consider temperatures in New York and San Francisco. Do the distributions of temperatures in these two cities differ in central tendency? In the summer New York is much warmer than San Francisco. In the winter, however, New York is much colder than San Francisco. On the average, the two cities have about the same temperature. Hence, their temperature distributions have about the same central tendency.

This fact is illustrated graphically in Figure 4.7. [Rather than showing the distribution of daily temperatures (365 numbers), Figure 4.7 shows the distribution of monthly means (12 numbers) for the two cities.] The medians of these two distributions are essentially equal. However, the distributions are markedly different in their *spread*. The technical term for spread is *variability*. So we say that New York and San Francisco differ in the spread or variability of their temperatures.

Range

Just as there are several measures of the central tendency of a distribution, so there are several measures of the variability of a distribution. The simplest measure is called the *range*. The range is the difference between the largest and smallest scores in a distribution. The range of the New York temperatures is $(85.2 - 38.5) = 46.7$, or about 47. (These numbers are found in Figure 2.8 but they can be estimated roughly from Figure 4.7.) Similarly, the range of the San

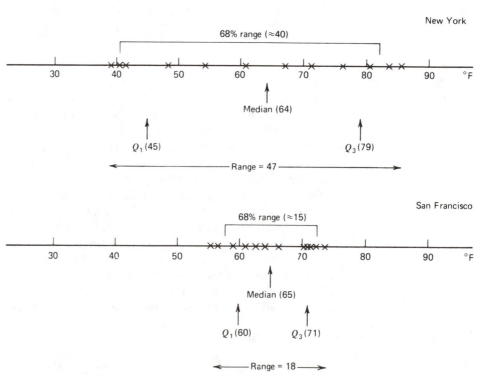

Figure 4.7 Monthly temperatures in New York and San Francisco. (For data, see Figure 2.8.)

Francisco temperatures is $(73.6 - 55.3) = 18.3$, or about 18. It is easy to see that the New York range is much greater than the San Francisco range.

Interquartile Range

Another way to measure the variability of the temperatures is to consider the first and third quartiles. If they are close together, then the variability is small; if they are far apart, then the variability is great. The first and third quartiles for the two cities are indicated in Figure 4.7 by the symbols Q_1 and Q_3. Remember that the quartiles divide the scores into quarters, so that for 12 scores, Q_1 has $12/4 = 3$ scores below it and Q_3 has 3 scores above it. Even without doing precise calculations, we can easily see that Q_1 and Q_3 are further apart in the New York distribution than in the San Francisco distribution. In order to get a number to represent the variability, we take the difference between Q_1 and Q_3. This difference $(= Q_3 - Q_1)$ is called the *interquartile range*. For New York, it is $(79 - 45) = 34$; for San Francisco, it is $(71 - 60) = 11$.

The 68% Range

The interquartile range extends from the 25th to the 75th percentile. For reasons that will be given later in the chapter, it is convenient to define another measure of variability, one that extends from the 16th to the 84th percentile. Since $(84 - 16) = 68$, we call this measure of variability the *68% range*. The 68% ranges for the two cities are marked in Figure 4.7. How were these ranges precisely located? They were *not* precisely located! The important idea is to be able to locate the 68% range *approximately*. Think of the 68% range as including about two-thirds of the scores in the middle of the distribution.[4] This leaves about one-sixth of the scores in each tail. Since one-sixth of 12 is two, there are two scores outside the 68% range at each end of the univariate scatterplots in Figure 4.7. We have indicated in the figure that the 68% ranges are *approximately* 40 (about $82 - 42$) for the New York temperatures and 15 (about $72 - 57$) for the San Francisco temperatures.

A Comparison of the Three Measures

The three measures of the variability of the temperature distributions agree in indicating that the New York temperatures are two-and-one-half to three times as variable as the San Francisco temperatures. This result also agrees with our intuitive knowledge of the temperatures in the two cities: the temperatures are fairly constant in San Francisco but vary widely in New York.

The range is clearly the easiest to compute: simply take the difference between the largest and smallest score. Its principal disadvantage is that the range is very sensitive to extreme scores. If there is one unusually large or one unusually small score, the range will be very large, even though the bulk of the scores may be only moderately variable. In order properly to represent the variability of the bulk of the scores, it is better to use the interquartile range or the 68% range. Changes in extreme scores have little or no effect on these two measures. The interquartile range is, in effect, the 50% range. Hence, it will always be somewhat smaller than the 68% range. The 68% range is preferable only because it is naturally related to the measure we will soon discuss, the

---

[4] Since $2/3 = 66.7\%$,. or about 67%, it would seem to be more accurate to call this interval the 67% range instead of the 68% range. Since we only want aproximate values, there is no essential difference. Furthermore, the number 68% has a special significance in statistical theory (see Section 7.3).

standard deviation. Otherwise, the interquartile range is just as good a measure of variability as the 68% range.

The interquartile range and the 68% range are more difficult to compute than the range, because percentiles must be calculated if precise values of these measures of variability are desired. We will rarely have to calculate these percentiles precisely, however, since usually we will estimate these percentiles from graphed frequency distributions.

4.3 The Variability Principle

This section initiates the description of the most important measure of variability, the standard deviation, which is related to the 68% range. The relationship, although only approximate, is extremely useful in estimating the standard deviation of most sets of data and most distributions. The formal definition of the standard deviation is given in the Section 4.4. The definition is a complex formula that does not show very obviously how the standard deviation measures the spread or variability of a set of scores.

The 68% range itself is a good measure of the variability of a set of scores. The standard deviation is approximately equal to *half* the 68% range. This result, with its qualifications, can be stated as a principle.[5]

P4: The Variability Principle
The standard deviation is approximately one-half the 68% range. This approximation is quite accurate for unimodal distributions that are not extremely skewed. The principle applies to both empirical and idealized distributions.

Look at Figure 4.8, which shows the frequency distribution of 20 scores. The 68% range is 22 and is marked on the graph. Half the 68% range is 22/2, or about 11. We indicate this value by the bar extending from the middle of the 68% range to its right boundary. (For convenience, we call the middle of the 68% range the "center". It is neither the mean nor the median exactly, unless the distribution is symmetric. However, for the purpose of the variability principle, which only states an approximation, we can use any convenient value in the center of the distribution or in the center of the 68% range.)

The above procedure for getting an approximate value of the standard deviation may seem complicated. It is not. To obtain the approximate standard deviation, look at the distribution, roughly locate the boundaries of the 68% range, and divide this range in two. It is as simple as that. No calculations are necessary. The estimated standard deviation found in this way can be used to check whether a standard deviation calculated from the complex formula of Section 4.4 is reasonable.

The variability principle applies both to empirical distributions derived from data (as in Figure 4.8) and to idealized distributions. Several idealized distributions are shown in Figure 4.9. A symmetric unimodal distribution is shown in Figure 4.9a. The 68% range extends from about 16 to 38, a distance of 22. Hence, we estimate that the standard deviation is about 22/2 = 11.

---

[5] The idea for the variability principle comes from W. Mendenhall and M. Ramey, *Statistics for Psychology*, Belmont, Calif.: Wadsworth, 1973, in which this principle, in a different form, is called the empirical rule. Since the rule applies to any distribution, theoretical or empirical, a more general name is used here.

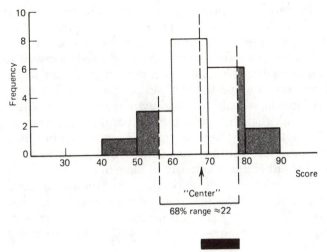

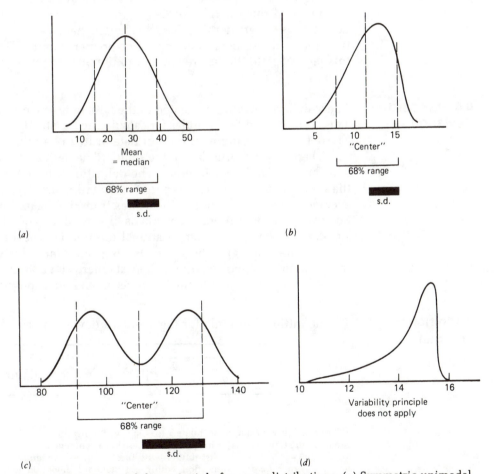

Figure 4.8 The variability principle for an empirical distribution. (For the data on which this distribution is based, see Table 4.4.)

Figure 4.9 The variability principle for some distributions. (a) Symmetric unimodal. (b) Moderately skewed unimodal. (c) Bimodal. (d) Extremely skewed.

The variability principle is most accurate for symmetric unimodal distributions. However, the principle is still quite accurate for nonsymmetric distributions, provided that they are not very skewed. For example, the principle would apply quite well to the moderately skewed unimodal distribution shown in Figure 4.9b. The 68% range extends from about 8 to 16. Since 16 − 8 = 8, we estimate that the standard deviation is 8/2, or about 4.

A bimodal distribution is shown in Figure 4.9c. Although at first sight you might think that the variability principle would not apply here, it actually works well. Rather than stating the precise conditions under which the variability principle applies, we will usually restrict its use to unimodal distributions that either are symmetric or have only moderate skew.[6] The fact that the variability principle also works for some bimodal distributions (such as the one shown in Figure 4.9c) is just an added bonus. Many distributions of real data and most theoretical distributions used by statisticians are unimodal with little or no skew (see Figures 4.8, 4.9a, 4.9b). The variability principle applies to such distributions.

Finally, Figure 4.9d shows a distribution to which the variability principle certainly does not apply. The distribution is too skewed. Even for this distribution, however, we can make a useful statement about the standard deviation. The range of this distribution is about 6.0. We can be sure that the standard deviation is not greater than half this range: that is, about 3.0. For any distribution, *the standard deviation cannot be greater than half the range.* (Note that this fact refers to the full range, not to the 68% range.)

4.4 Standard Deviation

The standard deviation is related to the 68% range by the variability principle. This section provides formulas for calculating the standard deviation precisely. These formulas apply to any set of data or to any empirical distribution.

There are two formulas for the standard deviation: the *definitional* formula and the *computational* formula. The definitional formula shows directly that the standard deviation is a measure of variability: namely, the square root of the average squared deviation of the scores from the mean. For most sets of data, however, the definitional formula is ill-suited to actual calculation of the standard deviation; the computational formula is used instead. The formulas are equivalent and give the same answer. [The standard deviation described in this section is called the *unadjusted* standard deviation; its symbol is s′ ("s-prime"). An adjustment to the formulas given here is described in Chapter 11.]

Definitional Formula

The definitional formula for the standard deviation is

$$s' = \sqrt{\frac{\Sigma(y - \bar{y})^2}{n}} \quad \text{(Definitional formula for s.d.)}$$

[6] The variability principle is related to some generalizations of Tchebychef's inequality. For example, the Gauss-Winkler inequality implies that a symmetric unimodal distribution must have at least 5/9 = 55.5% of the distribution within ±1 standard deviation of the center. On the other hand, in order for the proportion within ±1 standard deviation to greatly exceed 68%, the distribution must have extremely long tails. See H. J. Godwin, *Inequalities on Distribution Functions,* London: Griffin (1964) for a detailed technical discussion.

This formula directs us to compute the deviation, $y - \bar{y}$, of each score (y) from the mean $(\bar{y})$. Each of the deviations is squared and summed. The sum is divided by the sample size, n, to give the mean squared deviation. The standard deviation is the square root of the mean squared deviation.

Example 1	Twenty scores are listed in the first column of Table 4.4. The mean of these scores is 67.3 and the deviations of each score from this mean are given in the second column. The deviations, some of which are positive and some of which are negative, must sum to zero; it is wise to compute this sum as a check on the calculation of the mean and the deviations. (Recall, from Section 3.2, that the mean is the balance point of the distribution.)

The squares of the deviations, and their sum, are shown in the last column of the table. The standard deviation is, then,

$$s' = \sqrt{\frac{\Sigma(y - \bar{y})^2}{n}}$$

$$= \sqrt{\frac{2282.20}{20}}$$

$$= \sqrt{114.11} = 10.682228 = 10.7 \text{ (rounded)}$$

Notice that the standard deviation is rounded to one decimal place, following the principle of usefulness. The original scores are recorded to the nearest integer, and it is useful to report the standard deviation to one or two digits more than the original data. But note that the *intermediate results* have not been rounded. *You should never round*

TABLE 4.4 Sums Required for the Definitional Formula

SCORE $= y$	DEVIATION $=$ DISTANCE FROM MEAN $= y - \bar{y} = y - 67.3$	SQUARED DEVIATIONS $= (y - \bar{y})^2$
60	$-\ 7.3$	53.29
78	$+10.7$	114.49
68	$+\ 0.7$	0.49
69	$+\ 1.7$	2.89
43	-24.3	590.49
75	$+\ 7.7$	59.29
64	$-\ 3.3$	10.89
71	$+\ 3.7$	13.69
51	-16.3	265.69
69	$+\ 1.7$	2.89
78	$+10.7$	114.49
57	-10.3	106.09
83	$+15.7$	246.49
64	$-\ 3.3$	10.89
73	$+\ 5.7$	32.49
51	-16.3	265.69
62	$-\ 5.3$	28.09
69	$+\ 1.7$	2.89
78	$+10.7$	114.49
83	$+15.7$	246.49
	Sum = 0.0 $\Sigma(y - \bar{y})$	Sum = 2282.20 $\Sigma(y - \bar{y})^2$

numbers until the final result has been obtained. When you encounter more-complex calculations, your final answer may be incorrect if this rule is not followed.

The frequency distribution of these 20 scores is shown in Figure 4.8. We noted in Section 4.3 that the standard deviation of this distribution is about 11. Hence, the standard deviation we have computed is reasonable.

Computational Formula

The preceding formula for the standard deviation is called the *definitional* formula because it clearly shows that the standard deviation is computed from the deviations of the scores from the mean. It is easy to see that the result of this formula will be a measure of variability, since it depends on the distances of the scores from the mean. Although the *computational* formula does not reflect so clearly that it is a measure of variability, it gives *exactly* the same result as the definitional formula.

The computational formula for the standard deviation is

$$s' = \sqrt{\frac{\Sigma y^2 - (\Sigma y)^2/n}{n}} \quad \text{(Computational formula for s.d.)}$$

Two sums are required in this formula: the sum of the scores, Σy, and the sum of squares of the scores, Σy^2.

Example 2

The 20 scores of Example 1 are shown again in Table 4.5. The sum of the scores is shown at the bottom of the first column. The squares of each score appear in the second column, with their sum at the bottom. These two sums, and $n = 20$, are all that the computational formula requires.

The standard deviation is

$$s' = \sqrt{\frac{\Sigma y^2 - (\Sigma y)^2/n}{n}}$$

$$= \sqrt{\frac{92868 - (1346)^2/20}{20}}$$

$$= \sqrt{\frac{92868 - 90585.80}{20}}$$

$$= \sqrt{\frac{2282.20}{20}}$$

$$= \sqrt{114.11} = 10.682228 = 10.7 \text{ (rounded)}$$

The result is exactly the same as obtained from the definitional formula.

TABLE 4.5 Sums Required for the Computational Formula

SCORE $= y$	SQUARED SCORES $= y^2$
60	3600
78	6084
68	4624
69	4761
43	1849
75	5625
64	4096
71	5041
51	2601
69	4761
78	6084
57	3249
83	6889
64	4096
73	5329
51	2601
62	3844
69	4761
78	6084
83	6889
Sum = 1346 Σy	Sum = 92868 Σy^2

Deciding Which Formula to Use

The two formulas give the same result, but in the definitional formula the numerator is $\Sigma(y - \bar{y})^2$ and in the computational formula it is $\Sigma y^2 - (\Sigma y)^2/n$. The fact that the two formulas must produce the same result can be shown by an algebraic proof.[7]

For most purposes, you should use the *computational* formula. The definitional formula is difficult to use unless the mean, $\bar{y}$, is a "simple" number: that is, exactly an integer or (as in Example 1) exact to one decimal place. (The

---

[7] Here is a proof of this relationship.

$$\Sigma(y - \bar{y})^2 = \Sigma(y^2 - 2y\bar{y} + \bar{y}^2) \tag{1}$$
$$= \Sigma y^2 - 2\Sigma y\bar{y} + \Sigma\bar{y}^2 \tag{2}$$
$$= \Sigma y^2 - 2\bar{y}\Sigma y + n\bar{y}^2 \tag{3}$$
$$= \Sigma y^2 - 2\left(\frac{\Sigma y}{n}\right)(\Sigma y) + n\left(\frac{\Sigma y}{n}\right)^2 \tag{4}$$
$$= \Sigma y^2 - 2\frac{(\Sigma y)^2}{n} + \frac{(\Sigma y)^2}{n} \tag{5}$$
$$= \Sigma y^2 - \frac{(\Sigma y)^2}{n} \tag{6}$$

In step (1) the square is expanded. In step (2) the summation of each term is given separately. Since $\bar{y}$ is a constant in the summation in the second term, it can be taken outside the sum as a multiplier in step (3). Similarly, $\bar{y}^2$ is a constant summed n times in the third term, giving $n\bar{y}^2$ in step (3). In step (4), $\bar{y}$ is replaced by $(\Sigma y/n)$ in the second and third terms. The expressions are simplified in step (5), showing that the second and third terms are the same except for the multipliers; this results in a final simplification, in step (6).

answer obtained from the definitional formula will be inaccurate if a rounded value of the mean is used.)

The Variance

One other measure of variability that is sometimes used is called the *variance*, defined simply as *the square of the standard deviation*. We have already calculated it for the 20 numbers given in Tables 4.4 and 4.5, since the variance is the number obtained before the square root is taken. The symbol for the variance is s'^2 indicating directly that the variance is the square of the standard deviation. For the 20 numbers, s'^2 is 114.11 or 114.1.

The variance is frequently used by mathematical statisticians in preference to the standard deviation, because square roots are difficult to handle mathematically. We will also encounter variances in Chapter 17 in the method called the "analysis of variance". However, the standard deviation is easier to understand, since it measures the spread of scores in a natural way, as shown in the variability principle. The variance, on the other hand, is a squared measure and is difficult to interpret. Suppose that the 20 scores of Table 4.4 represented weights in kilograms. The standard deviation then would be 10.7 *kilograms*. This is easy to interpret. However, the variance is 114.1 *squared kilograms*. It is hard to interpret a "squared kilogram". Hence our preference for the standard deviation over the variance, even though we must calculate the variance before getting the standard deviation.

4.5 Summary

Percentiles are used to indicate the location of a score in a distribution of scores. Two basic techniques were described: finding the percentile corresponding to a given score and finding the score corresponding to a given percentile. Going from a score to a percentile is called a *forward* problem and going from a percentile to a score is called a *backward* problem. In order to compute percentiles precisely, we must compute *cumulative frequencies* and, in particular, the *cumulative frequencies below midpoint*. A cumulative frequency may be represented as the *area* below a given score in a frequency distribution.

The *median* is the 50th percentile in a distribution. The *first* and *third* *quartiles* are the 25th and 75th percentiles, respectively.

In order to measure the *variability* or *spread* of a distribution, we compare the location of various percentiles of the distribution. Three such measures are the *range*, the *interquartile range*, and the *68% range*.

By means of the *variability principle* we can relate the *standard deviation* to the *68% range*. The standard deviation is calculated precisely by either a *definitional formula* or a *computational formula*. The calculation can be checked by an estimate of the standard deviation obtained from half the 68% range.

Another measure of variability, commonly used in mathematical work, is the *variance*, which is simply the square of the standard deviation.

Exercises

The following data sets are used in the questions.

Data Set *A*: 73, 56, 39, 47, 88, 89, 74, 49, 58, 66, 91, 63.

Data Set *B*:

SCORE	FREQUENCY
57	3
56	4
55	7
54	12
53	29
52	15
51	6
50	1

Data Set *C*: 4.1, 8.1, 6.6, 8.8, 6.3, 3.7, 7.3, 7.7, 6.6, 8.2, 5.7, 5.3, 5.9, 6.0, 4.9, 7.0, 6.3, 6.4, 6.1, 5.7, 4.9, 5.8, 7.9, 6.1.

Data Set *D*: 10, 16, 8, 9, 14, 15, 12.

1 (Section 4.1)

(a) Find the percentiles (to the nearest integer) of the scores 47 and 88 in Data Set *A*.

(b) Find the percentile (to the nearest integer) of the score of 55 in Data Set *B*.

2 (Section 4.1)

(a) What score in Data Set *A* is closest to the 80th percentile?

(b) What score in Data Set *B* is closest to the 35th percentile?

3 (Section 4.1)

(a) Find the three quartiles in Data Set *C*.

(b) What is the percentile of a score of 4.9 in Data Set *C*?

(c) What is the median of Data Set *C*?

4 (Sections 4.2 and 4.3)

(a) Draw a histogram of Data Set *B*. Show on the graph the four measures of variability: range, interquartile range, 68% range, and standard deviation. (*Note:* The last three measures should be determined only approximately, from the graph. Do not carry out any detailed calculations.)

(b) Use a graph or scatterplot to show the four measures of variability of Data Set *C*. (*Note:* The range is easy to calculate and the interquartile range can be determined from your answer to 3(*a*). However, determine the 68% range and the standard deviation only roughly. Do not carry out any detailed calculations.)

5 (Section 4.4)

(a) Compute the variance and the standard deviation of the scores in Data Set *D*. Do the calculation in *two* ways: first, use the definitional formula; second, use the computational formula.

(b) Compute the variance and the standard deviation of the scores in Data Set *A* by the computational formula.

REVIEW CHAPTER A

Note: This chapter may be omitted without loss of continuity, although tree diagrams, briefly introduced here, are extensively used in Chapter 5.

The two main topics of this chapter—a summary of some of the material in the preceding chapters and an introduction to the use of tree diagrams—may at first sight appear to be unrelated. The connection between these two topics arises from the fact that tree diagrams can also be very useful in summarizing a set of concepts and in showing the relationships among them. Tree diagrams will be used extensively in later work in probability theory.

To introduce the technique of tree diagrams, some simple examples from everyday life are given in the next section. In Section A.2, tree diagrams are used to summarize the previous chapters. In Section A.3, the detailed characteristics of tree diagrams are described more formally.

A.1 Tree Diagrams in Everyday Life

Situations involving a series of decisions can be represented by a special kind of *tree diagram* called a *decision tree*. Consider a simple example from everyday life. Suppose you are trying to decide what to do in the coming summer. You think of various possibilities: getting a job, taking a trip, taking courses, doing work related to your future career, etc. It would be helpful to put some of the possibilities in a decision tree such as the one shown in Figure A.1. Note that at the left of the tree there is a general statement, "summer plans"—the overall purpose of the tree. Note, too, that this "tree" is on its side; if you rotate the picture 90° anticlockwise, you will see a rough similarity to a real tree.

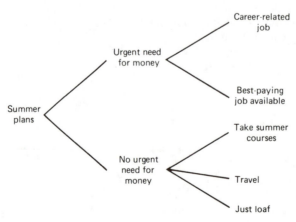

Figure A.1 Decision tree for summer plans.

Following "summer plans", the tree divides into two branches corresponding to two situations: you have enough money or you do not have enough money. In this way the tree shows clearly that in order to decide what to do in the summer, you must first consider the state of your finances. The several options shown at the next level of the tree depend on this state. If you need money, you must decide between a job that is related to your future career or one that simply pays you as much as possible. If you are not short of money, however, you have three choices according to the diagram: take summer courses, travel, or just loaf.

Let us take another example. Suppose you have $20 and the chance to buy two $10 tickets in a lottery that has several prizes, each worth $1 million. A decision tree representing the various alternatives is shown in Figure A.2. The first set of branches corresponds to a decision you must make: you can buy two tickets, one ticket, or no ticket. The branches that extend from these alternatives correspond to the outcome of the lottery, not to a decision you make. If you buy two tickets, you could win $2 million, $1 million, or nothing. These three possibilities are drawn from "buy two tickets", as they all are possible if you buy two tickets. If you buy one ticket, however, there are only two possibilities: win $1 million or win nothing.

If you do not buy a ticket at all (the lowest branch of the tree), you can win nothing, of course. Note that the result "win nothing" is explicitly shown following the choice "do not buy any tickets" in order to make this level of the tree complete. The column of "final assets", at the right of the diagram, allows you to compare the various outcomes. Note that if you bought only one ticket, you would retain $10, and that if you do not take part in the lottery, your assets would amount to $20.

At this point you might try to draw a decision tree for some decision you are currently facing. Are you going to stay home tonight or go out to see a movie? Should you sell your car or not? You are entertaining relatives from out of town and the weather forecast says that it may rain; should you take them to a baseball game or to the theater?

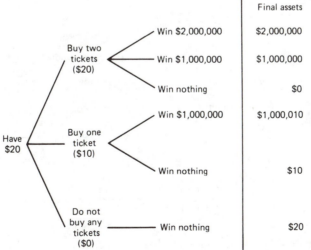

Figure A.2 Decision tree for lottery.

A.2 A Summary of Chapters 2, 3, and 4

This section summarizes some of the topics of Chapters 2, 3, and 4 by means of special tree diagrams.[1] When you have to make a choice among several techniques, these diagrams show you what the choices are. However, these diagrams are even more important for study and review purposes, since they allow you to see a chapter as a whole and to explore the relationships among the different topics.

Consider first the overview of Chapters 2, 3, and 4 in Figure A.3. The general topic of these chapters, "summarization of data", is given on the left of the figure, at the base of the tree. In general, the summarization principle (Section 2.3) guides all summarization. Three basic techniques of summarization have been described so far: distributions, measures of central tendency, and measures of variability. These techniques form the three main branches of the tree.

At the next level of the tree, each of these topics is broken down into subtopics. The divisions are still quite coarse, and the tree could be extended much further. Because the extended tree would be very cumbersome, however, the three main topics are shown in separate subtrees illustrated in Figures A.4, A.5, and A.6.

The subtree in Figure A.4 shows some of the topics of Chapter 2. Distributions are classified as either frequency or relative-frequency distributions. Distributions are either grouped or ungrouped and may be displayed in a table or in a graph. There are, therefore, eight possible distributions, shown clearly by the tree diagram. The arrow indicates the path that ends at "grouped relative-frequency graph". Each of the specific distributions is reached by a

---

[1] Most of the tree diagrams used in this section have been drawn to emphasize a choice among alternatives at each level. Such diagrams lead naturally to tree diagrams representing probabilistic situations. In summarizing any topic or chapter, however, it may be more useful to develop diagrams that have arrows connecting various parts of the diagram and thus show the relationships of topics. Since such complex diagrams do not have the shape of trees, they are not emphasized here.

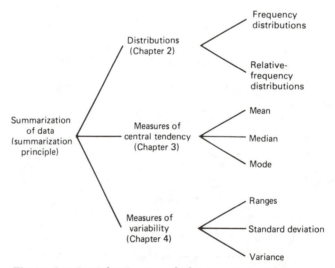

Figure A.3 Partial summary of Chapters 2, 3, and 4.

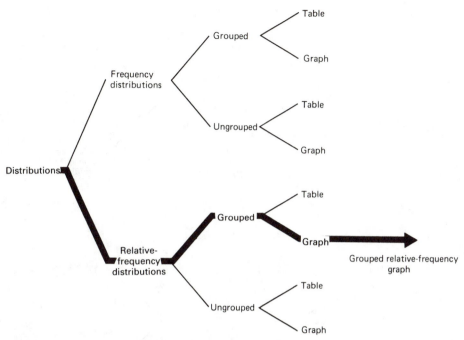

Figure A.4 Partial summary of Chapter 2.

different path through the tree. Try to follow the path that gives an "ungrouped frequency-distribution table".

There are, of course, many other topics in Chapter 2, including the distinctions between polygons and histograms, between discrete and continuous variables, between idealized and empirical distributions, and between various shapes of distributions. You should try to include some of these distinctions in Figure A.4. At some point your diagram will get too complicated, however. Several moderately simple diagrams are usually more helpful than one enormously complicated diagram.

Some topics of Chapter 3 are summarized in Figure A.5. The term "understanding" is used in the diagram to include the principle of reasonableness and some of the techniques that can be used to obtain approximate values of the mean and median. In particular, your understanding of the mean should be enhanced by thinking of the mean as a balance point and by considering the effect of extreme scores on it. (Notice that we have changed the treatment of the branches of the summary tree. The branches are no longer *choices* among alternatives, but *aspects* of one topic.)

Finally, some of the main topics of Chapter 4 are summarized in Figure A.6.

**A.3
Characteristics
of Tree
Diagrams**

The general notion of a tree diagram should be clear to you by now. The preceding sections have emphasized their usefulness in representing everyday decisions (decision trees) and in summarizing some of the topics of this book (summary trees). This section provides a more formal description of tree dia-

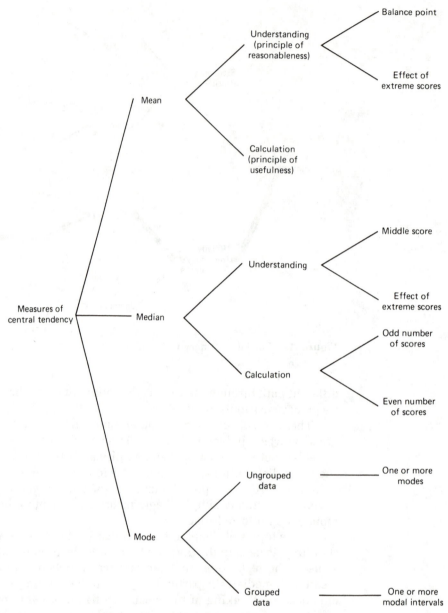

Figure A.5 Partial summary of Chapter 3.

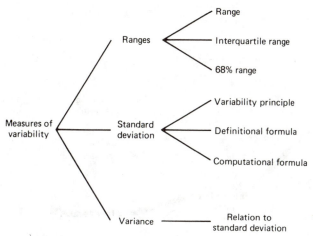

Figure A.6 Partial summary of Chapter 4.

grams. This information will prepare you for the probability theory of the next chapter.

In this text, tree diagrams are used primarily to illustrate situations in which a series of choices is made. Since the use of tree diagrams is of great help in understanding probability theory, the general characteristics of tree diagrams will be stated as a principle.

P5: Characteristics of Tree Diagrams

A series of choices or possibilities can be represented by a tree diagram. A set of choices, among which one choice is to be made, is called a *stage* of the diagram. The individual choices at any one stage are called *branches* of the tree. A complete series of choices, one choice at each stage, is called a *path* through the tree. The tree diagram as a whole (all the paths) represents all possible series of choices.

The key terms—stage, branch, and path—are illustrated in Figure A.7. The tree diagram for "summer plans" is the same as in Figure A.1. The base of the tree describes the whole tree ("summer plans"). Then there are two stages or levels of the tree corresponding to the two decisions or choices that must be made. Since the first decision is about the state of finances, the state of finances is the first stage of the tree. As there are two possible financial states, two branches are shown at this stage. If there were four possible financial states to be considered, there would be four branches at this stage. For example, the four branches might be "have less than $2,000", "have $2,000–$4,000", "have $4,000–$6,000", and "have over $6,000". The four branches at the first stage would have these four labels.

The second decision is represented by the second stage of the diagram, which includes the possible summer activities. There are five particular activities, which are shown as branches at the second stage. Note carefully that two of these activities ("career-related job" and "best-paying job available") begin from the top branch of the first stage, since they are considered pos-

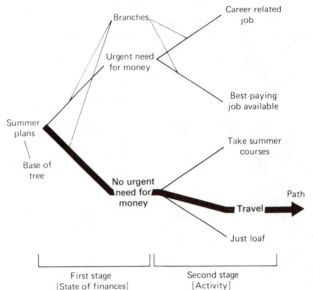

Figure A.7 Terminology for tree diagrams.

sibilities only if there is an urgent need for money. The remaining three activities begin from the bottom branch of the first stage.

The branches represent the possibilities at a particular stage. A complete series of choices is represented by a complete path through the tree, beginning at the base of the tree and continuing all the way to its outermost branches. There are five such paths in the tree, representing all possible series of choices. For example, the series of choices "no urgent need for money" and "choose to travel" is represented by the arrow shown in the tree. Notice that this path is composed of two branches, one at each stage of the tree. *A path is always composed of one branch for each stage of the tree.* If there were six stages in the tree, each path would be composed of six branches.

A path was drawn in Figure A.4 to represent a grouped relative-frequency graph. The tree diagram in Figure A.4 has three stages. In the first stage, the choice is between frequencies and relative frequencies; in the second stage, between grouped and ungrouped frequencies; and in the third stage, between presenting the distribution in a table or presenting it in a graph. The grouped relative-frequency graph is represented by one possibility at each stage.

Note that the tree diagram in Figure A.4 differs in an important way from the summer-plans tree. At the second stage of Figure A.4 the two possibilities at the top part of the tree ("grouped" and "ungrouped") are the same as those at the bottom of the tree. In the summer-plans tree, the branches of the second stage are different at the top and the bottom. In the next chapter, most tree diagrams for probabilistic events look more like Figure A.4 than Figure A.7; the possibilities (branches) at any stage are the same regardless of the particular branch chosen at the preceding stage.

Drawing tree diagrams is partly an art and partly a science. There is often more than one correct diagram for a given situation, but there is usually an appropriate level of detail to put in a diagram. Consider the two diagrams in Figure A.8. Each represents the situation of tossing two coins and one die. In

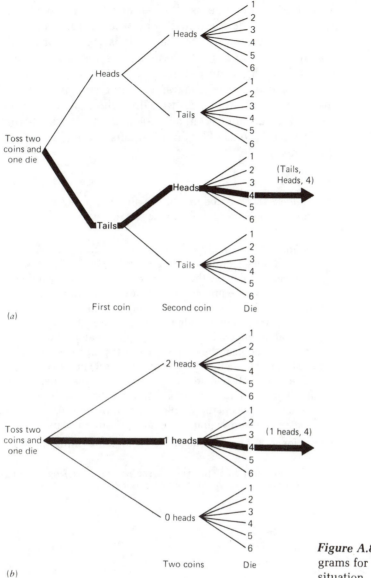

Figure A.8a there are three stages to the tree, representing the results of the first coin, the second coin, and the die. The particular result of a "tails" on the first coin, a "heads" on the second, and a "four" on the die is shown in the figure. In Figure A.8b this same result is shown in a different way. Here the two coins are placed together in a single stage and the result of tossing these two coins is indicated by the number of "heads" (2, 1, or 0). The path shows the same particular result as before—one "heads" (out of two coins) and a "four" on the die—but Figure A.8b is slightly more concise than Figure A.8a. One form of the diagram is sometimes more useful than another in considering a particular situation.

A.4 Summary

Tree diagrams can be used to represent everyday situations (*decision trees*) or to summarize the relationship between various topics in a chapter (*summary trees*). Tree diagrams are used extensively in probability theory; the general characteristics of such diagrams are stated as a principle. Each diagram has *stages*, *branches*, and *paths*.

(The use of tree diagrams is not confined to statistics and probability theory. In the field of scientific investigation called *problem solving*, to which psychologists, computer scientists, and others are making contributions, a major technique used by both humans and machines is the construction of tree diagrams.[2])

Exercises

1 (Section A.1) Describe in words a situation in everyday life and draw a tree diagram summarizing some of the choices in the situation. The diagram should have two or three stages.

2 (Section A.2) In each part of this question you are asked to sketch two distributions. You need not generate actual data for the graphs. The purpose of this question is to see if you can emphasize the difference between the two distributions in your sketch.
(a) Sketch the frequency distributions of IQ in two groups of persons. In one of the groups, all the persons have similar IQs. In the other, they have very different IQs. Label the axes, and include a scale on each IQ axis. Select the most significant or the principal difference between the graphs. In at most one or two sentences, describe this *difference* between the graphs, using appropriate statistical terms.
(b) Sketch the frequency distributions of IQ of two other groups of persons. In one of these groups, all persons have low IQs; in the other group, they all have high IQs. Label the axes and include an IQ scale. Select the most significant or the principal difference between the graphs. In at most one or two sentences, describe this *difference* between the graphs, using appropriate statistical terms.

3 (Section A.3) Look at the two tree diagrams in Figure A.9.

[2] For more information on this subject, see Wayne A. Wickelgren, *How to Solve Problems: Elements of a Theory of Problems and Problem Solving*. San Francisco: Freeman (1974).

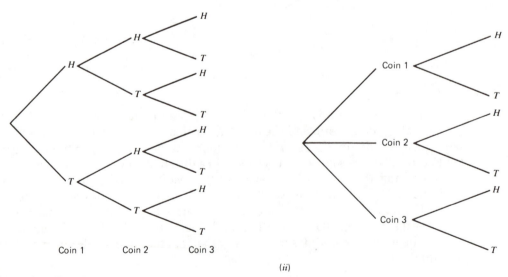

(i)   *(ii)*

Figure A.9 Exercise 3.

(a) Which of the two tree diagrams represents all the possible results of tossing three coins? Explain why the other diagram is incorrect, according to the definition of tree diagrams given in the chapter.

(b) Copy the correct diagram on your answer sheet and mark an example of a stage, a branch, and a path on it.

(c) Describe an experiment for which the incorrect diagram is the correct diagram.

4 (Section A.3) Look at the two tree diagrams in Figure A.10. Consider the two situations:

(a) A coin is tossed, then a die is tossed.

(b) A coin is tossed. If it comes up "heads", a die is tossed. If it comes up "tails", a second coin is tossed.

Identify which tree diagram represents (a) and which represents (b).

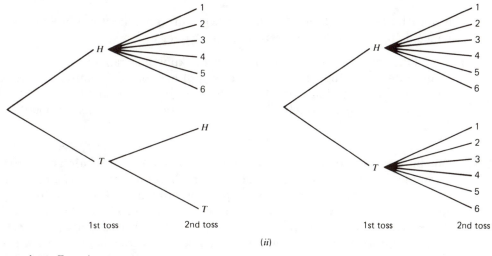

(i)   *(ii)*

Figure A.10 Exercise 4.

5 RANDOM SAMPLING AND PROBABILITY

Note: All of Chapter 5 may be omitted without loss of continuity. Some of the concepts in this chapter are repeated in Chapter 6.

Statistical inferences depend on what is known as random sampling. This chapter describes the basic process of sampling from populations. Since any consideration of random sampling requires a prior definition of probability, the theory of probability, which is central to statistics, is discussed in detail here.

5.1 Populations, Samples, and the Sample Space

One purpose of statistics is to draw conclusions about a population by studying a sample drawn from that population. A *population* can be thought of as a complete set—usually very large—of persons, animals, or things. A *sample* is a subset—usually small—selected or drawn in some way from the population.

Populations and samples can be represented by tree diagrams (Section A.3). A population may be represented by a one-stage tree diagram. A sample of size *n* is represented by a path through an *n*-stage tree. (For example, a sample of size four is represented by a path through a tree that has four stages.) Each of the *n* stages of the tree represents the selection of one person (or animal or thing) from the population. The full *n*-stage tree represents all possible samples of size *n* from the population. This set of all possible samples is called the *sample space*.

Example

Consider a country with a population of 1 million adults. You want to determine the proportions of males and females in this population. If males and females are equally numerous in the population, the proportion of males would be 0.500 and the proportion of females would be 0.500. Unequal numbers of the two sexes would yield proportions different from 0.500, although the two proportions would still add to 1.000.

Obviously, to obtain an accurate estimate of the proportions of males and females in the population, we would have to study a fairly large sample of persons from the population—say, 100 persons. The proportions of males and females found in that sample would then constitute estimates of the proportions in the population. For example, if there were 53 males and 47 females in a sample of 100, the sample proportion of males would be 0.53 and the sample proportion of females would be 0.47. These numbers are our best estimates of the population proportions.

A one-stage tree diagram can be used to represent this population of 1 million people (Figure 5.1). Each branch of the tree represents one person in the population. Several of the persons are identified in the diagram and each person's sex—the characteristic in which we are interested—is given. Notice that the diagram is drawn only in

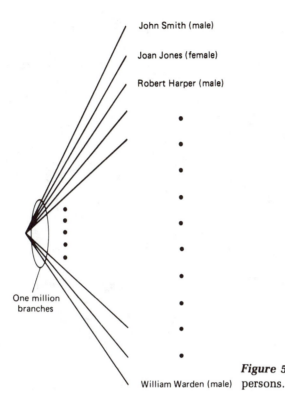

John Smith (male)

Joan Jones (female)

Robert Harper (male)

One million branches

William Warden (male)

Figure 5.1 A population of 1 million persons.

part, since it would be difficult to show 1 million branches. Furthermore, we presumably cannot label the 1 million branches with names and sexes; if we could do so, we would know all that we needed to know about the entire population, and therefore would not need a sample. Tree diagrams, although helpful in conceptualizing situations, seldom need to be complete and fully accurate representations of them.

Next, we want to represent a sample from the population. We do not know which sample will be drawn or selected in our study. *Therefore, we must represent all possible samples.* If you think about it, you will see that for a population of 1 million there are almost countless possible samples of size 100. How can we possibly list or represent all these possible samples?

The possible samples are represented in *a tree diagram with one stage for each person in the sample.* For samples of size two, there are two stages to the tree, as in Figure 5.2*a*. The first person can be anyone in the population, so the first stage is identical to the population tree in Figure 5.1. The second person in the sample can be anyone in the population except the first person chosen. So if in Figure 5.2*a* the first person selected is J. S., the selected second person can be J. J., R. H., etc.—that is, any one of the remaining 999,999 persons in the population. Similarly, if the first person selected is J. J., the second person selected can be any one of the remaining 999,999 persons. Since the first stage of the tree has 1 million branches and for *each* of these branches there are 999,999 branches at the second stage of the tree, there are 1,000,000 × 999,999 = 999,999,000,000 *paths* in the whole tree! Each path in the tree represents a possible sample. One of these samples (J. S. and W. W.) is shown in Figure 5.2*a*.

The tree for *n* = 2 is huge, but bigger diagrams are to come. Consider the tree in Figure 5.2*b*, which provides a mere hint of the possible samples of size 100. Try to visualize this tree. Each path through the tree consists of 100 branches, one for each stage of the tree. Each such path represents a possible sample from the population.

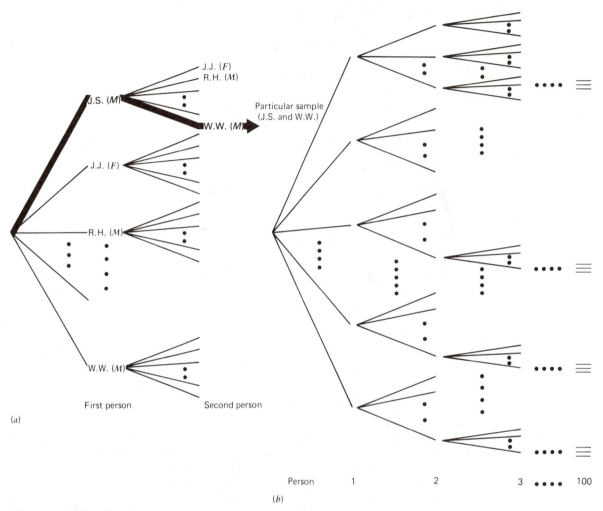

Figure 5.2 Sampling from a population of 1 million persons. (a) n = 2. (b) n = 100.

The next section will demonstrate how such a large tree diagram can be greatly reduced in size without losing the information that is essential for representing the sample space.

5.2 Reducing the Size of the Tree Diagram

The tree diagram illustrated in Figure 5.1 is so large because it attempts to show every person in the population as a separate branch. But we are interested only in each person's sex; individual identities are not essential. We can simplify the tree diagram for the population by having only two branches, one for each sex. This will also simplify the tree diagram for the sample space.

In a *full sample space* illustrated in Figure 5.2 all persons are shown as separate branches at each stage of the tree. In a *reduced sample space*, there is one branch for each different value of the variable being studied. (See Examples 1 and 2, below.) If the variable is sex, there would be two branches at each stage; if the variable is a rating of "sociability" and the rating can be any of five values, there would be five branches at each stage. In this way the sample space for any *discrete* variable having only a small number of values can be represented in a small tree diagram.

Many variables, however, are *continuous*, not discrete. A continuous variable has an infinite number of values. The device of having the branches represent only different values of a variable is not effective in reducing the size of the sample space for continuous variables. Example 3 suggests how to reduce the sample space for such variables.

Example 1

The example introduced in Section 5.1 can be simplified if we represent only the two sexes in each stage of the tree diagram of the reduced sample space. This step is taken in Figure 5.3a, which shows explicitly all possible samples of size four. The male (M) branch at the first stage represents all males in the population, and the female (F) branch

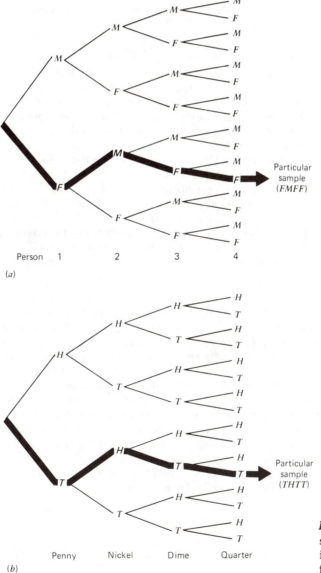

Figure 5.3 Two reduced sample spaces. (a) Selecting persons. (b) Tossing four coins.

represents all females. The first stage of the tree represents the first person in the sample, the second stage the second person, and so on. At each stage the person chosen may be male or female.

When the tree for the sample space is as small as that in Figure 5.3a, it is easy to count the number of paths in the tree and, hence, the number of samples in the reduced sample space. By direct count, there are 16 paths in the tree. (The number 16 can also be obtained arithmetically. There are two branches at the first stage. For each of these branches there are two branches at the second stage. This makes $2 \times 2 = 4$ paths through the first two stages. There are $4 \times 2 = 8$ paths through the first three stages and $8 \times 2 = 16$ paths through all four stages. In short, the number of paths or samples in the sample space is $2 \times 2 \times 2 \times 2 = 16$.)

The smallness of this number of samples (16) does not invalidate the previous count of millions of samples in the full sample space. Note that we have changed our definition of what counts as a "different sample". The particular sample (*FMFF*) singled out in Figure 5.3a represents *all* samples whose first member is female, second member is male, and third and fourth members are females. We count all such samples as just one sample, *FMFF*.

Example 2

When four coins are tossed, there are a large number of possible results. The reduced sample space is shown in Figure 5.3b. The four stages of the tree are labeled by the names of the four different coins, but the stages could be called "coin 1", "coin 2", "coin 3", and "coin 4", if, say, four pennies or four dimes were being tossed. Each coin can land on either heads (*H*) or tails (*T*); therefore, there are two new branches splitting off at each stage of the tree.

Note that the structures of trees for selecting four persons (Figure 5.3a) and for tossing four coins (Figure 5.3b) are identical. In both, a very large population has been reduced to only two possibilities. In the former, the reduction is made simply by considering the selected person's sex. In the latter, the reduction is made merely by considering whether the coin falls heads or tails. We could greatly expand the possibilities by noting where the coin fell on the table, the orientation of the coin, etc. So in many situations the population, and therefore the sample space, can be greatly simplified if we consider only one aspect of the individual members of the population.

Example 3

Let us return to the population of 1 million adults. Suppose that we are interested in the heights rather than the sexes of the adults. Figure 5.4a shows a tree diagram of the full sample space for samples of size two. The names of the persons are not given, but each person's height (in centimeters) is shown in the diagram. At the first stage, one person out of 1 million is selected. For each person selected at the first stage, another person is selected at the second stage. So far, there is no significant difference from the sample spaces shown in Figure 5.2. However, it is not possible to collapse the tree into manageable size by considering only different values of the variable, as was done in Figure 5.3 In Figure 5.3 the variable was *discrete* (there are only two values of the sex variable), whereas in Figure 5.4 the variable being studied is *continuous* (almost any height is possible between, say, 126 and 219 cm).[1] What we can do is to group the heights into a frequency distribution. In Figure 5.4b the heights at each stage have been grouped into five intervals. There are now 25 ($= 5 \times 5$) samples in the reduced sample space.

---

[1] The distinction between discrete and continuous variables was discussed in Section 2.4.

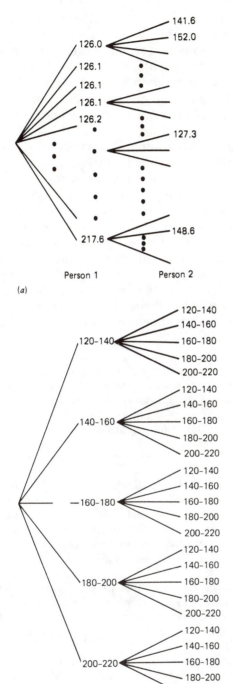

Figure 5.4 Sampling from a population of 1 million heights. (a) Full sample space. (b) Reduced sample space.

5.3 Adding Probabilities to the Tree Diagram

The purpose of many statistical studies is to determine what proportion of the persons or objects, in a given population, possess a certain characteristic. For example, what is the proportion of females in the population? or what is the proportion of persons in the population who can learn a certain task in 10 minutes? In order to determine proportions such as these, we must draw a *random sample* from the population. *A random sample is a sample in which every member of the population has an equal chance (probability) of being included.*[2]

In order to define a random sample, it is necessary to introduce the term "chance", or "probability". Everyone intuitively understands probability; we have all tossed coins, bought lottery tickets, or wondered whether it will rain. It is quite difficult, however, to *define* probability. The concept of probability is several centuries old, and yet there is no commonly accepted definition. We will examine two definitions of probability: the *theoretical* definition and the *empirical* definition. From the theoretical definition we can easily compute probabilities and add them to tree diagrams. Some examples will be given in this section. The empirical definition, introduced in Section 5.4, essentially provides a way to check whether the theoretical probabilities are correct.

According to the *theoretical definition of probability, the probability of a characteristic is the proportion of persons (or objects) having that characteristic in the population.*

Example 1

If there were 1,000 balls in a bag and a single ball were drawn at random, then the probability of selecting any specific ball in the bag would be 1/1,000. This value follows from the definition of a random sample given above. Every member of the population of 1,000 balls has an equal chance of being selected. Giving each ball a probability of 1/1,000 ensures that the sum of the probabilities is 1.00, which indicates that it is certain that some ball will be selected.

Example 2

In Figure 5.1 the population of 1 million persons was shown in a one-stage tree diagram. If persons are selected from this population at random, then the probability of selecting any single person is 1/1,000,000. These probabilities could be indicated on each branch of the tree. We found it more convenient to reduce the population tree to two branches only—one for males and the other for females. Each branch in that tree would be labeled with a probability of $\frac{1}{2}$, if the proportions of males and females in the population were each $\frac{1}{2}$.

This method of labeling a tree is extended to samples of size four in Figure 5.5a. The probability that the first person in the sample is male is $\frac{1}{2}$, and the probability that the first person is female is $\frac{1}{2}$. These probabilities are indicated on the branches at the first stage. At the second stage of the tree—for the second person in the sample—the probability of each sex is again $\frac{1}{2}$.

There is a slight inaccuracy in Figure 5.5a. The probabilities on the branches for the second, third, and fourth stages are not exactly $\frac{1}{2}$. If the first person in the sample is male, there will be 499,999 males and 500,000 females left in the population. Therefore the

---

[2] Although this definition probably provides the best way to introduce random sampling, it is defective in two ways. First, it is really a definition of *simple* random sampling. More generally, a sample is random if every member has a specified probability of being included; these probabilities do not have to be equal. Second, for some populations, random sampling must be defined in terms of specified probabilities of each value of a *variable*, rather than of individual members of the population. Again, in this case the probabilities do not have to be equal.

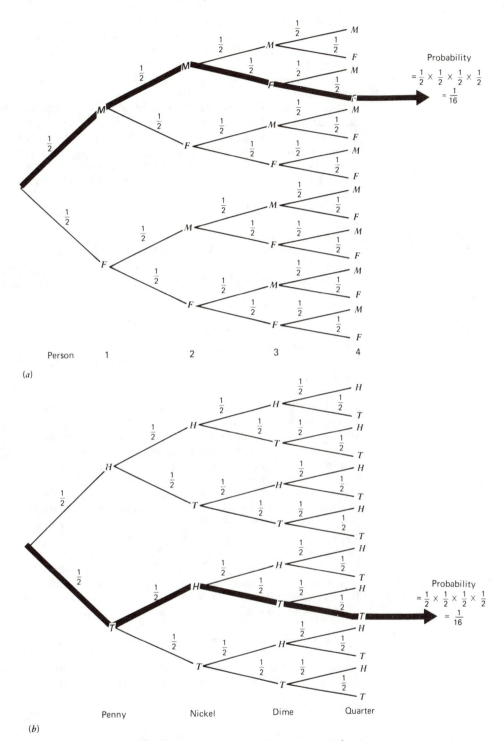

Figure 5.5 Adding probabilities to the tree diagram. (a) Selecting persons.
(b) Tossing four coins.

probability of M on the top branch at the second stage of the tree should be 499,999/999,999, and not $\frac{1}{2}$. Similar changes should be made for the other branches. However, these corrections are so slight that they are usually not necessary.

Example 3

The sample space for tossing four coins is shown again in Figure 5.5b, to which the probabilities of each branch have been added. The assumption has been made that all the coins are *fair* (that is, that there is an equal chance that they fall heads or tails). Hence, the probability marked on the two branches for each coin is $\frac{1}{2}$. We are assuming that in the population of tosses of the penny, for example, half the tosses will be heads and half will be tails. It then follows, by the theoretical definition of probability, that the probability that the penny will fall heads is $\frac{1}{2}$ and that the probability that it will fall tails is also $\frac{1}{2}$.

The Probability of a Sample

There is an additional reason for writing the probability on each branch of the tree diagram. By using these probabilities, we can very easily compute the probability of any sample in the sample space. *To get the probability of a sample, multiply the probabilities on the branches in the path representing the sample.*

Example 4

Let us return to the selecting of persons in Figure 5.5a and consider the probability of drawing the indicated sample *MMFF*. Following the above rule for getting the probability of a sample. we find that the probability of sample *MMFF* being selected is $\frac{1}{2} \times \frac{1}{2} \times \frac{1}{2} \times \frac{1}{2} = \frac{1}{16}$.

Explanation If we imagine samples of size four being drawn a large number of times, then we can say that in about half the samples the first person selected will be male and in about half the samples the first person will be female. Now consider the second person selected for the sample: again, in half the samples the second person will be male and in half the samples that person will be female. This is true whether the first person was a male (top branch) or a female (bottom branch). Accordingly, in the half of the samples in which the first person is male, the second person will be male in half of *those* samples. Hence the first *and* second persons will both be male in $\frac{1}{2} \times \frac{1}{2} = \frac{1}{4}$ of the samples. In short, the probability of *MM* is $\frac{1}{4}$. In the same way one can argue that the probability of *MF*, *FM*, and *FF* are each $\frac{1}{2} \times \frac{1}{2} = \frac{1}{4}$.

Following the same logic, the probability that the third person will be female is $\frac{1}{2}$, so in half the samples in which the first two persons are male (*MM*), the third person will be female (*F*). This means that the sequence *MMF* will occur in $\frac{1}{4} \times \frac{1}{2} = \frac{1}{8}$ of the samples. Finally, the sequence *MMFF* will occur in $\frac{1}{8} \times \frac{1}{2} = \frac{1}{16}$ of the samples. This confirms the probability of the sample *MMFF* shown in Figure 5.5a.

Example 5

The probability of any sample of four coins can be worked out in the same way—by multiplying probabilities along a path in the tree (Figure 5.5b). The particular sample *THTT* is shown in the figure. The probability of this sequence is $\frac{1}{2} \times \frac{1}{2} \times \frac{1}{2} \times \frac{1}{2} = \frac{1}{16}$. Try to give an explanation, like that given in Example 4, justifying the multiplying of probabilities here.

5.4 Empirical Definition of Probability

The theoretical definition of probability states simply that the probability of any characteristic is the proportion of persons or objects having that characteristic in the population. But how can this definition help us determine a probability for a population about which we do not have full information?

The mathematical theorem called the *law of large numbers* states that the relative frequency of occurrence of a characteristic in a random sample approaches the probability of that characteristic as the sample size becomes large. This theorem confirms our intuition: since probability, defined theoretically, is the proportion or relative frequency of a characteristic in the population, the relative frequency of a characteristic in a very large random sample should be very close to the value of the probability.

Having obtained the relative frequency in a very large random sample, we have justification, from the law of large numbers, to assert that the number obtained for the relative frequency *is* the value of the probability of the characteristic. We are using a second definition of probability called the *empirical definition of probability: the probability of a characteristic is the relative frequency of that characteristic in a very large random sample.*[3]

Consider a characteristic that has probability p. From the empirical definition of probability, we can say that the relative frequency of the characteristic in a random sample is expected to be close to p, if the sample size, n, is large enough. In short,

$$\text{Expected relative frequency} = p$$

Since the relative frequency = frequency/n, we can write

$$\text{Expected frequency} = n \times p$$

The empirical definition of probability and the calculation of the expected frequency of a characteristic in a sample of size n are illustrated in the following two examples.

Example 1

Consider a large population consisting of an equal number of the 10 digits: 0, 1, 2, . . . , 9. By the theoretical definition of probability, each digit has a probability of 1/10, since the proportions of each digit in the population are 1/10. Now suppose that digits are randomly drawn from this population, one at a time. Table A5 in the Appendix gives a list of digits drawn in this random fashion. The digits in the table are grouped in sets of five, but the table has been constructed so that each digit in the table is a *random digit* drawn from a population consisting of an equal number of the 10 digits.

Using this table, we can illustrate the law of large numbers and the empirical definition of probability. Consider the probability that the digit drawn is a 3. We want to show that, as the number of digits drawn increases, the relative frequency of 3s approaches the probability of a 3, which is 1/10 or 0.1000. Table 5.1 summarizes the results of drawing 5,000 digits. [These are the 5,000 digits in the first 10 columns of Table A5, read downward, column (set of five digits) by column. Table 5.1 gives the results for each block of 250 digits: that is, for each half column of Table A5.]

Look at the *block-by-block analysis* given in columns 2, 3, and 4 of Table 5.1. The number of 3s in each of the 20 blocks is shown in column 3. The ratio of the number of 3s to the number random digits in each block (250) is given in column 4. This ratio is simply the relative frequency of 3s in each block. For example, in the first block there are nineteen 3s, giving a relative frequency of 19/250 = 0.0760—quite a bit less than 0.1000, the theoretical probability. In the second block there are thirty-one 3s, giving a relative

---

[3] Alternatively, this definition can read *the probability of a characteristic is the* long-run *relative frequency of that characteristic.* The term "long-run" means that as the random samples get larger and larger, the relative frequencies in the samples approach the probability.

TABLE 5.1 Sampling from a Population of Digits

BLOCK NUMBER	BLOCK-BY-BLOCK ANALYSIS			CUMULATIVE ANALYSIS		
	DIGITS	THREES	RATIO	DIGITS	THREES	RATIO
1	250	19	.0760	250	19	.0760
2	250	31	.1240	500	50	.1000
3	250	28	.1120	750	78	.1040
4	250	31	.1240	1000	109	.1090
5	250	33	.1320	1250	142	.1136
6	250	25	.1000	1500	167	.1113
7	250	26	.1040	1750	193	.1103
8	250	29	.1160	2000	222	.1110
9	250	30	.1200	2250	252	.1120
10	250	22	.0880	2500	274	.1096
11	250	28	.1120	2750	302	.1098
12	250	27	.1080	3000	329	.1097
13	250	17	.0680	3250	346	.1065
14	250	18	.0720	3500	364	.1040
15	250	28	.1120	3750	392	.1045
16	250	34	.1360	4000	426	.1065
17	250	18	.0720	4250	444	.1045
18	250	31	.1240	4500	475	.1056
19	250	17	.0680	4750	492	.1036
20	250	29	.1160	5000	521	.1042

frequency = 31/250 = 0.1240, which is larger than 0.1000. As you scan down the column of relative frequencies (ratios), you can see that they are sometimes less than 0.1000 and sometimes greater than 0.1000, with no discernable pattern. (In the sixth block, the relative frequency is exactly 0.1000.)

A pattern does appear in the *cumulative analysis* provided in columns 5, 6, and 7. The cumulative (total) number of digits that have been sampled by the end of each block is shown in column 5. For example, by the end of the second block, 500 digits have been sampled. The cumulative number of 3s by the end of each block is shown in column 6. By the end of the second block there are 19 + 31 = 50 threes. The cumulative relative frequency is shown as a ratio in column 7. By the end of the second block, this ratio is 50/500 = 0.1000, which fortuitously equals the theoretical probability. By the end of the next block, the cumulative relative frequency has changed to 0.1040, and so on.

The last column of Table 5.1 shows that the cumulative relative frequency changes quite rapidly at first and increases to 0.1136. It then decreases slightly and oscillates for a while. But as the cumulative number of digits reaches 4,000 the relative frequency declines to 0.1065; and by 5,000 digits, it falls to 0.1042. If the number of digits drawn were increased to 10,000, or to 100,000, the relative frequency undoubtedly would be even closer to the theoretical value of 0.1000. This outcome is assured by the mathematical theorem (the law of large numbers) and by the careful construction of the table so that the digits are truly random. Table A5 is part of a much larger table. If 100,000 digits of this larger table were studied, and if the relative frequency of 3s were not very close to 0.1000, then we would have evidence that the random process used to produce the table was defective or that the probability (proportion) of 3s in the population was not 1/10.

Example 2

Consider an IQ test that has been standardized in such a way that 34% of the population have scores between 100 and 115. Using the theoretical definition of probability, we can say that the probability that a randomly selected person from this population will obtain a score between 100 and 115 is 0.34, or about one chance in three. How would this probability be reflected in a sample of 200 persons? From the empirical definition of probability, we expect the relative frequency of persons with IQs between the given limits to be approximately 0.34. So we would expect about $0.34 \times 200 = 68$ persons to have IQs in this range. We would not be surprised if the actual number fell between about 65 and 75, giving relative frequencies between 0.325 and 0.375. (Chapter 12 will show how to specify the limits within which we can expect the sample relative frequency to fall.)

The proportions of persons in the population with IQ scores in certain intervals are given in the following table.

IQ RANGE	PROBABILITY
Below 85	0.16
85 to 100	0.34
100 to 115	0.34
Above 115	0.16
	1.00

The frequencies and relative frequencies in a sample of 200 persons might be

IQ RANGE	FREQUENCY	RELATIVE FREQUENCY
Below 85	27	0.135
85 to 100	72	0.360
100 to 115	70	0.350
Above 115	31	0.155
	200	1.000

The relative frequencies and the corresponding probabilities are similar, and they would become more and more similar as the sample size increases.

5.5 Types of Random Processes

Terms such as "random digit" and "random process" were used in Section 5.4 without formal definition. These terms are difficult to define, and a precise definition will not be attempted here. A *random process* may be thought of as a process with various possible results, each of which is unpredictable and each of which has a certain probability. All the probabilities of the various results do not have to be equal.

The distinction between random processes in which the results all have the same probability and those in which the results have different probabilities is the distinction between *fair* and *biased* random processes. This distinction is illustrated in Example 1. Another distinction, between *sampling without replacement* and *sampling with replacement*, is illustrated in Example 2.

Example 1

When a single die is tossed, it can come up either 1, 2, 3, 4, 5, or 6. We say the die is *fair* if each of these results has equal probability: that is, $\frac{1}{6}$. Think of a large, hypothetical population of possible tosses. Theoretically, the proportion of 1s in this population is $\frac{1}{6}$; the proportion of 2s is $\frac{1}{6}$; and so on. We have the following probabilities for each result.

FACE OF DIE	PROBABILITY
1	$\frac{1}{6}$
2	$\frac{1}{6}$
3	$\frac{1}{6}$
4	$\frac{1}{6}$
5	$\frac{1}{6}$
6	$\frac{1}{6}$

The probability of the face coming up 5, for example, is $\frac{1}{6}$. This means that if we toss the die randomly, the chances are 1 in 6 that the die will come up 5. Suppose we toss the die 100 times. We would expect that about $\frac{1}{6}$ of the tosses would be 5: that is, that about $\frac{1}{6} \times 100 = 16.67$ of the tosses would result in a 5. This does not mean that exactly 16 or 17 of the tosses must be 5. Only in the long run will the proportion of 5s be $\frac{1}{6}$. This is what is meant by the empirical definition of probability given in Section 5.4. So in 100 tosses of the die we might get twelve 5s, giving a relative frequency of 5s in this sample of 12/100 = 0.120. In 1000 tosses, however, we would expect the relative frequency of 5s to be closer to the population probability. For example, there might well be 163 results of 5 in this sample, giving a relative frequency of 163/1000 = 0.163.

But, you may ask, suppose the long-run relative frequency of 5 does not approach $\frac{1}{6}$? Does this mean that the empirical definition is not valid? No. It means that the die is not fair: that is, that the die is *biased*. Suppose that the relative frequency of 5s in a very large random sample is 0.150. This would be taken as evidence that the probability of 5 is 0.150 and hence that the die is biased. A die (or coin) is *fair* if all the possible results are *equally likely*. A die (or coin) is *biased* if the possible results have *unequal* probabilities of occurring.

Let us return to the sample of 100 tosses. The number of 1s, 2s, etc., might be as follows.

FACE OF DIE	FREQUENCY	RELATIVE FREQUENCY
1	21	0.21
2	15	0.15
3	17	0.17
4	17	0.17
5	12	0.12
6	18	0.18
	100	1.00

Notice that the relative frequencies are similar, but not identical, to the probabilities (each $\frac{1}{6}$) given earlier. As the sample size increases, the former will become more and more similar to the latter.

Example 2

The two definitions of probability can be contrasted by considering the drawing of balls from a bag containing a large number of colored balls. The bag contains 1,000 balls, of which 700 are red and 300 are black. If we shake the bag well and draw one ball, we can reasonably assume that this ball is drawn at random. In the population of balls, the proportion of red balls is 700 /1,000 = 0.700, so we can say, following the theoretical definition of probability, that the probability of drawing a red ball is 0.70. Similarly, the probability of drawing a black ball is 0.30.

Now, if we draw a second ball from the bag, the probabilities will no longer be 0.70 and 0.30, since one ball has already been drawn. Such *sampling without replacement* has some complications, so for simplicity we will usually consider that the first ball has been returned to the bag and that the bag is well shaken before the second ball is drawn. In this method of *sampling with replacement*, the probabilities remain as 0.70 and 0.30. The distinction between these two methods of sampling is less important than it might appear, because if the original bag contains a very large number of balls, the probabilities will change only very slightly in sampling without replacement, and such sampling is then equivalent for all practical purposes to sampling with replacement.

If we now draw a sample of 75 balls from the bag, replacing each ball after its color has been recorded, we can record the relative frequencies of red and black balls and compare them to the corresponding probabilities. The results might be

COLOR OF BALL	PROBABILITY	FREQUENCY	RELATIVE FREQUENCY
Red	0.70	49	0.653
Black	0.30	26	0.347
	1.00	75	1.000

The relative frequencies are close to, but not identical to, the probabilities. If the sample size were larger, we would expect, from the empirical definition of probability, that the relative frequencies would be closer to the population probabilities.

Suppose we did not know the relative proportions of red and black balls in the bag. By drawing a random sample of balls we can *estimate* the proportion of red balls in the bag from the relative frequency of red balls in the sample. If the sample is of size 75, the estimate will be close but not completely accurate. If we wish to have a more accurate estimate, we must draw a larger sample. Much of the rest of this book is about further details of this inference process. For example, one question examined later is exactly what sample size to use in order to achieve a certain accuracy in the estimate of the population probability.

5.6 The Binomial Distribution

The set of probabilities, one for each possible result of a random process, is called a *probability distribution*. One such probability distribution is the *binomial distribution*, which is worked out in detail in this section.

We can illustrate the binomial distribution by the following situation. Suppose that a population of persons has a choice between Party A and Party B in the next election. We draw a random sample of four persons from the population and wish to determine the probabilities that various numbers of these four persons will favor Party A. These probabilities will be worked out assuming that the proportion of persons in the population favoring Party A is 0.7—that is, 70%. Hence, the probability that a randomly selected person favors Party A is 0.7.

The Binomial Distribution from First Principles

The sample space for samples of size four is shown in Figure 5.6a. The tree diagram has four stages, one for each member of the sample. Each person in the sample can indicate that either Party A or Party B is favored. At each stage of the tree, therefore, two branches split off from the previous branch. In all, there are 16 paths through the tree, corresponding to the 16 possible samples in this sample space.

The figure is based on the hypothesis that the proportion of persons in the population who favor Party A is 0.70. Hence, the probability of each A branch is 0.7 and the probability of each B branch is 0.3. We can then interpret the tree as representing the selection of one sample of size four in the following way. Starting at the base of the tree, we select the first person. With probability 0.7 we will "go down" the A branch, and with probability 0.3 we will go down the B branch. We are then ready to select the second person. This person, too, has probability 0.7 of favoring Party A, and hence we have probability 0.7 of going down the A branch at the second stage and probability 0.3 of going down the B

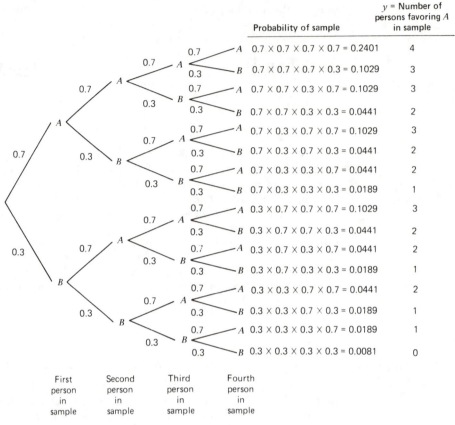

	Probability of sample	y = Number of persons favoring A in sample
	$0.7 \times 0.7 \times 0.7 \times 0.7 = 0.2401$	4
	$0.7 \times 0.7 \times 0.7 \times 0.3 = 0.1029$	3
	$0.7 \times 0.7 \times 0.3 \times 0.7 = 0.1029$	3
	$0.7 \times 0.7 \times 0.3 \times 0.3 = 0.0441$	2
	$0.7 \times 0.3 \times 0.7 \times 0.7 = 0.1029$	3
	$0.7 \times 0.3 \times 0.7 \times 0.3 = 0.0441$	2
	$0.7 \times 0.3 \times 0.3 \times 0.7 = 0.0441$	2
	$0.7 \times 0.3 \times 0.3 \times 0.3 = 0.0189$	1
	$0.3 \times 0.7 \times 0.7 \times 0.7 = 0.1029$	3
	$0.3 \times 0.7 \times 0.7 \times 0.3 = 0.0441$	2
	$0.3 \times 0.7 \times 0.3 \times 0.7 = 0.0441$	2
	$0.3 \times 0.7 \times 0.3 \times 0.3 = 0.0189$	1
	$0.3 \times 0.3 \times 0.7 \times 0.7 = 0.0441$	2
	$0.3 \times 0.3 \times 0.7 \times 0.3 = 0.0189$	1
	$0.3 \times 0.3 \times 0.3 \times 0.7 = 0.0189$	1
	$0.3 \times 0.3 \times 0.3 \times 0.3 = 0.0081$	0

First person in sample   Second person in sample   Third person in sample   Fourth person in sample

(a)

y	$P(y)$ = Sum of probabilities of all samples with given y	
0	0.0081	$= 1 \times 0.0081 = 0.0081$
1	0.0189 + 0.0189 + 0.0189 + 0.0189	$= 4 \times 0.0189 = 0.0756$
2	0.0441 + 0.0441 + 0.0441 + 0.0441 + 0.0441 + 0.0441	$= 6 \times 0.0441 = 0.2646$
3	0.1029 + 0.1029 + 0.1029 + 0.1029	$= 4 \times 0.1029 = 0.4116$
4	0.2401	$= 1 \times 0.2401 = 0.2401$
		1.0000

(b)

Figure 5.6 A binomial distribution. (a) Sample space. (b) Distribution of y.

branch. The selection of the third and fourth persons in the sample is represented by the figure in the same way.

The tree diagram in Figure 5.6a has certain characteristics that ensure that the probability distribution, which is derived from it, is the binomial distribution. We will call such a tree diagram a *binomial tree*.

A binomial tree has three characteristics (see Figure 5.7).

1 There are n stages in the tree, and each path must extend over all n stages.

2 At each stage there are two possible results, called A and B in the tree.

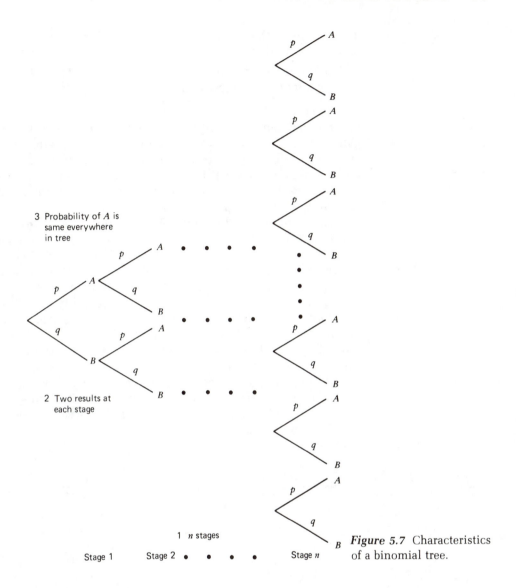

3 Probability of A is
same everywhere
in tree

2 Two results at
each stage

1 n stages

Stage 1   Stage 2 • • • •   Stage n

Figure 5.7 Characteristics of a binomial tree.

3 The probability of A is the same everywhere in the tree. This probability is denoted by p. Similarly, the probability of B is the same everywhere and is denoted by q. (The sum of p and q must be 1.)

Figure 5.6 satisfies these conditions: there are four stages; at each stage there are two possible results; the probability of A is 0.7 throughout the tree and the probability of B is 0.3 throughout.

Our aim is to determine the probability of each possible value of the number of persons favoring Party A in samples of size four. The steps we must follow include

1 Constructing the sample space: that is, the tree diagram. (Completed above.)

2 Computing the probability of each sample in the sample space: that is, the probability of each path in the tree diagram.

3 Determining the number of persons favoring Party A for each sample in the sample space. (Let y = number of persons favoring Party A.)

4 Combining the probabilities for all samples that have the same value of y and making a table of the probabilities.

The calculations for the second step are shown immediately to the right of the tree diagram in Figure 5.6a. The probability of any sample—that is, any path through the tree—can be found by multiplying the numbers on the branches of the path. Not all the samples in this example have the same probability, but note that there are only a few different probabilities (0.2401, 0.1029, 0.0441, 0.0189, and 0.0081). This fact will greatly simplify the calculations in the fourth step.

The third step is to determine the value of y for each sample. The first sample is $AAAA$. The number of persons favoring A in this sample is four—the number shown as y in the right hand side of Figure 5.6a. The next sample, $AAAB$, has three persons favoring A, so here $y = 3$. In the same way, the values of y are determined for each of the 16 samples in the sample space.

In the fourth step we find the total probability of getting $y = 0$, the total probability of $y = 1$, etc. Starting with $y = 0$, we look down the list of y for samples in the sample space with that value of y. For $y = 0$, there is only one such sample, with probability 0.0081—as indicated in the table in Figure 5.6b. For $y = 1$, there are four samples, each having probability 0.0189. So the total probability of $y = 1$ is simply $4 \times 0.0189 = 0.0756$. We similarly discover that all six samples with $y = 2$ have the same probability, and a total probability of $6 \times 0.0441 = 0.2646$. The table is completed in the same way for $y = 3$ and $y = 4$.

The resulting table or probabilities for each value of y can be abbreviated as follows.

y	$P(y)$
0	0.0081
1	0.0756
2	0.2646
3	0.4116
4	0.2401
	1.0000

This type of distribution occurs so frequently in statistics that it is given a special name: the *binomial distribution*. The prefix *bi*, meaning "two", describes the two branches at each stage of the tree diagram. (Keep in mind that some other distributions are not binomial distributions. If the tree diagram had three branches at each stage, the distribution would not be binomial.)

You may have noted two features of the calculation of the *binomial probabilities* (as the individual probabilities of a binomial distribution are known). First, it is not a coincidence that all the samples with the same value of the statistic y have the same probability. Consider the four samples that have $y = 1$: $ABBB$, $BABB$, $BBAB$, and $BBBA$. Each of these paths has probability 0.0189, since each path has one number 0.7 and three numbers 0.3, giving a product $0.7 \times 0.3 \times 0.3 \times 0.3 = 0.0189$. The only difference in the paths is the order in which these numbers appear. Each path has only one 0.7, since each has only one A. In the same way, each of these paths with $y = 1$ has three occurrences of 0.3, since there are three Bs in each path. A general property of sample spaces that lead to binomial distributions is that each probability can be calculated as the product of the number of samples with the specific value of y and the probability of any one such sample (since all the samples have the same probability). For $y = 1$, for example, the probability is $4 \times 0.0189 = 0.0756$, since there are four such samples and each sample has probability 0.0189.

The second feature of the calculation is that the *sum* of all the binomial probabilities is 1.0000. This should not come as a surprise, since a sample space includes all possible samples. One of these samples must occur in any actual experiment or survey. Hence, the probability of all samples together is 1.0000, indicating that it is certain that a sample does occur. This requirement[4] provides one way to check the calculation of a binomial distribution.

The preceding paragraphs, along with Figure 5.6, have illustrated one method for calculating the binomial distribution. This method may be called *working from first principles*. Two other methods are available. One of these methods uses a *formula* for the probabilities, and the other requires a *table* of the probabilities. Although the formula and the table are very useful in practice, you should keep in mind that the method of working from first principles always can be used. The formula method is described immediately below. The table method is described in Section 6.3.

The Binomial Formula

It is convenient to refer to the two possible results in a binomial tree as *success* and *failure*. For example, favoring Party A can be considered a success and favoring Party B a failure. Getting heads on a toss of a coin can also be considered a success. If we use this terminology, we can say that y is the *number of successes*, that p is the *probability of success*, and that q is the *probability of failure*.

Each binomial probability may be calculated as the product of two terms: the *number* of paths in a binomial tree with the desired number of successes, y; and the *probability* of one such path with y successes. The product of these two terms is the total probability of obtaining y successes in a sample of size n. The final result[5] for $P(y)$ is given by the following formula.

$$P(y) = \frac{n!}{y! \, (n - y)!} \times p^y q^{n-y} \quad \text{(Binomial formula)}$$

The symbol "!" in this formula means "factorial": $n!$ is defined as $n \times (n - 1) \times (n - 2) \times \ldots \times 3 \times 2 \times 1$. The terms $y!$ and $(n - y)!$ are defined in a similar way. Examples are given below. If y or $(n - y)$ is zero, we have the special case of 0! that is defined to be 1. In the term to the right of the multiplication sign, p is raised to the power y and q (the probability of failure $= 1 - p$) is raised to the power $(n - y)$.

The binomial formula is quite complex; it is a general formula and applies to any binomial tree. Here are some examples of its use.

---

[4] The precise restrictions on a probability distribution are given in Section 6.5.

[5] A proof of this formula may be found in Section 18.4.

Example 1

Let us find the probability that one person favors Party A in a sample of four persons, if the probability of a person favoring Party A is 0.70. This probability was worked out previously from the tree diagram in Figure 5.6. In this example, $n = 4$, $y = 1$, $p = 0.70$ and q (the probability of failure or favoring Party B) is $1 - 0.70 = 0.30$. The binomial formula then gives

$$P(1) = \frac{4!}{1! \, (4-1)!} \times (0.7)^1 \, (0.3)^{(4-1)}$$

$$= \frac{4!}{1! \, 3!} \times (0.7)^1 \, (0.3)^3$$

$$= \frac{4 \times 3 \times 2 \times 1}{1 \times 3 \times 2 \times 1} \times 0.7 \times 0.3 \times 0.3 \times 0.3$$

$$= \frac{24}{6} \times 0.7 \times 0.027$$

$$= 4 \times 0.0189$$

$$= 0.0756$$

This result agrees with Figure 5.6. Notice that the answer is the product of two terms, 4 and 0.0189—the number of paths with $y = 1$ and the probability of one such path, respectively.

In the remaining examples we will not write out each step of the calculation. The calculations can be simplified by noting that the largest of the factorials in the denominator of the first term can be canceled against part of the factorial in the numerator. In the calculation above, $3! = 3 \times 2 \times 1$ can be canceled against $4! = 4 \times 3 \times 2 \times 1$, leaving just 4 in the numerator (and $1! = 1$ in the denominator). Even if you use a calculator, this simplification can save time and decrease the chance of making an error.

Example 2

A fair coin is tossed 10 times. What is the probability that eight heads (and two tails) will be obtained? Here $n = 10$, $p = \frac{1}{2}$ (because the coin is fair), $q =$ probability of tails $= \frac{1}{2}$, and $y = 8$. The binomial formula gives

$$P(8) = \frac{10!}{8! \, (10-8)!} \times \left(\frac{1}{2}\right)^8 \left(\frac{1}{2}\right)^{(10-8)}$$

$$= \frac{10!}{8! \, 2!} \times \left(\frac{1}{2}\right)^8 \left(\frac{1}{2}\right)^2$$

With some practice, you should be able to write this last expression at once from the statement of the problem. Note that 8, the number of successes (heads), appears twice in the formula and that in a symmetric way, 2, the number of failures (tails), also appears twice. This example has a simplifying feature, since $p = q = \frac{1}{2}$. The two powers of $\frac{1}{2}$ can be combined to give

$$= \frac{10!}{8! \, 2!} \times \left(\frac{1}{2}\right)^{10}$$

$$= \frac{10 \times 9}{2 \times 1} \times \left(\frac{1}{2}\right)^{10} \qquad \text{(after cancellation of 8! into 10!)}$$

$$= 45 \times 0.0009765$$

$$= 0.0439425$$

$$= 0.0439, \text{ rounded to four decimal places}$$

Example 3

Let us repeat the last example but calculate instead the probability of 8 *or more* heads in 10 tosses of a fair coin. We have to find the probability of 8 heads, the probability of 9 heads, and the probability of 10 heads. These three probabilities are then added together.

$$P(8) + P(9) + P(10)$$

$$= \frac{10!}{8! \; 2!} \times \left(\frac{1}{2}\right)^8 \left(\frac{1}{2}\right)^2 + \frac{10!}{9! \; 1!} \times \left(\frac{1}{2}\right)^9 \left(\frac{1}{2}\right)^1 + \frac{10!}{10! \; 0!} \times \left(\frac{1}{2}\right)^{10} \left(\frac{1}{2}\right)^0$$

$$= \frac{10 \times 9}{2 \times 1} \times \left(\frac{1}{2}\right)^{10} + \frac{10}{1} \times \left(\frac{1}{2}\right)^{10} + \frac{1}{1} \times \left(\frac{1}{2}\right)^{10} \qquad \text{(note that 0! is 1)}$$

$$= 45 \times 0.0009765 + 10 \times 0.0009765 + 1 \times 0.0009765$$

$$= 0.0439425 + 0.0097650 + 0.0009765$$

$$= 0.0546840$$

$$= 0.0547, \text{ rounded to four decimal places}$$

Example 4

Subjects in a taste experiment are given three foods to taste. They are told to identify which of the three is different from the other two. Subjects have one chance in three of identifying the correct food, even if there is very little real difference in the foods. There are two possible results for each subject: the subject can succeed (identify the food that is different from the other two) or fail (select one of the two other foods). Here n is the number of subjects in the experiment, $p = \frac{1}{3}$ is the probability of succeeding, and $q = \frac{2}{3}$ is the probability of failure. The probability of succeeding is the same throughout the binomial tree if the subjects are tested individually and do not communicate with each other before completing the experiment.

What is the probability that at least five of six subjects choose the correct food?

$$P(5) + P(6) = \frac{6!}{5! \; 1!} \times \left(\frac{1}{3}\right)^5 \left(\frac{2}{3}\right)^1 + \frac{6!}{6! \; 0!} \times \left(\frac{1}{3}\right)^6 \left(\frac{2}{3}\right)^0$$

$$= \frac{6}{1} \times \frac{2}{729} + \frac{1}{1} \times \frac{1}{729}$$

$$= 0.0164609 + 0.0013717$$

$$= 0.0178326$$

$$= 0.0178, \text{ rounded to four decimal places}$$

5.7 Summary

Populations may be represented by one-stage tree diagrams. A *sample* of size n from a population is represented by a path through an n-stage tree. The n-stage tree represents the *sample space*.

The tree diagrams for *full sample spaces* usually are very large. However, in a *reduced sample space* each branch represents a value of a variable rather than a unique individual in the population.

There are two definitions of *probability*: the *theoretical definition* and the *empirical definition*. Probabilities can be added to the branches of a tree diagram. The probability of a particular sample can be calculated by multiplying the probabilities along its path through the tree. The basis of the empirical definition of probability is the *law of large numbers*.

One particular probability distribution is the *binomial distribution*, which arises from sample spaces that are represented in *binomial trees*. Two ways to compute binomial probabilities are from first principles using the binomial tree and from the binomial formula.

Additional terms discussed in this chapter include *random process*, a *fair* die or coin, a *biased* die or coin, *sampling without replacement*, and *sampling with replacement*.

Exercises

1 (Section 5.1) A jar contains six marbles—two are red and four are green. A person draws a marble at random from the jar (i.e., she shakes the jar thoroughly and selects one marble without looking at the jar). The color of the marble is noted, the marble is returned to the jar, and a marble is again drawn at random. (The second marble can, of course, be the same as the first.)

(a) Suppose that the marbles are numbered 1, 2, 3, 4, 5, and 6, as well as each having a color. Sketch a tree diagram to represent the sample space of this experiment. Represent each of the six marbles at each of the two draws.

(b) How many different samples are there in this sample space?

2 (Section 5.2)

(a) Sketch a tree diagram to represent a reduced sample space for the experiment of drawing two marbles in Question 1. Only the color of each marble is noted in the reduced sample space.

(b) How many different samples are there in the reduced sample space?

(c) Suppose that the experiment is changed to consist of *five* draws from the jar, rather than *two*. (The drawn marble is always returned to the jar before the next draw.) Sketch the sample space.

(d) How many different samples are there in this reduced sample space?

3 (Section 5.3)

(a) Use the theoretical definition of probability to determine the probabilities of drawing a red marble and a green marble in the experiments described in Question 1 and 2. Add these probabilities to the tree diagram for five draws (Question 2c).

(b) Mark all those paths in the tree diagram that represent a sample with two red and three green marbles, in *any* order. (The paths *RGRGG* and *GGRRG* are two such paths.) How many such paths are there?

(c) Calculate the probability of each of the samples with two red and three green marbles.

(d) Sum the probabilities found in (c) to obtain the probability of getting a sample with two red and three green marbles, in any order.

4 (Section 5.4) Consider an experiment of drawing *one* marble from a jar of six marbles, two of which are red and four of which are green.

(a) What is the probability of drawing a red marble? Of drawing a green marble? Use the theoretical definition.

(b) To determine the probabilities of a red and of a green marble using the *empirical* definition, you must actually carry out the experiment, say, 50 times. (An even larger number is desirable—you may wish to do the experiment with another student and pool your results.) There are two methods you may use to do the experiment. Choose one of these methods and state on your answer sheet which one you choose.

(i) Get a jar and six marbles, two of one color and four of another. Beads or buttons can also be used, but all six marbles, beads, or buttons must be similar, except for color. It is very important to shake the jar thoroughly before each draw.

(ii) Use the random number table (Table A5 in the Appendix). Each draw of a marble is represented by one digit in the table. In order to represent the two colors in their correct proportions, give the digits the following significances.

1, 2	—red
3, 4, 5, 6,	—green
7, 8, 9, 0	—not used (just skip over these digits)

Start at a "random" place in the table. (Do not start at top-left corner.) You can proceed across or down from the starting place. State on your answer sheet the location where you started and the direction in which you proceeded in order that your answer may be checked, if necessary.

Carry out the experiment 50 or more times using either method **(i)** or method **(ii)**. Report in a list the color of the marble drawn at each trial: that is, give a list of 50 (or more) Rs and Gs, in the order that you obtained them. Then report the relative frequency of R and G, and hence your estimated probability of R and G using the empirical definition.

5 (Section 5.5) Which terms, defined in Section 5.5, accurately describe the sampling you carried out in Question 4?

6 (Section 5.6) The probability that a pregnant woman over 45 years of age will bear a child with Down's syndrome is about 1/65. Consider three such women and their children.

(a) Represent all the possible results for the three children (Down's syndrome or normal) in a tree diagram.

(b) Is this tree a binomial tree? Explain.

(c) Work out the probabilities that none, one, two, or three of the children will have Down's syndrome.

7 (Section 5.6) Work out each probability in this question on your calculator. Round your answers to four decimal places.

(a) A multiple-choice quiz has 10 questions. Each question has four alternatives. If a student is completely unfamiliar with the subject matter, what is the probability that he or she gets exactly six answers correct, by guessing?

(b) Suppose that there are only two (instead of four) alternatives to each of the 10 questions on the quiz. What is the probability of getting at least six answers correct, by guessing?

6 STATISTICAL INFERENCE AND THE DISTRIBUTION OF THE STATISTIC

Note: In order to understand this chapter, it is not necessary to have read Chapter 5. A few terms extensively discussed in Chapter 5 are redefined here as required. Chapter 6 is essential for later chapters.

A statistical inference about a population is made on the basis of information in a random sample drawn from that population. This chapter describes in more detail certain aspects of populations and samples. A population can be described by one or more numbers, called parameters. The corresponding number in a sample is called a statistic. The parameter and the statistic are related by a probability distribution called the distribution of the statistic. The characteristics of such distributions are described at length.

The binomial distribution, one type of distribution of a statistic, is described here and used to illustrate two types of statistical inference: hypothesis testing and estimation.

6.1 Statistical Inference

The subject of statistics is divided into two main branches; descriptive statistics and inferential statistics. *Descriptive statistics* is a set of techniques for summarizing data. Some of the techniques of descriptive statistics were presented in Chapters 2, 3, and 4. *Inferential statistics* is a set of techniques for making inferences about a population from observations of a sample drawn from that population. In this definition, a *population* is any complete set—usually very large—of persons, animals, or things. A *sample* is a subset—usually small—selected or drawn in some way from the population.

Figure 6.1a shows, in simplified form, how statistical inference is carried out. The dashed arrow in the figure represents the drawing (selecting) of a particular sample from the population. This sample could be the focus of either a *survey* or an *experiment*.

A typical *survey* is a study of a sample drawn from an existing population such as all the people living in a city or country, all university students, or all learning-disabled adults. For example, a survey of learning-disabled adults might be conducted in order to determine whether such persons are more frequently male or female. A sample of such adults would be selected and the number of males and females counted. The ratio of males to females in the sample would be used as the basis of an inference about the corresponding ratio in the complete population.

An *experiment*, on the other hand, is a study of an artificial population created by the experimenter. For example, a group of subjects might be given a

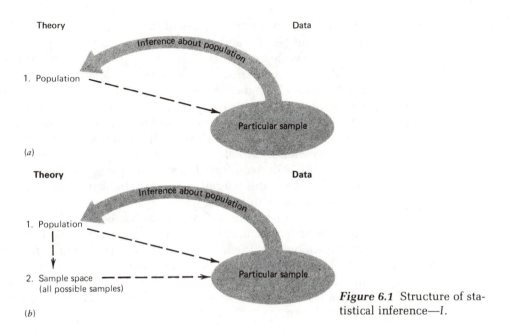

Theory   Data

Inference about population

1. Population

Particular sample

(a)

Theory   **Data**

Inference about population

1. Population

2. Sample space
(all possible samples)

Particular sample

(b)

Figure 6.1 Structure of statistical inference—*I*.

certain drug in order to see if the drug has an effect on the number of hours of sleep. The population in such a study consists of all persons who could be given the drug. Since most people do not take the drug, the population does not really exist and, hence, may be considered to be artificial. (An actual study usually would employ more than one drug, so that the effect of the different drugs could be compared; there would be several populations and several samples, one sample drawn from each population. The design of such experiments is described further in Section 11.3.)

Figure 6.1*a* also depicts a large arrow labeled "inference about population" that runs back from the sample to the population. Making an inference is coming to a conclusion. From the sample we come to a conclusion about some aspect of the population from which the sample was drawn. If we are studying the ratio of males and females, we can make a statement about the ratio in the population from the observed ratio in the sample. The inference is not guaranteed to be correct, however. Whether it is correct or not depends on many things: how the sample was selected, how large the sample is, whether the calculations are accurate, etc. In the chapters to come, the factors that influence the correctness of the conclusion are discussed extensively.

The two sides of Figure 6.1*a* are labeled "theory" and "data", respectively. The data side is quite clear. After making measurements on the persons in the sample (each person's height, IQ, hours of sleep, etc.), we have numbers or data that we process in some way to come to a conclusion about the population. But why do we label the population side of the figure as "theory"? We do not know everything there is to know about the population we are studying. If we did, we would not need to conduct a survey or experiment. We have some ideas or theories about the population: we think that the drug has no effect, or that the population has an equal number of males and females. But these are just theories. There are many other possibilities. The proportion of males in the population may be 0.48 or 0.49 rather than 0.50. Because we are therefore uncertain about the true nature of the population, we develop certain theories about it and test these theories by means of an experiment or a survey.

In carrying out a statistical inference, we first of all consider the sample space. The *sample space* is the set of all possible samples of a given size that can be drawn from a given population. If in our survey or experiment we draw a sample of 10 persons from a population, then the sample space is a list of all possible samples of 10 persons from the population. Such a list is usually very large. We will see later in this chapter that a statistical inference is actually based on what is called "the distribution of the statistic". The distribution of the statistic is derived from the sample space. Hence, it is important to understand the role of the sample space in statistical inference, even though we will not usually have to list the full sample space.

In Figure 6.1b, the sample space has been added to the basic drawing in Figure 6.1a. The samples in the sample space are purely theoretical (and hence are shown on the "theory" side of the figure); we only draw one sample in the actual study. Notice also the dashed arrow extending from the sample space to the particular sample. Instead of thinking of the particular sample as being drawn from the population, we can think of it as having been drawn from the sample space.

The Importance of Random Samples

The description of statistical inference given above, and summarized in Figure 6.1b, still is incomplete. We can make an inference about a population only if we draw a *random* sample. A random sample is a sample in which every member of the population has an equal probability of being included. Most readers will have an intuitive understanding of probability as "chance". A random sample is a sample in which every member of the population has an equal chance of being included. (See Chapter 5 for an extensive discussion of probability.)

Why is a random sample so important? It is obvious that the sample used to make an inference about a population should be *representative* of the population. As far as possible, the sample should be similar in all characteristics to the population. There is essentially only one way to ensure that a sample is representative: the sample must be a random sample.

Suppose that you want to determine the proportion of females in a population; you stand on a busy intersection in the center of a city and count how many males and females you see. Of the 174 persons in this sample 78 are female and the sample proportion is $78/174 = 0.448$. Is 0.448 a good estimate of the population proportion? No. Not only is there no sound basis for the inference that the population proportion of females is 0.448, but there are many reasons to believe that the sample is not *representative* of the population. People seen at the intersection are, presumably, shoppers, executives, retired persons, etc.; presumably, they do not include children, teachers, night-workers, etc. Furthermore, how is it possible to get an accurate count just by standing on a corner and counting heads? You can probably think of other ways in which this sample is defective and fails to represent the population.

The sample should be as similar to the population as possible. If it is similar, the proportion of females in the sample will be close to the proportion of females in the population. On the other hand, if the sample consists mainly of people of certain types (shoppers, executives, etc.), the proportion of females in the sample may not be close to the population proportion. Hence, it is important that the sample be as representative of the population as possible.

The best way to get a representative sample is to get a random sample.[1] In Figure 6.2*a* the basic figure illustrating the structure of statistical inference has been changed by replacing a "particular sample" by a "random sample".

Additional terms have been added to the figure. On the data side of the figure, the new terms are "sample distribution" and "relative frequencies". In

[1] A random sample is not guaranteed to be a representative sample. There is a certain chance that the random sample will be a deviant sample and not representative of the population.

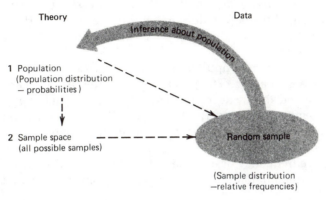

(a)

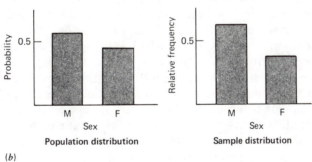

(b)

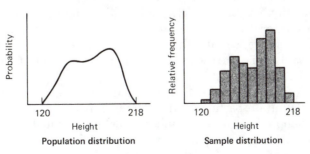

(c)

Figure 6.2 Structure of statistical inference—II. (*a*) . (*b*) Discrete case. (*c*) Continuous case.

any experiment or survey a certain variable is studied (for example, sex or height). The *relative frequencies* of the values of the variable are determined from the sample data. The set of relative frequencies is called the *sample distribution*. Examples of sample distributions are shown on the data side of Figures 6.2b and c.

Two analogous terms have been added to the population on the theory side of Figure 6.2a: "population distribution" and "probabilities". The *probability* of a value of a variable is the proportion of persons in the population who have that characteristic. A set of probabilities, one probability for each value of the variable, is called a *population distribution*. A population distribution is therefore a *probability distribution* (that is, a distribution of probabilities) and is often called the population probability distribution. (It is important to remember that there are other probability distributions that are not population distributions.)

Examples of population distributions are shown on the theory side of Figures 6.2b and c. The probabilities in the discrete case (Figure 6.2b) can be interpreted simply as the proportion of persons in the population who are male and female, respectively. It is more difficult to interpret the population distribution graph shown for the continuous variable height in Figure 6.2c. For the present, think of the population distribution of heights as giving the relative frequency of each height in the population. A more detailed explanation of the population distribution of a continuous variable is given in Chapter 7.

6.2 Parameters and Statistics

One purpose of statistics is to make inferences about a population from which a random sample has been drawn. Section 6.1 showed that a population is characterized by a *population distribution* that gives the probability of each possible value of the variable which is being studied. A sample is characterized by a *sample distribution* that gives the relative frequency of occurrence of each of these values in the sample. If the sample is random, each relative frequency approximates the corresponding population probability, and the approximation gets better as the sample size is increased.

Statistical inferences are inferences about characteristics of the population distribution called parameters, or population parameters.[2] *A (population) parameter is a number that describes a population distribution.* Examples of parameters are the population mean and standard deviation. The population mean is a number that measures or describes the central tendency of the population distribution. The population standard deviation is a number that measures or describes the variability of the population distribution. In a statistical inference we infer the value of a population parameter such as a mean or standard deviation. (See Section 6.6 for definitions of the population mean and standard deviation.)

A statistical inference is based on a random sample drawn from the population. In particular, the inference is based on a number computed from the sample data. This number is called a statistic or a sample statistic. *A (sample) statistic is a number that describes a sample distribution.* Note the close similarity of the definition of a statistic and the definition of a parameter.

[2] Exception: a special class of inferences called nonparametric. See Section D.5.

A parameter is to a population what a statistic is to a sample. The sample mean and standard deviation are statistics[3] that describe, respectively, the central tendency and the variability of the sample distribution. (The sample mean, previously called the mean, was defined in Section 3.2. The sample standard deviation, previously called the standard deviation, was defined in Section 4.4.) From statistics such as these we infer the values of the population parameters.

Example 1

The mean is the most commonly studied parameter of a population distribution. The mean income of a population of adults could be studied by drawing a random sample of adults and computing the mean income of the sample. The mean income of the population is a parameter; the mean income of the sample is a statistic. It will be shown later that the sample mean is closely related to the population mean when the sample is randomly drawn from the population. Hence, the sample mean can be used to make an inference about the population mean.

Example 2

Dice are used in casinos as a random device. If a die is fair, the probability of each of the six sides showing up will be 1/6 or 0.1667. In particular, the probability of the side 5 is 1/6. The probability itself can be considered as a population parameter, since it is a number that describes the population distribution. (Note that a parameter does not have to describe *fully* a population distribution; the mean does not fully describe the distribution either, since it measures the central tendency but not the variability of the distribution.)

In a sample of 75 tosses of the die, the 5 might come up 18 times. The relative frequency of the 5 in the sample is therefore $18/75 = 0.240$. The sample relative frequency can be used to judge whether the population probability of the 5 is really 1/6, and hence whether the die is really fair. Here the sample relative frequency is a statistic used to make an inference about the probability of the 5, which is a parameter.

This example actually mentions two statistics. The first is the number of times that the 5 is tossed in the sample of 75 tosses; the value of this statistic is 18. The second statistic is the relative frequency of the 5 in this sample; the value of this second statistic is 0.240. In this case the second statistic is more closely analogous than the first to the population parameter (probability of the 5). We will shortly see, however, that the first statistic, the number of times that an event occurs, is more commonly studied in discrete variable problems.

Structure of Statistical Inference

Look at Figure 6.3a, in which the diagram used to explain statistical inference in Section 6.1 has been augmented by "parameter" and "statistic". The figure emphasizes that a parameter is a characteristic of a population. On the other hand, a statistic is a characteristic of a sample and hence is computed from actual sample data. The inference about the population, indicated by the large arrow, usually is based on a value of the statistic computed in a sample. The inference is about the value of the population parameter.

These relationships are illustrated by the example in Figure 6.3b. Suppose that a certain political party has obtained 70% of the votes cast in each election during the past 10 years. Because of the death of its leader, however, the party

---

[3] The word "statistics" has two meanings, one singular and one plural. Its singular meaning is "a body of knowledge or set of techniques": for example, we can discuss the principles of statistics. On t' e other hand, when "statistics" is used as a plural word, as here, it simply means that we are discussing more than one statistic: that is, more than one number describing a sample.

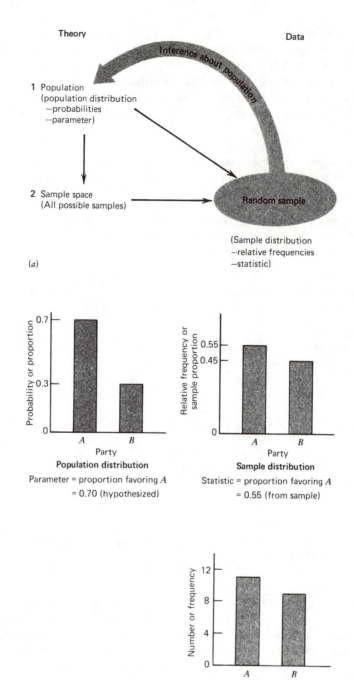

Figure 6.3 Structure of statistical inference—III. (a) .
(b) Example.

suspects that it will not be able to attract so large a proportion of votes in the next election. A poll is to be conducted to determine whether the proportion of voters favoring this party has changed.

In the figure, the party is called Party A and it is assumed that there is only one other party, Party B. The population distribution also is shown, under "theory", with the proportion favoring Party A = 0.70 and the proportion favoring Party B = 0.30. The parameter of interest in this population is the proportion favoring Party A. If there is no change in the proportion favoring Party A, this parameter will be 0.70.

The result from a random sample of 20 persons is depicted on the "data" side of the figure. Two sample distributions are shown, one of relative frequencies and the other of frequencies. The two distributions provide the same information because one distribution can be obtained from the other when the size of the sample is known.

The statistic based on the relative frequency distribution is the (sample) *proportion* of persons favoring Party A; the value of this statistic is 0.55. The statistic based on the frequency distribution is the *number* of persons favoring Party A; the value of this statistic is 11. Either statistic can be used to make an inference about the population proportion of persons favoring Party A. How this inference is made is outlined in Section 6.3.

6.3 The Distribution of the Statistic

The proportion of persons favoring Party A, in the sample of 20 persons described in the last section, was 0.55. This number differs from the proportion of 0.70, which is based on the elections of the last 10 years. We *hypothesize* a value of 0.70 for the population parameter. Does the difference between the sample statistic of 0.55 and the hypothesized population parameter of 0.70 allow us to conclude that the population proportion has decreased this year? No, it does not. The sample size is small (only 20). Even if the true population proportion is 0.70, it is quite possible in such a small sample to get a proportion of 0.55. In order to conclude that the population proportion has changed from 0.70, we have to show that the results obtained in the sample were very unlikely under that hypothesis.

The argument of the last paragraph is complex; several chapters of this text are devoted to elaborating it fully. The argument does make clear, however, that we have to evaluate the likelihood of the sample we obtained. In fact, we have to determine how likely every possible sample is. Recall from Section 6.1 that the set of all possible samples is called the sample space. So we have to obtain the probability of every sample in the sample space! This is a large job. To simplify it, we usually obtain the probabilities of the various values of the statistic. The number of persons favoring Party A in our sample was 11. It could have been 12, 10, 15, etc.—in fact, any number from 0 to 20. We must determine the probability of each of these possible values of the statistic in order to see if our obtained result of 11 is unusual or not.

The table of probabilities for each possible value of the statistic is called the *distribution of the statistic*.[4] Figure 6.4a shows how the distribution of the

---

[4] Many, but not all, statistics books call this distribution the *sampling distribution*. Even after an entire course in statistics, however, many students misunderstand this term, which is very difficult to distinguish from *sample* distribution. A sample distribution, like a population distribution, is a distribution of a variable. A sampling distribution, on the other hand, is a distribution of a statistic; that is its most important characteristic. Hence the decision to call it by the descriptive name—the distribution of the statistic.

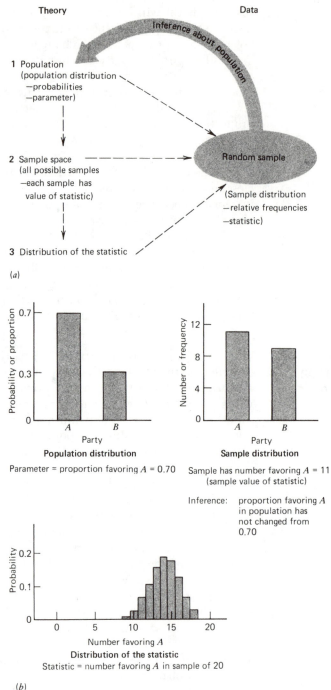

Figure 6.4 Structure of statistical inference—IV. (a) (b) Example.

statistic fits into the overall structure of statistical inference. This figure completes the series of four continuously augmented figures begun in Section 6.1. It shows that each sample in the sample space has a value of the statistic. Even though we obtain only one random sample in an actual study, and therefore only one value of the statistic, we can conceive of all possible samples (the sample space) and therefore can compute a statistic for each of these samples.

Because each of these samples has a certain probability, the value of the statistic for that sample has that probability. When all these probabilities are combined, the resulting probability distribution is the distribution of the statistic. Note that the distribution of the statistic is located on the theory side of the figure. Even though a statistic is computed from sample data, the distribution of the statistic is a theoretical notion, because we are considering all theoretically possible samples and the value of the statistic for each.

Finally, note the arrow connecting the distribution of the statistic with the random sample itself. The value of the statistic obtained in the random sample can be thought of as drawn from the distribution of the statistic. The probability of this value can be determined directly from the distribution. We then make the inference about the population parameter on the basis of this probability, in a manner still to be explained.

The Binomial Distribution and the Binomial Table

One type of distribution of the statistic is called the *binomial distribution*. Binomial distributions arise in situations with the following characteristics.

1 There are *two* possible results when any member of the population is sampled. These results may be called *success* and *failure*, respectively. (In the above example, favoring Party *A* may be considered "success" and favoring Party *B* may be considered "failure".)

2 A random sample of size *n* is drawn from the population.

3 Each member of the population has the same probability $= p$ of being a success and the same probability $= q$ of being a failure. ($p + q = 1$.) (In the above example, $p = 0.70$ and $q = 0.30$.)

4 The results for the *n* members of the sample must be independent.[5]

The *number of successes* in the sample of size *n* is the statistic of interest. We denote the number of successes by *y*. In the above example, *y* is the number of persons favoring Party *A*; in the actual sample, $y = 11$. Each value of *y* has a certain probability, called a *binomial probability*. The set of these probabilities, one probability for each possible value of *y*, is called the binomial distribution for the particular values of *n* and *p*. There is a different distribution for different values of *n* and *p*. (Of course, when *p* changes, *q* also changes, since their sum is 1.)

Example 1

One situation in which the binomial distribution arises is when a coin is tossed 10 times. Here the population consists of the infinite set of all possible tosses of the coin. Success may be defined as the coin falling heads and failure as the coin falling tails. The number

[5] This list of characteristics is equivalent to the list in Section 5.6. The fourth characteristic, independence, means roughly that the result for one member of the sample is not affected by, or related to, the result for any other member. See Chapter 18 for a more formal definition of independence.

of tosses is $n = 10$. If the coin is *fair* (not biased), the probability that any given toss will result in a heads is $p = 1/2 = 0.50$. The probability of tails is $q = 1/2 = 0.50$. The number of successes, y, is the number of heads in 10 tosses of the coin.

It is certainly true that the *expected* number of heads in 10 tosses of the coin is $10 \times 0.5 = 5$. We expect to get 5 heads and 5 tails in 10 tosses of the coin. However, we are not guaranteed to get *exactly* 5 heads (and 5 tails). We might get 4 heads or 6 heads, or, in fact, any number of heads from 0 to 10. Of course, it is most likely (probable) to get $y = 4$, 5, or 6 heads in 10 tosses of the coin, but the other values of y from 0 to 10 are also possible, although less probable. The precise probabilities of the possible values of y are given by the binomial distribution for $n = 10$ and $p = 0.5$.

Table A1 in the Appendix gives binomial distributions for various values of n and p. Figure 6.5 shows a section of Table A1, including several distributions for $n = 10$. The

n	y	.01	.05	.10	.20	.30	.40	.50	.60	.70	.80	.90	.95	.99	y
9	0	914	630	387	134	040	010	002	0+	0+	0+	0+	0+	0+	0
	1	083	299	387	302	156	060	018	004	0+	0+	0+	0+	0+	1
	2	003	063	172	302	267	161	070	021	004	0+	0+	0+	0+	2
	3	0+	008	045	176	267	251	164	074	021	003	0+	0+	0+	3
	4	0+	001	007	066	172	251	246	167	074	017	001	0+	0+	4
	5	0+	0+	001	017	074	167	246	251	172	066	007	001	0+	5
	6	0+	0+	0+	003	021	074	164	251	267	176	045	008	0+	6
	7	0+	0+	0+	0+	004	021	070	161	267	302	172	063	003	7
	8	0+	0+	0+	0+	0+	004	018	060	156	302	387	299	083	8
	9	0+	0+	0+	0+	0+	0+	002	010	040	134	387	630	914	9
10	0	904	599	349	107	028	006	001	0+	0+	0+	0+	0+	0+	0
	1	091	315	387	268	121	040	010	002	0+	0+	0+	0+	0+	1
	2	004	075	194	302	233	121	044	011	001	0+	0+	0+	0+	2
	3	0+	010	057	201	267	215	117	042	009	001	0+	0+	0+	3
	4	0+	001	011	088	200	251	205	111	037	006	0+	0+	0+	4
	5	0+	0+	001	026	103	201	246	201	103	026	001	0+	0+	5
	6	0+	0+	0+	006	037	111	205	251	200	088	011	001	0+	6
	7	0+	0+	0+	001	009	042	117	215	267	201	057	010	0+	7
	8	0+	0+	0+	0+	001	011	044	121	233	302	194	075	004	8
	9	0+	0+	0+	0+	0+	002	010	040	121	268	387	315	091	9
	10	0+	0+	0+	0+	0+	0+	001	006	028	107	349	599	904	10
11	0	895	569	314	086	020	004	0+	0+	0+	0+	0+	0+	0+	0
	1	099	329	384	236	093	027	005	001	0+	0+	0+	0+	0+	1
	2	005	087	213	295	200	089	027	005	001	0+	0+	0+	0+	2
	3	0+	014	071	221	257	177	081	023	004	0+	0+	0+	0+	3
	4	0+	001	016	111	220	236	161	070	017	002	0+	0+	0+	4
	5	0+	0+	002	039	132	221	226	147	057	010	0+	0+	0+	5
	6	0+	0+	0+	010	057	147	226	221	132	039	002	0+	0+	6
	7	0+	0+	0+	002	017	070	161	236	220	111	016	001	0+	7
	8	0+	0+	0+	0+	004	023	081	177	257	221	071	014	0+	8
	9	0+	0+	0+	0+	001	005	027	089	200	295	213	087	005	9
	10	0+	0+	0+	0+	0+	001	005	027	093	236	384	329	099	10
	11	0+	0+	0+	0+	0+	0+	0+	004	020	086	314	569	895	11

Figure 6.5 Using the binomial table.

**TABLE 6.1 The Binomial Distribution
for $n = 10$, $p = 0.50$**

y	$P(y)$
0	.001
1	.010
2	.044
3	.117
4	.205
5	.246
6	.205
7	.117
8	.044
9	.010
10	.001

distribution for $n = 10$ and $p = 0.50$ is the column of numbers that is marked in the figure. Note that the distribution is located by the value of n in the leftmost column and the value of p at the top of the table. The values in the marked column are the binomial probabilities—one probability for each value of y (ranging from 0 to 10) given on the left. (A decimal point is understood to precede each probability.) This binomial distribution is repeated in Table 6.1.

Note that the binomial probabilities are denoted $P(y)$, so that the probability of 0 heads is $P(0)$, the probability of 1 head is $P(1)$, and so on. If it is desired to find the probability that the number of heads is greater than, say, 7, the notation would be $P(y > 7)$, which has the value $P(8) + P(9) + P(10) = 0.044 + 0.010 + 0.001 = 0.055$.

Example 2

A coin is not fair if the probability of its falling heads is not 0.5 (and, therefore, the probability of its falling tails is also not 0.5). Such a coin is called a *biased* coin. Suppose we have a coin that tends to fall tails and that the probability of tails is 0.7. What is the probability of getting exactly 8 tails in 16 tosses of the coin?

In this example, "success" is defined as falling tails. The probability of success is the probability of tails—that is, $p = 0.7$. The binomial distribution for $n = 16$ and $p = 0.7$ can be found in Table A1. The required probability is $P(8) = 0.049$.

The probability of getting *at least* 8 tails is $P(y \geq 8)$. This probability can be found from the table by adding $P(8) + P(9) + \ldots + P(16) = 0.049 + 0.101 + \ldots + 0.003 = 0.974$. It can also be found by noting that the probability of getting at least 8 tails is 1 minus the probability of getting less than 8 tails.

$$\begin{aligned} P(y \geq 8) &= 1 - P(y < 8) \\ &= 1 - \{P(0) + P(1) + \ldots + P(5) + P(6) + P(7)\} \\ &= 1 - 0.026 \\ &= 0.974 \end{aligned}$$

Example 3

Let us return to the example of favoring either Party A or Party B. Suppose a random sample of 10 persons is selected and that the proportion of persons in the population who favor Party A is 0.70. This means that the probability, p, that any randomly selected person will favor Party A is 0.70. The binomial distribution of probabilities that 0, 1, . . ., 10 persons in the sample will favor Party A is given in Table A1 for $n = 10$ and $p = 0.70$.

The required section of Table A1 also is shown in Figure 6.5. The binomial distribution is in the column headed $p = 0.70$ and in the rows for $n = 10$. What is the chance that five or fewer persons will favor Party A in the sample? We require

$$
\begin{aligned}
P(y \leqq 5) &= P(0) + P(1) + P(2) + P(3) + P(4) + P(5) \\
&= 0.000 + 0.000 + 0.001 + 0.009 + 0.037 + 0.103 \\
&= 0.150
\end{aligned}
$$

Note that probabilities less than 0.001 are listed in the table as "0+". These probabilities are not precisely zero, but are 0.000 to three decimal places.

Example 4

We are now ready to consider the random sample of 20 persons, 11 of whom favor Party A and 9 of whom favor Party B. The sample is drawn from a population that is hypothesized to have 70% favoring Party A and 30% favoring Party B. These percentages, expressed as proportions, form the population distribution shown in Figure 6.4b. The distribution of the statistic, y, the number of persons favoring Party A, is a binomial distribution with $n = 20$ and $p = 0.70$. This distribution is found in Table A1 and is also shown in Table 6.2 (see Section 6.4). Each value of y from 0 to 20 has a certain probability. These probabilities are graphed in Figure 6.4b as the distribution of the statistic. Note that the most likely number of persons favoring Party A is 11, 12, 13, 14, 15, 16, or 17. The other values of this statistic are quite unlikely.

The "data" side of the figure shows the sample result of 11 persons favoring Party A. Since this result is included in the set of results, listed above, that are quite likely when the population is 0.70, we conclude, or make the inference, that the proportion favoring Party A has not changed from 0.70. This, in brief, is how a statistical inference about a parameter is made on the basis of a sample statistic. Of course, many details have been left out of the argument. Some of these details will be explained further in this chapter, but others must wait for later chapters.

6.4 Making Inferences from the Binomial Distribution

One purpose of statistics is to make inferences about a population from which a random sample has been drawn. A number of the terms and techniques required for such inferences have been developed in this chapter. This section illustrates in more detail how statistical inferences are carried out.

First, review Figure 6.4, which shows the relationships among the basic terms. We can summarize how a statistical inference is carried out by the following principle.

P6: The Basic Principle of Statistical Inference
In order to make an inference about a population parameter from a statistic computed from a random sample, we must consider the distribution of that statistic. The distribution of the statistic is derived from the probabilities of all samples in the sample space. The inference about the population parameter is made on the basis of (1) the distribution of the statistic and (2) the value of the statistic in the sample.

There are two types of statistical inference. In *hypothesis testing*, we decide, on the basis of the observed value of a statistic, whether to accept or reject a hypothesis or conjecture about the value of a population parameter. In *estimation*, we make a guess or an estimate of the value of a population parameter, on the basis of the observed value of a statistic. In both hypothesis testing and estimation, we must know the distribution of the statistic in order to make an inference. An example of each type of inference follows.

Hypothesis Testing

Let us return again to the study of the proportion of persons favoring Party A. This example was introduced in Section 6.2 by supposing that Party A had the support of 70% of the voters in each election during the last 10 years. The question was whether or not this proportion of support still held. In hypothesis testing, we set up a hypothesis on the basis of this conjecture about the population proportion favoring Party A. The hypothesis is $p = 0.7$. (Here we set up just one hypothesis; in later chapters we will usually set up two competing hypotheses.)

A random sample of 20 persons is drawn from the population; there are 11 persons in the sample who favor Party A and 9 persons who favor Party B. Can we conclude that the population proportion is unchanged at 0.7, or must we conclude that the proportion favoring Party A has changed?

At first glance it seems obvious that we should conclude that the population proportion has changed, since the sample proportion is $11/20 = 0.55$, which is clearly less than 0.70. However, this conclusion is not correct. We hardly expect to get a sample proportion of *exactly* 0.70 when the population proportion is 0.70. The sample proportion is a statistic, and this statistic has a distribution under the assumption or hypothesis that $p = 0.7$. There is some probability of getting a sample proportion of 0.55 even when the population proportion is 0.70.

It is conventional to work with the distribution of the *number* of persons favoring Party A rather than with the *proportion* favoring Party A. This distribution is the binomial distribution for $n = 20$ and $p = 0.7$; it can be found in Table A1 and also in Table 6.2. We see that the most likely number of persons

TABLE 6.2 Distribution of the Statistic
y, the Number of Persons
Favoring Party A, under the
Hypothesis that $p = 0.7$

y	$P(y)$
0	.000
1	.000
2	.000
3	.000
4	.000
5	.000
6	.000
7	.001
8	.004
9	.012
10	.031
11	.065
12	.114
13	.164
14	.192
15	.179
16	.130
17	.072
18	.028
19	.007
20	.001
	1.000

favoring Party A is 14; this result has a probability of 0.192, which is less than one chance in five.

The probability of obtaining the observed number, 11, of persons favoring Party A is 0.065, or about one chance in 16. Is this so unlikely that we should reject the hypothesis on which these probabilities are based—namely, the hypothesis that $p = 0.7$? The statistician's answer is no—this is not a small enough probability. Since 11 persons could, with probability 0.065, favor Party A when the true proportion is 0.7, we conclude that the proportion has not changed from 0.7.

Many details of this argument will be explained in Chapters 9 and 10. Two of these details will be mentioned here without any rationale, so that you can make statistical inferences in simple problems. First, in the above example, the relevant probability for deciding if the observed statistic is likely or unlikely under the hypothesis is not the individual probability $P(11)$ but the *cumulative* probability $P(y \leqq 11)$. The latter probability is the probability of the observed statistic, 11, and more extreme values: that is, values further away from the center of the distribution. In the above example, $P(y \leqq 11) = 0.065 + 0.031 + 0.012 + 0.004 + 0.001 + 0.000 + \ldots = 0.113$. Second, for the present we will define a small probability as a probability less than 0.05. Since $P(y \leqq 11)$ is greater than 0.05, it is not small by this definition. This is why we decided to accept the hypothesis that $p = 0.7$.

Suppose, on the other hand, that only 10 persons favored Party A in the random sample of 20 persons. From Table 6.2 we see that $P(y \leqq 10) = 0.031 + 0.012 + 0.004 + 0.001 + 0.000 + \ldots = 0.048$, which is less than 0.05. Hence, the observed result of 10 persons favoring Party A is sufficiently unlikely that we can conclude that the population proportion favoring Party A has changed from 0.7.

Estimation

The previous example of hypothesis testing may seem unnatural. If 11 persons in the random sample of 20 favor Party A, the proportion of persons in the sample favoring Party A is $11/20 = 0.55$. Surely, we should conclude that the population proportion, p, is also 0.55. This type of inference is called estimation.

The difference between hypothesis testing and estimation can be described as follows. In hypothesis testing, we set up a hypothesis about the population parameter. The hypothesis is set up before the study is done. On the basis of the sample statistic, we decide whether to accept or reject the hypothesis. In the preceding example the hypothesis was $p = 0.7$. Given the sample result, $y = 11$, we accepted this hypothesis.

In estimation, on the other hand, we consider all possible values of the population parameter, rather than just one possible value. On the basis of the sample statistic, we seek the best guess or estimate of p. It seems reasonable that the best estimate would be $11/20 = 0.55$.

Although statisticians have accepted this argument, they have found ways to define more precisely what is meant by the term "best estimate". In this example it is fairly obvious what the best estimate is. In examples involving other population parameters the best estimate is not always so obvious. Chapter 12 will provide a definition of a best estimate. For now, we simply should note that the definition of the best estimate requires a consideration of the distribution of the statistic y. Hence, the basic principle of statistical inference, given earlier, is still followed. Part of this principle stated that the inference "is made on the basis of (1) the distribution of the statistic and (2) the value of the statistic in the sample". The distribution of the statistic is as important in estimation as it is in hypothesis testing.

When the estimation procedure is described in more detail in Chapter 12, we will see that the average or mean of the distribution of a statistic is required. The mean of a probability distribution resembles very closely the concept of the mean of a frequency or sample distribution. We will also require the standard deviation of the distribution of the statistic. The techniques for computing the mean and the standard deviation of a probability distribution are developed in the next two sections. This discussion will prepare the way for a more detailed analysis of estimation (and hypothesis testing) in later chapters.

6.5 Random Variables and Discrete Probability Distributions

The distinction between two types of probability distributions has been stressed repeatedly in the earlier sections of this chapter. These types are the *population distribution* and the *distribution of the statistic*. Refer back to Figure 6.4 for a review of these two distributions. Both these distributions are probability distributions, since both give the probability of each possible value of a variable.

The two distributions shown graphically in Figure 6.4b are listed numerically in Table 6.3. The population distribution in Table 6.3a gives the proba-

TABLE 6.3 Two Probability Distributions

(a) A POPULATION DISTRIBUTION			(b) A DISTRIBUTION OF A STATISTIC	
PARTY	x	$P(x)$	y	$P(y)$
B	0	$0.3 = q$	0	.000
A	1	$0.7 = p$	1	.000
		1.0	2	.000
			3	.000
			4	.000
			5	.000
			6	.000
			7	.001
			8	.004
			9	.012
			10	.031
			11	.065
			12	.114
			13	.164
			14	.192
			15	.179
			16	.130
			17	.072
			18	.028
			19	.007
			20	.001
				1.000

bilities (that is, proportions in the population) of the two parties. Notice some details of this table. A number, called x, has been assigned to each party. Party A is assigned the value $x = 1$, and Party B is assigned the value $x = 0$. This way of assigning numbers to a variable that has only two possible values is standard in statistics. Party A is given the value or code "1" since it is the party on which most of our previous discussion has been focused. If a person favors Party A, that person is considered a "success", and is coded "1". On the other hand, the favoring of Party B is considered "failure" and is coded "0". Codes "1" and "2" could have been used instead, but there are mathematical reasons for the choice of "0" and "1". Finally, note that since it is usual to list the values of a variable from lowest to highest in a table, Party B, with code "0", is listed first, and Party A, with code "1", is listed second.

A variable that has numeric values and has probabilities associated with each value, such as x in Table 6.3a, is called a *random variable*. A random variable is simply a numeric variable that has an associated probability distribution. Much of statistics involves the study of random variables.

The probabilities of the values of this random variable are $P(0)$ and $P(1)$. In the previous discussion of the binomial (Section 6.3), these two values of the population distribution were given the symbols q (probability of 0 or failure) and p (probability of 1 or success), respectively. Here $P(0) = q = 0.3$ and $P(1) = p = 0.7$.

Another random variable is y, the number of successes in a random sample of 20 persons. The distribution of y is given in Table 6.3b. The random variable y has values extending from 0 to 20, and the probability of each value is given in the table.

These two probability distributions, along with many other, similar ones, are known as *discrete* probability distributions, because the random variable takes on a set of discrete values. (Recall the definitions of discrete and continuous variables given in Section 2.4.) A *continuous random variable* is a random variable whose values include all the numbers within a certain range. A *discrete random variable* is a random variable whose values have gaps between them: that is, not all values are possible in the range of values. The variable y is a discrete random variable, since the values are 0, 1, 2, . . . , 20 and values such as 5.62 are not possible. Continuous random variables are described in Chapter 7.

It is important to distinguish carefully between the *values* of a random variable (such as 0, 1, 2, etc.) and the *probabilities* of these values (such as 0.065, 0.130, etc.). There is no restriction on the numeric values a random variable may have, except that a discrete variable has gaps in its range and a continuous variable does not. However, there are definite restrictions on the probabilities of a random variable, or (what comes to the same thing) the probabilities in a probability distribution. *The probabilities in a probability distribution must each be nonnegative; must each not exceed 1.0; and must sum, in total, to 1.0.*[6] You will see that the two distributions in Table 6.3 satisfy

[6] These are reasonable restrictions on a probability distribution. Probability, or chance, is by its nature a positive or zero quantity. Second, a probability of 1.0 indicates certainty and, hence, cannot be exceeded. Third, the set of values of the random variable includes all those values that could occur; since one of these must occur the probability of one of them occurring (the sum of their probabilities) must be exactly 1.0—a certainty.

these restrictions. The following two distributions however, do not satisfy the restrictions.

x	P(x)		x	P(x)
1	0.4		5	0.6
2	0.5		7	−0.4
3	0.4		8	0.5
			9	0.3

The first distribution is not a legal probability distribution, because the probabilities add to 1.3, not the required 1.0. The second distribution has a negative probability that is not legal, even though the probabilities add to 1.0. Note that the *values* of the random variable in these examples differ from those in Table 6.3 but are quite acceptable as values of a discrete random variable. The value $x = 6$ does not occur in the second distribution, but there is no requirement for all integers in a certain range to be included. Another way of looking at this situation is to think of $x = 6$ as having probability zero ($P(6) = 0.0$).

This section has defined random variables in general, and discrete random variables in particular. The probability distribution of a discrete random variable is known as a discrete probability distribution.

6.6 The Mean and the Standard Deviation of a Discrete Random Variable

The mean and the standard deviation of a set or sample of numbers can be defined by a formula and can be related to the distribution of the numbers. The mean is the balance point of the distribution (Section 3.2) and the standard deviation is approximately equal to half the 68% range of the distribution (Section 4.3). These properties also apply to distributions of random variables.

Consider the simple probability distribution shown in Figure 6.6. The distribution satisfies the requirements of a probability distribution: all the probabilities are between zero and one and the sum of the probabilities is one. The distribution is symmetric and unimodal. Terms such as "symmetric", "unimodal", etc., which describe frequency distributions, also describe probability distributions.

Where is the balance point of this distribution? Obviously, at 7.0—the middle or center of this distribution. This number, 7.0, is the mean of the random variable y (or, what comes to the same thing, the mean of the probability distribution). For sets or samples of numbers, the symbol $\bar{x}$ or $\bar{y}$ is used for the mean. For a random variable we will use a new symbol—μ, the Greek letter m. The symbol μ is pronounced "mu". (It is very important to pronounce "mu" to yourself every time you see this symbol. A number of Greek letters are introduced in this text. If you do not attach a different name to each, you will not be able to tell them apart. If you think of each as just a "squiggle" or "funny letter" or "Greek letter", you will confuse a number of different concepts.)

So the mean of y is $\mu = 7.0$. What is the standard deviation of y? From the distribution, we can estimate that 68% of the distribution lies, roughly, between 6 and 8. This is a width of 2, giving an estimated standard deviation of $2/2 = 1$, by the variability principle. The symbol for the standard deviation of a

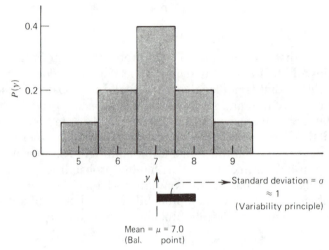

y	$P(y)$
5	0.1
6	0.2
7	0.4
8	0.2
9	0.1
	1.0

Figure 6.6 Mean and standard deviation of a probability distribution.

random variable is σ, the Greek letter s. It is pronounced "sigma". (Recall that the symbol Σ, used in summations, is also called "sigma". The symbol Σ is the uppercase Greek S, whereas σ is the lowercase letter.)

Mean of a Random Variable

These are, of course, formulas for precisely calculating the mean and standard deviation of any random variable y. The formula for the mean, μ, is

$$\mu = \Sigma y P(y) \quad \text{(Mean of a random variable } y\text{)}$$

This formula says that to get the mean of a random variable y, you take the sum (Σ) of the products of each value of y times the probability of that value. The mean of the variable y given in Figure 6.6 is computed as follows.

$$\mu = \Sigma y P(y)$$
$$= 5 \times 0.1 + 6 \times 0.2 + 7 \times 0.4 + 8 \times 0.2 + 9 \times 0.1$$
$$= 0.5 + 1.2 + 2.8 + 1.6 + 0.9$$
$$= 7.0$$

This answer confirms the value found from the balance point of the distribution.

Be sure to distinguish carefully between the formula for the sample mean ($\bar{y}$) and the formula for the mean of a random variable (μ).

$$\bar{y} = \frac{\Sigma y}{n} \qquad \text{versus} \qquad \mu = \Sigma y P(y)$$

The most common error made in calculating μ is to attempt to divide $\Sigma y P(y)$ by n. But what is n for a probability distribution? There is no n involved in a probability distribution! You should treat these two formulas as quite distinct—one is for the mean of a sample, whereas the other is for the mean of a random variable with a probability distribution.

Standard Deviation of a Random Variable

The standard deviation, σ, of a random variable is found by taking the square root of the variance of the random variable, σ^2. The formula for the variance of a random variable y is

$$\sigma^2 = \Sigma(y - \mu)^2 P(y) \quad \text{(Variance of a random variable } y\text{)}$$

This formula says that to get the variance of a random variable y, you must first compute the deviation of each value of y from the mean (μ). The variance is the sum of the products of the square of each deviation times the probability of each y.

The formula for the standard deviation is then

$$\sigma = \sqrt{\Sigma(y - \mu)^2 P(y)} \quad \text{(Standard deviation of a random variable } y\text{)}$$

The variance and standard deviation of the variable y of Figure 6.6 are computed as follows.

$$\sigma^2 = \Sigma(y - \mu)^2 P(y)$$

$$= (5 - 7)^2 \times 0.1 + (6 - 7)^2 \times 0.2 + (7 - 7)^2 \times 0.4 + (8 - 7)^2 \times 0.2 + (9 - 7)^2 \times 0.1$$

$$= (-2)^2 \times 0.1 + (-1)^2 \times 0.2 + (0)^2 \times 0.4 + (+1)^2 \times 0.2 + (+2)^2 \times 0.1$$

$$= 4 \times 0.1 + 1 \times 0.2 + 0 \times 0.4 + 1 \times 0.2 + 4 \times 0.1$$

$$= 0.4 + 0.2 + 0.0 + 0.2 + 0.4$$

$$= 1.2;$$

$$\sigma = \sqrt{1.2} = 1.10$$

So the estimate of 1.0 made from the variability principle is very close to the exact value of 1.10.

Bernoulli and Binomial Distributions

In Section 6.5, we considered two probability distributions for the study of the proportion of persons favoring Party A. Examples of these distributions are given in Table 6.3 and graphed in Figure 6.7. The first distribution is a population distribution. The random variable, called x, has only two values, 0 and 1, with probabilities q and p, respectively. This simple probability distribution is sometimes called a *Bernoulli distribution*. A Bernoulli distribution is the distribution of a random variable that has only two values. Therefore a

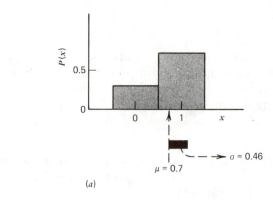

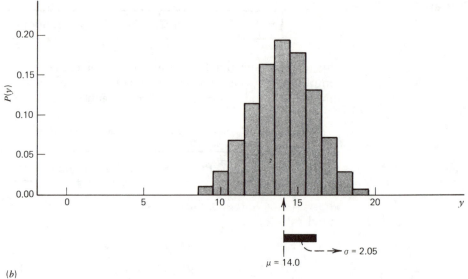

Figure 6.7 Two probability distributions. (a) A population distribution (Bernoulli distribution with $p = 0.7$). (b) A distribution of a statistic (binomial distribution with $p = 0.7, n = 20$).

Bernoulli distribution has a mean, a variance, and a standard deviation that can be calculated by the formulas given above. When $p = 0.7$, the calculations are (note that in this case the formulas are written in terms of x rather than y)

$$\mu = \Sigma xP(x)$$

$$= 0 \times 0.3 + 1 \times 0.7 = 0.7;$$

$$\sigma^2 = \Sigma(x - \mu)^2 P(x)$$

$$= (0 - 0.7)^2 \times 0.3 + (1 - 0.7)^2 \times 0.7$$

$$= 0.49 \times 0.3 + 0.09 \times 0.7$$

$$= 0.147 + 0.063$$

$$= 0.210;$$

$$\sigma = \sqrt{0.210} = 0.46$$

The values of μ and σ are shown in Figure 6.7a; they are reasonable on the basis of the mean being the balance point and the standard deviation being about half the 68% range.

It is possible to show algebraically that the mean, variance, and standard deviation have the following values for a Bernoulli distribution with any value of p.

$$\mu = p; \quad \sigma^2 = pq; \quad \sigma = \sqrt{pq} \quad \text{(Bernoulli distribution)}$$

For a general probability distribution, the general formulas given earlier for μ, σ^2, and σ must be used; but for the particular case of a Bernoulli distribution, the simpler formulas may be used, since they give the same result as the general formulas.

The second distribution in Table 6.3 and Figure 6.7 is a binomial distribution. A binomial distribution is a particular case of a distribution of a statistic, which itself is just a particular case of a probability distribution. Hence, the general formulas for μ, σ^2, and σ may be used. The lengthy calculations are not given here. The results, for the binomial distribution with $p = 0.7$ and $n = 20$, are

$$\mu = 14.0; \quad \sigma^2 = 4.20; \quad \sigma = 2.05$$

The values of μ and σ are shown in Figure 6.7b. They are reasonable.

As in the case of the Bernoulli distribution, it can be shown the general formulas for μ, σ^2, and σ reduce to the following simple formulas for the binomial distribution with any p, q, and n.

$$\mu = np; \quad \sigma^2 = npq; \quad \sigma = \sqrt{npq} \quad \text{(Binomial distribution)}$$

It can be confusing to have three sets of formulas for μ, σ^2, and σ. In the case of the three formulas for μ, the first $[\mu = \Sigma yP(y)]$ is a general formula that applies to *any* probability distribution of a random variable. Since it is general, it applies also to the Bernoulli and binomial distributions. But for these two special distributions, the general formula reduces to the simpler formulas given above. The same distinction applies to the three formulas for σ^2 and to the three formulas for σ.

Population Mean and Standard Deviation

In Section 6.2, the population mean and standard deviation were used as examples of parameters. Every population has a population distribution. The mean of this distribution—that is, the mean of the random variable having this distribution—is, by definition, the *population mean*. Similarly, the standard deviation of the population distribution is the *population standard deviation*.

The symbols μ and σ are used for the population mean and standard deviation, respectively, just as they are used for the mean and standard deviation of a random variable. This can be confusing. Note that the mean and standard deviation of a random variable that has a population distribution are the population mean and standard deviation. However, the mean and standard deviation of a random variable that is a statistic are simply the mean and standard deviation of a statistic and are *not* parameters of a population.

6.7 Summary

A statistical inference is an inference about a *population parameter*, which is a number that describes a *population distribution*. The inference is based on the value of a *sample statistic*, which is a number that describes a *sample distribution*. The link between the population parameter and the sample statistic is provided by the *distribution of the statistic*. A particular distribution of a statistic is the *binomial distribution*.

There are two types of statistical inference: *hypothesis testing* and *estimation*. The *basic principle of statistical inference* applies to each.

Probability distributions are distributions of *random variables*. A random variable may be *discrete* or *continuous*. The mean, variance, and standard deviation of a discrete random variable may be calculated by general formulas. These formulas have special forms for the *Bernoulli distribution* and for the binomial distribution.

Exercises

1 (Section 6.1) Consider an experiment in which one marble is drawn from a jar of six marbles, two of which are red and four of which are green. The experiment is repeated 50 times; after each draw, the marble is returned to the jar and the jar is shaken thoroughly. There are 15 red marbles and 35 green marbles drawn.

(a) Draw a graph of the *sample distribution* of marble color obtained in the 50 draws.

(b) Draw a graph of the *population distribution* of marble color.

(c) Suppose that you did not know the population distribution of marble color. What characteristic must the sample have before you can make an inference about the unknown population distribution from the results obtained in the sample?

2 (Sections 6.1, 6.2, and 6.3) In a certain country the proportion of male births is 0.45 and the proportion of female births is 0.55. Hence, the probability of a male birth is 0.45 and the probability of a female birth is 0.55. The probability distribution of the number of males in 10 consecutive births is given by the following table.

NUMBER OF MALES	PROBABILITY
0	0.0025
1	0.0208
2	0.0763
3	0.1664
4	0.2384
5	0.2340
6	0.1596
7	0.0746
8	0.0229
9	0.0042
10	0.0003

On one particular day, 10 babies were born—six male and four female.

(a) Draw graphs of

   (i) The population distribution of the sex of a baby.

   (ii) The sample distribution of the sex of a baby.

   (iii) The distribution of the statistic: y = number of males.

(b) State one parameter and its value. State two statistics and their values.

3 (Section 6.3) Determine the following probabilities from the binomial table (Table A1).

(a) The probability that a pregnant woman will have twins is 0.01. In three consecutive births in a hospital, what is the probability that one of the women gives birth to twins? What is the probability that none of the births are twins?

(b) A multiple-choice quiz has 10 questions. Each question has five alternatives. If a student is completely unfamiliar with the subject matter, what is the probability that he or she gets exactly six answers correct, by guessing?

(c) Suppose that there are only two (instead of five) alternatives to each of the 10 questions on the quiz. What is the probability of getting exactly six answers correct, by guessing? What is the probability of getting at least six answers correct, by guessing?

(d) Suppose that y is the number of successes in a study and that y has a binomial distribution. The sample size is 25 and the probability of success is 0.6. Find $P(15 \leq y \leq 18)$ and $P(y < 16)$.

4 (Section 6.4) A study is conducted to see if professional musicians can tell whether a piano sonata is played by a man or by a woman. Each musician listens to two recordings of the same piece, one performed by a man, the other by a woman. The musician has to judge which recording is which. If it is really impossible to make this discrimination, the probability that a musician makes the correct identification is 0.5.

(a) Suppose 25 musicians are tested in this way and that 18 correctly identified the pair of recordings by sex (and that the other seven incorrectly identified them). Is this sufficient evidence to infer that a musician can tell whether a man or a woman played the sonata?

(b) Estimate, from the results given in **(a)**, the proportion of musicians (in the population from which these musicians were selected) who can correctly identify the sex of the performer of a sonata.

(c) Repeat **(a)** and **(b)**, but suppose that the results of the study were that 16 of the musicians correctly identified the sex of the performer.

5 (Sections 6.5 and 6.6) Consider the following probability distribution.

x	$P(x)$
4	0.2
5	0.1
6	0.2
7	0.5

(a) Does this distribution satisfy the requirements of a probability distribution? Explain.

(b) Calculate the mean, variance, and standard deviation of this distribution.

(c) Consider the following two distributions: the Bernoulli distribution for $p = 0.2$ and the binomial distribution for $p = 0.2$ and $n = 5$.

  (i) Show each distribution in a table and a histogram.
  (ii) Compute the mean and standard deviation of each distribution.
  (iii) Show, from the graphs, that your answers to **(ii)** are reasonable.

7 CONTINUOUS DISTRIBUTIONS

Chapter 6 discussed discrete distributions. This chapter discusses continuous distributions. The basic concepts of population parameter, sample statistic, population distribution, sample distribution, and distribution of the statistic apply to both discrete and continuous distributions.

The most important continuous distribution is the normal distribution. Much of this chapter deals with that distribution. Probabilities of a normally distributed population can be calculated using a table. Even when the population distribution is not normal, the distribution of the sample mean is approximately normal and probabilities can therefore be calculated easily.

The two basic types of statistical inference, hypothesis testing and estimation, were introduced in Chapter 6. Further examples are given in this chapter, using the distribution of the sample mean.

7.1 Standard Scores

In Section 4.1 the *percentile* was introduced. The percentile of a score describes that score's position in a distribution. Another way to describe the position of a score is to compute the *standard score*. The standard score and the percentile provide related, but not identical, information about a score's position. The percentile gives the percentage of scores that fall below the given score. *The standard score states how far the given score is from the mean of the distribution, in standard deviation units.* (Recall, from Section 4.3, that the standard deviation is a measure of the variability of a distribution. The standard deviation is approximately half the 68% range of a distribution. The standard deviation can also be calculated by a formula, of course.)

Standard scores are much easier than percentiles to compute. A simple formula relating the original score and the standard score is given later in this section. First, however, consider an example showing the relationship between scores, percentiles, and standard scores.

Suppose a test marked out of 300 points has been given to a very large class of students. The distribution of scores on this test is shown in Figure 7.1a.[1] The mean of the distribution is 110 (the balance point of the distribution) and the standard deviation is 30 (about half the 68% range). Bob's score and John's score are also shown in the figure. Bob's score is 120, putting him at the 60th percentile of the distribution; hence, 60% of the area of the distribution is below his score. John's score is 168, at the 80th percentile; hence, 80% of the scores are below 168, or 20% are above 168, as shown. (Recall, from Section 2.6, that relative area can be interpreted as relative frequency.)

---

[1] For the purpose of this example and this section, it is not important to consider whether the test score is a continuous or discrete variable. The vertical axis on the graph is labeled "frequency"; the exact nature of this axis will be clarified in Section 7.2.

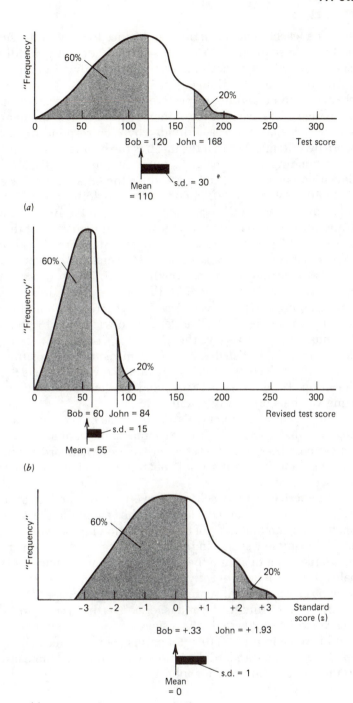

Figure 7.1 Standard scores.

The teacher who gave the test decides to convert the scores, originally out of 300, to a score out of 150. This is easily done by dividing each score by 2. These revised scores have the distribution shown in Figure 7.1b. Note that the shape of the distribution is unchanged. Bob's score and John's score have changed—to 60 and 84, respectively—but their location in the distribution is unchanged. The percentiles of their scores are also unchanged. The mean and standard deviation are each divided by 2, corresponding to the division of each score by 2. Remember that the mean is the balance point; the mean is now at 55, corresponding to the balance point of the new distribution. The standard deviation is also changed, to 15, since the 68% range now extends from about 40 to 70—a width of 30, in contrast to the former width of 60.

Now go back to the original test scores in Figure 7.1a. Bob's score of 120 is 10 points above the mean. As the standard deviation of the scores is 30, Bob's score is $10/30 = 0.33$, or one-third of a standard deviation above the mean. John's score of 168 is 58 points above the mean, which is $58/30 = 1.93$ or almost two standard deviations above the mean. These observations become more significant if we repeat them for the revised test scores, which are equal to the original scores divided by 2. Here Bob's score is 60 and the mean is 55, so he is 5 points above the mean. The standard deviation is 15, so Bob is $5/15 = 0.33$ of a standard deviation above the mean. *Bob's position with respect to the mean is the same in both distributions if we measure this position in standard deviation units.* Similarly, John's score on the revised scale is 84, which is 29 above the mean. He is $29/15 = 1.93$ standard deviation units above the mean, the same value as originally calculated.

In summary, any person's score can be converted to a score that measures how far that person is from the mean in standard deviation units. These converted scores are called *standard scores*. Standard scores have the property that the same standard score is obtained whether the test score or the revised test score is used.[2]

Suppose that every score is converted to a standard score. The distribution in Figure 7.1c is obtained. Note that once again the shape of the distribution is unchanged and that most of the distribution lies between standard scores of -3 and $+3$. Very few scores in this or any other distribution are further than three standard deviations, in either direction, from the mean. Also note that scores below the mean have *negative* standard scores, whereas scores above the mean have *positive* standard scores. The mean itself has a standard score of zero.

Let us now find the formulas for converting a score to a standard score, and vice versa. We converted John's score of 168 to a standard score by first noting that his score is $168 - 110 = 58$ points above the mean and then dividing this result by 30, the standard deviation. We will now denote a standard score by z, so the computation of z for John is

$$z = \frac{168 - 110}{30} = \frac{58}{30} = 1.93$$

[2] The original score can be multiplied or divided by any number without modifying the standard score. Furthermore, a constant can be added to or subtracted from the original scores without modifying the standard scores. In general, the standard scores are unchanged when the *scale* or the *origin* of the original scores is changed (see below).

In general, the formula is

$$z = \frac{y - \text{mean}}{\text{s.d.}} \quad \text{(Conversion of score } y \text{ to standard score } z\text{)}$$

This formula instructs us to take the original score y, subtract the mean, and divide the result by the standard deviation. The result is the standard score, which will often be called the *z-score*, for short.

This formula also works for scores below the mean. A score of 90 has a z-score of

$$z = \frac{90 - 110}{30} = \frac{-20}{30} = -0.67$$

This result is negative, indicating that the score is two-thirds of a standard deviation *below* the mean.

Suppose now that we wish to reverse the process. If a standard score is 1.5, what is the original test score? The z-score of 1.5 indicates that the score is 1.5 standard deviations above the mean. Since the standard deviation is 30, the score is $1.5 \times 30 = 45.0$ points above the mean. The mean itself is 110. Hence, the score in question is $110 + 45 = 155$.

The same procedure works for negative z-scores. Suppose that $z = -2.5$. Then this score is $(-2.5) \times 30 = -75.0$ points from the mean, the negative sign indicating that the score is *below* the mean. Hence, the score is $110 - 75 = 35$. Note carefully that 75 is subtracted from the mean here, since the z-score is negative.

The general formula for computing a score from a z-score is

$$y = \text{mean} + (z \times \text{s.d.}) \quad \text{(Conversion of z-score to } y\text{)}$$

This formula is just an algebraic rearrangement of the previous formula for converting y to a z-score. Repeating the calculations in the two preceding paragraphs, the formula gives

$$y = 110 + 1.5 \times 30 = 110 + 45 = 155$$

and

$$y = 110 + (-2.5) \times 30 = 110 - 75 = 35$$

Note that in the second example, the negative z-score automatically ensures that the product is subtracted from the mean.

Effects of Change of Scale and Origin

A simple way to change a set of scores is to multiply or divide each score by a constant number. This change to the original scores is called a *change of scale*. When each score is divided by 2, the scale is "shrunk" by a factor of 2. If, instead, each score is multiplied by a constant—say, 5—the scale is changed again, but now by an expansion factor of 5. Clearly the mean and standard deviation are increased by a factor of 5, too. In general,

1 When the scale of a set of scores is changed by multiplying or dividing by a constant factor, the mean and standard deviation of the changed scores are changed (multiplied or divided) by this same factor.

Another simple way to change a set of scores is to add a constant to, or subtract a constant from, all the scores. This change is called a *change of origin*. For example, the teacher who gave the test scored out of 300 might decide that the scores are too low and add 50 to each score. This change obviously would add 50 to the mean of the scores. However, the effect on the standard deviation is different. The variability of the distribution, as measured by the 68% range, does not change. The distribution has been shifted, but not shrunk or expanded. Hence there is no change in the standard deviation. In general,

2 When the origin of a set of scores is changed by adding or subtracting a constant amount, the mean of the changed scores is increased or decreased by the same amount, but the standard deviation is unchanged.

Finally, we can note that

3 When the scale or the origin of a set of scores is changed, there is no change in the characteristic shape of the distribution of the scores.

The *shape* of a distribution means those properties of the distribution that are distinct from the central tendency and variability of the distribution. As described above, a change of scale of a distribution does change the mean and standard deviation of the distribution. However, the overall or characteristic shape—whether unimodal or bimodal, symmetric or nonsymmetric, etc.—is unaffected. This means that we can talk about the shape of a distribution without necessarily being specific about its mean and its standard deviation. This property becomes very important in discussing the normal distribution (see Section 7.3). The normal distribution has a characteristic shape that remains unchanged through any changes of scale or origin.

Figure 7.2 illustrates and summarizes the effects of change of scale and origin on the mean, standard deviation, and shape of a distribution.

Properties of Standard Scores

Two properties of standard scores are listed below. These properties should be apparent from the example and from Figure 7.1, discussed earlier.

1 The mean of a set of standard scores is 0 and the standard deviation is 1. (See Figure 7.1c—the balance point is at zero and the 68% range extends from -1 to $+1$, implying a standard deviation of 1.)

2 When the scale or the origin of a set of scores is changed, there is no change to the standard scores. (This property was illustrated by Bob's score and John's score.)

It is the second property that makes standard scores so important.

7.2 Properties of Continuous Distributions

Continuous variables were introduced in Chapter 2 by noting that such variables are, essentially, idealized forms of observed variables. No observed variable can be, strictly speaking, continuous. But a discontinuous, or discrete, observed variable can be approximated by a continuous variable. Furthermore, continuous variables are easier to handle mathematically than are discrete

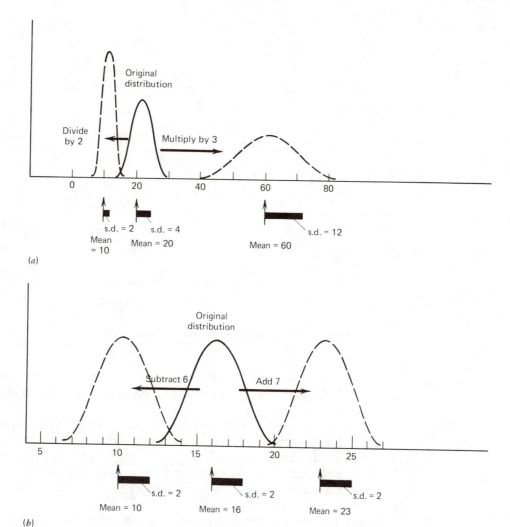

Figure 7.2 Effects of change of scale and origin. (*a*) Change of scale (mean and standard deviation change, shape unchanged). (*b*) Change of origin (mean changes, standard deviation and shape unchanged).

ones. In statistics, therefore, many variables and their distributions are continuous.

The mathematics of continuous distributions will not be given in this text, because it requires some knowledge of calculus. A few features of continuous distributions will be explained in this section by relating them to analogous features of discrete distributions.

A classification of distributions is given by the tree diagram of Figure 7.3. There are two types of *discrete distributions*: *empirical* and *probability* distributions. Empirical distributions arise from data collected in a study; such distributions must be discrete, since no measuring instrument is infinitely precise. Probability distributions of a discrete random variable (for example, the Bernoulli and binomial distributions) were discussed in detail in Chapter 6.

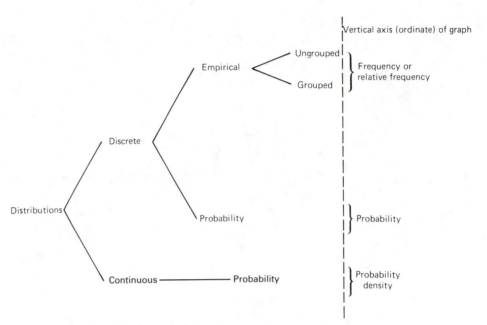

Figure 7.3 A classification of distributions.

There is only one type of *continuous distribution*: a *probability distribution*. These continuous probability distributions form the main subject of this chapter.

The distinction between discrete and continuous probability distributions is illustrated in Figure 7.4. In the discrete distribution the variable or score can have only integer values: 70, 71, etc. Each of these values has a probability, which can be read from the vertical axis in Figure 7.4a. In the continuous distribution of Figure 7.4b, on the other hand, the variable can have any value from about 65 to 100, including noninteger values such as 81.32. Except for the fact that the distribution is smooth, you might think that the graph could be read in the same way as the discrete graph in Figure 7.4a. Consider a score of 71. In the discrete graph you can read the probability directly from the vertical scale. In the continuous graph the vertical scale can also be read, *but the number read there is not a probability*. How could it be a probability? You would read a probability for a score of 71, another probability for 71.1, another probability for 71.135, etc. Every score would have a probability and there are an infinite number of scores! These probabilities would add up to a huge number, much greater than 1. But probabilities must sum to 1 (Section 6.5). The paradoxical conclusion from this argument is that *the probabilities of individual values of a continuous variable are zero*. Since there is an infinite number of values, the individual values cannot have any probability.

This paradox can be resolved by considering *intervals* of scores. Let us first return to the discrete distribution of Figure 7.4a. The probability of an interval such as 76–78 can be calculated in two ways. The first way: from the vertical axis, read the heights of the three bars for the scores of 76, 77, and 78; the sum of these three probabilities is the probability of the interval. The second way: measure the *area* of each of the three bars and sum these areas (the area of a rectangle equals its width times its height); also measure the total area of all the

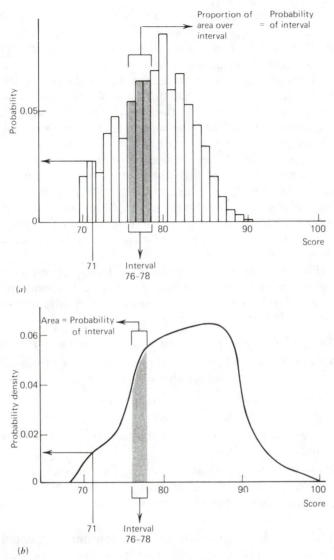

Figure 7.4 Discrete and continuous probability distributions. (*a*) Discrete. (*b*) Continuous.

bars in the distribution; the probability of the interval 76–78 is the *ratio* of the sum of the areas of the three bars to the total area. It should be obvious that the first way is much easier than the second, but that the two ways give the same answer.[3] In summary, *the probability of an interval equals the ratio of the area above that interval to the total area of the distribution*. (Section 2.6 demonstrated that the *relative frequency* of an interval equals the ratio of the area above that interval to the total area of a *frequency* distribution. Both relative frequency and probability can be determined from the *relative area* above an interval of frequency and probability distributions, respectively.)

[3] A formal proof of this equivalence depends on the fact that all the bars have the same width.

For discrete distributions, this second way of computing the probability of an interval is an *optional* way. But for continuous distributions, it is the *only* way. Consider the continuous distribution in Figure 7.4*b*. The probability of the interval 76–78 is given by the ratio of the area above that interval to the total area of the distribution. Continuous distributions are drawn so that the total area is 1.0. This means that the ratio of the area above the interval to the total area is simply equal to the area above the interval. Therefore *the probability of an interval of scores of a continuous variable is the area of the distribution above that interval.*

The area under the continuous distribution measures a probability. On the other hand, the ordinate of the distribution measures *relative probability*. Clearly, the greater the ordinate of the distribution above a given score, the more probable the score. So even though the probability of any particular score is zero, the distribution shows the relative probability of any particular score. The technical term used by statisticians for relative probability is *probability density*.

The following list summarizes the above properties of continuous distributions and adds some additional ones that correspond closely to properties of discrete distributions. All these properties are illustrated by the normal distribution, described in Section 7.3.

1 The probability of a particular value of a continuous random variable is zero.

2 The probability of an interval of values of a continuous random variable is the area of the distribution above that interval.

3 The total area of a continuous distribution is 1.0, so that the total probability is 1.0.

4 The ordinate of a continuous probability distribution is probability density or relative probability.

5 Percentiles have the same meaning for continuous as for discrete distributions. The percentile of a score is the percentage of scores below that score. The percentile is found from the area of the distribution below (to the left of) the score.

6 The mean and standard deviation of continuous random variables are calculated from complex formulas requiring the use of calculus. However, the interpretation of the mean as the balance point and of the standard deviation as approximately half the 68% range still applies.

7 Continuous random variables can be converted to standard scores using the formula of Section 7.1, if the mean and standard deviation are known. The properties of standard scores listed there apply to continuous as well as to discrete variables.

8 The effects of change of scale and origin listed in Section 7.1 apply to continuous random variables.

7.3 The Normal Distribution

The most important distribution in statistics is the *normal distribution*, the distribution of a continuous random variable. Three examples of the normal distribution are shown in Figure 7.5. All normal distributions have the characteristic "bell" shape and are symmetric and unimodal. The distributions extend to infinity in both directions but the probability density, as measured by

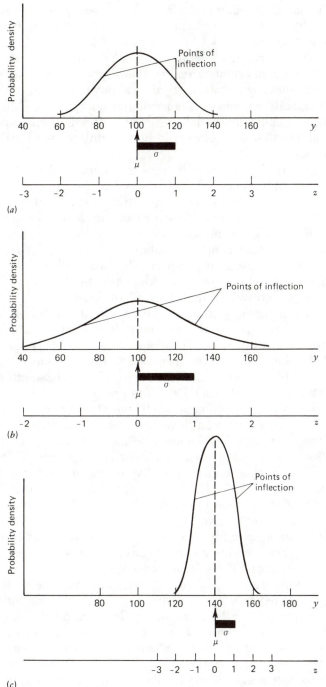

Figure 7.5 Three normal distributions. (a) $\mu = 100$, $\sigma = 20$. (b) $\mu = 100$, $\sigma = 30$. (c) $\mu = 140$, $\sigma = 10$.

the height of the curve above the axis, becomes very small when the distribution is extended beyond the confines of the figure: that is, beyond ±3 standard deviations.

Many distributions besides normal distributions are unimodal and symmetric. It is important to realize that a normal distribution has a particular shape as given by a mathematical formula.[4] Only distributions that satisfy this formula, and hence have this particular normal shape, qualify as normal distributions. Other unimodal, symmetric distributions may be approximately normal, but they are not exactly normal unless they satisfy the mathematical formula.

The three normal distributions in Figure 7.5 have the same characteristic shape but differ in central tendency and variability. Since the distributions are symmetric, the mean, μ, of each distribution falls at the center of symmetry, which is also the location of the median and mode. The distributions in Figures 7.5a and 7.5b have the same mean but differ in variability. Recall that the standard deviation is approximately half the 68% range. The standard deviations, σ, are 20 and 30, respectively. Since all distributions of continuous variables have a total area of 1.0, the wide distribution in Figure 7.5b is not as high as the narrow distribution in Figure 7.5a.

The standard deviation of a normal distribution can be determined, by eye, in another way. Each of the distributions, being unimodal, decreases in height on either side of its peak. The slope of the curve becomes steeper and steeper until a point of maximum steepness is reached. Once this point, called the *point of inflection*, is passed, the curve becomes less and less steep. There is a point of inflection on each side of the peak. *In a normal distribution, each of the two points of inflection is exactly one standard deviation from the mean, so that the distance between them is two standard deviations.*

The distribution in Figure 7.5c has a greater mean, 140, and a smaller standard deviation, 10, than the other two distributions. The figure shows that the mean is greater because the distribution is located farther to the right (at greater values of the variable, y) than the other two distributions. The figure shows that the standard deviation is smaller because the distribution is narrower than the other two distributions.

The three normally distributed random variables in Figure 7.5 are each called y. Each variable y can be converted to a standard score or a *standard normal variable* z by the usual formula, $z = (y - \text{mean})/\text{s.d.}$, which in this case may be written $z = (y - \mu)/\sigma$. When this is done, the horizontal scale for each distribution extends from roughly -3 to $+3$. The distribution actually extends from minus infinity to plus infinity, but beyond the range of -3 to $+3$ the probability density is almost zero. The mean of each distribution is at $z = 0$ and scores one standard deviation above and below the mean have standard scores of $+1$ and -1, respectively.

The distribution of the standard normal variable z has been redrawn to a different scale in Figure 7.6. This distribution is called the *standard normal*

---

[4] The *standard* normal distribution, described later in the section, has the formula

$$f(z) = \frac{1}{\sqrt{2\pi}} e^{-z^2/2}$$

where e is the base of natural logarithms. The formula for the general normal distribution, with mean μ and standard deviation σ, can be derived from the above formula.

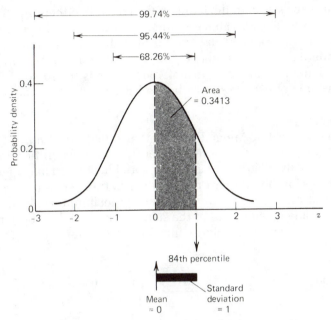

Figure 7.6 The standard normal distribution.

distribution. The total area of the distribution is 1.0. The area above the interval from z = 0 to z = 1 is shown in the figure. This area is 0.3413, to four decimal places. Since the distribution is symmetric, the area above the interval from z = −1 to z = +1 is 2 × 0.3413 = 0.6826, or 68.26% of the total area. Here is the origin of the variability principle of Chapter 4! For the normal distribution the standard deviation is *exactly* one-half of the 68.26% range. For other distributions this relationship is only approximate.

[*Caution.* The symbol z has two meanings. In Section 7.1, z is used as the symbol for a standard score: that is, a score computed from z = (y − mean)/ s.d. Such a z-score necessarily has a distribution whose mean is 0 and whose standard deviation is 1. In this section, z is the symbol for a z-score that is a standard normal variable: that is, a variable that has a standard normal distribution. This distribution has a mean of 0 and a standard deviation of 1, but, *in addition*, has the normal shape. Standard scores (z-scores) always have mean = 0 and standard deviation = 1, but their distribution does not have to be normal.]

The area above the interval from z = 0 to z = 2 is 0.4772 and the area from z = 0 to z = 3 is 0.4987. These areas, and others, are listed in a table described in Section 7.4. If we double each of these numbers and use the figure of 68.26% (worked out above), we obtain the following important results.

DISTANCE FROM MEAN	PERCENT OF AREA
± one s.d.	68.26%
± two s.d.	95.44%
± three s.d.	99.74%

Although these numbers were obtained from the standard normal distribution, they apply to any normal distribution. The area above an interval of values within one standard deviation of the mean (in either direction) is 68%, or about two-thirds. The area above an interval of values within two standard deviations of the mean is about 95%, and the area above an interval of values within three standard deviations is about 99.7%. (See Figure 7.6.) Clearly, there is not much area left out of the total area of 100%. This is the reason why normal distributions are not usually drawn beyond ±3 standard deviations of the mean.

7.4 Using the Normal Table

Table A3 in the Appendix is the basic table for finding areas of the normal distribution. It is therefore used for finding probabilities (represented by areas) of normally distributed random variables. The table gives areas and probabilities for the *standard* normal distribution. Areas and probabilities for other normal distributions can be found by techniques explained below. Furthermore, only certain areas of the standard normal distribution are given in Table A3. These areas are above intervals that extend from the mean of the dis-

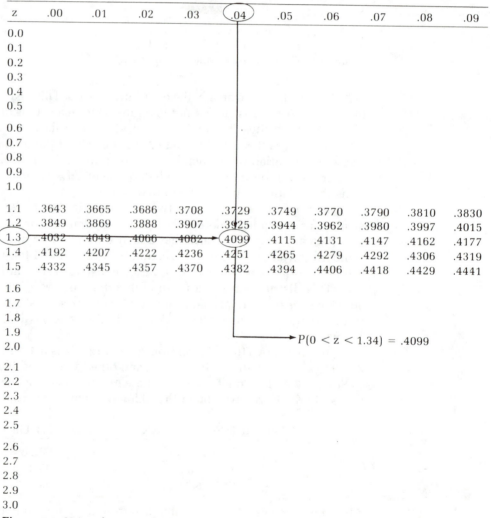

z	.00	.01	.02	.03	.04	.05	.06	.07	.08	.09
0.0										
0.1										
0.2										
0.3										
0.4										
0.5										
0.6										
0.7										
0.8										
0.9										
1.0										
1.1	.3643	.3665	.3686	.3708	.3729	.3749	.3770	.3790	.3810	.3830
1.2	.3849	.3869	.3888	.3907	.3925	.3944	.3962	.3980	.3997	.4015
1.3	.4032	.4049	.4066	.4082	.4099	.4115	.4131	.4147	.4162	.4177
1.4	.4192	.4207	.4222	.4236	.4251	.4265	.4279	.4292	.4306	.4319
1.5	.4332	.4345	.4357	.4370	.4382	.4394	.4406	.4418	.4429	.4441
1.6										
1.7										
1.8										
1.9										
2.0										
2.1										
2.2										
2.3										
2.4										
2.5										
2.6										
2.7										
2.8										
2.9										
3.0										

$P(0 < z < 1.34) = .4099$

Figure 7.7 Using the normal table.

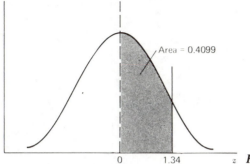

Figure 7.8a

tribution to positive values of z. Other areas of the distribution can be found by simple calculations also explained below.

Example (a)

Find the area above the interval from z = 0 to z = 1.34. This area is sketched in Figure 7.8a and given directly in Table A3. A portion of the table is reproduced in Figure 7.7. The number 1.34 is found on the borders of the table. Part of the number (1.3) is found on the left border; the other part (.04) is found on the top border. The row and column defined by these two points on the borders cross at the value .4099, which is the required area.

Notation for Intervals and Probabilities

In order to simplify the presentation of the examples in this section, it is important to have a suitable notation. We will be required to calculate, or will be given, the *area* above a certain *interval* of values of a variable. In the previous example the interval of values of z extended from 0 to 1.34. This interval will be denoted by (0 < z < 1.34), which may be read as "the interval of z from 0 to 1.34" or "0 (zero) less than z less than 1.34" or "z is between 0 and 1.34".[5] If the variable is y, the interval will be written in a similar way—for example, (20 < y < 40).

The area above the interval (and under the normal distribution) always represents a probability. Hence we will use the symbol P followed by the interval for this probability. The previous example would be summarized by P(0 < z < 1.34) = 0.4099. This is to be read as "the probability that 0 (zero) is less than z is less than 1.34 is 0.4099" or "the probability that z is between 0 and 1.34 is 0.4099".

Forward Problems in Standard Scores

In Section 4.1 a distinction was made between *forward* and *backward* problems. In forward problems, we are asked to find the percentile for a given score. In backward problems, we are asked to find the score for a given percentile. To generalize this distinction, we define a forward problem as any problem requiring the calculation of an area or probability from a given interval of scores. Conversely, a backward problem requires the calculation of an interval of scores that has the given area or probability. Four examples of forward problems follow.

[5] It may be thought that 0 and 1.34 should be included in the interval, so that it would be written (0 ≤ z ≤ 1.34). This interval has the same area and probability as (0 < z < 1.34), since the probability of z being exactly equal to 0 or 1.34 is zero. Hence, the notation (0 < z < 1.34) can be used for simplicity.

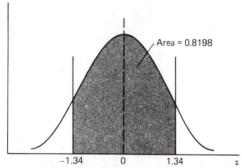

Figure 7.8b

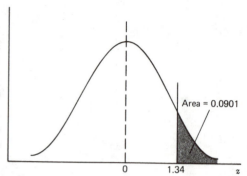

Figure 7.8c

Example (b)

Find the probability that z is between −1.34 and +1.34. The required probability may be written $P(-1.34 < z < +1.34)$. The required area is sketched in Figure 7.8b. Since the standard normal distribution is symmetric, the area to the left of $z = 0$ is the same as the area to the right. From Table A3, in Example (a), we already know that the area to the right is 0.4099. Hence, the required probability is $2 \times 0.4099 = 0.8198$.

Example (c)

Find the probability that z is greater than 1.34. This probability is usually written $P(z > 1.34)$, although it could be written $P(1.34 < z < \infty)$, where the symbol "∞" means "infinity". The required area is sketched in Figure 7.8c. This area is *not* given directly in Table A3, which only gives areas above intervals from $z = 0$ to a positive value of z. However, if the required area is added to the area above the interval $(0 < z < 1.34)$, the total will be exactly 0.5, since the two areas combined are the area to the right of the mean and median (one-half the total area). In symbols, $P(0 < z < 1.34) + P(z > 1.34) = 0.5$. Therefore, $P(z > 1.34) = 0.5 - P(0 < z < 1.34) = 0.5 - 0.4099 = 0.0901$.

Example (d)

Find the probability that z is between 1.5 and 2.5. This probability, $P(1.5 < z < 2.5)$, is sketched in Figure 7.8d. This area cannot be found directly in Table A3. However, the required area is the *difference* between two areas that can be found in the table: $P(1.5 < z < 2.5) = P(0 < z < 2.5) - P(0 < z < 1.5) = 0.4938 - 0.4332 = 0.0606$.

Example (e)

Many forward problems can be expressed as questions about percentiles. For example, what is the percentile for a z-score of −0.7 in a standard normal distribution? The required area, sketched in Figure 7.8e, obviously is less than one-half; hence, the percentile is less than the 50th. Several steps are needed to get the area. First, the area is

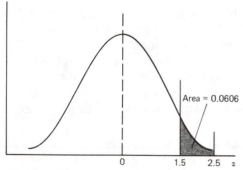

Figure 7.8d

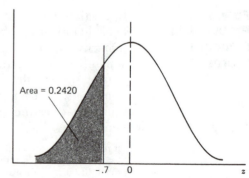

Figure 7.8e

one-half minus the area above the interval from -0.7 to 0—in symbols, $P(z < -0.7) = 0.5 - P(-0.7 < z < 0)$. Now, the area above the interval from -0.7 to 0 is, by symmetry, the same as the area above the interval from 0 to $+0.7$, which can be found in Table A3. Hence, the required area is $0.5 - P(0 < z < +0.7) = 0.5 - 0.2580 = 0.2420$. The percentile of a z-score of -0.7 in a standard normal distribution is the 24.20th, or the 24th to the nearest integer.

Backward Problems in Standard Scores

In a backward problem, we are given an area and have to find an interval of scores which has the given area above it.

Example (f)

Find the value of z such that the area above the interval, from $z = 0$ up to this unknown value, is 0.4495. The given area is sketched in Figure 7.8f. We must look for the number 0.4495 in the *body*, not the borders, of Table A3. Remember that this problem is a backward problem. We are given the area and have to find the z. If we search for 0.4495 in the body of Table A3, we find it in the row labeled "1.6" and the column labeled ".04". Hence, the required z is 1.64. The interval $(0 < z < 1.64)$ has the given area of 0.4495 above it.

Example (g)

What is the 40th percentile of the standard normal distribution? The given area is sketched in Figure 7.8g. It is clear that the required z will be negative, since the given area of 0.40 is less than half the total area and half the area extends up to the mean of $z = 0.0$. Should we look for 0.40 in the body of Table A3? No—the body of the table gives areas above intervals from $z = 0.0$ up to specific positive values of z. In the sketch, the given area is labeled A. The area above the interval from the unknown z to $z = 0$ is labeled B. This area must be $0.10 = 0.50 - 0.40$, since A and B together equal half the area. Now, the area labeled C is also 0.10, by symmetry, and this area is above an interval extending from 0 to a positive z. If we can find this positive z, the required 40th percentile simply will be the negative of this z. So we look up 0.10 in the body of Table A3. We cannot find this value exactly. The closest number we find is 0.0987, corresponding to a z of 0.25. Hence, the 40th percentile is approximately -0.25.

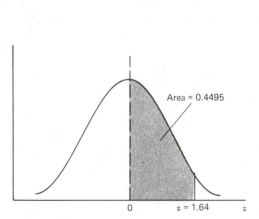

Figure 7.8f

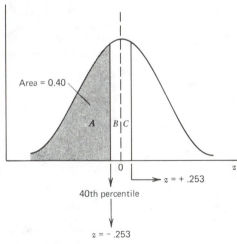

Figure 7.8g

In most backward problems, the exact area cannot be found in the body of Table A3. In the above example, we simply took the closest value. We might attempt to do better by finding, in this case, a value of z between 0.25 and 0.26, since the areas for these two values straddle the required 0.10. It is easier, however, to use a table specially constructed for backward problems. Table A4 in the Appendix gives the value of z corresponding to certain areas. In the above example the table gives directly a value of $z = 0.253$ for an area of 0.10. Hence, the 40th percentile is, more accurately, -0.253.

Example (h)

Many problems in statistics require that we find the value of z that cuts off a certain small area in the tail of the normal distribution. For example, what value of z cuts off an area of 0.001 in the upper tail of the normal distribution? (See Figure 7.8h). Table A4 gives the answer directly. We simply look for 0.001 in the second column of the table, labeled "Area in tail", and find the corresponding value of $z = 3.090$ in the third column. This answer can, of course, be checked in Table A3. There we find $P(0 < z < 3.09) = 0.4990$, which, by subtraction from 0.5, confirms that 3.09 cuts off the required 0.001 in the upper tail.

Forward Problems in Original Scores

The standard normal distribution is not the only normal distribution. Usually, the variables studied in surveys and experiments do not have mean = 0 and standard deviation = 1. Such variables will be called "original scores", in contrast to the standard scores we have been discussing. Forward problems in original scores require the calculation of the area or probability of an interval of original scores. The solution involves two steps: (1) converting the original scores to standard scores and (2) calculating the area over the interval of standard scores.

Example (i)

Suppose that the heights of a population of adult men are normally distributed with mean = 160 cm and standard deviation = 10 cm. What proportion of the men have heights between 145 cm and 165 cm, and in a sample of 800 men, how many would be expected to have heights in that interval? (See Figure 7.8i.) The ends of the interval, 145 and 165, must be converted to standard scores by the formula $z = (y - \mu)/\sigma$. For $y = 145$, the formula gives $z = (145 - 160)/10 = -1.5$; for $y = 165$, the formula gives $z = (165 - 160)/10 = 0.5$. This completes the first step. In the second step, we must find the area above the interval $(-1.5 < z < 0.5)$. This interval is shown on the separate scale below the graph in Figure 7.8i. The area is the sum of two parts, each of which can be found in Table A3: $0.4332 + 0.1915 = 0.6247$. Hence, a proportion of 0.6247, or about 62%, of the men have heights between 145 and 165 cm.

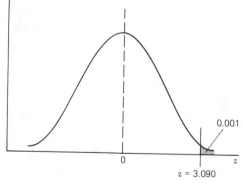

0.001

0

$z = 3.090$

z

Figure 7.8h

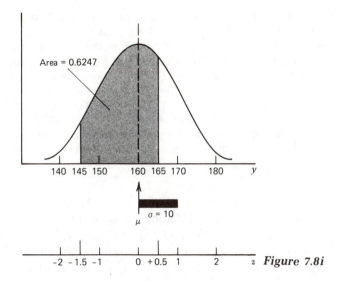

Figure 7.8i

The number of men in a sample of 800 men who would be expected to have heights between 145 and 165 cm is $0.6247 \times 800 = 499.76$ or 500 men.

Backward Problems in Original Scores

Backward problems in original scores require the calculation of an interval of scores that have a specified area or probability. The solution involves two steps: (1) finding the interval of standard scores that has the specified area and (2) converting this interval to an interval of original scores.

Example (j)

Suppose that the reaction times of subjects in a perceptual experiment are normally distributed with mean = 1,700 msec and standard deviation = 250 msec. We want to select, for another experiment, subjects who have times in the middle 80% of the

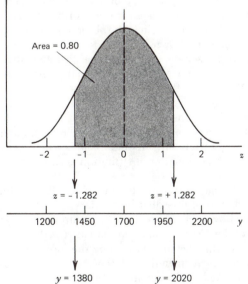

Figure 7.8j

distribution. What reaction times have this property? The first step is to find an interval of z-scores, centered on the mean, which has area = 0.80 (Figure 7.8j). We must find a z such that the area above the interval from 0 to that z is 0.80/2 = 0.40. The backward normal table, Table A4, directly gives us this z, which is 1.282. Hence, the required interval of z-scores is $(-1.282 < z < +1.282)$. In the second step, each of these z-scores is converted to an original score by the formula: $y = \mu + (z \times \sigma)$. For $z = -1.282$, this formula gives $y = 1,700 + (-1.282) \times 250 = 1,700 - 320.5 = 1,380$, to the nearest integer. For $z = +1.282$, the formula gives $y = 1,700 + 1.282 \times 250 = 1,700 + 320.5 = 2,020$. Hence, the required interval of original scores is $(1,380 < y < 2,020)$, and subjects should be selected for further experimentation who have times between 1,380 and 2,020 msec.

7.5 The Central Limit Theorem

The structure of statistical inference was outlined in Chapter 6. Figure 6.4 summarized the various distributions and their relationships. One of the distributions is the distribution of the statistic. Under certain conditions the statistic will be normally distributed. This important result is known as the central limit theorem. Before we state the theorem as a principle, let us look at three examples illustrating the theorem.

Example 1— Normal Population

Some population distributions are normal. When this is the case, the statistical treatment is particularly simple. Consider a population that is normally distributed with parameters $\mu = 110$ and $\sigma = 20$. (See Figure 7.9a.) A random sample drawn from this population might have the sample distribution shown in the figure. The sample mean, $\bar{y}$, computed from these data is 108.6.

The third distribution required for statistical inference is the distribution of the statistic. In Section 6.4 the statistic was y, the number of successes, and the distribution of the statistic was the distribution of y. Here, the statistic is $\bar{y}$, the sample mean, so we refer to the required distribution as the distribution of $\bar{y}$ (or as the distribution of the sample mean). Recall how this distribution would be computed in principle. The sample space for samples of size 25 drawn from the population would be listed. This list would include all possible samples of this size. Then, for each of the samples in the list the mean, $\bar{y}$, would be computed. Finally, a probability distribution would be constructed, giving the probability (more precisely, the probability density) of each possible value of the statistic, $\bar{y}$. The result of this process is the distribution of the statistic $\bar{y}$, shown in Figure 7.9a. One possible sample mean is the observed mean of 108.6. Its position in the distribution of the statistic is shown.

The distribution of the sample mean has three characteristics, the first two of which are not very surprising: (1) the statistic, $\bar{y}$, is normally distributed; (2) the mean of this distribution is 110, the same as the mean of the population distribution, μ. The mean of the distribution of the statistic might also have the symbol μ, but in order to keep it distinct from the population mean, we will give it the symbol $\mu_{\bar{y}}$. The second characteristic states that $\mu_{\bar{y}}$ is in fact equal to μ. (Warning: Even though these two means have the same value, they are conceptually distinct. That is why we have given them different symbols. The mean of the population distribution is μ; the mean of the distribution of the statistic $\bar{y}$ is $\mu_{\bar{y}}$.)

Now let us turn to the third characteristic of the distribution of the sample mean: (3) the standard deviation of this distribution is 4, which is the population standard deviation, 20, divided by $\sqrt{25}$, where 25 is the sample size $(20/\sqrt{25} = 20/5 = 4)$. The standard deviation of the distribution of the statistic $\bar{y}$ will be given the symbol $\sigma_{\bar{y}}$. The third characteristic can then be summarized by the formula $\sigma_{\bar{y}} = \sigma/\sqrt{n}$; in words, the standard deviation of the distribution of the statistic $\bar{y}$ is equal to the population standard deviation divided by the square root of the sample size.

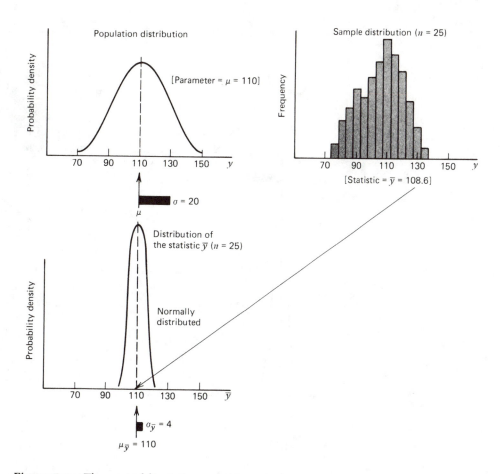

Figure 7.9a The central limit theorem: Normal population.

Why is the distribution of the statistic $\bar{y}$ so much less variable than the distribution of the population itself? The distribution of the statistic is derived from the sample space, as described above. Each sample in the sample space consists of 25 scores drawn from the population with mean = 110 and standard deviation = 20. In most samples, about half of the 25 scores will be above 110 and half below 110. This will make the sample mean near 110. Consider what would be necessary for a sample in the sample space to have a mean of, for example, 130, which is one standard deviation above the population mean. This would require that about half the scores would have to be above 130 and half below 130. But it would be rather unlikely that many scores above 130 would be drawn in a random sample. In a normal distribution the proportion of scores more than one standard deviation above the mean is 0.1587 (from Table A3). The chance of a sample having *half* its scores that large is extremely small. Hence, the probability of sample means very far from 110 is very small. It can be shown mathematically that the standard deviation of the distribution of $\bar{y}$ is equal to the population standard deviation divided by the square root of the sample size. This is the third characteristic listed above.

Example 2—Nonnormal Continuous Population

The three characteristics of the distribution of the statistic $\bar{y}$ might appear to depend on the fact that the population in Example 1 is normally distributed. This is not true. The second example uses a population that is continuously, but not normally, distributed. (See Figure 7.9*b*.) This skewed distribution has parameters $\mu = 95$ and $\sigma = 15$. The

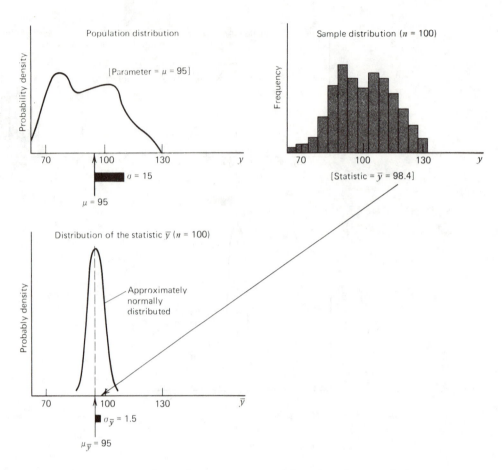

Figure 7.9b The central limit theorem: Nonnormal continuous population.

sample distribution for a particular random sample of size 100 is shown; the statistic, $\bar{y}$, for this sample has the value 98.4.

It is important to remember that on the whole the sample distribution and the population distribution will be *quite similar* and will get more and more similar as the sample size increases. The relative frequency of a score (or an interval of scores) becomes closer and closer to the probability of that score as the sample size increases.

On the other hand, the distribution of the statistic $\bar{y}$ is *quite different* from the population distribution. Sample means tend to be near the population mean. A sample mean is very unlikely to be far from the population mean. Hence, the mean of the distribution of the statistic $\bar{y}$, shown in Figure 7.9b, is equal to the population mean of 95, but its standard deviation is much smaller than the population standard deviation. The standard deviation of the distribution of the sample mean is 1.5 in this case. From the formula, $\sigma_{\bar{y}} = \sigma/\sqrt{n} = 15/\sqrt{100} = 15/10 = 1.5$.

What is the shape of the distribution of the sample mean in this example? One of the most surprising results in statistics is that the shape is approximately normal, even though the population distribution is quite nonnormal! It is not possible to give a simple explanation or proof of this fact. The third example in this section will, perhaps, make the result plausible. For now it should be noted that the distribution of the statistic $\bar{y}$ is not exactly normal, but rather is *approximately normal*. However, when the sample size is more than 10, in most cases the approximation to normality is so close that we can treat the distribution of $\bar{y}$ as normal for all practical purposes. So the distribution of the sample mean is approximately normal; however, the mean and standard deviation of this distribution are *exactly* given by the formulas $\mu_{\bar{y}} = \mu$ and $\sigma_{\bar{y}} = \sigma/\sqrt{n}$.

The purpose of the preceding two examples has been to show that when the population distribution is continuous, the distribution of the sample mean is normal if the population is normal, and approximately normal if the population is nonnormal. But the approximation in the second case is so good that we can usually treat the distribution as normal and use Table A3 to compute probabilities of sample means falling in specified intervals. In the third example, we turn to a population that has a discrete distribution. Even here the distribution of the sample mean is normal.

Example 3— Discrete Population

Note: If you have not read Chapter 5, you should skip this example and jump to the statement of Principle P7, the central limit theorem.

Consider a population in which the variable has only three possible values (1, 2, and 3), each of which is equally probable. This population might arise in an experimental task that could be learned in 1, 2, or 3 trials. Suppose that an equal number (proportion) of persons learn the task in 1, 2, or 3 trials. Such a population is shown in Figure 7.9c.

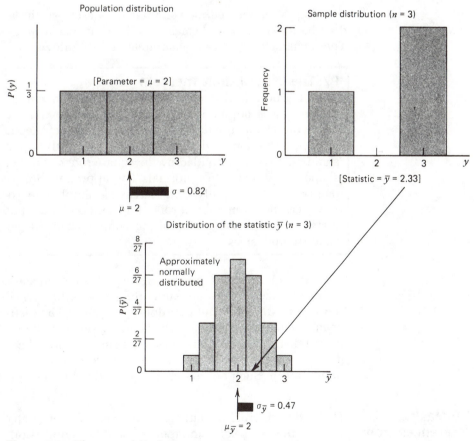

Figure 7.9c The central limit theorem: Discrete population.

The mean of this population is obviously 2.0, by symmetry, and the standard deviation can be calculated to be 0.82, a reasonable value by the variability principle.

The sample distribution for a particular sample (with scores 1, 3, 3) is shown in the figure. The mean of this sample is 2.33.

A small sample size ($n = 3$) has been chosen so that the distribution of the statistic $\bar{y}$ can be calculated easily. The calculations are shown in Figure 7.10. In Figure 7.10a the mean of each of the 27 samples is calculated. Each of these samples has the same probability: $\frac{1}{3} \times \frac{1}{3} \times \frac{1}{3} = \frac{1}{27}$. The table in Figure 7.10b summarizes the number of samples with each value of $\bar{y}$. Since each sample has probability $= \frac{1}{27}$, the probability of each value can be obtained by multiplying the number of samples by $\frac{1}{27}$. The result is the distribution of $\bar{y}$ given in both a table and a graph. This graph has been reproduced in Figure 7.9c.

Notice that the graph is similar to a normal distribution even though it is a discrete distribution. (The random variable, $\bar{y}$, has only seven different values.) The population certainly does not resemble a normal distribution, and the sample size is only three. When the sample size is greater than 10, the distribution of the sample mean is very close to a normal distribution even though, strictly, it is only approximately normal.

The mean and standard deviation of the distribution of $\bar{y}$ can be worked out quite easily from the basic formulas of Section 6.6. The values are $\mu_{\bar{y}} = 2.0$ and $\sigma_{\bar{y}} = 0.47$. These values confirm the formulas: $\mu_{\bar{y}} = \mu = 2.0$ and $\sigma_{\bar{y}} = \sigma/\sqrt{n} = 0.82/\sqrt{3} = 0.82/1.732 = 0.47$.

Each of the preceding examples has described three characteristics of the distribution of the sample mean: its shape, its mean, and its standard deviation. These characteristics can be summarized as follows.

P7: The Central Limit Theorem

The distribution of the sample mean, based on random samples of size n drawn from a population with mean μ and standard deviation σ, has the following characteristics: the mean, $\mu_{\bar{y}}$, is exactly equal to the population mean, μ; the standard deviation, $\sigma_{\bar{y}}$, is exactly equal to the population standard deviation divided by the square root of the sample size, $\sigma/\sqrt{n}$; the shape is approximately normal. The approximation of the shape to normality improves rapidly with increasing sample size, so that, for $n > 10$, the shape can be taken to be normal. Furthermore, if the population is normally distributed, the distribution of the sample mean is exactly normal, even for small sample sizes.

The importance of the central limit theorem in statistical work cannot be overemphasized. We rarely know the shape of the distribution of the population we are studying. But that does not matter! The distribution of the sample mean, for random samples whose size is greater than 10, is very close to a normal distribution. We need this distribution, together with the actual mean of our randomly drawn sample, in order to make an inference about the population mean. Such an inference is illustrated in Section 7.6.

7.6 Making Inferences from the Normal Distribution

The basic principle of statistical inference was presented in Section 6.4. This section illustrates some inferences about the population mean μ. The statistic on which the inferences are based is the sample mean $\bar{y}$. The basic principle states that in order to make an inference about μ from the statistic $\bar{y}$ computed

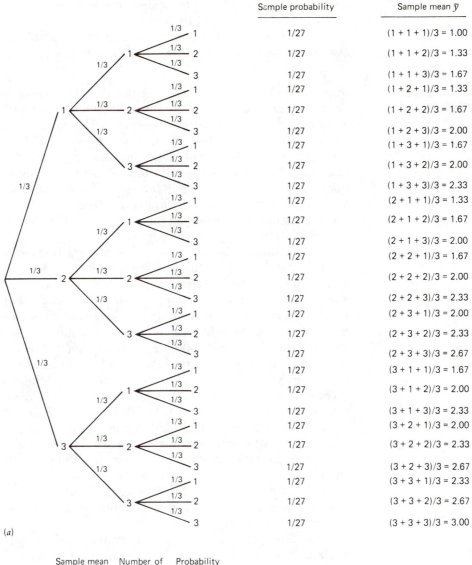

Sample probability	Sample mean $\bar{y}$
1/27	$(1 + 1 + 1)/3 = 1.00$
1/27	$(1 + 1 + 2)/3 = 1.33$
1/27	$(1 + 1 + 3)/3 = 1.67$
1/27	$(1 + 2 + 1)/3 = 1.33$
1/27	$(1 + 2 + 2)/3 = 1.67$
1/27	$(1 + 2 + 3)/3 = 2.00$
1/27	$(1 + 3 + 1)/3 = 1.67$
1/27	$(1 + 3 + 2)/3 = 2.00$
1/27	$(1 + 3 + 3)/3 = 2.33$
1/27	$(2 + 1 + 1)/3 = 1.33$
1/27	$(2 + 1 + 2)/3 = 1.67$
1/27	$(2 + 1 + 3)/3 = 2.00$
1/27	$(2 + 2 + 1)/3 = 1.67$
1/27	$(2 + 2 + 2)/3 = 2.00$
1/27	$(2 + 2 + 3)/3 = 2.33$
1/27	$(2 + 3 + 1)/3 = 2.00$
1/27	$(2 + 3 + 2)/3 = 2.33$
1/27	$(2 + 3 + 3)/3 = 2.67$
1/27	$(3 + 1 + 1)/3 = 1.67$
1/27	$(3 + 1 + 2)/3 = 2.00$
1/27	$(3 + 1 + 3)/3 = 2.33$
1/27	$(3 + 2 + 1)/3 = 2.00$
1/27	$(3 + 2 + 2)/3 = 2.33$
1/27	$(3 + 2 + 3)/3 = 2.67$
1/27	$(3 + 3 + 1)/3 = 2.33$
1/27	$(3 + 3 + 2)/3 = 2.67$
1/27	$(3 + 3 + 3)/3 = 3.00$

(a)

Sample mean $\bar{y}$	Number of Samples	Probability $P(\bar{y})$
1.00	1	1/27
1.33	3	3/27
1.67	6	6/27
2.00	7	7/27
2.33	6	6/27
2.67	3	3/27
3.00	1	1/27
	27	1

(b)

Figure 7.10 The distribution of the statistic $\bar{y}$ for a discrete population.

from a random sample, we must know the distribution of $\bar{y}$. This distribution is given by the central limit theorem.

As in Section 6.4, we will consider one example of each basic type of statistical inference—hypothesis testing and estimation.

Hypothesis Testing

Suppose we have designed an experiment in which subjects say words that are flashed onto a screen. By means of a microphone and suitable electronic apparatus, we can measure the time interval between the beginning of the flash and the moment the subject starts to say the word. This time interval is called the reaction time. A theory of word processing predicts that for the population of persons from which we will draw our subjects, the reaction times will be distributed with a mean of 1,500 msec and a standard deviation of 200 msec. We plan to test this theory by drawing a random sample of 100 subjects from the population and measuring their reaction times. Our hypothesis states that the population mean is 1,500 msec.

After running the experiment, we make a frequency distribution of the reaction times and find that the sample mean, $\bar{y}$, has a value of 1,550 msec. What can we conclude about the predicted value, for the population mean, of 1,500 msec? Clearly, the sample mean is not exactly equal to the predicted population mean. But we would never expect exact equality, since the sample means for various random samples from a population with mean 1,500 msec are distributed around 1,500 msec—some below and some above. This distribution is the distribution of the sample mean, whose shape, central tendency, and variability are specified by the central limit theorem. From that theorem, we know that if the population mean, μ, is 1,500 and the population standard deviation, σ, is 200, the distribution of the sample mean will be normal with mean $\mu_{\bar{y}} = 1,500$ and standard deviation $\sigma_{\bar{y}} = 200/\sqrt{100} = 200/10 = 20$. (Note that nowhere in this discussion have we had to mention the shape of the population distribution; for a sample size of 100, the distribution of the sample mean will be very close to normal, whatever the shape of the population distribution.) The hypothesized population distribution (with arbitrary shape) and the resulting distribution of the sample mean are shown in Figure 7.11.

The figure shows us that the observed sample mean of 1,550 msec is quite unlikely to occur. To compute how unlikely it is, we will compute the probability that the sample mean is 1,550 msec *or greater*: that is, $P(\bar{y} \geqq 1550)$. (A complete rationale for using this probability as a measure of how unlikely the observed result will be given in Chapters 9 and 10.) This probability, which is the same as $P(\bar{y} > 1550)$ since the probability that $\bar{y}$ is exactly equal to 1,550 is zero, may be worked out using the techniques of Section 7.4. We convert 1,550 to a z-score using the mean of 1,500 and the standard deviation of 20. Note carefully that the standard deviation of the distribution for which we are computing a probability is 20, not 200. The z-score is $(1,550 - 1,500)/20 = 50/20 = 2.50$. Hence,

$$P(\bar{y} > 1550) = P(z > 2.5) = 0.5 - 0.4938 = 0.0062$$

using Table A3. The observed sample mean of 1,550 is so unlikely to occur that the probability of such a mean, or a greater mean, is only 0.0062. Rather than think that such an unlikely event occurred, we prefer to conclude that the

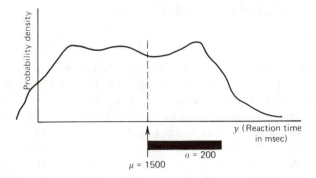

(a)

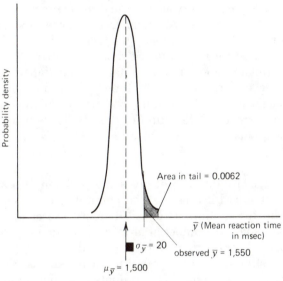

(b)

Figure 7.11 Testing a hypothesis. (*a*) Population
distribution. (*b*) Distribution of sample mean, $\bar{y}$
($n = 100$).

original hypothesis—that the population mean is 1,500—is incorrect. The
conclusion is that the population mean is not 1,500. This is the statistical
inference we make from the study we have described.

Let us change the example so that the opposite conclusion is made.
Suppose the sample mean is 1,529 msec. The z-score is $(1{,}529 - 1{,}500)/20 =$
$29/20 = 1.45$. Hence,

$$P(\bar{y} > 1529) = P(z > 1.45) = 0.5 - 0.4265 = 0.0735$$

This probability, although small, is not as small as 0.05, the usual criterion for a
small probability. The observed sample mean is not so unlikely, under the
hypothesis that the population mean is 1,500, that we should reject that
hypothesis. We therefore conclude that the population mean is 1,500. (For the
present we treat "failing to reject the hypothesis" and "accepting the hypoth-

esis" as equivalent. The distinction between these two conclusions is described in Chapter 10.)

Estimation

Many experiments are conducted without a theory that makes a specific prediction about a population parameter, such as the prediction that $\mu = 1,500$ in the above example. Upon completion of such an experiment, we may wish to *estimate* the population mean, rather than to accept or to reject a specific hypothesis about its value. Such an estimate is easily made. If the sample mean is 1,550 msec, the best estimate for the population mean will be 1,550 msec. Because the definition of a "best" estimate (see Chapter 12) requires knowledge about the distribution of the statistic $\bar{y}$, the central limit theorem is just as important for estimation as it is for hypothesis testing.

The distribution of the statistic $\bar{y}$ comes into estimation in another way. It usually is not satisfactory to give just an estimate of, say, 1,550 msec for a parameter. We know that this number is not absolutely accurate; it might be too large or too small. It is customary to attach a "plus or minus" to every estimate. In the present example, we might report the estimate as $1,550 \pm 40$, meaning that the estimate population mean falls somewhere between, 1,510 and 1,590. The number 40 is twice $\sigma_{\bar{y}} = 20$, the standard deviation of the distribution of the sample mean. The "plus or minus", or error of the estimate, is often taken to be twice the standard deviation of the distribution of the sample mean. (For detailed explanation, see Chapter 12.)

7.7 Summary

A *standard score*, or *z-score*, gives the position of a score in a distribution. Although standard scores may be computed for discrete variables, they are most useful for continuous ones. Standard scores apply both to data and to random variables. A number of *properties of standard scores* and the effects of *change of scale and origin* were discussed.

The distinctions and similarities between *discrete* and *continuous distributions* were elaborated. The ordinate in a continuous distribution is *probability density*; the probability of an interval is represented by the *area* of the distribution over that interval.

The *normal distribution* is the most important distribution in statistics. Any normal probability can be calculated by using two tables of the *standard normal distribution*.

The normal distribution is important because the *central limit theorem* states that the distribution of the sample mean is approximately normal with specified mean and standard deviation. This theorem provides the basis for statistical inferences of both types: *hypothesis testing* and *estimation*.

Exercises

1 (Section 7.1) Here are 16 scores, which we will call x: 34, 37, 33, 35, 34, 32, 34, 35, 36, 34, 33, 34, 31, 33, 34, 35.

(a) Compute the mean and standard deviation of these scores. (You can save time in these calculations by constructing an ungrouped frequency distribution of x and computing Σx and Σx^2 by the formulas

$$\Sigma x = \Sigma x f(x) \text{ and } \Sigma x^2 = \Sigma x^2 f(x)$$

In these formulas, $f(x)$ is the frequency of scores with value x.)

(b) Convert each different value of x to a standard score.

(c) Suppose that each score x is converted to a score y by subtracting 11 and dividing by 2 (i.e., $y = (x - 11)/2$). What are the values of y and what are their standard scores?

(d) Compute the percentiles of each x score and each y score.

(e) The scores x are equally spaced. Are either the standard scores or the percentiles equally spaced?

(f) What distribution would a set of scores have to have in order for the percentiles to be equally spaced?

2 (Sections 7.2 and 7.3) A variable x is distributed normally with mean = 50 and standard deviation = 10.

(a) What proportion of the area of this distribution falls between values of

    (i) 40 and 60.

    (ii) 20 and 80.

    (iii) 30 and 50.

    (iv) 40 and 80.

(b) The variable x is standardized by subtracting 50 and dividing by 10.

    (i) What are the mean, standard deviation, and shape of the distribution of the standardized variable?

    (ii) Find those intervals of the standardized variable that have the same proportion of area as the four intervals of x in **(a)**.

3 (Section 7.4) In the following problems, z is a variable that has the standard normal distribution and y is a variable that is normally distributed and has mean = 50 and standard deviation = 10. (Sketches of the required areas are very useful for solving these problems.)

(a) Find $P(0 < z < 1.86)$.

(b) Find $P(-1.26 < z < +2.37)$.

(c) Find the first quartile of z.

(d) Find the interval of z-scores, centered on zero, that includes 90% of the probability of z.

(e) Find the 60th percentile of z.

(f) Find $P(46 < y < 49)$.

(g) Find the third quartile of y.

(h) Find the interval of y, centered on the mean of y, that includes 75% of the probability of y.

(i) Find $P(y \le 55)$.

(j) Find $P(y = 55)$.

4 (Section 7.4) In the following problems, z is a variable that has a standard normal distribution and y is a variable that is normally distributed and has mean = 100 and standard deviation = 15. (Sketches of the required areas are very useful for solving these problems.)

(a) What is the probability that z is greater than 1.35?

(b) What is the probability that z is between 1.73 and 2.73?

(c) What are the three quartiles of z?

(d) Find the interval of z scores, centered on zero, which includes 80% of the probability of z.

(e) What is the 83rd percentile of z?

(f) What is the 17th percentile of z?

(g) What is the probability that y falls between 99 and 106?

(h) Find the interval of values of y, centered on the mean of y, which includes 50% of the probability of y.

(i) What is the probability that y is less than 115?

(j) What is the probability that y is at least 110?

5 (Section 7.4) Suppose that scores on a certain test are normally distributed with mean

= 50 and standard deviation = 20. The passing score on the test is 30. What proportion of students pass the test?

6 (Section 7.4) The time that it takes to complete a certain skilled task is normally distributed with mean = 45 minutes and standard deviation = 10 minutes. How long must subjects be given to do the task if you want 80% of the subjects to complete the task in the allotted time?

7 (Section 7.5) The distribution of weights of a population of adult men is skewed to the right. The mean of this distribution is 65 kg and its standard deviation is 8 kg. Random samples of size 70 are drawn from this population and the mean of each sample is calculated.

(a) What are the shape, mean, and standard deviation of the distribution of the sample means?

(b) What is the probability that a sample mean will fall between 60 and 70 kg?

(c) The standard deviation calculated in (a) is called the *standard error*. What is the probability that a sample mean will fall between (65 − one standard error) and (65 + one standard error)? What is the probability that it will fall between (65 − two standard errors) and (65 + two standard errors)? Use these answers to explain the answer in (b).

8 (Section 7.5) Suppose that income is normally distributed with mean = $8,000 and standard deviation = $1,000. A random sample of 50 persons is drawn from the population. What is the probability that the mean income of the sample will be less than $7,900?

9 (Section 7.6) A certain achievement test has a mean of 100 and a standard deviation of 12 in a national population. The school board in a city wants to determine if in its schools the test has the same mean. A random sample of 75 students in the city is tested and the sample mean is 98.23.

(a) Test the hypothesis that the city population mean is 100.

(b) Estimate the city population mean.

8 RELATIONSHIPS BETWEEN TWO VARIABLES

The studies used as examples in the preceding chapters share a common characteristic: each has only one variable (political party favored, result of a coin toss, male height, or time to respond to a flashed word, etc.).

Most research studies involve two or more variables. The purpose of such studies is to investigate the relationships among the variables. Consider, for example, a study of the political parties favored by males and by females. A natural question to ask is whether the favored political party is related to the sex of the respondent. Actually, two questions can be asked about the relationship between two variables: (1) Is there a relationship between the variables? and (2) If there is a relationship, how large is it? In the first four sections of this chapter several definitions are given of the relationship between two variables. These definitions provide a criterion for answering the first question. In Sections 8.5 and 8.6, several measures of the size of the relationship between two variables are introduced. These measures can be used to answer the second question.

8.1 The Study of Relationships

Consider a study of the intelligence of males as measured by an IQ test. Such a study of one variable could determine the shape of the distribution of IQ, its mean, and its standard deviation. Although these facts might be useful for checking the standardization of the test, such a simple study would hardly be considered a research or scientific study.

Suppose, however, that samples of both males and females are studied. The mean IQs for males and females could be compared. A difference between these means could then be interpreted in terms of some theory of the development of intelligence. This study could be augmented by the addition of other variables. For example, each subject could be judged as belonging to a certain social class; then, the mean IQs for persons of different social classes, and different sexes, could be compared. This study would now involve three variables: sex, social class, and IQ. An interesting result in the study might be that the difference between the two sexes in IQ occurs only in the lower social classes; there might be no difference in the higher classes. These results might then be explained by a sociological theory.

In Chapter 6, statistics was defined as a set of techniques for making inferences about a population from observations of a sample drawn from that population. A number of examples of such inferences were presented in Chapters 6 and 7. In each of these examples there was a single variable and the inference was about a parameter: either μ, the population mean; or p, the proportion of successes in the population. Most statistical inferences are, however, about the relationship between two (or more) variables. The purpose of statistics can therefore be generalized as follows.

P8: Definition of Statistics
Statistics is a set of techniques for making inferences about a single variable or about the relationship between two or more variables from observations of a sample drawn from a population.

As you read this chapter, keep in mind the following qualifications and comments.

Inferential and Descriptive Statistics The definition emphasizes statistics as *inferential*; as noted in Section 6.1, statistics has two main branches: descriptive statistics and inferential statistics.

Population/Sample Distinction The definition stresses the distinction between a sample and a population. We observe a sample and draw an inference about a population that we cannot directly observe. This chapter, however, will not discuss how to draw inferences about the relationship between two variables; its purpose is simply to define what a relationship is and how it is measured. A relationship can be defined and measured in either a sample or a population; the definition and measurement are similar in either case. For simplicity, all the examples in this chapter are of samples. Remember that the same ideas may be applied to populations. Further, remember that when a relationship is found and measured in a sample, you must still make an inference about the relationship in the population. How this inferential step can be taken is described in later chapters.

Science and Statistics A statistical inference about a relationship is not the same thing as a scientific theory. Earlier in this section a "theory of the development of intelligence" and a "sociological theory" were mentioned. These are scientific theories that do not depend directly on statistics. However, a scientific theory predicts that the variables of the theory will be related in certain ways. For example, females may be predicted to have higher intelligence than males, as measured by an IQ test, because of the difference in attitude of parents and teachers toward females and males. This is a prediction of the relationship between the variables IQ and sex. This relationship can be investigated by statistical techniques. From the data in a sample, an investigator would attempt to find out if the predicted relationship actually exists in the population. This inference would then have a bearing on whether or not the theory about intelligence is accepted.

8.2 Two-group Studies

Many studies in psychology and other social sciences involve the comparison of two groups. In some studies the groups occur naturally. Examples of *naturally-occurring groups* are males and females, high- and low-income persons, white and black persons, etc. The variable distinguishing the groups (sex, income, race, etc.) is called a *naturally-occurring variable*. In other studies, each group is identified by the *condition* under which the group is studied. Since most studies of this kind are experiments, the groups are called *experimentally-defined groups*. Examples include a group rewarded on every trial and a group rewarded on every second trial, a group tested in a darkened room and another tested in a bright room, a group given instructions intended to create moderate anxiety and another whose instructions are not anxiety-arousing. The variable or condition distinguishing the groups (frequency of reward, illumination of room, type of instructions) is called an *experimentally-defined variable*.

The distinction between naturally-occurring and experimentally-defined groups can be used to distinguish surveys and experiments (Section 6.1). *Surveys* that compare two groups of subjects usually are studies of naturally-occurring groups. Subjects in a survey seldom are assigned to different conditions. The difference between the two groups of subjects usually is based on a naturally-occurring variable such as sex, nationality, economic or social status, etc. *Experiments* that compare two groups of subjects usually are studies of experimentally-defined groups. Subjects are selected for study and are assigned to one of two conditions. The difference between the two conditions is based on an experimentally-defined variable such as type of reinforcement or level of illumination.

The variable that distinguishes the groups is known as the *independent variable*. It may be either a naturally-occurring or an experimentally-defined variable. Often, the purpose of the study is to see whether subjects who differ on the independent variable also differ on a second variable, called the *dependent variable*.[1] In a survey, the dependent variable usually is another characteristic of the subjects. If the independent variable is sex, the dependent variable might be the response to a question about income. The purpose of the survey might be to see if males and females differ in income. In an experiment, the dependent variable usually is a test or measure of performance. For example, subjects might be tested on a perceptual task after having spent 20 minutes in either a dark or a bright room.

The following example of a two-group study shows how to graph the results of a two-group study and how to see, from the graph, whether the independent and dependent variables are related. This graphical technique can be used when the independent variable is experimentally-defined, as in the example, and when it is naturally-occurring.

Example

Following the development of a new method of teaching arithmetic to fourth-grade children, a comparison of the new method with the old is carried out. Twenty children are selected for the study—10 assigned to a special class in which the new method will be taught and 10 assigned to a regular class using the old method. After three months of instruction, all students are given the same test. The score on this test is a measure of how well the students have learned arithmetic.

The experimentally-defined *independent* variable in this study is type of class (method of instruction). The two values of this variable are "regular class" and "special class". The *dependent* variable is the test score. We will assume that values of this

---

[1] Particularly in surveys, the designation of one variable as independent and the other as dependent often is made arbitrarily. The terms "independent" and "dependent" arise from the notion of cause: many studies seek to determine whether the independent variable causes the dependent variable. In statistics, however, we usually are limited to determining whether two variables are related or not. The definition of a relationship between two variables is given below.

TABLE 8.1 Data from a Two-group Study

REGULAR CLASS	SPECIAL CLASS
59	76
51	62
69	69
65	65
61	67
72	60
53	74
70	64
63	73
57	70
Means 62.0	68.0

variable range from 0 to 100. The results from the study (Table 8.1) can be shown in a *bivariate scatterplot* (see Figure 8.1a). This scatterplot is an extension of the univariate scatterplot of Section 2.5. Here we have two variables—the independent and dependent variables—but the independent variable has only two values. For each of its two values, "regular class" and "special class", a univariate scatterplot of the test scores is drawn in the figure in the vertical direction. Note also that the independent variable is drawn on the abscissa (horizontal axis) and the dependent variable is drawn on the ordinate (vertical axis). This is the usual convention.

The test scores for the two classes have different means: the mean of the regular class is 62 and the mean of the special class is 68. The difference between these two means indicates that there is a *relationship* or an *association* between test score and type of class. When the independent variable has only two values, we say that *there is a relationship or an association between two variables if the means of the dependent variable are different for the two values of the independent variable*.[2] If the two distributions have the same mean, we say that there is no association or no relationship between the two variables.

Distributions can be plotted in several ways. The scatterplot of Figure 8.1a is appropriate when the number of scores is quite small. For larger sample sizes a grouped frequency distribution must be used. The distributions for two groups of sizes 60 and 85 are illustrated in Figure 8.1b. Here the polygon form of frequency distribution is used, since the two distributions overlap and their histograms would be hard to distinguish. Again, we can see that the means of the distributions differ. In these data, therefore, test score and type of class are associated.

Although the graph in Figure 8.1b is easy to interpret, the independent and dependent variables no longer form the two axes of the graph. In Figure 8.1b the *dependent*, not the independent, variable is on the abscissa. The independent variable is on neither axis; the two values of the independent variable ("regular class" and "special class") are simply identifiers of the two distributions in the figure. What is the variable on the ordinate? The ordinate is *frequency*, which is neither the independent nor the dependent variable. This can be confusing, as it is customary to have the independent variable on the abscissa and the dependent variable on the ordinate.

In order to restore the independent and dependent variables to the abscissa and

---

[2] A more general definition of relationship is sometimes used: there is a relationship if the *distributions* of the dependent variable are different for the two values of the independent variables. In most cases, however, the difference between the distributions is a difference in their means. (Two distributions with the same means but different variabilities would indicate a relationship between the two variables by this more-general definition, but not by the simpler definition.)

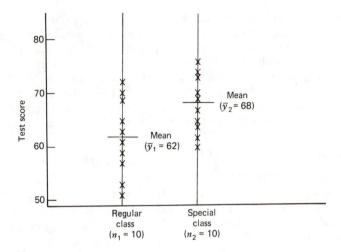

(a)

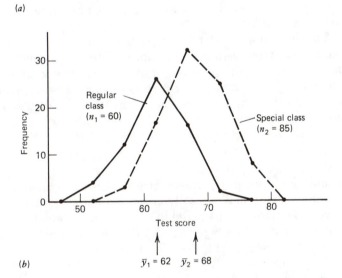

(b)

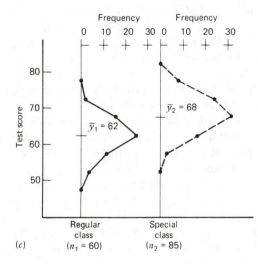

(c)

Figure 8.1 Graphing the
results of a two-group
study. (a) Scatterplot.
(b) Frequency distribution.
(c) Modified scatterplot.

ordinate, respectively, we can draw a modified scatterplot (Figure 8.1c). This scatterplot is similar to Figure 8.1a except that the univariate scatterplot for each class is replaced by a polygon drawn on its side. It helps in interpreting this graph to imagine a scatter of points along each of the two vertical lines. The density of the scatter is given by the frequency indicated by the polygon. (An alternative way to think of this graph is to imagine the two polygons folded out of the plane of the paper and standing vertically on the page.) It is easy to see that test score is associated with type of class, since the means of the distributions of test scores are different for the two types of class.

The three graphical methods of Figure 8.1 can be applied to many studies. In Figure 8.1 the independent variable is an experimentally-defined variable. Similar graphs can be drawn if the independent variable is a naturally-occurring variable. In the case of sex, for example, there would be one distribution for males and another for females. These graphical methods can also be extended to studies with three, four, and even more groups defined by values of the independent variable. There would then be three, four, or more distributions in the graphs.

The definition of relationship or association can be extended to cover this more general case. *Definition of relationship or association between two variables: Consider the distributions of the dependent variable, one distribution for each value of the independent variable. There is a relationship or association between the two variables if the means of these distributions are different. Conversely, there is no relationship or no association between the two variables if the means of these distributions are identical.*

8.3 Correlational Studies

In some studies the *independent variable* has more than just a few values. Such a variable usually is a naturally-occurring variable such as age, IQ, or height. The variable will be considered to be continuous, even though in practice only a discrete set of values is recorded. The *dependent variable* also is continuous. Studies of the relationship between two continuous variables are called *correlational studies*.

Consider a study of the relation between IQ and test score in a sample of fourth-grade children. The IQ of each child is measured, and the score on a test at the end of the school year is recorded. Because a child's IQ is a more enduring property of the child than his or her test score, IQ would be considered to be the independent variable and test score the dependent variable, and not vice versa.

In a correlational study a relationship is defined in the same way as it is in a two-group study. Consider the distributions of the dependent variable, one distribution for each value of the independent variable. There is a relationship or association between the two variables if the means of these distributions are different. Conversely, there is no relationship or no association between the two variables if the means of these distributions are identical.

This definition is more difficult to apply to a correlational study than to a two-group study, since the independent variable is continuous and therefore has a large number of possible values. In some cases, however, it is possible to determine whether there is a relationship by looking at the scatterplot. The definition of relationship will be applied in this section by grouping the values of the independent variable into just two categories. Section 8.4 will show how

Example

Suppose that a sample of 50 father-son pairs is studied to see if there is a relationship between the height of the father and the height of the son. Is there a tendency for tall fathers to have tall sons and short fathers to have short sons (and hence is there a relationship between father's height and son's height)? Or are the sons of both tall and short fathers about the same height (and hence is there no relationship between father's height and son's height)? The scatterplot of fathers' heights and sons' heights can help to answer these questions. In Figure 8.2a each father-son pair is indicated by a cross. The horizontal position of the cross indicates the height of the father and the vertical position indicates the height of the son.

The scatterplot shows quite clearly that the taller fathers tend to have tall sons and the shorter fathers tend to have short sons. Hence, there is a relationship between the heights of fathers and sons. If we want to investigate this relationship in more detail, we first change the scatterplot to look like a two-group scatterplot. This change will mean, however, that much of the information about the relationship will be lost. Section 8.4 will show how to extract more information from the scatterplot.

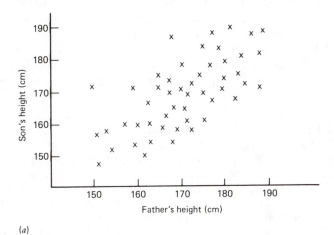

(a)

(b)

Figure 8.2 Graphing the results of a correlational study. (a) Scatterplot. (b) Modified scatterplot (two categories).

The scatterplot is changed into a two-group format in Figure 8.2b. The change is made by grouping all fathers who are short (less than 170 cm) into one category and all fathers who are tall (more than 170 cm) into a second category. Then two distributions of son's heights are drawn: the distribution of heights of sons of short fathers and the distribution of heights of sons of tall fathers. The means of these two distributions are different (167 cm and 175 cm, respectively), and hence son's and father's heights are related.

8.4 Regression Functions

We have seen that we must compare the means of the distributions of the dependent variable (for different values of the independent variable) in order to decide whether or not the two variables are related. These means form what is called the regression function: *a regression function is a curve or line connecting the means of the distributions of the dependent variable*. There is one mean for each value of the independent variable. When all the means of the distributions are equal, the regression function is a horizontal straight line and the two variables are not related. When the means of the distributions are not all equal, the regression function is not a horizontal straight line and the two variables are related. We can, therefore, use the regression function to determine whether or not two variables are related.

Example 1 shows how, when the independent variable has only two values, the regression function may be used to determine whether or not there is a relationship. Example 2 illustrates this approach when the independent variable is continuous.

Example 1

Consider first a study of the relationship between IQ and sex. The scatterplot is shown in Figure 8.3a. (If the samples of females and males were larger, a frequency distribution for each sample would be drawn instead of the univariate scatterplot.) The two distributions appear to have similar shapes and variabilities. The biggest difference between the distributions occurs in their central tendencies. The means are 103 and 95 for females and males, respectively. The means of each distribution are joined by a line, which serves to draw attention to the two points marking the means.

Since the two means are different, there is a relationship between IQ and sex. In this study, females tend to have higher IQs than males. Note that this is just a (central) tendency, as there is considerable overlap in the female and male distributions. In this study, some males have higher IQs than some females but overall the females have higher IQs than the males. This is what is meant by a relationship between IQ and sex.

What would the results of the study look like if there were no relationship between the two variables? Look at Figure 8.3b—not only are the shapes and variabilities similar, but the means are the same. Hence there is no relationship between IQ and sex. On the average, the IQs of the two sexes are the same.

The lines connecting the means of the dependent variable in Figures 8.3a and 8.3b are called *regression functions*. [The terms *regression* (used as a noun) and *regression curve* are also used.] A regression function is a curve or line connecting the means of the distributions of the dependent variable. In a two-group study the regression function is very simple: it is just a line connecting two means. The regression function in correlational studies is more complex (see Example 2).

The regression function can be used to determine whether or not two variables are related. From Figure 8.3 it is clear that *when the regression*

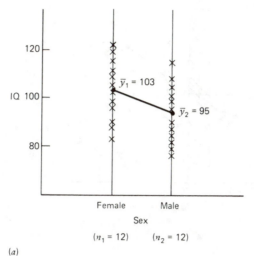

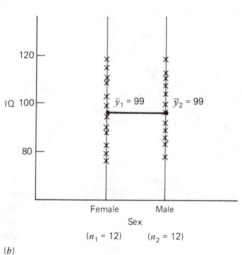

Figure 8.3 Regression function in a two-group study. (a) Relationship between IQ and sex. (b) No relationship between IQ and sex.

function is a horizontal straight line, the two variables are not related; when the regression function is not a horizontal straight line, the two variables are related.

Example 2

This example continues the study of fathers' and sons' heights begun in the last section. The scatterplot of the 50 father-son pairs is redrawn in Figure 8.4. We want to find the distributions of son's height for specific values of father's height so that we can compare the means of these distributions. This is not so easy to do as it would be in a two-group study. The first attempt, in Figure 8.2b, resulted in only two distributions. This, however, is not satisfactory—too much information has been lost. We would like to have a distribution for each possible father's height (the distribution for fathers of 150 cm, 151 cm, 152 cm, etc.), but the lack of sufficient data clearly makes this impossible: there are at most one or two fathers at each of these heights. The solution is to group the heights of fathers into intervals that are not as wide as the two intervals in Figure 8.2b, nor so narrow that each contains just a few fathers. For these data a good choice would be intervals such as 151–155, 156–160, etc.

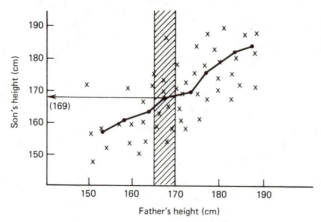

(a)

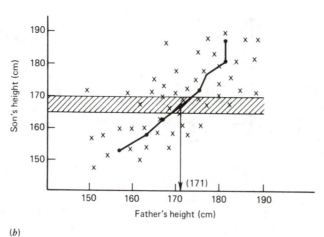

(b)

Figure 8.4 Regression functions in a correlational study. (a) Regression of son's height on father's height. (b) Regression of father's height on son's height.

After the intervals for father's height (the independent variable) have been selected, we identify all the crosses that belong to each interval. The nine father-son pairs in which the father is between 166 and 170 cm are shown in the shaded panel of Figure 8.4a. The mean height of these nine *sons* (dependent variable) is computed as 169 cm, indicated by a dot above the midpoint (168) of the 166–170 interval of father's height. In the same way, we compute and plot the mean of the heights of sons whose fathers have heights falling in the other intervals.

The resulting points are joined by short lines, giving a *regression function*. This is the regression function when son's height is treated as the dependent variable and father's height is treated as the independent variable. In order to indicate which variable has been treated as the dependent variable and which as the independent variable, it is customary to call this regression function by the expression *the regression of son's height on father's height.*[3] Note the order of the variables in this expression! In general, the expression is "the regression *of* (dependent variable) *on* (independent variable)". (The means of the dependent variable are plotted against the *midpoints of the intervals* of the independent variable.)

[3] Alternatively, "the regression *function* of son's heights on father's heights." In this expression, however, the word "function" usually is left out.

Is son's height related to father's height? Certainly it is, since the means of the different distributions are different. This is clearly indicated by the fact that the regression function is *not* a horizontal straight line. (If it were a horizontal straight line, all the means would be equal, which would indicate that the two variables were not related or associated.) The *rising* regression function in Figure 8.4*a* indicates that on the whole, as father's height increases the son's height also increases.

In some studies it is not clear which variable should be considered independent and which dependent. In the study of fathers' and sons' heights this is fairly clear, since the fathers reach their full heights before the sons. Even in this study, however, it is of interest to reverse the roles of the independent and dependent variables. This can be done by replotting the data with son's height on the abscissa and father's height on the ordinate and then following the above procedure.

An alternative procedure is to keep the original scatterplot and to proceed with the independent variable on the ordinate and the dependent variable on the abscissa. The scatterplot in Figure 8.4*b* is identical to that in Figure 8.4*a*. Now, however, it is the sons whose heights have been divided into intervals and the fathers whose mean heights have been computed. The eight father-son pairs in which the son is between 166 and 170 cm are shown in the shaded panel of Figure 8.4*b*. The mean height of these eight *fathers* (dependent variable) is 171 cm. This calculation has been repeated for each interval and the resulting means joined by short lines, thus producing the *regression of father's height on son's height*. Note carefully the reversal of wording from the previous regression function, which was "the regression of son's height on father's height".

The new regression function can be thought of as a "backward" regression: we find the mean height of the fathers whose sons have a particular height (or have a height in a narrow interval). When the sons are between 166 and 170 cm tall, for example, the fathers are, on the average, 171 cm tall. The backward regression should be contrasted with the original "forward" regression in which we found the mean height of the sons whose fathers have a particular height (or a height in a narrow interval). When the fathers are between 166 and 170 cm tall, for example, the sons are, on the average, 169 cm tall.

Although the data in the previous example are fictitious, the regression functions are typical. Two regression functions can always be computed for a scatterplot of two continuous variables. If the two variables are called x and y, one regression function is the regression of y on x and the other is the regression of x on y. *These two regression functions usually are not the same.*

Beginning in Section 8.6 we will be studying situations in which the regression function is a *straight line*. Such situations are special or idealized, and are by no means universal. *There is no requirement that all regression functions be straight.* Although the regression functions in Figure 8.4 are approximately straight, their deviations from straightness could well be real. There is no mathematical, biological, or other reason why the means of a dependent variable must always fall on a straight line. A straight-line regression represents either a special case or an approximation of a curved regression function by a straight line.

8.5 Measures of Relationship for Two Groups

The relationship between IQ and sex is indicated in Figure 8.3*a* by the difference in the mean IQ of the two sexes. The relationship between IQ and sex is sometimes described as an *effect of sex on IQ*. Although the term "effect" derives from a causal interpretation of a relationship, the term is used in statistics to describe any relationship, whether causal or not.

Figure 8.3b shows no relationship between IQ and sex because there is no difference in the mean IQ of the two sexes. Therefore, we say that there is *no effect of sex on IQ* in this study. (Alternatively, we can say that sex does not have an effect on IQ.)

Preceding sections have shown how to determine, from a graph, whether or not there is a relationship between two variables: that is, whether or not one variable has an effect on the other. We now describe how to state precisely the *size of the effect*. For example, if there is an effect of sex on IQ, how large is this effect? A numeric measure of the size of an effect is called a *measure of the relationship* between two variables.

The simplest measure of a relationship is the difference between the two means.

$$\text{Effect size} = \bar{y}_2 - \bar{y}_1$$

In Figure 8.3a, the effect size is $(95 - 103) = -8$. The negative sign indicates that males (group 2) have lower IQs than females (group 1). Usually, as in this case, the numbering of the groups is arbitrary and the negative sign is dropped: that is, the effect size is *usually* taken as the *absolute value*[4] of the difference between the two group means. We say that, in this study, the effect size is eight IQ points. This information is more useful than simply saying that there is a relationship between sex and IQ.

There is no effect of sex on IQ in Figure 8.3b. This is confirmed by the effect size, which is $(99 - 99) = 0$.

In summary, a positive (or negative) effect size indicates that the independent variable has an effect on the dependent variable. Conversely, an effect size of zero indicates that the independent variable does not have an effect on the dependent variable.

The effect size as measured by the difference between the means is quite useful in discussing the data in Figures 8.3a and 8.3b. However, it is not wholly satisfactory in instances in which the *scale* and *origin* of the dependent variable change. Recall from Section 7.1 that changes of scale and origin produce changes in the mean of the variable. Suppose that in Figure 8.3a every IQ score is divided by 10, producing a new measure of IQ that we will call IQA. Using this new measure, which is just as good a measure of intelligence as the original IQ, the mean for males is $95/10 = 9.5$ and the mean for females is $103/10 = 10.3$. (Section 7.1 stated that if the scale of a variable is changed by dividing by 10, the mean will be divided by 10.) The effect size, for the new variable IQA, is $(9.5 - 10.3) = -0.8$. The new effect size is the former effect size, -8, divided by 10.

Why is this a problem? An effect size of 8 sounds like a "big" effect, whereas an effect size of 0.8 sounds rather small. Yet these two effect sizes come from the same data! It is not possible to judge whether an effect size is big or small unless it has been *standardized*. The standardized effect size is computed by dividing the effect size by a standard deviation, s_p'.

---

[4] The absolute value of any number x is denoted by $|x|$. When x is positive, $|x| = x$. However, when x is negative, $|x|$ is a positive number and equals the original number with the negative sign dropped. So $|5| = 5$ and $|-6| = 6$.

$$\text{Standardized effect size} = \frac{\text{Effect size}}{s_p'}$$

The standard deviation in this formula, s_p', is computed from the two standard deviations, s_1' and s_2', of the two groups. This standard deviation is a particular kind of average of the two standard deviations; it is called the *pooled standard deviation*—hence the subscript "p", for "pooled".

When the two groups are of the same size, the formula for s_p' is quite simple.

$$s_p' = \sqrt{\frac{s_1'^2 + s_2'^2}{2}} \quad \text{(Equal sample sizes)}$$

This formula indicates that we compute the two variances, by squaring each standard deviation, and then compute their mean, which is then the pooled variance. We take the square root of the pooled variance to get the pooled standard deviation.

When the two sample sizes are unequal, the pooled standard deviation[5] is computed as follows.

$$s_p' = \sqrt{\frac{n_1 s_1'^2 + n_2 s_2'^2}{n_1 + n_2}} \quad \text{(Unequal sample sizes)}$$

Example 1

In Figure 8.3a, s_1' is 14.6 and s_2' is 15.2. Since the sample sizes are equal ($n_1 = n_2 = 12$), the simpler formula for s_p' can be used.

$$s_p' = \sqrt{\frac{(14.6)^2 + (15.2)^2}{2}} = \sqrt{\frac{213.16 + 231.04}{2}}$$

$$= \sqrt{\frac{444.2}{2}} = \sqrt{222.1} = 14.903$$

A useful check on the calculation of s_p' is that s_p' must always fall between the values s_1' and s_2'. This is true for both equal and unequal sample sizes. The value of 14.903 in this example satisfies the requirement, since it is greater than 14.6 and less than 15.2.

The standardized effect size for Figure 8.3a is, in absolute value, $8/14.903 = 0.54$. Since the standardized effect size is the effect size divided by the standard deviation, we can say that the observed effect size is 0.54, or slightly over one-half, a standard deviation. In other words, in this study, the females, on the average, scored about half a standard deviation higher in IQ than the males. This statement contains more information to a reader than a statement that the females scored, on the average, 8 IQ points higher than the males (since the number "8" has no significance by itself). The reader must know the standard deviation of IQs in order to decide whether this difference of 8 points is small or large. The reader can interpret a standardized effect size of 0.54 without knowing the value of the standard deviation.

---

[5] The pooled variance is simply a *weighted* average of $s_1'^2$ and $s_2'^2$, where the weights are n_1 and n_2, respectively. When the sample sizes are equal, these weights are equal, and the complex formula reduces to the simpler formula given for equal sample sizes.

Example 2

In Figure 8.3*b*, the effect size was zero, so the standardized effect size is also zero. The pooled standard deviation does not have to be calculated.

Example 3

Let us return to the rescaled IQ scores, called IQA above. These are the original IQ scores divided by 10, and the effect size was calculated to be 0.8. What is the standardized effect size? The standard deviations of IQA will be one-tenth the original standard deviations (see the rules about the effects of change of scale in Section 7.1). It can be shown algebraically, or by calculation, that s_p' will also be one-tenth of the original value. Hence, the new s_p' is 1.4903, and the standardized effect size is $0.8/1.4903 = 0.54$, just as before. This example shows one of the chief merits of the standardized effect size. It does not change when the scale of the variable is changed.

Example 4

The data in Figures 8.1*b* and 8.1*c* can be summarized as follows.

	GROUP 1 REGULAR CLASS	GROUP 2 SPECIAL CLASS
Mean	$\bar{y}_1 = 62$	$\bar{y}_2 = 68$
s.d.	$s_1' = 7.3$	$s_2' = 8.6$
Sample Size	$n_1 = 60$	$n_2 = 85$

The effect size is $(68 - 62) = 6$. The pooled standard deviation is

$$s_p' = \sqrt{\frac{60 \times (7.3)^2 + 85 \times (8.6)^2}{60 + 85}}$$

$$= \sqrt{\frac{60 \times 53.29 + 85 \times 73.96}{145}}$$

$$= \sqrt{\frac{9484.00}{145}} = \sqrt{65.4069} = 8.087$$

Hence, the standardized effect size is $6/8.087 = 0.74$, or about three-quarters of a standard deviation.

You should note that there is considerable overlap of the two distributions in Figures 8.1*b* and *c*. If the distributions overlap, the standardized effect size usually is less than 1.0. If the distributions do not overlap, the standardized effect size usually is greater than 1.0.

8.6 The Pearson Correlation

There are a number of ways to measure the relationship between two continuous variables. This section focuses on one measure, called the Pearson correlation (or simply "the correlation"), showing how to compute the correlation from the original data. In order to interpret the Pearson correlation by a regression function, the regression function must be a straight line or be approximated by a straight line. The Pearson correlation is, therefore, a measure of straight-line or linear relationship.

Correlation from Original Data

In a correlation study (Section 8.3) there are n pairs of scores (x, y). From these data we can compute the means ($\bar{x}$ and $\bar{y}$) and the standard deviations (s'_x and s'_y). The correlation, r, can be calculated by either of the following equivalent formulas.

$$r = \frac{\Sigma(x - \bar{x})(y - \bar{y})}{n\, s'_x\, s'_y} \quad \text{(Definitional formula)}$$

$$r = \frac{\Sigma xy - (\Sigma x)(\Sigma y)/n}{n\, s'_x\, s'_y} \quad \text{(Computational formula)}$$

The distinction between the two formulas for r is the same as the distinction between the definitional and computational formulas for the standard deviation, s' (Section 4.4). The *definitional* formula shows more clearly than the computational formula that the numerator of the correlation is a *sum of products of deviations*, $\Sigma(x - \bar{x})(y - \bar{y})$. For this reason, the Pearson correlation is sometimes called the *product-moment correlation*.[6] However, the definitional formula is not a practical way to compute the correlation, except for very special data sets with simple values of $\bar{x}$ and $\bar{y}$. The computational formula is almost always preferable.

In the unlikely event that we do not wish to compute the standard deviations and we want to simplify the formula for r as much as possible, we can use the following formula.

$$r = \frac{\Sigma xy - (\Sigma x)(\Sigma y)/n}{\sqrt{\Sigma x^2 - (\Sigma x)^2/n}\;\sqrt{\Sigma y^2 - (\Sigma y)^2/n}} \quad \begin{array}{l}\text{(Computational}\\ \text{formula without}\\ \text{standard deviations)}\end{array}$$

Example 1

Let us compute the correlation for a very small set of data. The data and the calculations using the computational formula are shown in Tables 8.2*a* and 8.2*b*. The two standard deviations are computed from the sums and sums of squares in the usual way. Note that a fifth sum, Σxy, is computed from the data. This sum of the products of each x times y can be computed on a single pass on most calculators, without recording individual products. For these data, the calculation is $\Sigma xy = (4 \times 16) + (7 \times 12) + (2 \times 10) + (5 \times 8) + (6 \times 18) + (3 \times 14) = 64 + 84 + 20 + 40 + 108 + 42 = 358$. The sums and standard deviations are then used in the computational formula to give a correlation of 0.200 for these data.

The definitional formula also can be used for data as simple as these. The calculations are illustrated in Table 8.2*c*. First, the means are calculated. Then the deviations of each x and each y from their respective means are calculated. The products of these deviations are shown in the last column of Table 8.2*c*. The sum of each column of deviations should be computed to check that the sum is zero. The sum of the products of the deviations, 7.0, is the numerator of the definitional formula for the correlation. The result is again $r = 0.200$. Note a major difference between the calculations using the two different formulas. When we use the computational formula, we do not have to write down intermediate products. When we use the definitional formula, in contrast, we have to write down the deviations and their products. The computational formula is therefore much easier to use, even for these simple data.

[6] A deviation from the mean is sometimes called a *moment* about the mean.

TABLE 8.2 The Pearson Correlation

(a) Data, Basic Sums, and Standard Deviations

x	y
4	16
7	12
2	10
5	8
6	18
3	14

$\Sigma x = 27 \qquad \Sigma y = 78$

$\Sigma x^2 = 139 \qquad \Sigma y^2 = 1084$

$\Sigma xy = 358$

$$s'_x = \sqrt{\frac{\Sigma x^2 - (\Sigma x)^2/n}{n}} = \sqrt{\frac{139 - (27)^2/6}{6}} = 1.7078$$

$$s'_y = \sqrt{\frac{\Sigma y^2 - (\Sigma y)^2/n}{n}} = \sqrt{\frac{1084 - (78)^2/6}{6}} = 3.4157$$

(b) Computational Formula

$$r = \frac{\Sigma xy - (\Sigma x)(\Sigma y)/n}{n\, s'_x\, s'_y} = \frac{358 - (27)(78)/6}{6 \times 1.7078 \times 3.4157} = \frac{7.00}{35.000} = 0.200$$

(c) Definitional Formula

$\bar{x} = (\Sigma x)/n = 27/6 = 4.5$

$\bar{y} = (\Sigma y)/n = 78/6 = 13.0$

x	y	$(x - \bar{x})$	$(y - \bar{y})$	$(x - \bar{x})(y - \bar{y})$
4	16	−0.5	+3.0	−1.5
7	12	+2.5	−1.0	−2.5
2	10	−2.5	−3.0	+7.5
5	8	+0.5	−5.0	−2.5
6	18	+1.5	+5.0	+7.5
3	14	−1.5	+1.0	−1.5
		Sum = 0.0	Sum = 0.0	Sum = 7.0

$$r = \frac{\Sigma(x - \bar{x})(y - \bar{y})}{n\, s'_x\, s'_y} \qquad \frac{7.0}{6 \times 1.7078 \times 3.4157} = \frac{7.0}{35.000} = 0.200$$

(d) Computational Formula without Standard Deviations

$$r = \frac{\Sigma xy - (\Sigma x)(\Sigma y)/n}{\sqrt{\Sigma x^2 - (\Sigma x)^2/n}\sqrt{\Sigma y^2 - (\Sigma y)^2/n}}$$

$$= \frac{358 - (27)(78)/6}{\sqrt{139 - (27)^2/6}\,\sqrt{1084 - (78)^2/6}} = \frac{7.00}{4.1833 \times 8.3666} = 0.200$$

The computational formula without standard deviations is illustrated in Table 8.2d. The formula requires only the five sums computed in Table 8.2a. However, the effort required to use this formula is similar to that required to use the basic computational formula.

Interpreting the Correlation

It can be shown that the values of the Pearson correlation, r, are restricted to the range $-1 \leq r \leq +1$. Values of r near $+1.0$ are indicative of a strong relationship between the two variables, x and y: that is, when x is large, y is also large; and when x is small, y is also small. Values of r near -1.0 are indicative of a strong (negative) relationship between x and y: that is, when x is large, y is small; and when x is small, y is large. Values of r near 0.0 (either positive or negative) are indicative of a weak relationship between x and y: that is, when x is large, y has both large and small values; and when x is small, y also has both large and small values.

Values of r that are neither near 0.0 nor near ± 1.0 are indicative of a moderate relationship. The values of r that separate a strong relationship from a moderate one (and that separate a moderate relationship from a weak one) are, of course, arbitrary. If we choose ± 0.7 and ± 0.3 for these boundaries, we will get the summary shown in Table 8.3.

The Pearson correlation is related to the regression function. Consider one of the regression functions—say, the regression of y on x. (The following discussion also applies to the regression function of x on y.) If the regression function is not straight, draw a straight line that best approximates the regression function. This line is called the *best-fitting straight line* or the *linear regression function* or the *regression line*. (For now, simply estimate by eye where this line is located. Its precise location is discussed in Section 16.3).

The regression line has a certain *slope* (ratio of its rise to its run). If the *line slopes upward*, the slope is positive and the correlation, r, is positive. A positive correlation indicates that the mean of the dependent variable increases as the independent variable increases. If the *line is horizontal*, it has zero slope and the correlation is zero. A zero correlation indicates that the mean of the dependent variable does not change as the independent variable increases. If

TABLE 8.3 Interpreting the Correlation

r	STRENGTH OF RELATIONSHIP	SPREAD OF POINTS ABOUT REGRESSION LINE
$r = -1.0$	Perfect negative	None
$-1.0 < r \leq -0.7$	Strong (negative)	Small
$-0.7 < r \leq -0.3$	Moderate (negative)	Moderate
$-0.3 < r < 0.0$	Weak (negative)	Large
$r = 0.0$	No (linear) relationship	Very large
$0.0 < r < 0.3$	Weak	Large
$0.3 \leq r < 0.7$	Moderate	Moderate
$0.7 \leq r < 1.0$	Strong	Small
$r = +1.0$	Perfect positive	None

Note: The boundaries of (± 0.7 and ± 0.3) between the intervals of r are arbitrary.

the *line slopes downward*, the slope is negative and the correlation is negative. A negative correlation indicates that the mean of the dependent variable decreases as the independent variable increases.

For example, consider a large number of used cars. The two variables "age of car" and "number of miles the car has been driven" are positively correlated, since the greater the age, the greater the mean number of miles driven. The two variables "age of car" and "age of owner" have, probably, a very small or zero correlation, since there is probably no change in the mean age of the owner for cars of different age. The two variables "number of miles the car has been driven" and "resale value" are negatively correlated, since the mean resale value decreases as the number of miles increases.

It may appear that the steeper the regression line, the greater the value of correlation: that is, the greater the absolute value of the slope, the greater the absolute value of the correlation ($|r|$). But this statement is true only if other characteristics of the data (such as the standard deviations) do not change. Therefore, the statement is of limited usefulness in estimating the value of r from the slope of the regression line, except that the *sign* of the correlation can be determined from the *sign* of the *slope*.

The points in a scatterplot have a certain degree of *scatter* or *spread* about the regression line. The smaller the spread, the greater the absolute value of the correlation (r is nearer to -1.0 or $+1.0$). Conversely, the more the data are spread about the line, the smaller the absolute value of the correlation (nearer to 0.0). The degree of spread for different values of r is summarized in Table 8.3.[7]

Example 2	The scatterplots of six small sets of data are shown in Figure 8.5. (The data of Table 8.2 are plotted in Figure 8.5c.) In Figure 8.5a, the six points fall exactly on a straight line. It is therefore clear that the best-fitting straight line goes through these points. There is no spread of the points about the straight line, which implies the highest possible correlation: namely, $r = 1.00$.

Consider next the data in Figure 8.5b. Here the data certainly do not fall on a straight line. As there are only six points, it is not reasonable to group the data into intervals of x in order to compute means of y and hence the regression function. In any case, we want the best-fitting straight line to this regression function. It can be shown mathematically that the best-fitting straight line to the regression function is exactly the same as the best-fitting straight line to the original data themselves. Hence, we can draw the linear regression function by eye, by attempting to draw that line that is closest to all the points. Such a line is shown in Figure 8.5b.

Clearly, this line has less slope than the line in Figure 8.5a, and the points are more scattered about it. The standard deviations of x and y are, in fact, the same in the two figures. (All the data sets in Figure 8.5 have the same s'_x and the same s'_y.) Hence, the correlation is lower in Figure 8.5b. A calculation confirms that r is 0.657.

In Figure 8.5c the best-fitting straight line has even less slope. The scatter or spread of points about this line is greater than it was in Figure 8.5b. This indicates that the correlation is even lower (nearer zero). This correlation was found to be 0.200 in Table 8.2.

---

[7] The relation of spread (or scatter) to r depends on size of s'_y (for the regression of y on x) and, therefore, is also subject to the condition "if other characteristics of the data do not change", just as the relation of slope to r does. However, it does seem to be easier to estimate values of r from the spread than from the slope.

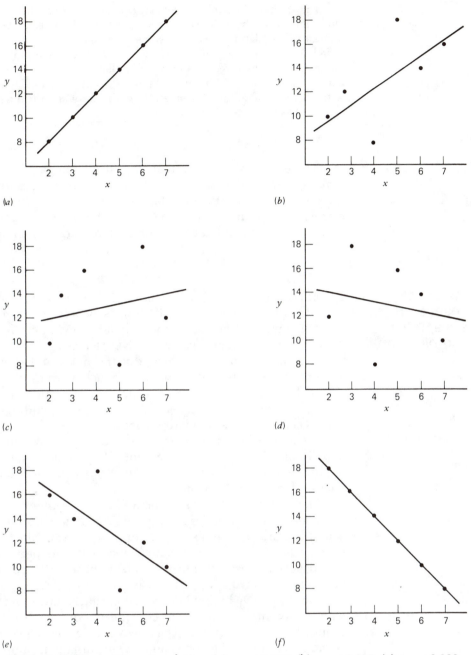

Figure 8.5 Interpreting a correlation. (a) r = +1.000. (b) r = +0.657. (c) r = +0.200. (d) r = − 0.200. (e) r = −0.657. (f) r = −1.000.

The straight lines in Figure 8.5d, 8.5e, and 8.5f are different. They have negative slope, indicating that on the average y decreases as x increases. A negative slope implies a negative correlation. When the slope is small and the scatter great (Figure 8.5d), the correlation is near zero. By calculation, r is −0.200 here. In Figure 8.5e the slope is moderate, as is the spread of points about the line, so the correlation is somewhat more negative (r = −0.657). Finally, in Figure 8.5f the points fall exactly on a straight line, with no spread about the line at all. Here the correlation is the most negative possible: r = −1.000.

These figures illustrate that the correlation can be roughly estimated by looking at the spread or scatter of points about the best-fitting straight line. If the spread is great, the correlation is near zero. If the spread is small, the correlation is near either +1.0 or −1.0. Positive correlations occur when the line is increasing and negative correlations occur when the line is decreasing.

Effects of Change of Scale, Origin, and Direction

The effects of change of scale and origin on the mean and standard deviation were described in Section 7.1. If all the scores are multiplied (or divided) by a constant number, the mean and standard deviation are also multiplied (or divided) by that same constant number. Hence, a change of scale in the scores produces a corresponding change in the mean and standard deviation. If a constant number is added to (or subtracted from) all scores, the mean is changed in the same way, but the standard deviation is unchanged. Hence, a change of origin in the scores produces a corresponding change in the mean but no change in the standard deviation.

What effect do changes of scale and origin have on the Pearson correlation? None! The correlation is completely unaffected by changes of scale and origin. This is one of the virtues of the Pearson correlation as a measure of the relationship of two variables.

The sign of the correlation between two variables is changed if the direction of one of the variables is changed. The notion of the direction of a variable is best explained by an example. Suppose the variable "introversion" is studied. A scale of introversion is developed with the property that highly introverted persons receive high scores (say, near 100) and persons who are not introverted receive low scores (say, near 0). However, it would be equally satisfactory to give highly introverted persons a low score (near 0) and persons who are not introverted a high score (near 100). The direction of the variable is changed by this procedure. The new variable could be called "negative introversion," or simply "extraversion." One variable is simply the reverse of the other, and the scatterplots of these variables with another variable—for example, neuroticism—are identical except that one scatterplot is the mirror image of the other.

If introversion is positively correlated with neuroticism (the linear regression function slopes upward), extraversion is negatively correlated with neuroticism (the linear regression function slopes downward). The two correlations are exactly the same except for their signs. In other words, the correlations have the same absolute value.

Since the direction in which a variable is defined usually is arbitrary, the sign of the correlation of that variable with another variable is arbitrary to that extent. It is therefore important to keep in mind when interpreting any correlation that it is the *absolute value* of the correlation that indicates the strength of the linear relationship between the variables. The *sign* of the correlation indicates whether one variable increases or decreases when the other variable is increased, and this sign will change if the direction of one of the variables is changed.

8.7 Summary

Many *surveys* and *experiments* are studies of the *relationship* between two variables. The two variables are often called the *independent variable* and the *dependent variable*. Two types of independent variable can be distinguished: *naturally-occurring* and *experimentally-defined*.

When the independent variable has only two values, the study is a *two-group study*. The results of such a study can be displayed in a *scatterplot*. The presence or absence of a relationship (or *association*) between the two variables is defined in terms of the two distributions in this scatterplot.

The definition of a relationship can be extended to studies in which the independent variable has more than two values or is continuous. A study in which both the independent and the dependent variables are continuous is called a *correlational study*. The data from such a study can also be displayed in a scatterplot. If the regression function is a horizontal straight line, there is no relationship between the variables.

It is important to measure the size of the relationship between two variables in addition to determining that there is a relationship. For two groups, two such *measures of relationship* can be defined: the *effect size* and the *standardized effect size*. The *pooled standard deviation* is required for the latter measure.

In correlational studies the most common measure of relationship is the *Pearson correlation*. This correlation measures the degree of *linear relationship* between the two variables. It is interpreted by considering the *best-fitting straight line* or *linear regression function*. The correlation is unaffected by changes in the scale and origin of the variables, but the *sign of the correlation* changes when the *direction* of one of the variables changes.

Exercises

1 (Sections 8.1 and 8.2) Describe in a few sentences a study that is reported in either a published paper or a textbook.
   (a) Identify the variables whose relationship is investigated in the study.
   (b) Identify an independent variable and a dependent variable and state whether the independent variable is a naturally-occurring or an experimentally-defined variable.

2 (Section 8.2)
   (a) In a sleep-deprivation study, two groups of subjects are studied. One group, of 10 subjects, is prevented from sleeping during a 24-hour period. The other group, of 12 subjects, is allowed to sleep a normal amount during a 24-hour period. At the end of the 24 hours all subjects are tested on a vigilance task. The number of errors that each subject makes is listed below.

DEPRIVED GROUP	NORMAL GROUP
8	4
4	7
12	2
5	3
11	6
9	7
6	1
9	0
10	4
8	3
	3
	7

Does there appear to be a relationship between experimental condition and performance on the vigilance task? Explain.

(b) The annual incomes of persons who went to college and who did not go to college are compared by studying 400 persons aged 35. Of these 400 persons, 100 went to college and 300 did not go to college. The frequency distributions for each group are given below.

INCOME	COLLEGE	NON-COLLEGE
$0–$ 5,000	7	21
$5,001–$10,000	15	45
$10,001–$15,000	25	75
$15,001–$20,000	31	93
$20,001–$25,000	17	51
over $25,000	5	15

Does there appear to be a relationship between level of education and annual income? Explain. (*Hint:* A difference in the height of two distributions alone does not indicate that there is a relationship. Distributions can be compared more readily by comparing *relative* frequency distributions.)

3 (Sections 8.3 and 8.4) Here are the verbal and performance IQs of 40 subjects.

SUBJ.	VERB.	PERF.	SUBJ.	VERB.	PERF.	SUBJ.	VERB.	PERF.
1	79	74	15	97	93	28	101	107
2	78	85	16	97	97	29	102	106
3	87	79	17	92	100	30	108	103
4	82	84	18	95	106	31	103	116
5	82	85	19	90	105	32	100	118
6	84	90	20	97	105	33	117	88
7	89	92	21	109	79	34	114	97
8	99	77	22	105	82	35	112	92
9	96	76	23	104	85	36	118	101
10	96	87	24	104	97	37	117	104
11	97	81	25	106	92	38	115	113
12	90	81	26	102	95	39	121	106
13	94	92	27	103	104	40	120	116
14	93	90						

(a) Make a scatterplot of these data, using graph paper.

(b) Compute and draw regression functions of verbal IQ on performance IQ and of performance IQ on verbal IQ. Use intervals of width 10 (70–79, 80–89, etc.) for the intervals in which you compute the means that you need for the regression functions. Be sure to make clear which regression function is which!

4 (Section 8.5) This question is a continuation of Question 2.

(a) Compute the effect size for the data in Questions 2a and 2b.

(b) Compute the standardized effect size for the data in Question 2a.

5 (Section 8.6) This question is a continuation of Question 3. Draw, as best you can, the linear regression function that best approximates the regression function of verbal IQ on performance IQ that you drew in Question 3b. Using this linear regression function, estimate the correlation, r.

6 (Section 8.6) Here are two data sets.

DATA SET A		DATA SET B	
x	y	x	y
11	21	18	22
17	28	10	28
14	21	11	27
16	24	17	23
12	24	12	26
14	25	19	21
20	28	16	24
15	23	13	26
18	25	17	22
15	26	14	24

(a) Calculate the Pearson correlation in Data Set A.

(b) Calculate the Pearson correlation in Data Set B.

(c) Make scatterplots of the two sets of data and, by considering the linear regression function, show that the scatterplots are consistent with the calculated values of r.

REVIEW CHAPTER B

The general approach to analyzing a set of data is the same whether the basic variable is two-valued (Chapter 6) or continuous (Chapter 7). The similarities, rather than the differences, between these two analyses are emphasized in this chapter. The data in some studies can be analyzed in both ways: by using the binomial or by using the normal distribution. The two analyses provide related, but not identical, information. There is not, therefore, only one "correct" way to analyze a given set of data.

**B.1
Proportions
Versus Means**

In Chapters 6 and 7 a sharp distinction was made between analyses of a discrete variable (with two values) and a continuous variable. This distinction can be described as one between an *analysis of a proportion* and an *analysis of a mean*. However, the data from a single study can often be analyzed in either manner. Consider the following example.

A large mental hospital has begun a new program of giving some patients a certain psychoactive drug and then discharging them. The effectiveness of the program is to be evaluated by studying a random sample of discharged patients. (The method of evaluation used here is very simple, because the statistical techniques that have so far been described are simple. However, the example should be sufficient to illustrate the distinction between an analysis of a mean and an analysis of a proportion.)

Each discharged patient is to live in the community for a certain length of time. Some eventually will return to live in a mental hospital; others will remain in the community. Each patient in the population of patients who are discharged during the next six months will spend a certain length of time in the community before returning to a mental hospital. The population distribution of these times might have the form shown in Figure B.1a. (For simplicity in the description of this study, we ignore patients who die before returning to a mental hospital.) In this distribution, most patients have returned within five years of discharge; some, however, are still in the community after 10 years.

How can this study be completed if some of the patients will still be in the community after 10 years? One simple approach is to ignore these few people; the study is limited to those who return to a hospital within 10 years. In this approach the variable to be studied is the time, in years, before return, up to a maximum of 10 years. The distribution of the population on this variable is given in Figure B.1b. This population has a number of characteristics (parameters) that we might want to determine. For example, it has a mean, μ, and a standard deviation, σ. One measure of the effectiveness of the discharge program is the mean length of stay in the community before return to a hospital. In the figure this parameter, μ, is assumed to be 3.5 years. However, this number would be unknown before the study begins. From the random sample of patients an inference about this parameter can be made.

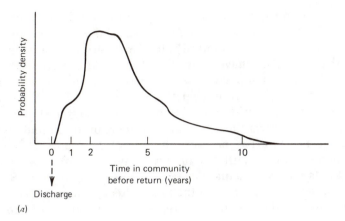

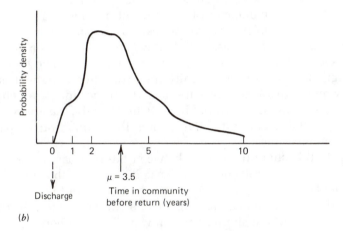

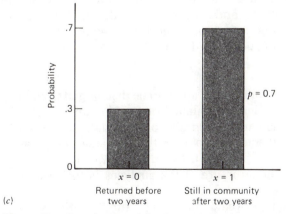

Figure B.1 Population of discharged patients. (*a*) Basic population. (*b*) Continuous variable (10-year study). (*c*) Discrete variable (2-year study).

A second approach is to divide all patients who are readmitted to hospital within 10 years into two groups: those who return within two years and those who are still in the community after two years. (One advantage of this approach is that we do not have to wait 10 years!) After two years each person in the population (or sample) can be assigned to one of these two groups. We can then calculate the proportion of patients who are still in the community after two years. The population proportion ($p = 0.7$, say) is shown in Figure B.1c. The figure also shows the population proportion of patients who returned before two years. These two proportions add to 1.0, of course. Note that the category of patients who are still out in the community after two years includes those patients who will actually stay out more than 10 years. Such patients are not excluded from the study if the data collection is completed after two years. (In this respect the populations represented in Figures B.1b and c are slightly different.)

The contrast we wish to draw is between (1) a study of a continuous variable and a parameter of its distribution, the mean, μ; and (2) a study of a discrete variable (with two values) and a parameter of its distribution, the proportion of successes, p. If data will be available for the full 10-year period we can look at the data from either perspective. The two parameters—mean length of stay in the community and proportion of patients still in the community after two years—measure related, but not identical, characteristics of the basic population shown in Figure B.1a. Usually, the greater the mean length of stay in the community, the greater the proportion of patients still in the community after two years.

In this study the choice between the two parameters would probably be based on practical considerations. If it is possible to keep records on the patients in the sample for 10 years, the *mean* would be preferable, since it does not depend on an arbitrary choice of a two-year cutoff. If it is planned to keep records only for a shorter time, however, the *proportion* would be preferable, since the data could be collected in two years.

So a continous variable can sometimes be studied in two ways. It can be analyzed directly and the *mean* of its distribution determined. Or the variable can be divided into two categories, labeled "success" and "failure," and the *proportion* of successes determined.

B.2 The Common Structure of Statistical Inference

A population of patients about to be discharged from hospital was described in Section B.1. Here, the data from a random sample of 25 such patients is analyzed. As emphasized above, such data can be analyzed in two different ways: as a discrete variable with two categories or as a continuous variable. Both analyses will be illustrated here. A comparison of the two analyses will show that the structure of the analyses is the same even though the details differ.

The time, in years, that each of the 25 patients spent in the community is given in Table B.1. These times range from a minimum of 0.1 years to a maximum of 8.0 years. The table shows that the proportion of patients staying in the community more than two years is 0.80, or 80%. On the other hand, if the times are averaged as they stand, without being grouped into two categories, the mean length of stay in the community is 3.63 years. These two basic results will now be studied further. Even though the two analyses are described in

Table B.1 A Sample of Twenty-five Patients

TIME IN THE COMMUNITY BEFORE RETURN (IN YEARS)

0.7	1.2	4.1	8.0	3.6
2.2	0.1	6.0	2.6	4.9
4.7	1.7	3.9	4.3	4.9
4.5	3.7	0.6	5.6	3.2
2.7	2.3	4.1	6.2	5.0

ANALYSIS OF PROPORTION

   Number of patients with time > 2 years = 20
   Proportion of patients with time > 2 years = 20/25 = 0.80.

ANALYSIS OF MEAN

   Mean time = 90.8/25 = 3.63 years

separate subsections, you should compare them to confirm that their structure is the same, preferably by comparing Figures B.2 and B.3.

Inference about a Proportion

Two theoretical distributions (the population distribution and the distribution of the statistic) are shown on the left side of Figure B.2. An empirical distribution (the sample distribution) is shown on the right. From the empirical distribution, we can make an inference about the population distribution.

Suppose we want to test the hypothesis that the proportion of successes (number of patients staying in the community more than two years) is 0.7. The population distribution shown in Figure B.2 is based on this hypothesis. From the hypothesis, we can work out the distribution of the statistic for samples of size 25. Here the statistic is the number of successes. Its distribution is the binomial distribution with $p = 0.7$ and $n = 25$.

In the actual sample (Table B.1) the number of successes, y, is 20. From this statistic, we can make an inference about the population proportion, p. We can proceed either by testing a hypothesis about p or by estimating p.

Hypothesis Testing If we wish to test the hypothesis that $p = 0.7$, we must calculate the probability that 20 or more patients will be successes in a sample of 25 patients if $p = 0.7$. (Note that it is the probability of 20 *or more*, since we always work out the probability of the observed result or *a more extreme* result: that is, a result further from the mean. Since the mean of the distribution of the statistic is $np = 25 \times 0.7 = 17.5$, values *more* than 20 are more extreme: that is, further from the mean than 20. For now, concentrate on the distinctions between hypothesis testing and estimation and between tests of means and tests of proportions, rather than on the details of these tests. An explanation of why this particular probability is calculated is given in Chapter 9.)

The probability of 20 or more successes can be found from the binomial table, Table A1. The probability is 0.192, which, though small, is not as small as 0.05, the conventional criterion of a small probability. We therefore note that a result of 20 successes (out of 25) is quite probable under the hypothesis that $p = 0.7$. Accordingly, since we have no reason to reject that hypothesis, we accept it. (For the present, we will treat "failing to reject the hypothesis" and

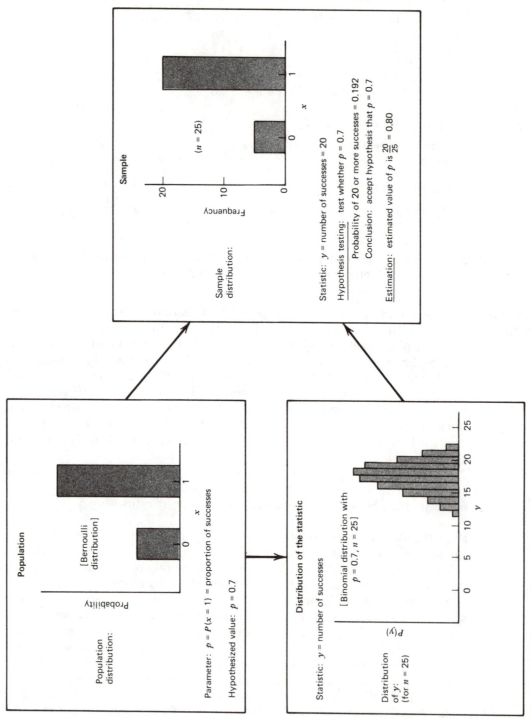

Figure B.2 Inference about a proportion.

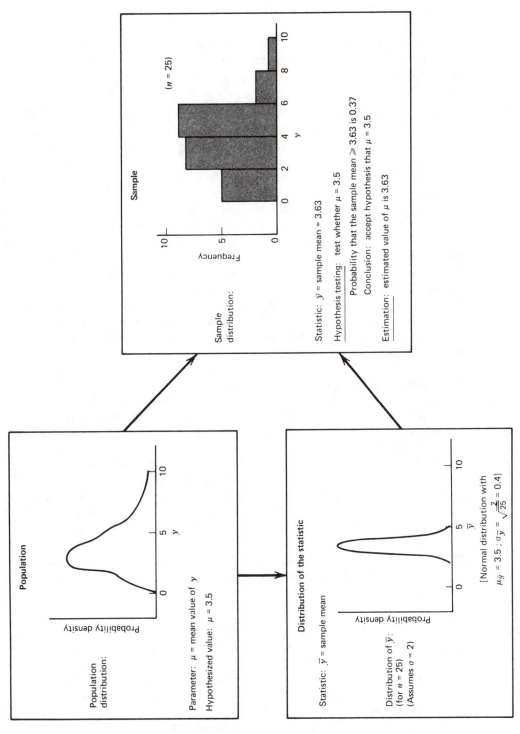

Figure B.3 Inference about a mean.

"accepting the hypothesis" as equivalent. The distinction between these two conclusions will be described in Chapter 10.)

Estimation The sample proportion, 0.80, is the best estimate of the population proportion. Therefore, our estimate of the proportion of patients who remain in the community after two years is 0.80, or 80%.

Inference about a Mean

The analysis of the mean in Figure B.3 closely parallels the analysis of the proportion in Figure B.2. In the figure the theoretical distributions (population distribution and distribution of the statistic) appear on the left and the sample distribution appears on the right. (The continuous variable has been grouped into intervals in the sample distribution.) Now the parameter is the population mean, μ, and the statistic is the sample mean, $\bar{y}$.

Suppose we want to test the hypothesis that the mean of the population distribution, μ, is 3.5. Assuming this hypothesis, we know, from the central limit theorem, that the distribution of the sample mean is approximately normal with mean = 3.5 and standard deviation = $\sigma_{\bar{y}} = \sigma/\sqrt{n}$. We must know (or assume) the value of the population standard deviation, σ, in order to calculate $\sigma_{\bar{y}}$. In this example, $\sigma = 2$; hence, $\sigma_{\bar{y}} = 2/\sqrt{25} = 0.4$.

In the actual sample (Table B.1) the mean length of stay in the community is 3.63. From this statistic, we can make an inference about the population mean, μ. We can proceed either by testing a hypothesis about μ or by estimating μ.

Hypothesis testing To test the hypothesis that $\mu = 3.5$, we first calculate the probability that the sample mean is 3.63 or greater. (Just as in the test of a proportion, it is the probability of 3.63 *or greater* that is calculated. Values greater than 3.63 are *more extreme*: that is, further from the mean than 3.63.)

The required probability is $P(\bar{y} \geq 3.63)$, where $\bar{y}$ has the normal distribution with mean = 3.5 and standard deviation = 0.4. To find this probability in Table A3, we must convert 3.63 to a standard score: $z = (3.63 - 3.5)/0.4 = 0.325$. From Table A3, $P(z \geq 0.325) = 0.5 - 0.13 = 0.37$. This probability is much greater than 0.05; the observed result is not improbable under the hypothesis that $\mu = 3.5$. We therefore accept this hypothesis.

Estimation The sample mean, 3.63, is the best estimate of the population mean. Therefore, our estimate of the mean length of stay in the community is 3.63 years.

A Note about Symbols

A careful comparison of Figures B.2 and B.3 shows that the symbol y is not used in the same way in the two figures. First note that in each figure there is a *basic variable* and a *statistic*. In the case of the proportion (Figure B.2), the basic variable, x, is two-valued—returned before two years *versus* still in community after two years—and the statistic, y, is the number of patients still in the community after two years. In the case of the mean (Figure B.3), the basic variable, y, is the time in the community before return and the statistic, $\bar{y}$, is the mean time in the community. The uses of the symbol y and the other symbols in these two figures can be summarized as follows.

	Proportion (B.2)	Mean (B.3)
Basic variable	x	y
Statistic	y	$\bar{y}$

Although other notational systems are feasible, this system is fairly conventional among statisticians.

B.3 Summary

In some research studies the basic variable is *discrete with two values.* (Examples: a coin can land either heads or tails; a person either does or does not volunteer to give blood.) In other research studies the basic variable is *continuous.* (Examples: the time taken to complete a task; the score on a 100-item quiz; an IQ.) Still other possibilities, such as a discrete variable with more than two values, are described in later chapters.

When the variable is discrete with two values, or categories, we can compute the *proportion* of scores in one of the categories. When the variable is continuous, two statistics can be computed: we can compute the *mean* of the scores or we can divide the variable into two categories and compute the *proportion* of scores in one of the categories. These two statistics provide related but not identical information about the population distribution of the continuous variable.

The basic steps in analyzing a mean or analyzing a proportion are the same. A statistic is computed from the sample distribution. If we are *testing a hypothesis* about a population parameter, we obtain the distribution of the statistic; this distribution is based on the hypothesized value of the parameter and the given sample size. One then computes the probability of the observed value of the statistic or a more extreme value. If this probability is less than a small number—say, 0.05—we reject the hypothesis. Otherwise, we accept it. If we are *estimating the parameter*, we use a suitable sample statistic as the estimate. For estimating the population mean, the suitable statistic is the sample mean; for estimating the population proportion, the suitable statistic is the sample proportion.

In the next six chapters the basic theory of hypothesis testing and estimation are discussed in greater detail. The parameters about which inferences are made will still be proportions and means. However, the analyses will be extended to studies that involve two proportions or two means. Then, beginning in Chapter 14, a number of other parameters and statistics are described.

Exercises

1 (Section B.1) Explain, briefly, how a study of the final-exam scores of a group of students can be analyzed in two ways: as an analysis of a proportion or as an analysis of a mean. (*Hint:* each student either passes or fails).

2 (Section B.2) For each of the five parts of this question do **(i)**, **(ii)**, and **(iii)**.

  (i) Identify the situation as involving *means* or *proportions.*

  (ii) Identify the analysis as being *estimation* or *hypothesis testing.*

  (iii) Carry out the analysis in order to answer the indicated question.

  (a) Five hundred children take an IQ test. The *sum* of their scores is 51,217. What is the mean IQ of the population of children from which the sample was randomly drawn?

(b) It is believed that a certain strain of rats should have an equal chance of turning left or right in a T-maze: that is, the probability of a rat turning right should be 0.5. In an experiment with a random sample of 23 of these rats, 18 turn right. Do the data support the belief that the rats have an equal chance of turning left or right?

(c) An IQ test is standardized to have mean = 100 and standard deviation = 15 in a certain population. A sample of 200 persons has mean = 101 on this test. Is it reasonable to suppose that this sample came from the standardization population?

(d) Twenty-seven persons out of a random sample of 46 persons pass a law school admission test. What is the probability of passing this test?

(e) A professor wants to compare two methods of instruction: lecture and Keller-plan. He divides his 50 students into two matched groups of 25 students each, so that each student in one group is matched with a student in the other group in terms of grade-point average. The scores of the 50 students on a final exam are shown below. Note that the scores of each matched pair are given together.

We have not yet studied how to analyze the data from two groups of subjects. However, these data can be analyzed by working with *difference scores*: that is, for each pair compute the difference (lecture minus Keller-plan) between the two scores. Be sure to retain the sign of this difference. In your analysis, assume that the population standard deviation of these difference scores is 8.0. Do you conclude that the mean final-exam scores of the two populations from which the groups were drawn are the same or different? (*Hint:* test whether the population mean of the difference scores is zero).

PAIR	LECTURE	KELLER-PLAN	PAIR	LECTURE	KELLER-PLAN
1	82	82	14	78	88
2	75	84	15	78	84
3	80	92	16	94	98
4	87	93	17	93	94
5	83	97	18	78	87
6	75	92	19	76	81
7	99	93	20	79	81
8	83	78	21	77	76
9	98	90	22	98	98
10	95	99	23	78	93
11	78	75	24	86	90
12	77	74	25	81	83
13	94	82			

9 HYPOTHESIS TESTING: THE BASICS

Hypothesis testing has been illustrated by several examples in the preceding chapters. This chapter provides a more-formal description of how hypotheses are tested.

In each of the examples, we assumed that there was just one hypothesis, and on the basis of the sample data the hypothesis was accepted or rejected. This is a simplified approach, because a hypothesis cannot be rejected except in favor of another hypothesis. This chapter shows how to include a second hypothesis and how to use sample data to decide which of the two hypotheses to accept.

We cannot be absolutely sure that our conclusion, made on the basis of sample data, is correct; we may conclude that one hypothesis is correct when in reality the other hypothesis is correct. Much of this chapter is concerned with evaluating the probability of making such an incorrect conclusion.

The distinction between tests of proportions and tests of means was emphasized in Review Chapter B. These tests are discussed in detail below. The test of a proportion is called the binomial test and the test of a mean is called the z-test.

9.1 The Binomial Test

The binomial test is a test of two hypotheses about the population proportion of successes, p. In this section each hypothesis is a hypothesis that p has a specific value—for example, $p = 0.6$. The hypotheses will be labeled H_0 and H_1. The two hypotheses might be H_0: $p = 0.5$ and H_1: $p = 0.7$. Hypotheses which state that a population parameter has a specific value are called *specific hypotheses* or *simple hypotheses*. In Section 9.3 the hypotheses will be about μ rather than p, but will still be specific: for example, H_0: $\mu = 100$. (In Chapter 10, nonspecific hypotheses will be introduced. Such a hypothesis is $p > 0.5$, which states that p has an interval of values rather than a specific value.)[1]

The decision between the two hypotheses is based on the number of successes, y, which are obtained in a binomial study. This statistic, y, has a probability distribution that depends on the sample size, n, and the probability of success, p. Therefore, the distributions are different for the two hypotheses,

---

[1] The approach to hypothesis testing taken in this chapter, based on two specific hypotheses, allows both α and β to be defined and calculated (see below). The following topics will not be considered until Chapter 10: testing a specific (null) hypothesis against a nonspecific (alternative) hypothesis, describing the decisions to be made as *rejecting* or *failing to reject* the specific (null) hypothesis instead of just accepting either H_0 or H_1, and two-tailed tests. Discussion of these topics has been postponed because they are easier for students to understand once the basic theory described in this chapter is understood. Finally, note that β rather than *power* ($= 1 - \beta$) is defined in this chapter, to emphasize the parallel definitions of α and β. The power of a test is introduced and described in detail in Chapter 13.

H_0 and H_1. Both distributions are binomial distributions—Figures 9.1a and 9.1b show the distributions for $n = 15$, H_0: $p = 0.5$, and H_1: $p = 0.7$.

The possible values of y are divided into two regions, called the *acceptance region for H_0* and the *acceptance region for H_1*. In Figure 9.1c the acceptance region for H_0 includes all values of y from 0 to 11, inclusive. If the observed value of y in the binomial study falls in this interval, the decision is "accept H_0" or "conclude that H_0 is true." In the figure the acceptance region for H_1 includes all values of y from 12 to 15, inclusive. If the observed value of y falls in this interval, the decision is "accept H_1" or "conclude that H_1 is true."

Let us consider an example before we examine how the acceptance regions are determined.

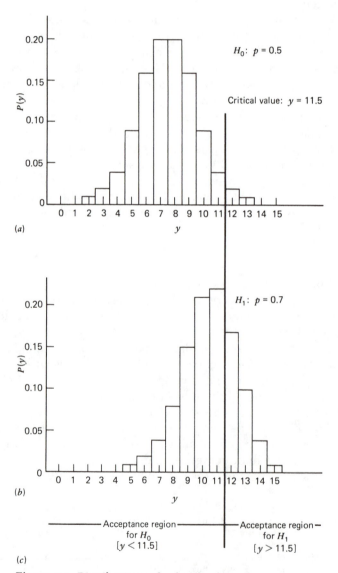

Figure 9.1 Distributions of y for two hypotheses.

Example 1

Learning experiments with rats are often conducted in a T-maze, which consists of a straight alley down which the rats run. At the end of this alley the rat must choose between two identical paths, one to the left and the other to the right. Rats that have had no experience in such a maze are called *naive* rats. One theory about naive rats holds that there is an equal chance that a naive rat will turn right or left at the choice point: in other words, the probability that a naive rat turns *right* is 0.5. This hypothesis about naive rats can be summarized as $H_0: p = 0.5$.

Another theory about naive rats states that they have an inborn tendency to turn right rather than left. This theory predicts that there is a 70% chance ($p = 0.7$) that a naive rat will turn right in the T-maze. The second hypothesis is therefore $H_1: p = 0.7$. Note that both hypotheses, H_0 and H_1, are specific hypotheses.

We try to determine which hypothesis is correct by drawing a random sample of rats. Each rat is run in the T-maze, and we record that 10 of the 15 rats in the experiment turn right. From Figure 9.1, we see that $y = 10$ falls in the acceptance region for H_0. We therefore accept H_0 and conclude that naive rats have a 50% chance of turning right.

Type I and Type II Errors

How are the acceptance regions determined? Roughly speaking, the acceptance regions are located so that the chance of making a wrong conclusion is as small as possible. This subsection describes what it means to make a wrong conclusion and how to calculate the probability of such a conclusion. The next subsection shows how to locate the acceptance regions.

In order to describe a conclusion as wrong or in error, we have to consider two cases. First, there is the case when H_0 is actually true. In that case, accepting H_0 is the correct conclusion and accepting H_1 is an error. Second, there is the case when H_1 is actually true. In that case, accepting H_0 is an error and accepting H_1 is the correct conclusion. The two kinds of errors are distinguished by calling them Type I and Type II errors, respectively.

A *Type I error* is made if H_1 is accepted when H_0 is in fact true.
A *Type II error* is made if H_0 is accepted when H_1 is in fact true.

To discover the probabilities of making these errors, given that the acceptance regions are those shown in Figure 9.1, we first consider the case that H_0 is true. The distribution of y is binomial with $n = 15$ and $p = 0.5$ (graphed in Figure 9.1a). The probability of a Type I error is the probability that y falls in the acceptance region for H_1. This probability is (using Table A1)

$$P(y \geqq 12) = P(12) + P(13) + P(14) + P(15)$$
$$= 0.014 + 0.003 + 0.000 + 0.000$$
$$= 0.017$$

A correct conclusion is made if y falls in the acceptance region for H_0. The probability of a correct conclusion is

$$P(y \leqq 11) = P(0) + P(1) + \cdots + P(10) + P(11)$$
$$= 0.000 + 0.000 + 0.003 + 0.014 + 0.042 + 0.092 + 0.153 +$$
$$0.196 + 0.196 + 0.153 + 0.092 + 0.042$$
$$= 0.983$$

Before we proceed to the case when H_1 is true, two new notations must be explained. The acceptance region for H_0 can be written as $(y < 11.5)$ instead of $(y \leq 11)$ and the acceptance region for H_1 can be written as $(y > 11.5)$ instead of $(y \geq 12)$. The advantage of this change is that a single number, 11.5, divides the values of y into two intervals. The intervals are unambiguously defined because y can never equal 11.5. We do not have to worry whether to write "$<$" or "$\leq$", since it does not matter; the interval $(y < 11.5)$ includes the same integer values of y as the interval $(y \leq 11.5)$. *The number that divides the possible values of y into two acceptance regions will be called the critical value of y.* In this example, values of y greater than the critical value are in the acceptance region for H_1, whereas values of y less than the critical value are in the acceptance region for H_0. (If you wish, you can think of 11.5 as being the upper real limit of the score of 11 and the lower real limit of the score of 12. However, the scores, y, in a binomial study are truly integers and are not continuous. Strictly speaking, the notion of real limits does not apply to them, even though it is convenient to use numbers such as 11.5 as critical values.)

The second notational change is to denote the probability of a Type I error as α (the Greek letter a, pronounced "alpha"). Since the probability of a Type I error and the probability of a correct conclusion (when H_0 is true) must sum to 1.0, we need only be concerned with one of these probabilities; the other can be found by subtraction from 1.0, if it is required. So we concentrate on α, the probability of a Type I error: *α is the probability of concluding that H_1 is true when H_0 is in fact true.* In the example we have been considering, $\alpha = P(y > 11.5) = 0.017$ and the probability of a correct conclusion $= 1 - \alpha = 0.983$. These probabilities are shown as branch probabilities on the first tree in Figure 9.2a.

We turn now to the case when the second hypothesis, H_1, is in fact true. In this case, the distribution of y is binomial with $n = 15$ and $p = 0.7$ (graphed in Figure 9.1b). If we conclude that H_0 is true when H_1 is in fact true, we are making a Type II error. The probability of a Type II error is called β (the Greek letter b, pronounced "beta"): *β is the probability of concluding that H_0 is true when H_1 is in fact true.* The probability that y falls in the acceptance region for H_0 is

$$\begin{aligned}
\beta &= P(y < 11.5) \\
&= P(0) + P(1) + \cdots + P(10) + P(11) \\
&= 0.000 + 0.000 + 0.000 + 0.000 + 0.001 + 0.003 + 0.012 + 0.035 + \\
&\quad 0.081 + 0.147 + 0.206 + 0.219 \\
&= 0.704
\end{aligned}$$

When H_1 is true, a correct conclusion is made if y falls in the acceptance region for H_1. The probability of a correct conclusion is simply $1.0 - \beta = 1.0 - 0.704 = 0.296$. This probability can also be worked out directly as

$$\begin{aligned}
P(y > 11.5) &= P(12) + P(13) + P(14) + P(15) \\
&= 0.170 + 0.092 + 0.031 + 0.005 \\
&= 0.298
\end{aligned}$$

(The discrepancy between this result and the value found from $1.0 - \beta = 1.0 - 0.704 = 0.296$ is caused by the rounding of the entries in Table A1.) The two

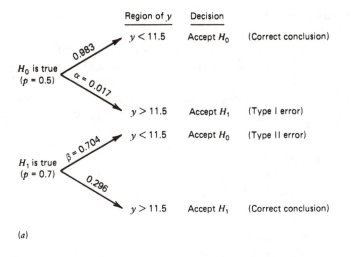

(a)

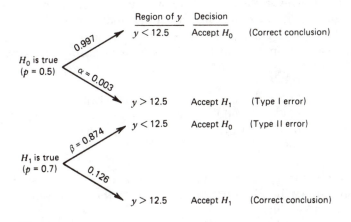

(b)

Figure 9.2 Decisions and their probabilities (binomial test, n = 15). (a) Critical value = 11.5. (b) Critical value = 12.5.

probabilities, β and the probability of a correct conclusion, are shown as branch probabilities on the second tree in Figure 9.2a.

Finding the Critical Value and the Acceptance Regions

It is easy to carry out the binomial test if the acceptance regions are given. How are the acceptance regions determined? Essentially, the problem reduces to finding the critical value, since the regions are on the two sides of the critical value. (Even after the critical value has been determined, we must be sure which side is the acceptance region for H_0 and which side is the acceptance region for H_1.)

At the beginning of the last subsection it was suggested that the acceptance regions, and therefore the critical value, are located so that the chance of making an error is as small as possible. We have seen that there are two kinds of error and two probabilities, α and β. Which probability should be made small? We show first that as one probability is made smaller, the other increases. We then describe the conventional procedure that makes the probability of a Type I error, α, small.

From Figure 9.1a it should be clear that α can be made smaller by *increasing* the critical value. Since α is the probability that y is in the acceptance region for H_1 (when H_0 is true), the probability can be decreased by reducing the size of this acceptance region. The size is reduced if the critical value of y is increased from 11.5 to, say, 12.5. The new probability of a Type I error is, using $p = 0.5$,

$$\alpha = P(y > 12.5)$$
$$= P(13) + P(14) + P(15)$$
$$= 0.003 + 0.000 + 0.000$$
$$= 0.003$$

which is smaller than the previous value of 0.017.

Using a critical value of 12.5 certainly reduces the probability of a Type I error. However, the probability of a Type II error is increased. This is clear from Figure 9.1b. Since β is the probability that y is in the acceptance region for H_0 (when H_1 is true), the probability is increased when the critical value is increased from 11.5 to 12.5. The new probability is, using $p = 0.7$,

$$\beta = P(y < 12.5)$$
$$= P(0) + P(1) + \cdots + P(11) + P(12)$$
$$= 0.000 + 0.000 + 0.000 + 0.000 + 0.001 + 0.003 + 0.012 + 0.035 +$$
$$\quad 0.081 + 0.147 + 0.206 + 0.219 + 0.170$$
$$= 0.874$$

which is larger than the previous value of 0.704. These new values of α and β are shown in Figure 9.2b, along with the corresponding probabilities of a correct conclusion.

It would be desirable to reduce the values of α and β simultaneously. However, a careful study of Figure 9.1 demonstrates that this goal is impossible to achieve. If the critical value is increased, then α decreases but β increases. Conversely, if the critical value is decreased—say, to 10.5 or 9.5—then β decreases but α increases.

Given this relationship between α and β, one approach to selecting a critical value would be to choose that critical value that gives the smallest sum of α and β. Although this approach has some merit, it is rarely used in statistics. The following approach is used instead: *select the critical value that makes α sufficiently small.* Essentially, we decide to concentrate on only one of the errors—namely, the Type I error—and make its probability sufficiently small.

How small should α be in order to be "sufficiently" small? The most common value of α is 0.05, but other standard values, such as 0.01, 0.005, 0.001, etc., are used. For example, suppose we want α to be no larger than 0.05 (that is, $\alpha \leq 0.05$). To find the critical value of y, we select from among those values of y that satisfy the condition $\alpha \leq 0.05$, that value which yields the smallest value of β. As we saw earlier, as α gets smaller, β gets larger, so we must find the critical value that makes α just less than 0.05. The values of α and β for three different critical values of y are given below.

Critical Value of y	α	β
10.5	0.059	0.485
11.5	0.017	0.704
12.5	0.003	0.874

From this table we see that a critical value of 11.5 satisfies $\alpha \leq 0.05$; for this critical value, $\beta = 0.704$. A critical value of 12.5 also satisfies $\alpha \leq 0.05$, but the value of β for 12.5 is larger than for 11.5. Therefore, we choose 11.5 as the most satisfactory critical value.

Beginning students of statistics sometimes have difficulty determining which side of the critical value is the acceptance region for H_0 and which is the acceptance region for H_1. The following rule should be applied.

Rule for locating the acceptance regions for a test with a specified condition on α:

Locate the acceptance region for H_1 in one of the tails of the H_0 distribution. Choose the *upper* tail if the H_1 distribution is located at higher values than the H_0 distribution; choose the *lower* tail if the H_1 distribution is located at lower values than the H_0 distribution. The critical value is chosen so that the probability of the tail of the H_0 distribution just satisfies the condition on α. The acceptance region for H_0 consists of all values of y that are not in the acceptance region for H_1.

Explanation The condition on α applies to probabilities in the H_0 distribution (and not the H_1 distribution), since a Type I error occurs when H_0 is true. Furthermore, it can be shown that β is as small as possible (keeping the condition on α) when the acceptance region for H_1 is in a tail of the H_0 distribution. Note carefully that the particular tail used depends on the relative locations of the two distributions (see the second sentence of the rule).

Example 2

Consider two hypotheses: H_0: $p = 0.2$ and H_1: $p = 0.6$. The two distributions of y for $n = 15$ are shown in Figure 9.3. The H_1 distribution is located at higher values than the H_0 distribution (the mode of H_1 is 9; the mode of H_0 is 3). Hence, the acceptance region for H_1 is in the upper tail of the H_0 distribution. If we specify $\alpha \leq 0.05$, we find from Table A1 that the critical value is 6.5. [$P(y > 6.5) = 0.014 + 0.003 + 0.001 + 0.000 = 0.018$, for $n = 15$ and $p = 0.2$.] The acceptance region for H_1 is ($y > 6.5$) and the acceptance region for H_0 is ($y < 6.5$). The value of β is 0.094. [$P(y < 6.5) = 0.000 + 0.002 + 0.007 + 0.024 + 0.061 = 0.094$ for $n = 15$ and $p = 0.6$.]

Example 3

Consider these two hypotheses: H_0: $p = 0.6$ and H_1: $p = 0.4$. Here the H_1 distribution is located at lower values than the H_0 distribution (because $0.4 < 0.6$). The two distributions of y for $n = 15$ are shown in Figure 9.4. Note that the mode of H_1 is less than the mode of H_0. The acceptance region for H_1 is in the lower tail of the H_0 distribution. If we again specify $\alpha \leq 0.05$, we find that the acceptance region for H_1 is ($y < 5.5$) and the acceptance region for H_0 is ($y > 5.5$). The actual value of α is 0.033 and the value of β is 0.596.

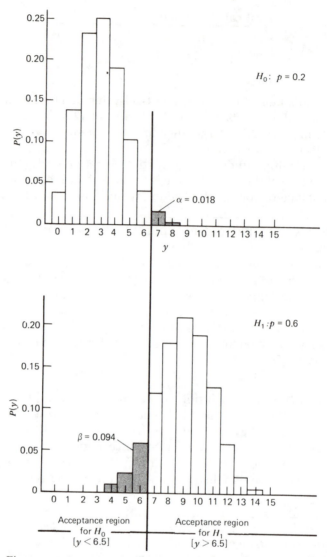

Figure 9.3 Locating the acceptance regions.

The Two Hypotheses Are Not Treated Equally

At the beginning of this section, the two hypotheses, H_0 and H_1, were treated equally: there is an acceptance region for each, the Type I and Type II errors are defined in a symmetric way, and the probabilities of these errors (α and β) are also defined symmetrically. We have found, however, that it is impossible to make both α and β small. We decided, arbitrarily, to make α small (less than a standard value such as 0.05, 0.01, etc.) and to choose the critical value so that β is as small as possible, subject to the condition on α. The net result is that the two hypotheses are no longer treated equally.

This result has the further effect that the two possible conclusions "accept H_0" and "accept H_1", based on the acceptance region in which the observed y falls, cannot be made with equal confidence. When y falls in the acceptance

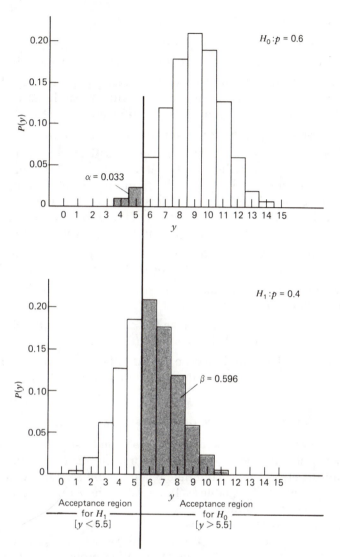

Figure 9.4 Locating the acceptance regions.

region for H_1, we can be quite confident in accepting H_1, since the probability that y falls in the acceptance region for H_1 when H_0 is true (i.e., α, the probability of a Type I error) is very small. We cannot have such confidence in the other possible conclusion (accept H_0) which occurs when y falls in the acceptance region for H_0, because β, the probability of a Type II error, is not usually small. Hence, we must be more cautious in accepting H_0 than in accepting H_1.

The preceding argument is subtle, and it may be difficult to grasp on first reading. For simplicity, the two conclusions "accept H_0" and "accept H_1" should be stated in this way throughout Chapter 9. Then, in Section 10.2, alternative and more proper ways to state the conclusions are introduced.

Summary of Steps

Carry out the binomial test as follows.

1 Set up two hypotheses about the population parameter p. In this chapter the two hypotheses are specific. (*Example*: H_0: $p = 0.5$ and H_1: $p = 0.7$.)

2 Choose a value for α. This value is usually quite small. Since the binomial distribution is discrete and only certain probabilities of a Type I error are possible, the restriction on α is stated as an inequality. (*Example*: $\alpha \leq 0.05$.)

3 Choose the sample size, n, for the study. (*Example*: $n = 15$.)

4 Find the critical value of y. Choose the critical value so that the probability in the tail of the H_0 distribution of y satisfies the restriction on α. The critical value falls in the upper tail if the H_1 distribution is located at higher values than the H_0 distribution. The critical value is located in the lower tail if the H_1 distribution is located at the lower values than the H_0 distribution. Both distributions of y are binomial distributions. (*Example*: critical value of y is 11.5.)

5 Locate the two acceptance regions. [*Example*: acceptance region for H_0 is ($y < 11.5$) and acceptance region for H_1 is ($y > 11.5$).]

6 Compute β. If β is larger than desired, α may have to be changed or the study modified. (*Example*: $\beta = 0.704$.)

7 Draw a random sample of size n. Carry out the study. Record the number of successes, y, observed in the study. (*Example*: $y = 10$.)

8 Make the decision, based on the acceptance region in which the observed y falls. (*Example*: accept H_0 and conclude that $p = 0.5$.)

9.2 The Cumulative Binomial Table

Note: This section may be omitted with no loss of continuity.

In order to calculate α and β, we have to sum individual probabilities. A special table has been prepared to simplify this calculation and to reduce the rounding errors that accumulate when a number of individual probabilities are added together. This table is Table A2 in the Appendix: the cumulative binomial probability table.

Cumulative probabilities are expressed by using the inequality symbols: $<$, $\leq$, $>$, $\geq$. The cumulative probability of values of y less than or equal to 3 may be expressed as $P(y \leq 3)$. The cumulative probability of an interval of values may be expressed in a similar fashion. For example, the cumulative probability of values of y between 8 and 13 is written as $P(8 < y < 13)$. If we want to indicate that $y = 8$ is included in this interval, the inequality is written as $P(8 \leq y < 13)$. The cumulative probability of values of y greater than 13, is expressed as $P(y > 13)$. Again, if we wish to include $y = 13$, the expression is written as $P(y \geq 13)$.

All these probabilities can be calculated from the binomial formula (Section 5.6) or from Table A1 (Section 6.3). But it is easier to use Table A2, which gives cumulative binomial probabilities for $n = 5, 10, 15, 20,$ and 25. For other values of n, you must continue to add the individual probabilities from Table A1.

The basic rules to remember when using Table A2 are

1 Use the *top* and *left* scales when you need a *lower* cumulative probability: that is, a probability of the form $P(y < a)$ or $P(y \leq a)$ for some a. For example, use the top and left scales to find $P(y < 8)$ or $P(y \leq 6)$.

2 Use the *bottom* and *right* scales when you need an *upper* cumulative probability: that is, a probability of the form $P(y > a)$ or $P(y \geq a)$ for some a. For example, use the bottom and right scales to find $P(y > 4)$ or $P(y \geq 10)$.

3 Choose the table for the correct n (5, 10, 15, 20, or 25).

4 Decide on the *real limit* of the interval for which you want the probability; this real limit will be a number like 4.5, 10.5, etc.

Even though Table A2 is a table of cumulative probabilities, it may also be used to get individual probabilities and probabilities of intervals of values of y. Examples of these uses are given below, along with examples of using the table to calculate α and β and to determine the critical value of y.

Cumulative Probabilities	Lower or upper cumulative probabilities of the binomial distribution may be determined directly from Table A2. When a lower cumulative probability is desired, the top and left scales of the table are used. When an upper cumulative probability is desired, the bottom and right scales are used.
Example (a)	Find $P(y \leq 5)$ when $p = 0.2$ and $n = 10$. Since the desired probability is a lower cumulative probability, the top and left scales of the table are used. The section of the table for $n = 10$ and $p = 0.2$ is shown in Table 9.1. The real limit of the interval is $y = 5.5$. The table shows that the cumulative probability of y less than or equal to 5.5 is 0.994. Therefore, $P(y \leq 5) = 0.994$.
Example (b)	Find $P(y \geq 12)$ when $p = 0.7$ and $n = 15$. Since the required probability is an upper cumulative probability, the bottom and right scales are used. The section of the table for $n = 15$ and $p = 0.7$ is shown in Table 9.1. The real limit of the required interval is $y = 11.5$. The table indicates that $P(y > 11.5) = 0.297$. Hence, the required probability is $P(y \geq 12) = 0.297$.
Example (c)	Find $P(y > 10)$ when $p = 0.7$ and $n = 15$. This probability is also an upper cumulative probability, and it can be found in the same section of the table as used in the previous example. The real limit of the interval is 10.5, since $y = 10$ is not included in the interval. The table shows that $P(y > 10) = 0.515$.
Individual and Interval Probabilities	Individual and interval probabilities are the difference between two cumulative probabilities. They can be determined from either lower or upper cumulative probabilities. For simplicity, the examples are worked out using lower cumulative probabilities: that is, using the top and left scales of Table A2. The alternative calculation for each example is given in a footnote.
Example (d)	Find $P(4)$ when $p = 0.5$ and $n = 15$. The required probability is the colored bar in the histogram of Figure 9.5. The real limits of the bar at $y = 4$ are 3.5 and 4.5. The required probability is found by subtracting the cumulative probability at 3.5 from the cumulative probability at 4.5. The section of the cumulative probability table for $n = 15$ and $p = 0.5$ is shown in Table 9.1. Since $P(y < 3.5) = 0.018$ and $P(y < 4.5) = 0.059$,

$$P(4) = P(y < 4.5) - P(y < 3.5)$$
$$= 0.059 - 0.018$$
$$= 0.041$$

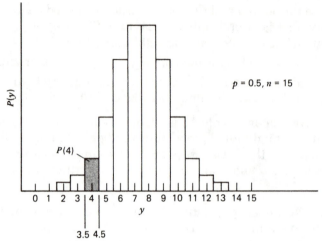

Figure 9.5 Example (d).

TABLE 9.1 Using the Cumulative Binomial Table

EXAMPLE (a)			EXAMPLES (b) AND (c)			EXAMPLE (d)		
		p						*p*
n	*y*	.20				*n*	*y*	.50
10	0.5	107	005	14.5		15	0.5	000
	1.5	376	035	13.5			1.5	000
	2.5	678	127	12.5			2.5	004
	3.5	879	297	11.5			3.5	018
	4.5	967	515	10.5			4.5	059
	5.5	994	722	9.5			5.5	151
	6.5	999	869	8.5			6.5	304
	7.5	1000	950	7.5			7.5	500
	8.5	1000	985	6.5			8.5	696
	9.5	1000	996	5.5			9.5	849
			999	4.5			10.5	941
			1000	3.5			11.5	982
			1000	2.5			12.5	996
			1000	1.5			13.5	1000
			1000	0.5	15		14.5	1000
			.70	*y*	*n*			
			p					

The value of $P(4)$ can also be found directly in Table A1, which gives it as 0.042. The discrepancy between the two values is due to rounding.[2]

Example (e)

Find $P(9 \leqq y \leqq 11)$ when $p = 0.7$ and $n = 15$. The required probability is shown in Figure 9.6. The sum of the probabilities of the three colored bars must be determined. The real limits of the interval are 8.5 and 11.5. From Table A2, using the top and left scales, we see that the two cumulative probabilities are $P(y < 8.5) = 0.131$ and $P(y < 11.5) = 0.703$. Hence,[3]

$$P(9 \leqq y \leqq 11) = P(y < 11.5) - P(y < 8.5)$$
$$= 0.703 - 0.131$$
$$= 0.572$$

Finding α and β

Finding α, given a critical value, requires finding a probability in the upper or lower tail of the H_0 distribution of y. The probability of a Type II error, β, is found from the H_1 distribution of y. Both probabilities can be taken directly from Table A2.

Example (f)

Consider a binomial study with $n = 25$, $H_0: p = 0.4$, and $H_1: p = 0.8$. If the critical value of y is 13.5, what are the values of α and β? The easiest way to solve problems of this kind is to draw a sketch such as the one in Figure 9.7. Two distributions are sketched in the figure, one for H_0 and one for H_1. The principal purpose of the sketch is to help you decide which tail of which distribution represents α and which tail of which distribution represents β. For this purpose the sketch does not have to be very accurate. The binomial distributions should be sketched as continuous distributions, for simplicity. The peak of the distribution should be located close to the mean (np). The mean of the H_0 distribution is $25 \times 0.4 = 10.0$ and the mean of the H_1 distribution is $25 \times 0.8 = 20.0$.

[2] The probability can also be found from upper cumulative probabilities as follows: $P(4) = P(y > 3.5) - P(y > 4.5) = 0.982 - 0.941 = 0.041$.

[3] The probability can also be found from upper cumulative probabilities as follows: $P(9 \leqq y \leqq 11) = P(y > 8.5) - P(y > 11.5) = 0.869 - 0.297 = 0.572$.

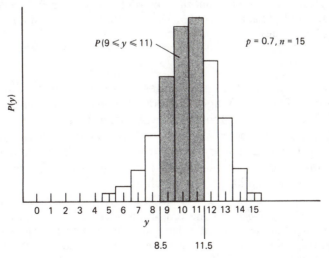

Figure 9.6 Example (e).

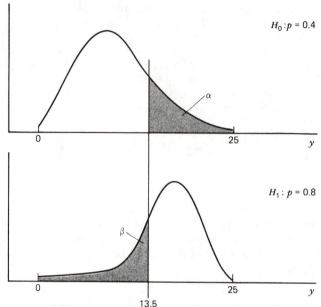

$H_0 : p = 0.4$

α

$H_1 : p = 0.8$

β

0   25 y

0   25 y

13.5

Figure 9.7 Example (f).

Once having sketched the distributions in this way, you locate the critical value (which, here, is in the upper tail of the H_0 distribution) and label the appropriate areas as α and β. Note that α is the area of the upper tail of the H_0 distribution, since the H_1 distribution is located at higher values than the H_0 distribution. The value of α, found by using the bottom and right scales for $n = 25$ and $p = 0.4$, is 0.078. The value of β, found by using the top and left scales for $n = 25$ and $p = 0.8$, is 0.002.

Example (g)

Consider a binomial study with $n = 15$, H_0: $p = 0.4$, and H_1: $p = 0.2$. Find the values of α and β if H_1 is accepted for values of y between 0 and 4, inclusive. A sketch of the two distributions is shown in Figure 9.8. The problem is similar to the preceding example except that the acceptance region for H_1 is now in the lower tail of the H_0 distribution. The given information about the values of y that are in the acceptance region implies that the critical value of y is 4.5. The two probabilities, α and β, can be found in Table A2. Their values are $\alpha = 0.217$ and $\beta = 0.164$.

Finding the Critical Value

Up to this point, all the examples in this section have had a common characteristic: in each, a single value or an interval of values of y was given and the probability of that value or interval had to be determined. In earlier chapters such problems were called *forward problems*. Suppose, instead, that we are given a probability and have to find an interval of values of y that has this probability. Such problems are known as *backward problems*. Finding the critical value is a backward problem.

Example (h)

A psychologist feels that the probability of a certain behavior is 0.5 (H_0). Another psychologist argues forcefully that p is actually 0.7 (H_1). An experiment is devised to determine which psychologist is correct. A sample size of 20 is planned. What critical value of y must be used if α is to be less than or equal to 0.10?

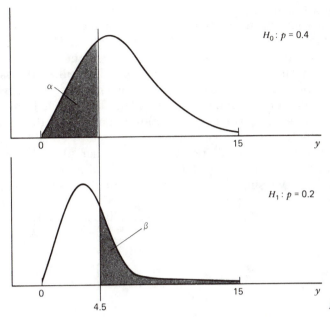

$H_0: p = 0.4$

α

0   15 y

$H_1: p = 0.2$

β

0   15 y
4.5

Figure 9.8 Example (g).

The two distributions are sketched in Figure 9.9. The acceptance region for H_1 must be in the upper tail of the H_0 distribution, since the H_1 distribution is above the H_0 distribution. The unknown critical value of y must be chosen so that the area of the upper tail is $\alpha \leqq 0.10$. We use the bottom and right scales of Table A2 with $n = 20$ and $p = 0.5$. We look for the probability that is closest to but smaller than 0.10. The closest probability is 0.058, corresponding to a critical value of 13.5. In this experiment, therefore, values of $y \geqq 14$ are evidence for H_1. For this critical value, the value of β is 0.392.

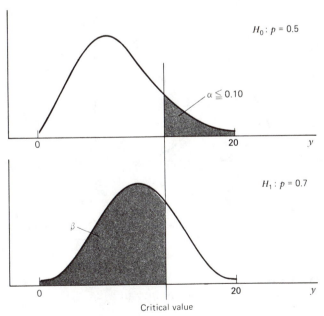

$H_0: p = 0.5$

$\alpha \leqq 0.10$

0   20 y

$H_1: p = 0.7$

β

0   20 y
Critical value

Figure 9.9 Example (h).

9.3 The z-test of One Mean

The z-test described in this section is a test of two hypotheses about the population mean, μ. Each hypothesis states that μ has a specific value. For example, the two specific hypotheses might be $H_0: \mu = 18$ and $H_1: \mu = 16$.

In order to make a decision between the two hypotheses, we draw a random sample from the population and calculate the sample mean, $\bar{y}$. The possible values of $\bar{y}$ are divided into two regions, called the acceptance region for H_0 and the acceptance region for H_1. The observed sample mean falls in one of these regions. The corresponding hypothesis is then accepted.

The two acceptance regions are located by considering the distribution of the statistic, $\bar{y}$, for all possible random samples of size n drawn from the population, and by considering the probabilities of Type I and Type II errors. There are two such distributions of the statistic, because there are two hypotheses about the population. An example, based on a sample size $n = 25$, is shown in Figure 9.10. Each of the two hypothesized population distributions,

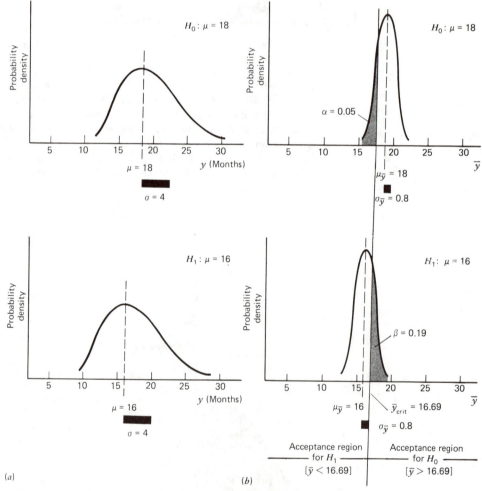

Figure 9.10 Distributions for the z-test of one mean. (a) Population distributions. (b) Distributions of the sample mean, $\bar{y}$ ($n = 25$).

for H_0 and H_1, that appear in Figure 9.10a is slightly skewed to the right, and each has a standard deviation, σ, of 4. The distributions of $\bar{y}$ for H_0 and H_1, are shown in Figure 9.10b. The central limit theorem (Section 7.5) states that these distributions are approximately normal with mean equal to the population mean (18 and 16, respectively) and with standard deviation equal to $\sigma/\sqrt{n} = 4/\sqrt{25} = 0.8$. The acceptance regions are shown in the figure. For the derivation of the critical value, 16.69, see below. The acceptance region for H_0 is the interval ($\bar{y} > 16.69$); if the sample mean in the study falls in this interval, the decision will be "accept H_0" or "conclude that H_0 is true". The acceptance region for H_1 is the interval ($\bar{y} < 16.69$); if the sample mean falls in this interval, the decision will be "accept H_1" or "conclude that H_1 is true".

Example 1

Children learn to talk at different ages. Consider the age at which a child first says a two-word sentence. For children in the general population (excluding retarded children) these ages have a distribution with a mean of 18 months and a standard deviation of 4 months.[4] Suppose a parents' organization develops a program designed to get children to talk earlier. During the program, parents of newborn babies are encouraged to talk more with their babies, read more to them, etc. The hope is that the children will, on the average, talk two months earlier than children in the general population. If this hope is realized, the distribution of ages at which children in this program say a two-word sentence will have a mean of 16 months.

An experiment is designed to see if the program is effective; on the basis of the experiment, a decision will be made between two hypotheses.

H_0: The program is not effective; the distribution of ages at which children in the program say two-word sentences is the same as the distribution in the general population: that is, $\mu = 18$ months.

H_1: The program is effective; the distribution of ages at which children in the program say two-word sentences has a mean of 16 months: that is, $\mu = 16$ months.

Note carefully that *both* hypotheses are about the population of children whose parents could *potentially* take part in the program; in other words, the hypotheses are about the population of all such children, not about the sample of children actually studied. We use the results from the sample to make an inference about the population of all children. The first hypothesis states that the mean age at which children in this population talk is 18 months. The second hypothesis states that the mean age at which children in this population talk is 16 months.

The two hypothesized population distributions are shown in Figure 9.10a. The distributions of the sample mean (for $n = 25$) are shown in Figure 9.10b, one for each hypothesis. We will explain below how the two acceptance regions shown in the figure are determined. Suppose that the data from the experiment have been collected and the sample mean age, $\bar{y}$, of the first two-word sentence has been calculated. If the sample mean were 17.52 months, H_0 would be accepted, since 17.52 is in the acceptance region for H_0: ($\bar{y} > 16.69$). We would conclude that the mean age of talking for the population of children who take the program is 18 months. In other words, the program is not effective.

---

[4] The parameters are approximately those of E. H. Lenneberg, "Speech development: its anatomical and physiological concomitants", in E. C. Carterette (ed.), *Brain function III: speech, language, and communication* (Berkeley and Los Angeles: University of California Press, 1966).

Finding the Critical Value and the Acceptance Regions

As stated above, the possible values of the sample mean, $\bar{y}$, are divided into two regions, called acceptance regions. It is obvious that a large value of the sample mean (say, more than 20) is strong evidence that H_0: $\mu = 18$ is true, because such large values are very unlikely when H_1: $\mu = 16$ is true. Conversely, a small value of the sample mean (say, less than 15) is strong evidence for H_1, because such small values are very unlikely when H_0 is true. So for large values of $\bar{y}$, we will accept H_0; and for small values of $\bar{y}$, we will accept H_1. But what if $\bar{y}$ has a value in the middle—say, about 17? We must find the boundary between the two acceptance regions. This boundary, called the critical value of $\bar{y}$, is abbreviated as $\bar{y}_{crit}$.

The critical value, $\bar{y}_{crit}$, is chosen so that the probability of accepting H_1 when H_0 is actually true (i.e., the probability of a Type I error) is a specified small number, α. When the H_1 distribution of $\bar{y}$ is located at lower values than the H_0 distribution (as in the example), the acceptance region of H_1 is in the lower tail of the H_0 distribution. When the H_1 distribution of $\bar{y}$ is located at higher values than the H_0 distribution, the acceptance region of H_1 is in the upper tail of the H_0 distribution. In either case, the area of this tail is α. The acceptance region of H_0 consists of all other values of $\bar{y}$. The area above this region, under the H_1 distribution, is β, the probability of accepting H_0 when H_1 is true (a Type II error).

Example 2

The values of $\bar{y}_{crit}$ and β for Example 1 are found as follows. The hypotheses are H_0: $\mu = 18$ and H_1: $\mu = 16$. The sample size is n = 25. Suppose we set $\alpha = 0.05$. The population distributions and the distributions of the sample mean, $\bar{y}$, are shown in Figure 9.10. The two errors and the two acceptance regions are summarized in Figure 9.11.

In order to find $\bar{y}_{crit}$, we first find its value in standard scores. Since the distribution of $\bar{y}$ is (approximately) normal, the standard score cutting off an area of 0.05 in the lower tail is −1.64 (from Table A3 or A4). Then, converting this standard score to a score with mean = 18 and standard deviation = 0.8, we get

$$\bar{y}_{crit} = 18 + (-1.64) \times 0.8 = 18 - 1.312 = 16.69 \text{ (rounded)}$$

(The formula on which this conversion is made can be found in Section 7.1. In general, if z_{crit} is the critical value in standard scores, then

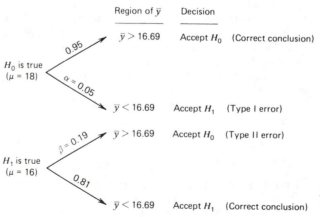

Region of $\bar{y}$	Decision	
$\bar{y} > 16.69$	Accept H_0	(Correct conclusion)
$\bar{y} < 16.69$	Accept H_1	(Type I error)
$\bar{y} > 16.69$	Accept H_0	(Type II error)
$\bar{y} < 16.69$	Accept H_1	(Correct conclusion)

H_0 is true ($\mu = 18$): 0.95, $\alpha = 0.05$

H_1 is true ($\mu = 16$): $\beta = 0.19$, 0.81

Figure 9.11 Decisions and their probabilities (z-test).

$$\bar{y}_{\text{crit}} = \mu_0 + (z_{\text{crit}} \times \sigma_{\bar{y}})$$

where μ_0 is the hypothesized mean when H_0 is true.)

The acceptance region for H_1 consists of all values of $\bar{y}$ less than $\bar{y}_{\text{crit}}$: ($\bar{y} < 16.69$). The calculation of $\bar{y}_{\text{crit}}$ ensures that the probability that $\bar{y}$ is in the region, when H_0 is true, is 0.05. On the other hand, the acceptance region for H_0 is ($\bar{y} > 16.69$). The probability that $\bar{y}$ is in this region, when H_1 is true, is β, which can now be calculated. First, convert 16.69 to a standard score. Since the probability we want is a probability when H_1 is true, the standard score must be computed using the mean and standard deviation of the H_1 distribution of $\bar{y}$.

$$z_\beta = \frac{16.69 - 16}{0.8} = \frac{0.69}{0.8} = 0.86$$

Second, find the area of the normal curve above the interval ($z > 0.86$). This area is 0.19, as found in Table A3. (In general, z_β can be calculated by the following formula.

$$z_\beta = \frac{\bar{y}_{\text{crit}} - \mu_1}{\sigma_{\bar{y}}}$$

Here, μ_1 is the hypothesized mean when H_1 is true.)

Summary of Steps

The above-described test of one mean is usually called the z-test of one mean. A standard score, or z-score, must be found in order to calculate $\bar{y}_{\text{crit}}$. To perform the z-test, proceed as follows.

1a Set up two hypotheses about the population parameter μ. In this chapter the two hypotheses are specific. (*Example:* H_0: $\mu = 18$ and H_1: $\mu = 16$.)

b Record the population standard deviation σ. The standard deviation must be known, or a value must be assumed for it. The standard deviation is assumed to be the same for each hypothesis about μ. (*Example:* $\sigma = 4$.)

2 Choose a value for α. The value should be quite small. Since the normal distribution is continuous, the probability of a Type I error can have any value between 0 and 1; hence, α may be specified exactly. (*Example:* $\alpha = 0.05$.)

3 Choose the sample size, n, for the study. (*Example:* n = 25.)

4 Find the critical value of $\bar{y}$. The critical value, called $\bar{y}_{\text{crit}}$, is chosen so that the area of the tail of the H_0 distribution of $\bar{y}$ equals the specified α. Choose the upper tail if the H_1 distribution is located at higher values than the H_0 distribution; choose the lower tail if the H_1 distribution is located at lower values than the H_0 distribution. Both distributions of $\bar{y}$ are (approximate) normal distributions with standard deviation = $\sigma_{\bar{y}} = \sigma/\sqrt{n}$. (*Example:* $\bar{y}_{\text{crit}} = 16.69$.)

5 Locate the two acceptance regions. [*Example:* acceptance region for H_0 is ($\bar{y} > 16.69$) and acceptance region for H_1 is ($\bar{y} < 16.69$).]

6 Compute β. If β is larger than desired, α may have to be changed or the study modified. (*Example:* $\beta = 0.19$.)

7 Draw a random sample of size n. Carry out the study. Record the sample mean, $\bar{y}$, actually observed in the study. (*Example:* $\bar{y} = 17.52$.)

8 Make the decision, based on the acceptance region in which the observed $\bar{y}$ falls. (*Example:* accept H_0 and conclude that $\mu = 18$.)

An Alternative Approach: The Test Statistic z_{obs}

The acceptance regions have been expressed above in terms of the sample mean, $\bar{y}$. Alternatively, we can convert $\bar{y}$ to a standard score and express the acceptance regions in terms of this standard score. The standard score is computed using the mean and standard deviation of the H_0 distribution of $\bar{y}$. This standard score is called z_{obs} because it is a z-score based on the "observed" sample mean, $\bar{y}$. The formula for z_{obs} is

$$z_{obs} = \frac{\bar{y} - \mu_0}{\sigma_{\bar{y}}} = \frac{\bar{y} - \mu_0}{\sigma/\sqrt{n}}$$

where μ_0 is the hypothesized mean when H_0 is true. The acceptance regions can be expressed very simply in terms of z_{obs}, once z_{crit} has been found. Recall that z_{crit} is the standard score that cuts off an area of α in the appropriate tail of the H_0 distribution. In the example of this section, $z_{crit} = -1.64$ for $\alpha = 0.05$. The acceptance region for H_0 is $(z_{obs} > -1.64)$, and the acceptance region for H_1 is $(z_{obs} < -1.64)$.

Essentially, what we have done is convert $\bar{y}$ to a z-score instead of converting z_{crit} to $\bar{y}_{crit}$. Since the new quantity, z_{obs}, is computed from the sample, it is a statistic. Because it is easy to compare the value of z_{obs} with a number found in Table A3 or A4, this statistical test is simple to carry out. The statistic z_{obs}, like many other statistics that will be introduced in later chapters, is called a *test statistic*. *A test statistic is a statistic that is used to make a decision between two hypotheses*. (Note that by this definition, $\bar{y}$ is also a test statistic. In the binomial test, y is the test statistic.)

The steps for carrying out the test of one mean by using z_{obs} parallel the steps outlined above for carrying out the test by using $\bar{y}$. Some steps are identical—these are marked "unchanged" in the list below. One step, the computation of β, cannot be done directly if z_{obs} is used, since $\bar{y}_{crit}$ must be calculated in order to calculate β. However, statistical tests are often carried out without calculating β.

To carry out the alternative approach to the test of one mean,

1a Set up two hypotheses about the population parameter μ. (Unchanged.)

b Record the population standard deviation σ. (Unchanged.)

2 Choose a value for α. (Unchanged.)

3 Choose the sample size, n, for the study. (Unchanged.)

4 Find the critical value of z_{obs}. The critical value, called z_{crit}, is the standard score that cuts off an area of α in the tail of the H_0 distribution of $\bar{y}$. Choose the upper tail if the H_1 distribution is located at higher values than the H_0 distribution; choose the lower tail if the H_1 distribution is located at lower values than the H_0 distribution. (*Example:* $z_{crit} = -1.64$.)

5 Locate the two acceptance regions. [*Example:* acceptance region for H_0 is $(z_{obs} > -1.64)$ and acceptance region for H_1 is $(z_{obs} < -1.64)$.]

6 This step is omitted. If the value of β is desired, steps 4, 5, and 6 from the previous summary must be used.

7 Draw a random sample of size n. Carry out the study. Record the sample mean, $\bar{y}$, actually observed in the study and compute z_{obs} from it. [*Example:* $\bar{y} = 17.52$ and $z_{obs} = (17.52 - 18)/0.8 = -0.60$.]

8 Make the decision, based on the acceptance region in which z_{obs} falls. (*Example:* accept H_0 and conclude that $\mu = 18$.)

Example 3

An experiment is conducted to decide between two hypotheses: H_0: $\mu = 100$ and H_1: $\mu = 108$. A sample of 50 subjects is studied; the sample mean is 105.68. What decision is made if the experimenter wants $\alpha = 0.001$ and the population standard deviation is assumed to be 12?

The hypotheses about μ and the values of σ, α, and n are given directly in the statement of the problem. The value of z_{crit} can be found most easily from Table A4. We require a tail probability of 0.001; the table gives a critical value of $z = 3.090$. We have to decide whether z_{crit} is $+3.090$ or -3.090. The mean of the H_0 distribution of $\bar{y}$ is 100, whereas the mean of the H_1 distribution is 108, which is greater than 100. Hence the *upper* tail of the H_0 distribution is the required tail. This means that $z_{crit} = +3.090$ and that the acceptance region for H_1 is ($z_{obs} > 3.090$). The acceptance region for H_0 is then ($z_{obs} < 3.090$).

The above steps are preliminary. We now turn to the data from the study. The sample mean of 105.68 is converted to z_{obs} as follows.

$$z_{obs} = \frac{\bar{y} - \mu_0}{\sigma/\sqrt{n}} = \frac{105.68 - 100}{12/\sqrt{50}} = \frac{5.68}{1.697} = 3.347$$

Since 3.347 is greater than 3.090, z_{obs} is in the acceptance region for H_1; hence, we accept H_1 and conclude that $\mu = 108$.

9.4 The Normal Approximation to the Binomial Distribution

(*Note:* This section may be omitted with no loss of continuity. Some of the results of this section are used in Section 12.5.)

We return now to the binomial distribution and the binomial test of a proportion. Cumulative binomial probabilities are required in order to carry out the binomial test. In Section 9.1 these cumulative probabilities were found by summing individual binomial probabilities from Table A1. In Section 9.2 the cumulative probabilities were found directly from Table A2. This section outlines a third, approximate, method for finding either individual or cumulative binomial probabilities. This method makes use of the standard normal distribution.

The normal approximation can be used for many of the combinations of n and p found in Tables A1 and A2. However, the approximation is most useful for values of n and p not in those tables. (Neither Table A1 nor Table A2 extend beyond $n = 25$, and Table A2 only includes $n = 5, 10, 15, 20$, and 25. Only certain values of p are included in these tables.) It must always be kept in mind, however, that the normal approximation is just that—an *approximation* to the exact binomial probabilities. Later in this section the accuracy of this approximation will be stated. If the approximation is not accurate enough, you can consult the more-extensive tables of the binomial distribution available in many libraries.

Although both the binomial and the normal distributions are unimodal, they differ in two important ways. First, the normal distribution is always symmetric; the binomial distribution is symmetric only when $p = 0.5$. (For other values of p, the binomial distribution is skewed: to the left if $p > 0.5$ and to the right if $p < 0.5$.) Second, the normal distribution is a distribution of a *continuous* variable, whereas the binomial is a distribution of a *discrete* variable. A binomial variable can have only the values 0, 1, 2, 3, etc., whereas a normal variable may take on any value from minus infinity to plus infinity and each value (such as 2.57, -3.145, etc.) can have any number of decimal places. How is it possible for the normal distribution to be an approximation to the binomial distribution when the distributions are so dissimilar? The method is best illustrated by an example.

Example 1

Look at the binomial distribution for $p = 0.7$ and $n = 15$ in Figure 9.12. A normal distribution has been superimposed on the binomial distribution. Note that the two distributions are similar but by no means identical. They differ in the two ways mentioned above: the normal is symmetric and the binomial is skewed (in this example); the normal is a smooth curve and the binomial is a series of bars.

Suppose we want to find $P(12)$. This probability is represented by the height of the shaded bar at $y = 12$. The bar extends from $y = 11.5$ to $y = 12.5$. The binomial probability also can be thought of as the *area* of the bar, since area is base $\times$ height. The base is 1.0 and the height is 0.170, so the area is $1.0 \times 0.170 = 0.170$.

In Figure 9.12, the area under the normal curve between the real limits 11.5 and 12.5 is very close to the area of the bar representing the binomial probability of $y = 12$. The top of the bar is not included in the normal area, but part of the normal area is not included in the bar—these areas are almost equal. So the binomial probability is approximately equal to the area under the normal curve. This is the essence of the technique of approximating the binomial distribution by the normal distribution.

Which normal distribution has been drawn in the figure? Recall that there are many different normal distributions, differing in their means (μ) and standard deviations (σ). A binomial distribution also has a mean and standard deviation, given by (Section 6.6):

$$\mu = np \text{ and } \sigma = \sqrt{npq}$$

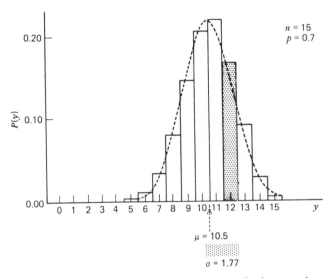

Figure 9.12 The normal approximation to the binomial distribution.

We use these values as the mean and standard deviation of the approximating normal distribution. The binomial distribution in Figure 9.12 has $\mu = 15 \times 0.7 = 10.5$ and $\sigma = \sqrt{15 \times 0.7 \times 0.3} = 1.77$; the approximating normal distribution has the same values for μ and σ.

We can now estimate the binomial probability, $P(12)$, using normal probabilities. To find the area under a normal curve between 11.5 and 12.5, we simply convert each of these numbers to z-scores.

$$z_1 = \frac{11.5 - 10.5}{1.77} = 0.56 \text{ and } z_2 = \frac{12.5 - 10.5}{1.77} = 1.13$$

We then find the area under a standard normal distribution between 0.56 and 1.13. Using Table A3, we find that this area is $0.3708 - 0.2123 = 0.1585$. This is our estimate of the binomial probability, $P(12)$. The exact value (from Table A1) is 0.170. Since our estimate is accurate to no more than two decimal places, we should round the original result of 0.1585 to 0.16 before reporting it. Even the second decimal place is out by one unit (0.16 instead of 0.17), but this digit is still worth reporting since it is close to the correct value. Of course, in practice we would not know the correct value!

The Accuracy of the Normal Approximation

Why is the answer obtained in the above example not more accurate? First, we approximated a bar by the area under a curve, and these two areas are not exactly equal. Second, we rounded the standard deviation to two decimal places and used this rounded value in the calculation of the z-scores. Third, we rounded the z-scores to two decimal places before looking them up in the normal table. (We would have done better if we had kept another decimal place in the z-scores and had interpolated in the normal table, but such a refinement usually is not necessary.) The main reason for the inaccuracy is the first—using an area under a curve to approximate a bar.

The accuracy of the normal approximation can be summarized by the following rules.[5]

(a) The approximation is accurate to two decimal places in all cases except those given in **(b)**, **(c)**, and **(d)**.

(b) If $p = 0.50$ and $n > 50$, the approximation is accurate to three, not two, decimal places.

(c) If $n < 5$, the approximation is poor and should *not* be used. Use Table A1 instead.

(d) If $p < 0.1$ or $p > 0.9$, the approximation is poor and should *not* be used. (Such extreme values of p are rarely encountered.)

How to Carry Out the Normal Approximation

1 Draw a sketch of the binomial distribution with the approximating normal distribution superimposed.

2 Calculate the mean and standard deviation of the binomial distribution, and hence of the approximating normal distribution. Check these values for reasonableness in terms of your sketch, and show them on the sketch.

3 Indicate the interval of values of y for which you need the probability. Use the real limits of the bars.

[5] These rules are based on the results of M. S. Raff, "On approximating the point binomial", *American Statistical Association Journal* 51 (1956), 293–303.

4 Shade in the area under the normal curve above this interval.

5 Calculate the z-scores corresponding to the ends of the interval. Check them against the sketch. If a z-score is positive, it should be to the right of the mean; if it is negative, it should be to the left.

6 Find the area under the normal curve between the two z-scores, using Table A3.

7 Round the area to two decimal places (or to three places if $p = 0.5$ and $n > 50$). The result is the approximate value of the binomial probability.

In Example 1 above, the required binomial probability is an individual binomial probability, $P(12)$, not a cumulative probability. The seven steps apply both to individual probabilities as in that example and to cumulative probabilities as in Examples 2 and 3 below.

Example 2

We will once again choose an example for which we "know the answer", so that we can see how accurate the normal approximation is. Let us take $p = 0.5$ and $n = 20$ and calculate $P(4 \leq y \leq 8)$.

First we make a sketch of the two distributions (Figure 9.13a). Second, we calculate the mean and standard deviation: $\mu = np = 20 \times 0.5 = 10$ and $\sigma = \sqrt{npq} = \sqrt{20 \times 0.5 \times 0.5} = \sqrt{5} = 2.24$. The mean and standard deviation are added to the sketch in Figure 9.13b. Note that the center of the distribution is at the mean and that about 68% of the distribution falls above the interval from one standard deviation below to one standard deviation above the mean.

Third, we show the limits of the desired interval. Since we want to include the bars for $y = 4$ and $y = 8$, we must take values of y from 3.5 (lower limit of the bar at $y = 4$) to 8.5 (upper limit of the bar at $y = 8$). Fourth, we shade in the area under the normal curve above this interval. Both these steps are shown in the next sketch, Figure 9.13c. (In practice, only one sketch is made—the sketch has been drawn in three stages to show the sequence in which it is drawn.)

Fifth, we calculate the z-scores that correspond to the ends of the interval.

$$z_1 = \frac{3.5 - 10}{2.24} = -2.90 \text{ and } z_2 = \frac{8.5 - 10}{2.24} = -0.67$$

Note that the mean (10) and the standard deviation (2.24), which were previously obtained, are used to calculate these z-scores.

Sixth, we find the area under the normal curve above the interval ($-2.90 < z < -0.67$). Because the normal curve is symmetric, this area will be the same as the area above the interval ($0.67 < z < 2.90$). The latter area can be found as the difference of two values in Table A3: $0.4981 - 0.2486 = 0.2495$.

Seventh and finally, we round this result to two decimal places, giving a probability of 0.25.

We can check how close the approximation is by finding the exact binomial probability from Table A2. The exact answer is $0.252 - 0.001 = 0.251$, which confirms the result obtained from the normal approximation. In this example, the result obtained from the normal approximation is accurate to 0.001 (three decimal places), but we would not know that this was the case without the exact table. The normal approximation is used when the exact table is unavailable.

Example 3

We wish to make a binomial test of two hypotheses: $H_0: p = 0.6$ and $H_1: p = 0.8$. We plan to use a sample size of 60. If the acceptance region for H_1 is the set of values of $y \geq 43$, what is the probability of a Type I error?

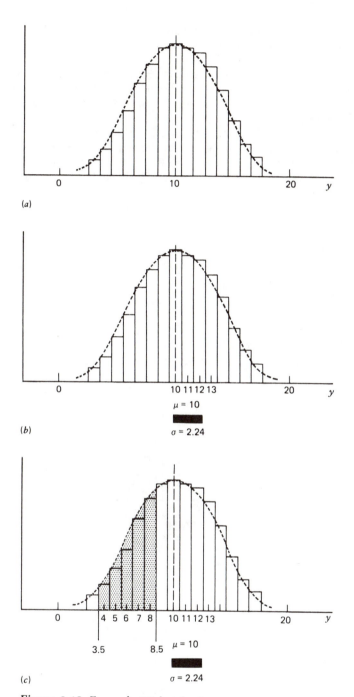

(a)

(b)

(c)

Figure 9.13 Example 2 (sketches).

We want to find the tail probability, $\alpha = P(y \geqq 43)$, when $p = 0.6$ and $n = 60$. We must use the normal approximation, since Tables A1 and A2 do not extend beyond $n = 25$. We cannot easily sketch the full binomial distribution, because there are too many bars. In any case, it is only necessary to sketch the normal distribution and a few bars in the vicinity of $y = 43$, since one end of the interval is located near this value. See Figure 9.14a, which also shows the mean ($\mu = np = 60 \times .06 = 36$) and the standard deviation ($\sigma = \sqrt{npq} = \sqrt{60 \times 0.6 \times 0.4} = 3.79$).

The critical value is 42.5. We show by the shading in the next sketch (Figure 9.14b) the area that we want to determine. Strictly speaking, we want to include all values of y up to 60.5 (the end of the bar at $y = 60$). But there is little error in taking the upper limit of the interval as $z = +\infty$ (plus infinity). The lower limit, corresponding to $y = 42.5$, has a z-score of

$$z = \frac{42.5 - 36}{3.79} = 1.72$$

Hence the required normal probability is $\alpha = P(z > 1.72) = 0.5 - 0.4573 = 0.0427$, using Table A3. The approximate binomial probability is 0.04, since we must round the result to two decimal places.

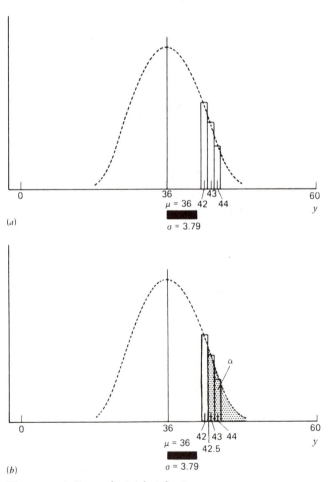

Figure 9.14 Example 3 (sketches).

Relation to Previous Topics

It is easy to calculate binomial probabilities by the normal approximation if the connection between this technique and other topics in this book is understood.

(a) *Real limits.* The notion that an interval of values of a discrete binomial variable can have real limits was introduced earlier in this chapter. Recall that critical values of the binomial test are numbers like 4.5, 7.5, etc. These numbers are used in Table A2 to find cumulative binomial probabilities.

(b) *Normal table.* When the limits of the interval have been converted to z-scores, the area is found in the normal table, using the methods of Section 7.4.

(c) *Reasonableness.* The concept of reasonableness was first introduced in Chapter 3. When using the normal approximation, you should first check that the values of μ and σ are reasonable by comparing them with the sketch you have drawn. Similarly, the calculated z-scores are compared with the sketch to see if they are reasonable.

(d) *Usefulness.* The concept of usefulness was also introduced in Chapter 3. The final result obtained from the normal approximation to a binomial probability is not useful unless the number of decimal places that are retained matches the accuracy of the approximation. If you report four decimal places when the accuracy is only two decimal places, your answer will not be useful—the last two digits are "junk".

(e) *Standard score formula.* The standard score, or z-score (Section 7.1), is

$$z = \frac{\text{Statistic} - \text{Mean}}{\text{Standard Deviation}}$$

It is important to distinguish between the standard score for the sample mean (in the z-test) and the standard score for the number of successes (in the normal approximation). In converting the statistic $\bar{y}$ to a z-score, the mean is μ_0, the hypothesized mean under H_0; and the standard deviation is $\sigma_{\bar{y}} = \sigma/\sqrt{n}$. The standard score is, therefore,

$$z_{\text{obs}} = \frac{\bar{y} - \mu_0}{\sigma/\sqrt{n}}$$

In converting the statistic y to a z-score, the mean is np and the standard deviation is $\sqrt{npq}$. The standard score is

$$z = \frac{y - np}{\sqrt{npq}}$$

Note that y in this formula is a real limit, i.e., a number like 4.5, 7.5, etc.

9.5 Summary

The *binomial test* is a test of one proportion; the *z-test* is a test of one mean. Each test is carried out by setting up two hypotheses, called H_0 and H_1, about the population parameter p (binomial test) or μ (z-test). Each hypothesis is a *specific hypothesis* or a *simple hypothesis*. The decision between the two hypotheses is made on the basis of the value of a sample statistic (called a *test statistic*). For the binomial test, the test statistic is y, the number of successes in the sample. For the z-test of one mean, the test statistic may be either $\bar{y}$, the sample mean, or z_{obs}, the sample mean converted to a z-score.

The possible values of the test statistic are divided into two regions, called the *acceptance region for H_0* and the *acceptance region for H_1*. The boundary between the two regions is called the *critical value of the test statistic*.

Two kinds of error can be made in a hypothesis test: a *Type I error* and a *Type II error*. The probabilities of these errors are called α and β, respectively. The critical value is chosen so that α is equal to a specified small number. The value of β can then be calculated.

Binomial probabilities can be found from the *cumulative binomial table* or from the *normal approximation to the binomial*. The latter technique is particularly useful when $n > 25$.

Exercises

1 (Section 9.1) A binomial test of the two hypotheses H_0: $p = 0.4$ and H_1: $p = 0.7$ is carried out. A sample of size 20 is used and the critical value of y is 10.5. What are the probabilities of Type I and Type II errors in this study? (Use Table A1.)

2 (Section 9.1) A binomial test of the two hypotheses H_0: $p = 0.8$ and H_1: $p = 0.6$ is carried out. A sample of size 25 is used. The test is to have $\alpha \leq 0.05$. Use Table A1 to answer the following questions.
   (a) What is the critical value of y for this test?
   (b) What are the acceptance regions for H_0; for H_1?
   (c) What is the value of β?

3 (Section 9.1) A handwriting expert claims that on the basis of a sample of handwriting, he can judge a person's sex with a 70% success rate (H_1). A skeptic thinks that the success rate is just at chance level, 50% (H_0). A test of the expert's skill is conducted on 15 samples of handwriting. The expert makes 11 correct identifications of sex. Make a decision between the two hypotheses. The test is to be conducted so that the probability of concluding that the expert really can judge a person's sex on the basis of handwriting, when he in fact cannot, is less than 1 in a 100.

4 (Section 9.2) Suppose that 60% of a certain population have a gene for red hair. What is the probability that in a random sample of 25 persons from this population,
   (a) At least 15 people will have the gene.
   (b) Between 10 and 16 persons, inclusive, will have the gene.
   (c) Exactly 10 persons will have the gene.
   (*Note:* Use Table A2 and show your work.)

5 (Section 9.2)
   (a) Find the probabilities in Question 1 using Table A2.
   (b) Find the critical value and β in Question 2 using Table A2.

6 (Section 9.3) A test of two hypotheses H_0: $\mu = 40$ and H_1: $\mu = 38$ is carried out using a sample of size 100. The population standard deviation is 3. The hypothesis H_1 is accepted for all sample means that are less than 39.4.
   (a) What are α and β for this test?
   (b) If the mean of 100 scores is 39.2, what conclusion do you draw?

7 (Section 9.3) A test of two hypotheses H_0: $\mu = 50$ and H_1: $\mu = 55$ is carried out using a sample of 64 subjects. The population standard deviation is 10. The probability of a Type I error is to be 0.05.
   (a) What is the critical value and what are the acceptance regions for this test? (Express in terms of $\bar{y}$.)
   (b) Express the critical value and the acceptance regions in terms of z_{obs}.
   (c) The sample has a mean of 51.3. What conclusion do you draw?

8 (Section 9.3) A program of dietary control for a group of 50 mentally retarded children is studied. Each child's IQ is determined twice, once before treatment and once after. The mean IQ difference (IQ after treatment minus IQ before treatment) is $\bar{y} = 3.20$. Test the hypothesis of no difference (H_0: $\mu = 0$) against the hypothesis that the program had an effect of 3 IQ points (H_1: $\mu = 3$). Assume that the population standard deviation of the difference scores is 6.0. Use $\alpha = 0.01$.

(a) Make the test using $\bar{y}$ as the test statistic.

(b) Make the test using z_{obs} as the test statistic.

(c) Compute the probability of a Type II error.

9 (Section 9.4) Find the two probabilities in Question 1 and the three probabilities in Question 4 by using the normal approximation to the binomial distribution.

10 HYPOTHESIS TESTING: FURTHER CONSIDERATIONS

The basic theory of hypothesis testing was developed in Chapter 9. In this chapter the theory is extended in several ways. Whereas in Chapter 9 both hypotheses were specific, here one of the hypotheses is specific but the other is composite, which means that more than one value of the parameter is included in the hypothesis. The specific hypothesis is called the null hypothesis and the composite hypothesis is called the alternative hypothesis. The reason for introducing this change is that in many situations it is not possible to state two specific hypotheses, as required by the theory of Chapter 9; however, it usually is possible to state a null and an alternative hypothesis. The change in the nature of the hypotheses has little effect on the procedure used to test them.

The test of a null hypothesis against an alternative hypothesis is often called a significance test. Several ways to report the result of such a significance test are described, and the distinction between the statistical significance of a result and its scientific significance is emphasized.

The theory of hypothesis testing as described in Chapters 9 and 10 leads to the division into two regions of the values of the test statistic. We decide between the two hypotheses by noting into which region the observed value of the test statistic falls. A variation on this technique is to compute a quantity called the "p-value" and to base the decision between the two hypotheses on it. The advantages and disadvantages of using the p-value are discussed.

Finally, a new test, the z-test of two independent means, is described.

10.1 The Null Hypothesis and a Composite Alternative Hypothesis

In Chapter 9 the hypothesis H_0 was tested against a *specific* hypothesis H_1 (for example, H_0: $p = 0.5$ versus H_1: $p = 0.7$; or H_0: $\mu = 100$ versus H_1: $\mu = 90$). In this chapter the first hypothesis, H_0, is tested against a *composite* hypothesis (for example, $p > 0.5$; or $\mu < 100$). *In a composite hypothesis the parameter has an interval of possible values*, rather than a specific value. This section shows that there is no change in the acceptance regions when we change the value of the parameter in the second hypothesis, H_1. Therefore, there would be no change in the acceptance regions if we consider H_1 to be composed of an interval of values. The terms "null hypothesis" and "alternative hypothesis" are then introduced, and two examples are given.

Acceptance Regions Do Not Depend on H_1

Consider a test of two specific hypotheses, H_0 and H_1, for given values of α and n. From this information the acceptance regions for H_0 and H_1 can be worked out. Now suppose that a different value of the parameter in H_1 is considered, keeping H_0, α, and n unchanged. It is a fact that the acceptance regions do not depend on the specific value of the parameter in H_1. This statement is true as long as the value of the parameter in H_1 stays on the same "side" of the value of the parameter in H_0. Hence, we can think of the parameter in H_1 as having an interval of possible values: in other words, the hypothesis is composite.

The property that the acceptance regions for H_0 and H_1 do not depend on the specific value of the parameter in H_1 is illustrated in Figures 10.1 and 10.2. Both figures illustrate a binomial test, but the property also applies to the z-test of one mean. In Figure 10.1, H_0 is $p = 0.5$ and H_1 is either $p = 0.6$ (in Figure 10.1b) or $p = 0.8$ (in Figure 10.1c). If we denote the values of the parameters in H_0 and H_1 by p_0 and p_1, respectively, then in this figure, p_1 is always greater than p_0. We can say that in this example, p_1 stays on the same side of p_0.

In Figure 10.1 the critical value of y is 14.5 for $\alpha \leq 0.05$ and $n = 20$. Since the critical value is chosen to make the probability in the tail of the H_0 distribution less than or equal to 0.05, the critical value does not depend on the H_1 distribution as long as the H_1 distribution is located at higher values than the H_0 distribution—which it is, since $p_1 > p_0$.

When the parameter in H_1 changes, there is one change in the test of H_0 versus H_1—the probability of a Type II error, β, changes. As shown in Figure 10.1, as p_1 increases, β decreases.

Figure 10.2 provides a second illustration of the unchanging acceptance regions; here, the value, p_1, of the parameter in H_1 changes but remains less than p_0. The acceptance region for H_1 is in the lower tail of the H_0 distribution, since the H_1 distribution is located at lower values than the H_0 distribution.

Null and Alternative Hypotheses

Since the acceptance regions do not depend on the specific value of the parameter in H_1 (except for the side on which the parameter falls), we can consider the parameter in H_1 as having an interval of possible values. In Figure 10.1, H_1 can be considered to be $p > 0.5$, a composite hypothesis. We are testing a specific hypothesis, $p = 0.5$, against a composite hypothesis, $p > 0.5$. The acceptance regions are the same whatever the particular value of p in the second hypothesis, provided that p is greater than 0.5.

By convention, when a specific hypothesis is tested against a composite hypothesis, the specific hypothesis is called the *null hypothesis* (H_0) and the composite hypothesis is called the *alternative hypothesis* (H_1). These names serve to remind us that we are no longer testing one specific hypothesis against another specific hypothesis, even though we do not change the symbols denoting the hypotheses.

The null hypothesis is just another name for H_0. The term "null" is used because in later work the value of the parameter in the null hypothesis often is zero. (One meaning of "null" is "zero".) In the binomial test the null hypothesis often is $p = 0.5$: if the null hypothesis is true, there is an equal chance of success and failure. Example 1 (see below) shows how such a hypothesis arises naturally in an experimental situation; however, other values of p can serve as a null hypothesis value. In the z-test of one mean the null hypothesis value of μ sometimes is determined from a previous standardization of a variable: for example, if we wanted to determine whether or not a particular population has the same mean IQ ($= 100$) as the population on which the IQ test was standardized, the null hypothesis would be $\mu = 100$.

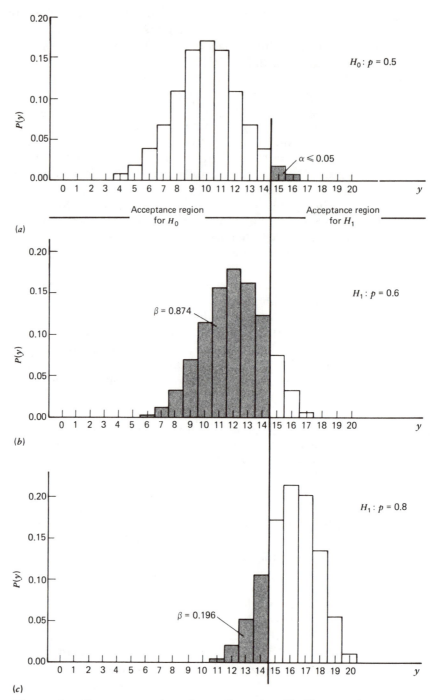

Figure 10.1 Acceptance regions do not depend on H_1 (if $p_1 > p_0$).

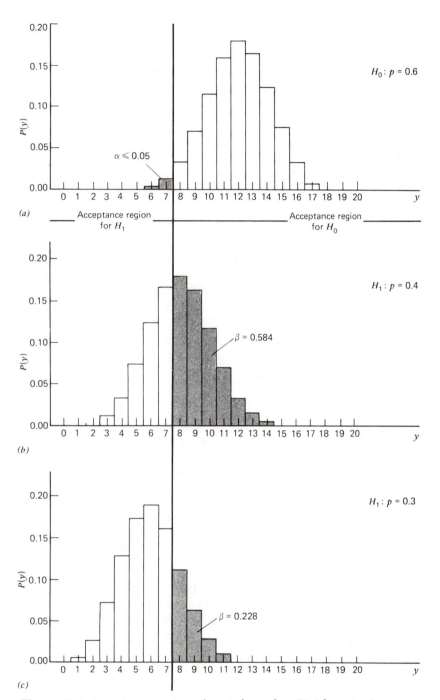

Figure 10.2 Acceptance regions do not depend on H_1 (if $p_1 < p_0$).

The alternative hypothesis is so named because it is an "alternative" possibility to the null hypothesis. For the present, we will consider only a *one-sided* alternative hypothesis: that is, the values of the parameter specified by the alternative hypothesis are on one side of (greater than) or on the other side of (less than) the null hypothesis value of the parameter, but not on both sides. Two-sided hypotheses, such as $p \neq 0.5$, will be discussed in Section 10.3.

Acceptance Regions and the Critical Value

How is the test of the two hypotheses, H_0 and H_1, made? Essentially, there is no change from the procedures followed for a test of two specific hypotheses, H_0 and H_1. The same *test statistic* and the same acceptance regions are used. The boundary between the two regions is called the *critical value* of the test statistic, as before.

In reading the following examples, note how the null and alternative hypotheses are determined from the description of each example. Also note that, as before, the acceptance regions are specified in terms of a *test statistic*, which is simply the statistic used to make a decision between two hypotheses. In the binomial test (Example 1), the test statistic is y, the number of successes. In the z-test of one mean (Example 2), the test statistic is $\bar{y}$, the sample mean; or, alternatively, z_{obs}, the sample mean converted to a z-score.

Example 1

Polygraphs, or lie detectors, are sometimes used to determine whether or not criminal suspects are telling the truth. There is considerable controversy about polygraphy. Let us design a very simple test of the effectiveness of the polygraphic technique.

We obtain polygraph records from 20 suspects, 10 of whom are known by other evidence to be lying and 10 of whom are known to be telling the truth. After deleting from them any information about the known truthfulness or untruthfulness of each suspect, we place these 20 records in a random order. We then call in a trained polygraph operator to interpret each record and to judge whether or not the record indicates that the suspect was lying. [In an actual study, it would be sensible, of course, to have a number of operators do the ratings. Also, in order for the test statistic to have a binomial distribution, we must make the (probably unreasonable) assumption that the operator's 20 ratings are independent of each other—see Section 6.3.]

We then compare the operator's judgments with the actual known status of the suspects. We count how many correct judgments the operator makes, out of the 20 records; the number of correct judgments is called y. If y is large, we will accept the hypothesis that the operator can discriminate between the records of truthful and untruthful suspects; if y is not large, on the other hand, we will accept the hypothesis that the operator cannot make the discrimination. In other words, the decision between the two hypotheses is based on y, and hence y is the test statistic.

Let us express the two hypotheses in terms of a parameter. The operator has a certain probability, p, of being correct about a given record. If the operator cannot make the discrimination between the two kinds of records, he is, in effect, guessing. For each record the probability of his being correct is 0.5 (and the probability of his being incorrect is 0.5). This hypothesis is the *null hypothesis*, H_0: $p = 0.5$. If the operator can discriminate in any way between the records, the probability of his being correct is greater than 0.5. Hence, the *alternative hypothesis* is H_1: $p > 0.5$. Notice that we do not require that he be able to make perfect discrimination ($p = 1.0$).

Suppose that we are skeptical about the polygraphic technique and will require strong evidence of its effectiveness before concluding that it actually is effective. In the language of hypothesis testing, this means that we want the probability of a Type I error, α, to be very small—say, less than or equal to 0.01. In other words, we want there to be only a very small chance of our concluding that the operator can discriminate between the two kinds of records when in fact he cannot.

The sample size, $n = 20$, has already been stated. We now have all the information necessary to determine the critical value of y and the acceptance regions for the two hypotheses. The distribution of the test statistic, y, when H_0 is true, is called "the H_0 distribution of y" (or simply "the H_0 distribution" when no ambiguity results). This is a binomial distribution with $n = 20$ and $p = 0.5$. The H_1 distribution of y cannot be specified exactly, but since $p > 0.5$, the H_1 distribution is located at higher values than the H_0 distribution of y. Hence, the critical value of y lies in the upper tail of the H_0 distribution of y and the acceptance region for H_1 consists of values of y greater than this critical value. From the binomial table, we see that the critical value is 15.5, corresponding to a probability of a Type I error of 0.006—less than the required limit of 0.01. The acceptance region for H_0 is $(y < 15.5)$, and the acceptance region for H_1 is $(y > 15.5)$.

Suppose that in the actual test the operator makes 14 correct identifications. This result falls in the acceptance region for H_0, and we will conclude that $p = 0.5$ and that the operator cannot make the discrimination between the two types of records. (See Section 10.2 for further discussion of how to state the conclusion of this study.)

Example 2   A test of mechanical aptitude has a mean of 70 and a standard deviation of 10 in a standardization population. We want to use the test in another population. Does the test have the same mean in the new population? We expect the test to have either the same mean or a lower mean. In other words, we know that people in the new population will perform, at best, at an average score of 70, but may score lower.

We select a random sample of size 80 from the new population and find that its mean score is 67.4. What conclusion should we draw if we want a 0.05 probability of concluding that the new population has a different mean when it in fact does not?

From the given information the hypotheses are $H_0: \mu = 70$ and $H_1: \mu < 70$. Let us use z_{obs} as the test statistic. The acceptance region for H_1 lies in the lower tail of the H_0 distribution of z_{obs}, since the H_1 distribution is located at lower values than the H_0 distribution. For $\alpha = 0.05$, the critical value of z is -1.64. Hence, the acceptance region for H_0 is $(z_{obs} > -1.64)$ and the acceptance region for H_1 is $(z_{obs} < -1.64)$.

The observed value of the test statistic is

$$z_{obs} = \frac{\bar{y} - \mu_0}{\sigma/\sqrt{n}} = \frac{67.4 - 70}{10/\sqrt{80}} = -\frac{2.6}{1.118} = -2.33$$

Hence, z_{obs} falls in the acceptance region for H_1, and we conclude that $\mu < 70$. The test of mechanical aptitude appears to have a lower mean in the new population than it has in the standardization population. (See Section 10.2 for further discussion of how to state the conclusion of this study.)

Type I and Type II Errors   The terminology of Type I and Type II errors is exactly the same for tests of two specific hypotheses (H_0 and H_1) and for tests of a specific hypothesis (H_0) against a composite hypothesis (H_1).

A *Type I error* is made if H_1 is accepted when H_0 is in fact true.
A *Type II error* is made if H_0 is accepted when H_1 is in fact true.

The probabilities of these errors are α and β, respectively.

α is the probability of concluding that H_1 is true when H_0 is in fact true.
β is the probability of concluding that H_0 is true when H_1 is in fact true.

The probability of a Type I error, α, can be calculated from the critical value, since the H_0 distribution of the test statistic is known; conversely, the critical value can be determined from a specified value of α. On the other hand, the probability of a Type II error, β, *cannot* be determined in a test of null hypothesis against an alternative hypothesis. As shown in the beginning of this section, the value of β depends on the specific value of the parameter in H_1; since H_1, the alternative hypothesis, now includes an *interval* of possible values of the parameter, the value of β is indeterminate. Section 10.2 shows that we must change the way we report the result of a test of two hypotheses to take into account our inability to calculate β.

10.2 Reporting the Result of a Significance Test

Hypothesis testing as described in Section 10.1 is the testing of a (specific) null hypothesis against a (composite) alternative hypothesis. The critical value and the acceptance regions are chosen so that the probability of a Type I error is quite small. If the observed result falls in the acceptance region for H_1, we say that the result is *significant* or that the result is *significantly different* from the result that would be expected if the null hypothesis were true. The hypothesis test itself is often referred to as a *significance test*.

It is important to remember that the two hypotheses in a significance test have quite different characters. The null hypothesis is a *specific* hypothesis, and the probability, α, of making a wrong decision when the null hypothesis is true is always kept very small. On the other hand, the alternative hypothesis is a *composite* hypothesis, and we cannot even calculate the probability, β, of making a wrong decision when the alternative hypothesis is true, let alone be certain that β is small.

Because the two hypotheses differ in character, the significance test commonly is thought of in the following way. The object of the test is to disprove or *reject the null hypothesis* (that is, to accept the alternative hypothesis). If the observed test statistic is far out in the tail of the H_0 distribution, the null hypothesis is rejected; the result is significant. If the observed test statistic is not far out in the tail, on the other hand, the null hypothesis is not rejected; the result is not significant. *As the probability of a Type II error, β, is not known to be small, we cannot say that the null hypothesis is accepted.*

When a significance test is thought of in this way, attention is focused on the null hypothesis. The acceptance region for H_1 is referred to as the *rejection region*. The acceptance region for H_0 is referred to simply as the *acceptance region*. In this new terminology and terms "rejection" and "acceptance" refer to the acceptance and rejection of the null hypothesis. If the test statistic falls in the rejection region, the decision is to reject the null hypothesis and accept the alternative hypothesis. If the test statistic falls in the acceptance region, on the other hand, the decision is to accept the null hypothesis; or rather, as pointed out above, the decision is usually to *not reject* the null hypothesis. (In certain circumstances, it is possible to accept the null hypothesis—see Section C.4.)

We can summarize the above discussion as follows.

	Acceptance region for H_0	Acceptance region for H_1
Former name		
New name	Acceptance region	Rejection region
Decision	Fail to reject H_0 (Sometimes, accept H_0)	Reject H_0 and accept H_1

Let us consider how the results of Examples 1 and 2 of the previous section would be reported using the new terminology. In Example 1 the hypotheses are

H_0: $p = 0.5$ and H_1: $p > 0.5$. The acceptance region is ($y < 15.5$) and the rejection region is ($y > 15.5$) for $\alpha \leq 0.01$. The observed result of 14 correct identifications out of 20 falls into the acceptance region. We conclude that the hypothesis that $p = 0.5$ cannot be rejected: there is not sufficient evidence to conclude that the operator can discriminate between the two kinds of records.

In Example 2 the hypotheses are H_0: $\mu = 70$ and H_1: $\mu < 70$. The acceptance region is ($z_{obs} > -1.64$) and the rejection region is ($z_{obs} < -1.64$) for $\alpha = 0.05$. In the study the test statistic ($z_{obs} = -2.33$) falls into the rejection region; hence, we conclude that the hypothesis that the mean is 70 can be rejected in favor of the hypothesis that the mean is less than 70. On the evidence of this study, the test of mechanical aptitude has a lower mean in the new population than it has in the standardization population.

Other Ways to Report the Result

Put simply, the purpose of a significance test is to get a significant result: that is, a result in the rejection region. A significant result also can be referred to as a *statistically significant result* or as a result that is significant at the 0.05 level (if $\alpha = 0.05$). We can also use an expression such as "the null hypothesis was rejected at the 0.05 level of significance". In Example 2 of Section 10.1 the sample mean is 67.4 and H_0 is 70; this result may be reported as "the sample mean was significantly less than the hypothesized mean". In other studies the observed result may be significantly more than the hypothesized value. Significant results are also reported as reliable or statistically reliable results.[1]

There also are many ways to report nonsignificant results. We may say that a result is not significant at the 0.05 level (0.01 level, etc.), or that the result is not significantly less than or not significantly more than a hypothesized value. Or the result may be described as not reliable. Another way to report a nonsignificant result is to say that the result is at chance level. By definition, a nonsignificant result falls into the acceptance region; such a result has a high chance of occurring if the null hypothesis is true. Hence, such a result is a chance result. [In Example 1 of Section 10.1, there were 14 successes in 20 tries. If the null hypothesis ($p = 0.5$) is true, the expected (mean) number of successes is $20 \times 0.5 = 10$. Since $y = 14$ falls into the acceptance region, we say that the result is at chance level.]

There are many other ways to report the result of a significance test. A summary of those mentioned above follows:

FAIL TO REJECT H_0	REJECT H_0
Nonsignificant result	Significant result
Result is not statistically significant	Result is statistically significant (statistically significant result)
Not significant at the 0.05 level (0.01 level, etc.)	Significant at the 0.05 level (0.01 level, etc.)
Not significantly less (more) than	Significantly less (more) than
Result is not reliable	Result is reliable (reliable result)
Result is not statistically reliable	Result is statistically reliable (statistically reliable result)
Result is at chance level	

[1] A significant result can occur only rarely if the null hypothesis is true, since the probability of such a result is α. By accepting the alternative hypothesis when a significant result is obtained, we are claiming that what occurred was not a rare chance, but that the alternative hypothesis is true and that if we repeated the experiment we would get a similar result. Hence, the result is reliable.

Statistical Significance and Scientific Significance

The goal of many research studies is to obtain a statistically significant result. Such a result means only that we can be reasonably sure that the null hypothesis is false. However, a statistically significant result is just that—a *statistical* result. The result applies only to the particular variables and to the particular population studied.

The variables in which psychologists and other scientists are interested are theoretical variables, not the actual measures in an experiment or survey. And the populations in which they are interested are difficult to sample from. We can say that psychologists are looking for *scientific significance*, not just statistical significance, in their work.

The distinction between these two kinds of significance can be stated generally as follows.

P9: Statistical Significance and Scientific Significance

The statistical significance of an observed result indicates which of two statistical hypotheses about a population is supported by the result, given a certain probability of a Type I error. The statistical hypotheses are expressed in terms of the variables measured in the study. The scientific significance of a result goes beyond a simple report of statistical significance. The justification of scientific significance requires much more than just statistical significance. The scientific significance is expressed in terms of more general or theoretical variables and may take into account the results of other studies.

Consider again Example 1 of Section 10.1, a study of the effectiveness of polygraphs. A psychologist working in this field would be interested in the relationship between mental variables such as "feeling of guilt" and physiological variables such as "electrical skin resistance". Since polygraph records are interpreted by an operator, the psychologist also would have to consider variables that are related to the ability of the operator to detect differences between the records of those who are lying and those who are telling the truth.

All the variables mentioned in the last paragraph are theoretical and quite different from the variable actually measured in the study. The measured variable indicates, for each record, whether the operator was correct or incorrect. Since this variable is so different from the theoretical variables, it is difficult to generalize from the result of one study on the effectiveness of a polygraph operator to any statement about the theoretical variables. A statistically significant result in the one study may assure us that the operator in that study is effective, but it tells us little more.

It is also clear that it is impossible to generalize the results of the study of one operator to the population of all possible operators. The only sampling done in this study is the drawing of a sample of records from the population of possible records—and even this sampling is difficult to carry out rigorously.

Principle P9 states that the scientific significance of a study may take into account the results of other studies. A statistically significant result in the study of a one polygraph operator may add to previous knowledge about the variables involved in polygraphy and allow the psychologist to make further claims about the theory. These claims would not be based solely on the statistical significance of the one result, but would be based on a series of studies.

In Example 2 of Section 10.1 a random sample was drawn from a new population to determine if the mean score on a test of mechanical aptitude is the same in the new population as it is in a standardization population. The observed mean is significantly lower than the mean in the standardization population. The result is statistically significant, but is it scientifically significant? The answer to this question would depend on the purpose of the study. Was the study designed to show that the test is biased in the new population? Or was the study designed to show that the measurement is not of mechanical aptitude? Considerations of this sort relate to the scientific significance of the result.

Consider two ways in which the results of the study in Example 2 could be reported: (1) the mean of the aptitude test is significantly lower in the new population than in the standardization population ($\alpha = 0.05$); (2) the aptitude test is a biased measure of mechanical aptitude for persons in the new population. The first statement is a direct statement of statistical significance: the level of significance is indicated. The reader of this report knows that the result is statistically reliable. The second statement is less statistical and goes beyond the first. This statement indicates that the theoretical variable of mechanical aptitude is not measured satisfactorily by the particular aptitude test used in the study.

10.3 Two-sided Alternative Hypotheses and Two-tailed Tests

In many situations, we may be able to specify the null hypothesis but not the direction in which the parameter will lie if the null hypothesis is false. In such situations, we can include in the alternative hypothesis *all* values of the parameter except the null-hypothesis value; such an alternative hypothesis is called a *two-sided hypothesis*, or a *nondirectional hypothesis*. For example, if H_0 is $p = 0.5$, the two-sided alternative hypothesis is H_1: $p \neq 0.5$. From the point of view of the previous section, the aim of a significance test is to disprove the null hypothesis. By using a two-sided instead of a one-sided alternative hypothesis, we are not commiting ourselves to the direction in which the parameter falls if the null hypothesis is false.

When H_1 is two-sided, the null hypothesis is rejected for both small and large values of the test statistic. Hence, the rejection region consists of two intervals of values, one interval in the lower tail of the H_0 distribution of the test statistic and the other interval in the upper tail of the H_0 distribution. The test is therefore called a *two-tailed test*. Recall that when H_1 is one-sided, the rejection region is in either the upper or the lower tail (but not in both), and the test is a *one-tailed test*.

Example 1 illustrates a two-sided alternative hypothesis for a test of a proportion. Example 2 illustrates a two-sided alternative hypothesis for a test of one mean. The section concludes with a discussion of the relative merits of one- and two-sided alternative hypotheses.

Example 1

We attempt to determine whether or not a coin is fair by tossing it 20 times. The probability that heads turns up on a single toss is p. Since we do not have a particular direction in mind for the coin's bias, a value of p less than 0.5 and a value of p greater than 0.5 are both reasonable alternatives to H_0. The hypotheses are, therefore, H_0: $p = 0.5$ (the coin is fair) and H_1: $p \neq 0.5$ (the coin is biased).

Figure 10.3 shows the distribution of the statistic y, the number of heads, when H_0 is true. We cannot show the precise distribution of y when H_1 is true, since H_1 is a

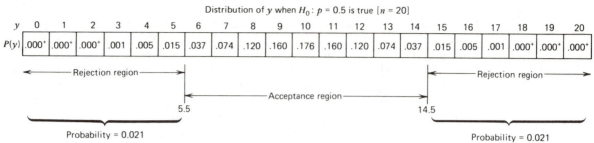

Figure 10.3 A two-tailed binomial test ($\alpha \leqq 0.05$).

composite hypothesis. However, when $p > 0.5$ the distribution of y will be concentrated at large values of y and when $p < 0.5$, the distribution of y will be concentrated at small values of y. Given these general locations of the distribution of y when H_1 is true, we should accept H_1 (and reject H_0) if the observed value of y is either very large or very small. These two intervals of values of y form the rejection region for the test.

The figure shows the acceptance and rejection regions for a 0.05 level test. You can check that the probability that y falls in the rejection region when H_0 is true, is 0.042. Note that the probability of each tail is 0.021. However, the probability of a Type I error is the probability of rejecting H_0 when it is true; since H_0 is rejected for both small and large values of y, we find α by adding the probabilities of the two tails: $\alpha = 0.021 + 0.021 = 0.042$.

The critical values, 5.5 and 14.5, can be found as follows. If α is to be less than or equal to 0.05, the probability of each tail must be less than or equal to $0.05/2 = 0.025$. We look in Table A1 or Table A2 for the largest lower-tail probability not exceeding 0.025. In the present example, the lower-tail probability is 0.021 for a critical value of 5.5. One tail of the rejection region is, therefore, ($y < 5.5$). The upper critical value can be found by looking for the largest upper-tail probability not exceeding 0.025 ($y = 14.5$ corresponds to a probability of 0.021). Hence, the upper tail of the rejection region is ($y > 14.5$). The values of y that are not in either tail form the acceptance region: ($5.5 < y < 14.5$).

We toss the coin 20 times and get 12 heads; this result is in the acceptance region. We fail to reject the hypothesis that the coin is fair. In other words the observed result, $y = 12$, is not significantly different from the expected number, 10, when H_0 is true. (Note that terms such as "not significantly greater than", "not significantly less than" usually are changed to "not significantly different from" when a two-sided alternative hypothesis is used.)

On the other hand, if the observed number of heads is 18, which falls in the rejection region, we can reject the hypothesis that the coin is fair and conclude that the coin is biased. The result, $y = 18$, is significantly different from the expected number, 10, when H_0 is true.

Example 2

Two-tailed tests are also appropriate for tests of one population mean. If we want to test $H_0: \mu = 100$ against $H_1: \mu \neq 100$, the rejection region for $\alpha = 0.01$ can be found as follows. If we use z_{obs} as the test statistic, we must find the critical value of z, called z_{crit}, such that $P(z_{obs} > z_{crit}) + P(z_{obs} < -z_{crit}) = 0.01$.

Note that the rejection region includes the two tails of the normal distribution. From the symmetry of the distribution it is clear that the tails have equal probabilities: that is, that each tail has a probability of $0.01/2 = 0.005$. The value of z_{crit} that cuts off a probability of 0.005 in the upper tail is 2.58 (found from the normal table, Table A3 or A4). Hence, the rejection region consists of the two intervals: ($z_{obs} < -2.58$) and ($z_{obs} > 2.58$). See Figure 10.4. The rejection region can be stated more briefly as ($|z_{obs}| > 2.58$).

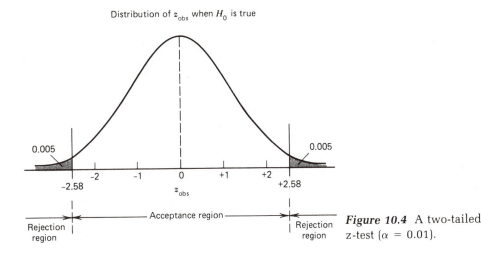

Figure 10.4 A two-tailed z-test ($\alpha = 0.01$).

The sample mean is 101.84 for a random sample of size 36. Assuming a population standard deviation of 5, the test statistic is

$$z_{obs} = \frac{\bar{y} - \mu}{\sigma/\sqrt{n}} = \frac{101.84 - 100}{5/\sqrt{36}} = \frac{1.84}{0.833} = 2.21$$

This value is not in the rejection region, and therefore the sample mean of 101.84 is not significantly different from the hypothesized mean of 100.

Should H_1 Be One- Or Two-sided?

In most situations the appropriate H_1 is nondirectional, or two-sided; therefore, the test should be two-tailed. A directional or one-sided H_1 can only be justified in a few, special situations; in these rare cases, a one-tailed test may be used.

Two-sided Alternative Hypothesis In many studies in psychology and other social sciences the hypotheses are not clearly formulated before the study is carried out, and the researcher usually will be pleased with a significant result in either direction. Both tails of the distribution of the test statistic must be included in the calculation of the probability of a Type I error. Hence, a two-tailed test is appropriate and the alternative hypothesis is two-sided.

Two-tailed tests can also be justified on the grounds that they are less sensitive than one-tailed tests to the *skewness* of the population distribution. The requirements for the validity of statistical tests are discussed in detail in Section C.3. One of the requirements of the z-test is that the population distribution is normal; when the distribution is skewed, however, the z-test is still quite satisfactory, provided that a two-tailed, not a one-tailed, test is used.

One-sided alternative Hypothesis The use of a one-sided H_1 and the corresponding one-tailed test can be justified in two situations. First, if a one-sided hypothesis is based on a well-worked-out theory that has some status in the literature, it may be used. The hypothesis would not be revised if the result of one experiment was in the "wrong" direction; the result would simply be interpreted as evidence for the null hypothesis. Second, if it is essentially impossible to obtain a result in one direction, the alternative hypothesis may be one-sided, in the other direction. For example, suppose a safety campaign is launched. It is hoped that the campaign will reduce the mean number of accidents per day; it is hard to imagine that the campaign would *increase* the mean number of accidents. The null hypothesis states that the mean is unchanged from its previous value; the alternative hypothesis states that the mean is less than the previous value—a one-sided hypothesis.

10.4 Summary of Steps in Hypothesis Testing

We have discussed several modifications to the basic approach to hypothesis testing introduced in Chapter 9: (1) the alternative hypothesis, H_1, is composite; (2) H_1 is either one- or two-sided; (3) the result of a significance test may be reported in one of several ways. The following list summarizes the eight steps to follow when carrying out a test of two hypotheses, H_0 and H_1. These steps correspond closely to those listed (for tests of two specific hypotheses) in Sections 9.1 and 9.3.

1a State two hypotheses: H_0 and H_1. The null hypothesis is a specific hypothesis about the population parameter. The alternative hypothesis is a composite hypothesis that is either one- or two-sided (usually two-sided). (*Example:* H_0: $\mu = 100$ and H_1: $\mu \neq 100$.)

1b Record the value of any additional parameter that must be known for the test to be carried out. This step is omitted for the binomial test but is required for the z-test of one mean. (*Example:* $\sigma = 5$.)

2 Choose a value for α. This value usually should be quite small. The value of α should be stated as an inequality if the distribution of the test statistic is discrete (as in the binomial test) and as an equality if the distribution is continuous (as in the z-test). (*Example:* $\alpha = 0.01$.)

3 Choose the sample size. (*Example:* $n = 36$.)

4 Find the critical value(s) of the test statistic. When H_1 is one-sided, there is one critical value selected, so that the probability in one tail of the H_0 distribution of the test statistic satisfies the restrictions on α. When H_1 is two-sided, there are two critical values selected, so that the sum of the probabilities in the two tails satisfies the restriction on α. (*Example:* critical values of z are $+2.58$ and -2.58.)

5 Locate the acceptance and rejection regions. [*Example:* acceptance region is $(-2.58 < z_{obs} < +2.58)$ and rejection region consists of the two intervals: $(z_{obs} < -2.58)$ and $(z_{obs} > +2.58)$.]

6 There is no step 6! This step is retained in the list as a reminder that when two specific hypotheses are tested, the probability of a Type II error, β, can be calculated. When a composite H_1 is tested against H_0, however, β cannot be calculated. We will return to this topic in Chapter 13.

7 Carry out the study and record the value of the test statistic. (*Example:* $z_{obs} = 2.21$.)

8 Make the decision, based on whether the test statistic falls into the acceptance or rejection region. There are various ways to state the conclusion—see Section 10.2. (*Example:* H_0 cannot be rejected; the sample mean does not differ significantly from the hypothesized mean of 100.)

10.5 The *p*-value   When the steps summarized in the preceding section are followed, the probability that the test statistic falls in the rejection region, if H_0 is true, is less than or equal to the prescribed α. Once the rejection region has been located, it is a very simple matter to compare the observed value of the test statistic with the rejection region and to make the appropriate decision: reject H_0, if the test statistic falls into the rejection region; fail to reject H_0, if it does not. This procedure may be called the *rejection region procedure.*

A slightly different procedure, called the *p-value procedure*, may be followed. In all instances, this procedure leads to exactly the same decision as the rejection region procedure. The principal reason for introducing the p-value procedure is that in some situations the p-value procedure is easier to apply than the rejection region procedure. However, there are some dangers in using the p-value procedure; these are explained at the end of the section.

In the p-value procedure it is not necessary to calculate the critical values (step 4 of the preceding section) or the rejection region (step 5). The p-value[2] is computed from the observed value of the test statistic by the following definition.

> The *p-value* of an observed value of a test statistic is the probability of that value plus the probability of more-extreme values. These probabilities are calculated from the H_0 distribution of the test statistic. "More-extreme" means "in the direction of the H_1 distribution".

The computed p-value is compared with the prescribed value of α. The decision is made according to the following rule.

- Reject H_0 if the p-value is less than or equal to α.
- Fail to reject H_0 if the p-value exceeds the prescribed α.

Figure 10.5 illustrates why the p-value procedure leads to the same decision as the rejection region procedure. The acceptance and rejection regions for a certain value of α are shown on the graph of the H_0 distribution of the test statistic. When the test statistic falls in the rejection region (Figure 10.5*a*), the p-value, which is the area in the tail beyond the observed value of the test statistic, is less than α. [Note that the H_1 distribution (not shown) must be located at higher values than the H_0 distribution, since the rejection region is in the upper tail of the H_0 distribution. Hence, more-extreme values of the test statistic (those in the direction of the H_1 distribution) are also in the upper tail of the H_0 distribution.] Similarly, when the test statistic falls in the acceptance region (Figure 10.5*b*), the p-value is greater than α.

The p-value procedure is illustrated here by three examples. In Example 1 the p-value procedure is applied to a binomial test with a one-sided alternative hypothesis. This example is based on Example 1 of Section 10.1, in which the rejection region procedure was used; a comparison of the two examples will show that the two procedures must give the same decision. In Example 2 the p-value procedure is applied to a z-test with a two-sided alternative hypothesis. This example is based on Example 2 of Section 10.3, and again the two

---

[2] Be sure to distinguish the p-value from p, the population probability of success in a binomial study. The two are quite different.

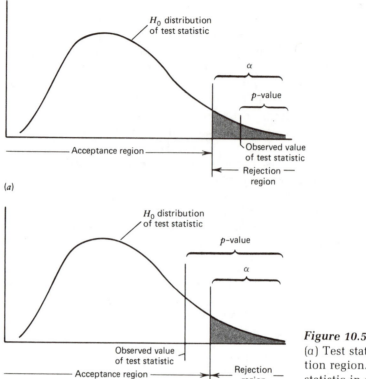

(a)

(b)

Figure 10.5 p-value.
(a) Test statistic in rejection region. (b) Test statistic in acceptance region.

procedures may be compared. Finally, in Example 3 the choice between one- and two-tailed p-values is discussed.

Example 1

In the polygraph study (Example 1 of Section 10.1) the hypotheses are $H_0: p = 0.5$ and $H_1: p > 0.5$. The sample size is 20 and $\alpha \leq 0.01$. The observed value of the test statistic is $y = 14$.

The p-value is computed from the H_0 distribution of y, a binomial distribution with $n = 20$ and $p = 0.5$. The p-value is the probability of the observed value, $y = 14$, plus the probability of more-extreme values. Since the H_1 distribution of y is concentrated at large values of y, more-extreme values are values greater than 14. Hence, the p-value $= P(y = 14) + P(y > 14) = P(y \geq 14) = P(y > 13.5)$. This probability is 0.058 (Table A1 or A2). A p-value computed in this way is referred to as a *one-tailed p-value*, since it is found from one tail of the H_0 distribution. It is appropriate to use only one tail because H_1 is one-sided. (A two-tailed p-value is illustrated in the next example.)

The decision between the two hypotheses is made by comparing the p-value with 0.01. Since the p-value exceeds 0.01, we cannot reject H_0; we conclude that there is no evidence that the operator can discriminate between the two groups of polygraph records. It is common in reports of the result of significance tests to add the p-value to the conclusion. This may be done in one of two ways. The exact p-value may be given: for example,

(i) There is no evidence that the operator can discriminate between the two kinds of polygraph records (p-value = 0.058).

Or the p-value may be incorporated in an inequality that indicates the level of significance (α) at which the test was carried out: for example,

(ii) There is no evidence that the operator can discriminate between the two kinds of polygraph records (p-value > 0.01).

In published reports, "p-value" is usually abbreviated as "p": for example, ($p = 0.058$). This abbreviation is not used in this text, because the p-value could be confused with the parameter p.

How is the p-value reported when H_0 is rejected? If the observed value of y is 16, instead of 14, the p-value $= P(y \geq 16) = P(y > 15.5) = 0.006$. Two ways to report the conclusion are

(iii) The operator can discriminate reliably between the two kinds of polygraph records (p-value $= 0.006$).

(iv) The operator can discriminate reliably between the two kinds of polygraph records (p-value < 0.01).

Note carefully that H_0 is rejected when the p-value is less than the prescribed α. In conclusion **ii**, on the other hand, H_0 was not rejected, since the p-value is greater than the prescribed α.

Example 2

Consider the hypotheses H_0: $\mu = 100$ and H_1: $\mu \neq 100$, as in Example 2 of Section 10.3. The other specifications are $\sigma = 5$, $n = 36$, and $\alpha = 0.01$. The sample mean, 101.84, is converted to a test statistic, $z_{obs} = 2.21$, as shown before.

What is the p-value of the test statistic? Recall that the p-value is the probability of the observed value plus the probability of more-extreme values. The probability that z_{obs} is exactly equal to 2.21 is zero, since z_{obs} is a continuous variable (Section 7.2). Values of z_{obs} greater than 2.21 are certainly more extreme than 2.21, and these values are in the direction of the H_1 distribution if μ is greater than 100. But μ may also be less than 100. Values of z_{obs} less than -2.21 are considered to be just as extreme as values greater than $+2.21$. The p-value is the probability of two tails of the H_0 distribution and is therefore called a *two-tailed p-value*. In this example,

$$
\begin{aligned}
p\text{-value} &= P(z_{obs} > 2.21) + P(z_{obs} < -2.21) \\
&= 2 \times P(z_{obs} > 2.21) \\
&= 2 \times (0.5 - 0.4864) \qquad \text{(from Table A3)} \\
&= 2 \times 0.0136 \\
&= 0.0272
\end{aligned}
$$

Since the normal distribution is symmetric, the two-tailed p-value is simply twice the probability in the one tail beyond the observed value of the test statistic. The p-value is rounded *up* to 0.03. (Note that p-values usually are rounded up, in order not to make them smaller than they really are—statistical procedures are often applied conservatively.)

Since the p-value is greater than 0.01, we cannot reject H_0. The conclusion could be reported in either of the following ways.

(i) The sample mean is not significantly different from 100 (p-value $= 0.03$).

(ii) The sample mean is not significantly different from 100 (p-value > 0.01).

In order to make it clear that a two-tailed p-value was computed, the parenthesized expression is often extended as follows: (p-value $= 0.03$, two-tailed) or (p-value > 0.01, two-tailed).

If the p-value turned out to be less than 0.01, the sample mean would be significantly different from 100. Its exact value, or the inequality (p-value < 0.01), would then be reported.

One- and Two-tailed *p*-values

As shown in Examples 1 and 2, a one-tailed p-value is computed when H_1 is one-sided and a two-tailed p-value is computed when H_1 is two-sided. The *one-*

tailed p-value is the probability of the observed value of the test statistic plus more-extreme values in the tail of the distribution.[3] The *two-tailed p-value* includes the probability of the values in both tails that are as extreme or more extreme than the observed value of the test statistic. Since the H_0 distribution of the test statistic usually is symmetric, the probabilities of extreme values in the two tails are the same. The two-tailed p-value therefore, can be easily obtained by doubling the one-tailed p-value.

> Two-tailed p-value = $2 \times$ one-tailed p-value

We will use this formula even when the H_0 distribution is not symmetric (for example, in the binomial test when H_0 is $p = 0.6$).

Example 3	In order to compare two methods of training, 24 pairs of subjects are selected. One member of each pair is trained by method A and the other member is trained by method B. After training, each subject is given a test and it is recorded for each pair whether the A member or the B member did better on the test. If there is no difference between the methods, the probability that the A member of a pair will do better on the test is 0.5; *while* if there is a difference between the methods, this probability will be different from 0.5. So the hypotheses are H_0: $p = 0.5$ and H_1: $p \neq 0.5$.

We count the number of pairs in which the A member does better, and call this number y. Under the null hypothesis, y will be binomially distributed with $p = 0.5$ and $n = 24$. The A member does better than the B member in 17 out of the 24 pairs: that is, $y = 17$. Since the alternative hypothesis is two-sided, we compute a two-tailed p-value, as follows.

The two-tailed p-value is found from Table A1 to be

$$
\begin{aligned}
\text{p-value} &= 2 \times P(y \geqq 17) \\
&= 2 \times \{P(y = 17) + P(y = 18) + \cdots + P(y = 24)\} \\
&= 2 \times \{0.021 + 0.008 + 0.003 + 0.001 + 0.000 + 0.000 + \\
&\quad\quad 0.000 + 0.000\} \\
&= 2 \times 0.033 \\
&= 0.066
\end{aligned}
$$

A value of α has not been specified in this example. It is customary to assume that $\alpha = 0.05$, if a smaller value has not been specified. The calculated p-value is greater than 0.05. We therefore conclude that there is not a significant difference between the two methods of training (p-value > 0.05, two-tailed).

The *one-tailed* p-value in this example *is* less than 0.05. If H_1 had been $p > 0.5$ rather than $p \neq 0.5$, the result would have been significant. If we had been able to predict confidently that method A would be superior to method B, before the experiment was carried out, we could have used the one-sided alternative hypothesis and a one-tailed p-value. It would be incorrect, however, to decide on the direction of the one-sided alternative hypothesis after completing the experiment.

Dangers in the p-value Procedure

Because the decision made from the p-value procedure is exactly the same as the decision made from the rejection region procedure, we can use whichever procedure is most convenient. However, there are two dangers in the use of the p-value procedure: (1) the p-value procedure encourages the practice of setting

---

[3] This statement assumes that the observed value of test statistic is in the direction of the H_1 distribution.

α *after*, rather than *before*, the data are collected; and (2) the computed p-value is all-too-often erroneously interpreted as a measure of the effect size.

If α is specified, the decision based on a comparison of the sizes of the p-value and α is exactly the same as the decision based on whether or not the test statistic falls in the rejection region. However, it is difficult for experimenters to set α, as there is no completely objective basis on which to do it. For this reason the custom has developed of computing the p-value and, if it is small, reporting the conventional value of α (0.05, 0.01, etc.) that is just greater than the p-value. For example, if the p-value is 0.037, a value of α = 0.05 will be chosen and the result reported as (p-value < 0.05). Or if the p-value is 0.006, α will be given the value 0.01 and the result reported as (p-value < 0.01). The largest α usually allowed in this procedure is 0.05. (If the p-value is greater than 0.05, the result usually is reported as nonsignificant.) In brief, experiments are carried out with an implied value of α = 0.05, but if the p-value turns out to be much smaller than 0.05, the conclusion is reported as if α had been 0.01 or 0.005 or 0.001, etc. This procedure is clearly misleading—α has been specified after the data were collected, rather than before.

A further difficulty with the p-value procedure is that it encourages the belief that smaller p-values are better. This belief is false! Chapter 8 introduced the effect size, which is a measure of the relationship between two variables. If the relationship is strong, there is a large effect size; if the relationship is weak, there is a small effect size. The objective of many studies is to find a relationship—particularly, a strong relationship—between two variables. It might seem that the smaller the p-value is, the greater the effect size and therefore the stronger the relationship between the two variables. However, as shown in detail in Chapter 13, the p-value also depends on the size of the sample, n, and other characteristics of the study. Only if the sample size and all these other characteristics are fixed can we claim that the smaller the p-value, the greater the effect size. If the sample sizes in two studies are different, a smaller p-value in one of the studies does not, by itself, indicate that the effect size in that study is larger than the effect size in the other one.

For these reasons, the p-value procedure, although convenient, must be used with care.

10.6 The *z*-test of Two Independent Means

The z-test is used in a number of situations. If we want to know whether the mean of one population has a given value, we would use the z-test of one mean. This section presents two populations and two means. The z-test of two independent means determines whether or not the two means are equal. As the name indicates, z-tests are based on a test statistic z_{obs} that is normally distributed. These tests have the characteristic that one or more additional parameters must be known or assumed in order for the test to be carried out; *in most situations these parameters are not known, and therefore the z-test is not often used in practice*. Other tests, called t-tests (which do not require that additional parameters be known) have been developed; these tests are discussed in Chapter 11. The z-test of two independent means is introduced here because the theory of this test is simpler than the theory of the corresponding t-test; it is easy to understand the t-test after having studied the z-test.

The z-test of two independent means is used to determine whether or not two population means are equal. If the two means are denoted μ_1 and μ_2, then the null hypothesis is $H_0\colon \mu_1 = \mu_2$ and the alternative hypothesis is $H_1\colon \mu_1 \neq \mu_2$. This alternative hypothesis is two-sided. It is also possible to use a one-sided alternative hypothesis: that is, $H_1\colon \mu_1 > \mu_2$ or $H_1\colon \mu_1 < \mu_2$.

Two random samples are drawn, one from each population. Since the two samples are drawn *independently*, the test is referred to as a test of two *independent* means. Roughly speaking, two samples are independent if there is no matching or pairing of the individuals in the two samples; this means that there is no connection between the scores of the two samples. (The distinction between independent and dependent samples is discussed in detail in Section 11.3.)

Each of the two populations has a distribution of scores. These distributions have means, μ_1 and μ_2, and standard deviations, σ_1 and σ_2. The means are unknown—the purpose of the test is to see if they are equal. In order to make the test, however, we must know the standard deviations. (Recall that in the z-test of one mean, the population standard deviation, σ, had to be known.)

In order to have a statistical test of H_0 and H_1, we must have a test statistic. Obviously, we should compare the two *sample* means if we wish to determine whether the *population* means are the same or different. We therefore consider the statistic: $\bar{y}_1 - \bar{y}_2$, the difference between the sample means. Clearly, the greater the difference between the population means, the greater the difference between the sample means, on the average. But we must know the distribution of the statistic in order to know how big the difference must be to prove that the population means differ.

The distribution of $\bar{y}_1 - \bar{y}_2$ can be shown to have the following properties.

$$\text{mean} = \mu_1 - \mu_2$$

$$\text{standard deviation} = \sqrt{\frac{\sigma_1^2}{n_1} + \frac{\sigma_2^2}{n_2}}$$

shape: approximately normal (or exactly normal if the populations are normally distributed)

In the formula for the standard deviation, n_1 and n_2 are the sizes of the samples drawn from the first and second populations, respectively. (Note carefully that although the mean of $\bar{y}_1 - \bar{y}_2$ is the *difference* of the two population means, the standard deviation of $\bar{y}_1 - \bar{y}_2$ is the square root of the *sum* of two terms.)

We can convert $\bar{y}_1 - \bar{y}_2$ to a standard score by subtracting its mean and dividing by its standard deviation.

$$z_{\text{obs}} = \frac{(\bar{y}_1 - \bar{y}_2) - (\mu_1 - \mu_2)}{\sqrt{\dfrac{\sigma_1^2}{n_1} + \dfrac{\sigma_2^2}{n_2}}}$$

If $H_0\colon \mu_1 = \mu_2$, or $(\mu_1 - \mu_2) = 0$, is true, then the formula simplifies to the *test statistic*.

$$z_{\text{obs}} = \frac{\bar{y}_1 - \bar{y}_2}{\sqrt{\dfrac{\sigma_1^2}{n_1} + \dfrac{\sigma_2^2}{n_2}}}$$

This test statistic is approximately normally distributed with mean = 0 and standard deviation = 1. Tables A3 and A4 can be used to find the probabilities of this statistic. In particular, if we are testing a two-sided alternative hypothesis and want $\alpha = 0.05$, the rejection region consists of the two intervals: ($z_{obs} < -1.96$) and ($z_{obs} > +1.96$). In brief, H_0 would be rejected if $|z_{obs}| > 1.96$. The test is two-tailed. On the other hand, if a one-sided alternative hypothesis such as $H_1: \mu_1 < \mu_2$ is tested, then the one-tailed rejection region is ($z_{obs} < -1.64$).

Example

Suppose we wish to compare the physical ability of eight-year-old boys and girls. We draw random samples of boys and girls and ask each child to make as many jumps, standing at one place, as possible. We wish to compare the mean number of jumps of boys and girls. Let us follow the steps in hypothesis testing, summarized in Section 10.4, and modify them as needed for the z-test of two independent means.

1a The hypotheses are $H_0: \mu_1 = \mu_2$ and $H_1: \mu_1 \neq \mu_2$, where μ_1 is the mean number of jumps by boys and μ_2 is the mean number of jumps by girls.

1b We must know two additional parameters, σ_1 and σ_2, in order to carry out the z-test. Suppose we know that $\sigma_1 = 7$ and $\sigma_2 = 8$.

2 Let us set a stringent value for the probability of a Type I error: $\alpha = 0.01$. In other words, we want the chance of incorrectly concluding that boys and girls have different means, when they actually have the same mean, to be very small—only one chance in 100.

3 There are two samples in this test; hence, two sample sizes are recorded: $n_1 = 37$, $n_2 = 53$.

4 The test statistic is z_{obs} and its critical values are -2.576 and $+2.576$ (Table A4).

5 The acceptance region is ($-2.576 < z_{obs} < +2.576$) and the rejection region consists of the two intervals: ($z_{obs} < -2.576$) and ($z_{obs} > +2.576$). These regions can be written more simply by using absolute values: the acceptance region is ($|z_{obs}| < 2.576$) and the rejection region is ($|z_{obs}| > 2.576$).

6 This step is omitted, as β cannot be calculated when one of the hypotheses is composite. (But see Chapter 13.)

7 The sample means in the study are $\bar{y}_1 = 51.6$ and $\bar{y}_2 = 53.7$. The observed value of the test statistic is

$$z_{obs} = \frac{\bar{y}_1 - \bar{y}_2}{\sqrt{\dfrac{\sigma_1^2}{n_1} + \dfrac{\sigma_2^2}{n_2}}} = \frac{51.6 - 53.7}{\sqrt{\dfrac{(7)^2}{37} + \dfrac{(8)^2}{53}}}$$

$$= \frac{-2.1}{\sqrt{1.324 + 1.208}}$$

$$= \frac{-2.1}{1.591}$$

$$= -1.320$$

8 The observed value of the test statistic falls in the acceptance region. The difference in the mean number of jumps of boys and girls is not significant.

Comment Steps 4 and 5 can be replaced by a computation of the p-value. The two-tailed p-value is $2 \times P(z < -1.320) = 2 \times 0.0934 = 0.1868$. The p-value is greater than 0.01, the specified α. Hence, the observed difference in means is not significant (p-value > 0.01).

10.7 Summary

Hypotheses may be either *specific* or *composite*. In this chapter we introduced the testing of a specific *null hypothesis* against a composite *alternative hypothesis*. The critical value of the test statistic and the acceptance regions are unchanged from Chapter 9. It is convenient, however, to change the names of the two acceptance regions (one for H_0 and one for H_1) to *acceptance region* and *rejection region*, respectively. Usually, we say that we *fail to reject* H_0 if the test statistic falls in the acceptance region and that we *reject* H_0 if the test statistic falls in the rejection region. There are, however, many other ways to state the result of a *significance test*. The importance of distinguishing between the *statistical significance* and the *scientific significance* of a result is stressed.

The alternative hypothesis may be either *one-sided* or *two-sided*. The test is, correspondingly, either *one-tailed* or *two-tailed*.

The *rejection region procedure* for carrying out a test of two hypotheses consists of eight steps. One of these steps, the calculation of the probability of a Type II error, can only be carried out if both hypotheses are specific, as in Chapter 9. Another way to test two hypotheses is the *p-value procedure*, based on the calculation of the *p-value*.

The difference between two population means may be tested by the *z-test of two independent means*.

Exercises

Several of the questions are tests of hypotheses. You may find the summary in Section 10.4 useful when answering those questions.

1 (Section 10.1) The purpose of this question is to draw, on the same page, three graphs like Figure 10.1 in the text but for the z-test instead of the binomial test. You may find it useful to refer to Section 9.3. Proceed as follows.

(a) Consider a hypothesis H_0: $\mu = 50$ and suppose that the population standard deviation (σ) is 10. Use $\bar{y}$ as the test statistic for a sample of size 25. What are the mean and standard deviation of the distribution of $\bar{y}$ when H_0 is true? Sketch a graph of this distribution at the top of a blank page. Show a scale on the abscissa and indicate the mean and standard deviation of the distribution in the usual way.

(b) Now consider the hypothesis H_1: $\mu = 53$ and suppose that the population standard deviation remains at 10. Draw a graph of the distribution of $\bar{y}$, when H_1 is true, below the graph for (a). Again show the mean and standard deviation.

(c) Compute the critical value of $\bar{y}$ for a test with $\alpha = 0.05$. Show this critical value as a vertical line extending from one graph to the other.

(d) Compute the value of β and mark it on the graph. (Also mark α.)

(e) Finally, draw a graph of the distribution of $\bar{y}$ if H_1 is changed to $\mu = 54$. Draw this graph below the other two. You need not calculate β for this hypothesis, but state whether it is greater or less than the β calculated in (d). How do you know?

(f) Mark the acceptance regions for H_0 and H_1. Do these regions (and the critical value) change when H_1 changes? Why?

2 (Section 10.1) A student completes a multiple-choice test that contains 20 questions. Each item has five alternatives, of which one is correct. If the student has no knowledge of the subject matter and guesses the answer to each question, then her chance of getting a particular question correct is 1/5. Suppose that she gets eight questions correct. We want to test the hypothesis that she is guessing (H_0) against the hypothesis that she has some knowledge of the subject matter and is doing better than guessing (H_1). What conclusion should be drawn if the probability of a Type I error is not to exceed 0.05? (Show the steps in your work.)

3 (Section 10.1) A school principal wants to determine whether the mean number of days that students attend his school is 150 or whether it is less than 150. He believes

that, in any case, the standard deviation of the distribution of days at school is 10. He draws a random sample of 20 student records and finds that the mean number of days in this sample is 137.6. What conclusion should he draw if $\alpha = 0.01$?

4 (Section 10.2) This question is designed to illustrate the use of the word "significant" (or "significantly") in the reporting of research. Find a report in a published paper or in a textbook that includes the word "significant" in the description of the result of a statistical test. (The test need not be a binomial or a z-test.)

(a) Quote the sentence that contains the word "significant".

(b) Give the complete reference to the paper or text, including page number.

(c) Does the sentence mean that H_0 cannot be rejected or that H_1 is claimed to be true?

5 (Section 10.3) In a test of ESP, one person tosses a coin 25 times and another person guesses the result of each toss. The hypothesis of no ESP (H_0) is the hypothesis that the probability of success on each toss is 0.5. The hypothesis of ESP (H_1) is the hypothesis that the probability is different from 0.5. This probability can be either greater than 0.5 (more successes than chance level) or less than 0.5 (fewer successes than chance level).

(a) Why is a probability less than 0.5 just as much evidence for ESP as a probability greater than 0.5? (This is a nonstatistical question, and a brief two- or three-sentence answer should suffice.)

(b) Suppose that the correct guess is made on 8 of the 25 tosses. What conclusion should be drawn if $\alpha \leq 0.05$?

6 (Section 10.3) For a new IQ test, it is important to determine whether the mean, in a certain population, is 100 or is different from 100. The standard deviation is known to be 15. A random sample of 50 persons is tested; the sample mean is 105.7. What conclusion should be drawn if $\alpha = 0.01$?

7 (Section 10.5)

(a) State the one-tailed p-value for the following observed values of the test statistic. State the p-value both exactly and as an inequality.

Example: $y = 16, n = 20, H_0: p = 0.5$; p-value $= 0.006$, p-value < 0.01.

(i) $y = 18, n = 25, H_0: p = 0.5$

(ii) $y = 2, n = 15, H_0: p = 0.5$

(iii) $y = 3, n = 10, H_0: p = 0.5$

(iv) $\bar{y} = 53, n = 25, \sigma = 10, H_0: \mu = 50$

(v) $\bar{y} = 71, n = 100, \sigma = 20, H_0: \mu = 76$

(b) State the two-tailed p-value (both exactly and as an inequality) for each of the observed values in (a).

8 (Section 10.5)

(a) Compute the exact p-value in Questions 2 and 6.

(b) Report these two p-values as inequalities using conventional values (e.g., 0.05, 0.01, etc.)

(c) Write the conclusions of Questions 2 and 6 in words using the p-values.

(d) What would be the p-value in Question 6 if the sample mean had been 102.6? Give an exact value and also an inequality.

9 (Section 10.6) In comparing two methods of teaching reading in grade two, two samples of children are studied. One sample of 75 children is taught by the conventional method (the control group). The other sample of 100 children is taught by a new method (the experimental group). After two months, each child is given a standard test of reading. It is known that the standard deviation (σ) on this test is 20, for either method of teaching. The sample means are 78.2 (control) and 83.5 (experimental). Carry out a two-tailed test of the hypotheses at the 0.01 level.

11 t-TESTS

The theory of hypothesis testing was presented in the last two chapters, in which three statistical tests were described in detail: the test of one proportion (the binomial test), the z-test of one mean, and the z-test of two independent means. Tests of means are particularly important in statistics; unfortunately, the z-tests have a major limitation—the population standard deviation must be known in order to calculate the test statistic.

This chapter introduces tests that do not have this limitation. The tests are called t-tests. The test statistics have t-distributions, rather than the normal distributions of the z-tests.

The t-tests may be used to study the relationship between two variables. Section 11.3 is devoted to showing how experiments are designed, so that this relationship is not obscured by the effects of other variables.

11.1 The t-distribution

Two z-tests have been described: the z-test of one mean (Section 9.3) and the z-test of two independent means (Section 10.6). In each test the test statistic, z_{obs}, has the standard normal distribution when H_0 is true. Consider the test of one mean. The hypotheses are H_0: $\mu = \mu_0$ and H_1: $\mu \neq \mu_0$, where μ_0 is a specific number. The formula for z_{obs} is

$$z_{obs} = \frac{\bar{y} - \mu_0}{\sigma/\sqrt{n}}$$

In this formula, $\bar{y}$ is the sample mean, μ_0 is the population mean if the null hypothesis is true, σ is the population standard deviation, and n is the sample size. This test statistic can be calculated only if σ is known. (The other quantities would be known: $\bar{y}$ is calculated from the data; μ_0 is given by the null hypothesis; n is the sample size.)

What can be done if σ is not known? The sample standard deviation can be calculated from the data and substituted into the above formula in place of σ. There is, however, a slight complication in calculating the standard deviation. In Section 4.4, the formula for the sample standard deviation was

$$s' = \sqrt{\frac{\Sigma y^2 - (\Sigma y)^2/n}{n}} \quad \text{(Unadjusted standard deviation)}$$

Note that the divisor in this formula is n, the sample size. For reasons given in detail in Chapter 12, it is not correct to use s' as the estimate of σ when inferences are made from samples to populations. The correct formula to use is

$$s = \sqrt{\frac{\Sigma y^2 - (\Sigma y)^2/n}{n - 1}} \quad \text{(Adjusted standard deviation)}$$

Note that the divisor in the new formula is $(n - 1)$, not n. The standard deviation, s, is called the *adjusted standard deviation*, to distinguish it from s', the *unadjusted standard deviation*. The adjusted standard deviation, s, is always greater than the unadjusted standard deviation, s'. [A final note: both of these formulas are *computational* formulas: in both cases, the definitional formulas are the same except that the numerator is replaced by $\Sigma(y - \bar{y})^2$.]

If we now replace σ by s in the above formula for z_{obs}, the formula for testing one mean becomes

$$t_{obs} = \frac{\bar{y} - \mu_0}{s/\sqrt{n}}$$

In this formula all quantities are known: $\bar{y}$ and s come from the sample; μ_0 is given by the null hypothesis; n is the sample size. As indicated, the new statistic is given the symbol t_{obs}, to distinguish it from z_{obs}.

In order to use t_{obs} as a test statistic, we must know its distribution when H_0 is true. It would be convenient if this statistic, like z_{obs}, were normally distributed, since then we could continue to use the normal table. Recall what we mean when we say that z_{obs} is normally distributed. Suppose that a great many samples are selected at random from a population with mean μ_0. If z_{obs} is calculated for each sample and a distribution of z_{obs} is constructed, the distribution will be normal. Unfortunately, if t_{obs} is calculated for each sample and a distribution of t_{obs} is constructed, the distribution will *not* be normal—it will be what is called the *t-distribution*.

If there were just one *t*-distribution, the situation would be quite simple. We would just need a table of the *t*-distribution similar to the table of the normal distribution. Rather than one *t*-distribution, however, there is an infinite number of *t*-distributions. Each *t*-distribution is labeled by an integer called its *degrees of freedom*. Chapters 14 and 17 explain why the unusual term "degrees of freedom" is used to label or index distributions such as the *t*-distribution. For now, it is best to think of the degrees of freedom as just a label distinguishing the different *t*-distributions. The term "number of degrees of freedom" is abbreviated *df*.

The number of degrees of freedom depends on the sample size. In the case of t_{obs} for one mean, the number of degrees of freedom is one less than the sample size: that is, $df = n - 1$. However, for other situations the formula for *df* is different. [Note that the denominator of s, the adjusted standard deviation, is also $(n - 1)$, the number of degrees of freedom.]

There is a *t*-distribution for every integer value of *df* from 1 to ∞ (infinity). Three distributions are shown in Figure 11.1. The distribution with $df = 1$ is the broadest (most variable); the distributions become narrower as *df* increases. The distribution with $df = \infty$ is identical to the standard normal distribution. The distributions with other values of *df* fall between the distributions for $df = 1$ and $df = \infty$. One such distribution, with $df = 9$, is shown in the figure.

To perform *t*-tests, we need a table of the *t*-distribution for each *df*. This is not practical. When *df* is greater than 20 or 30, there is little difference between the *t*-distribution and the normal distribution. Table A6 in the Appendix contains certain important percentiles for each *t*-distribution for $df = 1, 2, \ldots$, 30 and also for $df = 40, 60, 120$, and ∞ (equivalent to the normal distribution). The percentiles are the 90th, 95th, 97.5th, 99th, 99.5th, and 99.95th; each row of the table contains the percentiles for a particular *df*. The percentiles are, of

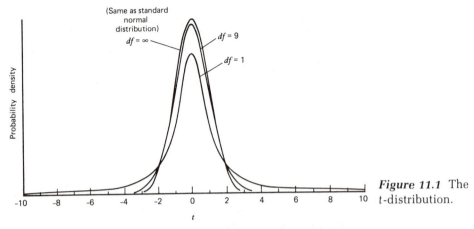

Figure 11.1 The *t*-distribution.

course, the critical values of *t*; for example, the 95th percentile is the critical value for $\alpha = 0.05$. The columns of Table A6 are headed by t_α: that is, a *t* subscripted by the pertinent α. For example, the column of the 95th percentiles is headed by $t_{.050}$, indicating that the values in that column are the critical values of *t* for $\alpha = 0.050$. (These are one-tailed critical values; see Section 11.2 for an example of two-tailed critical values.)

Example 1

Let us use the *t*-table to find the 95th percentile of the *t*-distribution with nine degrees of freedom. The line of the table for $df = 9$ is reproduced in Figure 11.2*a*.

The 95th percentile is that value below which 95% of the distribution falls; therefore, 5% of the distribution falls above the required value. See Figure 11.2*b* for a sketch of the distribution. To find the percentile we look for the column labeled $t_{.050}$, since 5% = 0.050. We find that the 95th percentile is 1.833.

The *t*-distribution is symmetric: therefore, a percentile below the mean or median (which is zero) is the negative of the corresponding upper-tail percentile. For example, the 5th percentile is -1.833. Figure 11.2*b* also shows the 1st and 99th percentiles (-2.821 and 2.821, respectively).

Example 2

Consider three questions about a random variable that is distributed as *t* with 9 *df*.

(a) What is the probability that the variable is greater than 2.13?

(b) What is the probability that the absolute value of the variable is greater than 2.13?

The answer to **(a)** can be seen in Figure 11.2*c*, in which the percentiles near 2.13 are shown. The required probability is the shaded area; it is greater than 0.025 and less than 0.05. The answer to **(a)** may be written

$$0.025 < P(t > 2.13) < 0.05$$

This formula states that the required probability lies between 0.025 and 0.05.

Since the *t*-distribution is symmetric, the probability that *t* is less than -2.13 is the same as the probability that *t* is greater than $+2.13$. Hence, the probability that the absolute value of *t* is greater than 2.13 is double the probability found in **(a)**. Doubling the limits found in **(a)**, we obtain

$$0.05 < P(|t| > 2.13) < 0.10$$

---

df	$t_{.100}$	$t_{.050}$	$t_{.025}$	$t_{.010}$	$t_{.005}$	$t_{.0005}$	df
9	1.383	1.833	2.262	2.821	3.250	4.781	9

(a)

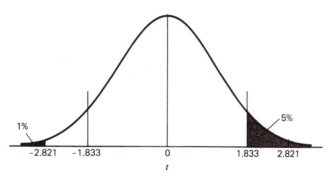

(b)

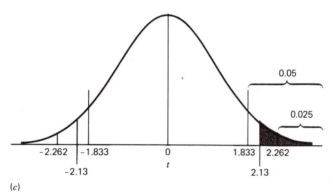

(c)

Figure 11.2 Using the t-table.

Variability of the t-distribution

Let us compare the two statistics, z_{obs} and t_{obs}. The formulas for the test of one mean are

$$z_{obs} = \frac{\bar{y} - \mu_o}{\sigma/\sqrt{n}} \text{ and } t_{obs} = \frac{\bar{y} - \mu_o}{s/\sqrt{n}}$$

When H_0 is true, z_{obs} has the normal distribution whereas t_{obs} has the t-distribution with $df = n - 1$. Consider the values of these statistics for many random samples. The values of z_{obs} for samples of a given size will differ for only *one* reason: the sample mean, $\bar{y}$, usually has a different value for each sample. All the other quantities in z_{obs} (that is, μ_o, σ, and n) are the same from sample to sample. Hence, the variability of z_{obs} is related to the variability of $\bar{y}$.

The values of t_{obs} will vary from sample to sample for *two* reasons: $\bar{y}$ usually has a different value for each sample, and s has a different value for each sample. Hence, the variability of the t-distribution is greater than the variability of the normal distribution. However, the variability of s decreases as the sample size and the number of degrees of freedom increase. For this reason, the variability of the t-distribution decreases as df increases, until the point is reached at which the t-distribution is indistinguishable from the normal distribution (say, when $df \geqq 30$).

11.2 The *t*-test of One Mean

When the population standard deviation, σ, is unknown, we use the t-test to determine whether or not a population has a specified mean, μ_0.

$$t_{obs} = \frac{\bar{y} - \mu_0}{s/\sqrt{n}} \quad \text{(Test of one mean)}$$

As noted in Section 11.1, this formula is the same as the formula for z_{obs} except that the adjusted sample standard deviation, s, replaces the population standard deviation, σ, in the denominator.

The steps used to carry out the t-test are the eight basic steps of Section 10.4. The t-test differs from the z-test in only two ways: the formula for the test statistic is different and the test statistic, when H_0 is true, has a t-distribution instead of a normal distribution. The t-distribution has $(n - 1)$ degrees of freedom. Critical values of the test statistic can be found from Table A6.

Example 1

Let us test the mean of a certain population by drawing from it a random sample of size 16. Suppose the hypotheses are $H_0: \mu = 30$ and $H_1: \mu > 30$. The data and the calculations are shown in Table 11.1. The following steps are used to carry out the statistical test.

TABLE 11.1 The *t*-test of One Mean

(a) Data

36.4, 31.3, 35.2, 29.3, 38.6, 36.0, 30.4, 26.3, 35.1, 28.3, 29.6, 33.3, 34.3, 34.9, 31.6, 34.8.

(b) Basic Calculations

$\Sigma y = 525.4$

$\Sigma y^2 = 17426.04$

$\bar{y} = \dfrac{\Sigma y}{n} = \dfrac{525.4}{16} = 32.8375$

$s^2 = \dfrac{\Sigma y^2 - (\Sigma y)^2/n}{n - 1}$

$\quad = \dfrac{17426.04 - (525.4)^2/16}{15} = \dfrac{173.218}{15} = 11.548$

$s = \sqrt{11.548} = 3.398$

(c) Calculation of Test Statistic

$t_{obs} = \dfrac{\bar{y} - \mu_0}{s/\sqrt{n}}$

$\quad = \dfrac{32.8375 - 30}{3.398/\sqrt{16}} = \dfrac{2.8375}{0.8495} = 3.34$

1a The hypotheses are H_0: $\mu = 30$ and H_1: $\mu > 30$. Here μ_0 is 30. Note that the alternative hypothesis is one-sided. (Refer to Section 10.3 for a discussion of one- and two-sided alternative hypotheses.)

1b No additional parameters are needed to carry out the t-test. (In the z-test it is necessary to specify the value of σ at this point. In the t-test this step is unnecessary.)

2 Let $\alpha = 0.01$.

3 The sample size is $n = 16$.

4 The critical value of the test statistic is found as follows. H_0 will be rejected when $\bar{y}$ is sufficiently larger than 30 to be improbable under H_0. Since "sufficiently larger" is defined here in terms of $\alpha = 0.01$, H_0 will be rejected when t_{obs} is larger than the 99th percentile of the t-distribution with $df = n - 1 = 16 - 1 = 15$. The critical value is, therefore, $t_{crit} = t_{.010} = 2.602$.

5 The acceptance region is ($t_{obs} < 2.602$); the rejection region is ($t_{obs} > 2.602$).

6 There is no step 6, since the probability of a Type II error, β, cannot be calculated.

7 The test statistic, found from Table 11.1, is $t_{obs} = 3.34$.

8 Since t_{obs} falls in the rejection region, we reject the hypothesis that the population mean is 30 (and accept the hypothesis that it is greater than 30).

Comments *(a)* If the alternative hypothesis were H_1: $\mu < 30$, the rejection region would be ($t_{obs} < -2.602$), for the given $\alpha = 0.01$. *(b)* In the p-value procedure, the observed value of the test statistic, 3.34, is compared with the percentiles of the t-distribution with 15 df (Table A6); since 3.34 exceeds the 99.5th percentile, $t_{.005} = 2.947$, the observed mean is significantly different from 30 (p-value < 0.005, one-tailed).

Example 2

In order to determine whether a population has a mean of 100, a random sample of size 25 is drawn from the population. The hypotheses are H_0: $\mu = 100$ and H_1: $\mu \neq 100$. The sample mean and (adjusted) standard deviation are $\bar{y} = 97.3$ and $s = 11.3$. Let us make the test with $\alpha = 0.05$. Note that the alternative hypothesis is two-sided, and therefore the test is two-tailed. The test statistic, t_{obs}, is distributed as t with $df = 25 - 1 = 24$ when H_0 is true. The critical values are the 2.5th and the 97.5th percentiles of this distribution: that is, -2.064 and $+2.064$. The null hypothesis is rejected if ($t_{obs} < -2.064$) or ($t_{obs} > 2.064$). Put more simply, the null hypothesis is rejected if ($| t_{obs} | > 2.064$).

From the sample data,

$$t_{obs} = \frac{\bar{y} - \mu_0}{s/\sqrt{n}} = \frac{97.3 - 100}{11.3/\sqrt{25}} = \frac{-2.7}{2.260} = -1.19$$

Since the absolute value of this result does not exceed 2.064, we conclude that the sample mean is not significantly different from 100.

Alternatively, if we use the p-value procedure, we find that the absolute value of the observed test statistic, 1.19, is smaller than 1.318, the 90th percentile of the t-distribution with 24 df. Hence, the probability of the observed value, or a more extreme value, is greater than $2 \times 0.10 = 0.20$. The sample mean is not significantly different from 100 (p-value > 0.20, two-tailed).

11.3 Experimental Design

Of the three t-tests discussed in this chapter, the t-test of one mean was the topic of Section 11.2. The other t-tests are called the t-test of two independent means and the t-test of two dependent means. The very important distinction between these two tests is discussed in this section; the t-tests themselves are described in Sections 11.4 and 11.5.

One of the purposes of statistical techniques is to make inferences about the relationship between two variables. (See Principle P8, in Section 8.1.) Usually, one of these variables is called the independent variable and the other the dependent variable. This section describes, at an elementary level, how to design an experiment to study the relationship between two variables. The simplest experiments have only two groups; accordingly, our consideration of experimental design will be restricted to such experiments.

Manipulated and Observed Variables

In Section 8.2 the groups in an experiment were described as *experimentally-defined* groups, since each group was identified by the condition under which the group is studied. In a particular learning experiment, for example, the subjects in one condition are reinforced on every trial whereas the subjects in the other condition are reinforced on every second trial. Or in a problem-solving experiment the conditions differ in the amount of information given to the subjects. In each of these experiments, the independent variable has two different values, and these values are the conditions: in the learning experiment the independent variable is frequency of reinforcement, and in the problem-solving experiment it is the amount of information provided. The independent variable may be called an experimentally-defined variable (as in Section 8.2) or a *manipulated variable*.

The dependent variable in an experiment may be called the *observed variable*, to distinguish it from the independent, or manipulated, variable. In the learning experiment, the observed variable is the number of trials the subject takes to learn the required task, or the time taken for learning. In the problem-solving experiment, the observed variable is the time taken to solve the problem, or a measure of the originality of the solution.

Irrelevant Variables

Experiments would be quite simple if they involved only two variables, the manipulated variable and the observed variable. But even if the experimenter plans to include only these two variables, there are always many other variables involved. These other variables are known as *irrelevant variables* (or *nuisance variables*, or *extraneous variables*). Unless the experiment is well designed, the irrelevant variables may make interpretation of the results difficult or even impossible.

Suppose that the mean number of trials to learn a task is much greater for a group that is reinforced on every second trial than for a group reinforced on every trial, and that this difference is statistically significant. Can we conclude that frequency of reinforcement has affected learning? In order to answer this question, we would have to know something about the irrelevant variables. Suppose that all the subjects in one group had learned a similar task in a previous experiment, whereas the subjects in the other group did not have this experience. The two groups would then differ in prior experience as well as in frequency of reinforcement. The variable "prior experience" is an irrelevant (or nuisance) variable. It is obvious that the experiment can be interpreted only if some way is found to equate the two groups on this variable.

An irrelevant variable is any variable that can affect or *confound* the relationship between the independent and dependent variables. In many experiments, variables such as age, sex, IQ, time of day, attitude toward the experiment, etc., can affect the relationship between the independent and dependent variables. The experiment must be designed, therefore, to minimize the effect of these variables. But it is impossible to completely eliminate the effect of all irrelevant variables. One of the essential skills of a good experimental scientist is the ability to identify the important irrelevant variables and to reduce their effect at least to the extent that the results of the experiment can be interpreted. (In some experiments the effect of an irrelevant variable cannot be reduced but can be taken account of in the analysis. In other words, the effect is reduced or eliminated in the *analysis*, rather than in the experiment itself. One such technique is called the analysis of covariance; this and related techniques are discussed in advanced books on experimental design.)

Let us now summarize the basic components of an experiment.

P10: The Basic Principle of Experimental Design

The relationship between an independent and a dependent variable can be investigated in an experiment involving two or more groups. Each group is defined by a condition: that is, by a value of the independent (manipulated) variable. The subjects are measured on the dependent (observed) variable. The effect of each irrelevant variable is minimized by *(a)* holding the variable constant for all subjects or *(b)* randomizing the effect of the variable or *(c)* matching the subjects on the variable.

Holding the Variable Constant The effect of an irrelevant variable can be eliminated by holding the variable constant for all subjects in the experiment. In a learning experiment, all subjects should have equal previous experience on the task. Usually, the subjects will not have had any previous experience on the specific experimental task, and therefore their previous experience on that task is equated. Many irrelevant variables are held constant by running an experiment in the same way for all subjects: the room is the same, the experimenter is the same, the apparatus is the same, etc.

Many variables—particularly, variables that are characteristics of the subjects—cannot be held constant. How can IQ, motivation, and fatigue, for example, all be held constant? The experimenter can select subjects with IQs of 122 in order to hold IQ constant, but such selectivity severely limits the number of potential subjects. Another difficulty with this strategy is that the results of the experiment could not be generalized to subjects with IQs other than 122. Furthermore, if the experimenter also tries to restrict the experiment to subjects who have the same motivation (however measured), the number of potential subjects will be even more limited. Clearly, we cannot hold constant all the individual characteristics of the subjects.

Randomizing the Variable The effect of a number of irrelevant variables can be minimized by assigning subjects at random to the two groups. Suppose that 40 subjects are selected for an experiment which has two conditions, *A* and *B*. Number the subjects from 00 to 39. Start at some point in the random-number table (Table A5) and record *pairs of* random digits. Write down each number (two digits) as it occurs, if the number is between 00 and 39, inclusive. Discard all numbers greater than 39. Once a number has been recorded, do not record the same number again. Eventually, a list of the 40 numbers from 00 to 39, in a random order, will be obtained. The first 20 numbers are the numbers of the subjects who should be assigned to condition *A*; the remaining numbers are the numbers of the subjects who should be assigned to condition *B*.[1]

How are the effects of the irrelevant variables minimized by this procedure? Consider the IQ of the subjects. It is very unlikely that all high-IQ subjects will be assigned to condition *A* and all low-IQ subjects to condition *B*. We expect that the average IQ will be about the same in the two conditions. Note carefully that the randomization procedure does *not* ensure that the two conditions—that is, samples—have *exactly* the same mean IQ. What has just been said for IQ applies to all the other characteristics of the subjects. The effect of each irrelevant variable is greatly reduced by the randomization procedure.

Matching the Variable The effect of an irrelevant variable can be reduced by matching the subjects, in pairs, on the variable. Consider again the 40 subjects in the learning experiment. Suppose that two of the subjects have IQ = 130, two subjects have IQ = 127, two subjects have IQ = 125, etc. In other words, suppose there are 20 pairs of subjects, each with the same IQ. (Such a perfect pairing of the subjects ordinarily is not possible, but it is always possible to pair the two persons with the highest IQ, then pair the next two highest persons, then the next two highest, etc.)

One member of each pair of subjects is assigned to condition *A*, the other to condition *B*. In this way the effect of IQ on the outcome of the experiment is reduced. We must still be careful in assigning the members of each pair to the conditions. It would not be satisfactory to assign the older person in each pair to condition *A* and the younger to condition *B*. Obviously, the results of the experiment then would be confounded by the irrelevant variable age. The effect of this variable, and of other irrelevant variables, should be eliminated by randomly assigning the members of each pair to the two conditions. The randomization is again carried out by a random-number table. The two members of each pair are listed in a specific order—say, alphabetically—and a random digit is selected from the table for each pair. If the digit is even, the first listed member of the pair is assigned to condition *A* and the other to condition *B*. If the digit is odd, the assignment is reversed.

It would be desirable to match the subjects on more than one variable, but this usually is difficult to do. However, the subjects in some experiments are matched, by nature, on several variables. Suppose identical twins are used as the subjects in an experiment. Identical twins are alike on many variables. An experiment in which identical twins are assigned at random, one to each of two conditions, minimizes the effects of a large number of irrelevant, genetically determined characteristics.

Since it is very difficult to find a sufficient number of identical twins for most experiments, another technique often used is to match each subject with himself or herself. In other words, each person selected for the experiment is a subject in condition *A* *and* a subject in condition *B*. In this way a number of

---

[1] This procedure can be speeded up in two ways. Only the first 20 random numbers (each between 00 and 39) need be obtained; these are the numbers of the subjects in condition *A*. The remaining subjects are then in condition *B*. Fewer numbers need be discarded if 40 is subtracted from numbers between 40 and 79; the resulting number is between 00 and 39 and can be used as a subject number. The numbers between 80 and 99 are still discarded. (Variations of this procedure are used when the number of subjects is different from 40.)

characteristics that are irrelevant are matched and their effects minimized. Although this solution seems perfect, there are drawbacks. Suppose a subject gets condition *A* first. After completing the task in condition *A*, the subject will be affected by his or her experience and will not approach the task in condition *B* in the same way as a subject who gets condition *B* first. For this reason it is not an advantage, in many experiments, to match subjects with themselves.

Independent and Dependent Groups

We have discussed the three ways in which the effect of an irrelevant variable can be minimized. Two experimental designs can be distinguished. In one type of design, called the *dependent-groups* design, some of the irrelevant variables are matched and the remaining variables are either randomized or held constant. In the other type of design, called the *independent-groups design*, there are no matched variables, so that all irrelevant variables are either randomized or held constant.

Note carefully that when used in this context, the terms "independent-groups" and "dependent-groups" are not connected with the terms "independent variable" and "dependent variable"; both types of design have an independent and a dependent variable. The difference between independent- and dependent-groups designs is that there are no matched variables in an independent-groups design and that there are one or more matched variables in a dependent-groups design. The names of these designs are chosen to indicate that the scores of the subjects in the two groups are unrelated or independent in an independent-groups design and that paired scores are related or dependent in a dependent-groups design. (See Chapter 18 for a more precise definition of independence and dependence.)

Other names are used for the dependent-groups design. The design is sometimes called a *related-groups* design or a *correlated-groups* design. Since the subjects are matched, the design may also be called a *matched-groups design*; it should be noted, however, that the subjects are matched, not just the groups. When subjects are matched with themselves, the design may be called a *repeated-measures* design. This name is chosen because each subject receives both conditions in the experiment and therefore is measured twice on the dependent variable.

Two types of independent-groups designs may be distinguished. One type is based on an independent variable that is manipulated by the experimenter: that is, the groups are *experimentally-defined*. All preceding designs in this section are of this type. In the other type, the independent variable is *naturally-occurring* and the experimenter cannot randomly assign the subjects to the groups. Consider a study that compares males and females on some task; the groups of males and females already exist in nature. Such a study is not a true experiment, since the independent variable is not manipulated; however, the statistical techniques for independent groups are still appropriate.

Example

An important topic studied by industrial psychologists is the selection of new employees. Two methods commonly used in selection are the application form and the interview. Let us design a study of these two methods.[2] Our interest is in the effect of the

[2] This example is based on D. H. Tucker & P. M. Rowe, "Consulting the application form prior to the interview: an essential step in the selection process", *Journal of Applied Psychology* 62 (1977), 283-287.

information in the application form on the time taken by the interviewer to make an initial decision about the applicant.

We plan to conduct a number of interviews. There are two conditions: in one, the interviewers see the completed application form before each interview; in the other, they never see the form and conduct the interviews without it. The independent variable is, therefore, "presence or absence of application form information". In each interview the time required by the interviewer to make the initial decision about the applicant is recorded. Thus, time is the dependent variable.

What are some of the irrelevant variables? Here we will consider two types of irrelevant variables: characteristics of the person being interviewed (the interviewee) and characteristics of the interviewer. We can eliminate the effect of many of the *interviewee* variables by stipulating that the same person be interviewed by all the interviewers; hence, these variables are held constant. We must be careful, though: even though the same person is interviewed in all the interviews, he changes from interview to interview. He becomes more experienced, and perhaps bored with the experiment. But we will not pursue this difficulty any further.

Let us randomize the characteristics of the *interviewer* by assigning the interviewers at random to the two conditions. We have, then, an independent-groups design. Half the interviews are conducted by interviewers who have seen the application form; the remaining interviews are conducted by interviewers who have not.

Consider an alternative to the independent-groups design. Suppose we know that the principal variable affecting the length of the interview is the experience of the interviewer: the more experience, the shorter the time needed before making a decision in an interview. In other words, the length of the interview is negatively correlated with experience. (The topic of correlation was discussed in Section 8.6.) We can match the interviewers on the experience variable. The two interviewers of longest experience are paired; one is selected at random for the application-form condition; the other is assigned to the no-application-form condition. The interviewers who are third and fourth in length of experience are paired and randomly assigned to the two conditions, and so on. In this way, one interviewer variable, length of experience, is matched and the other interviewer variables are randomized. The design is now a dependent-groups, or matched-groups, design.

A Comparison of the Two Designs

Two ways to reduce the effects of irrelevant variables are by randomizing and by matching. In practice, the effects of these variables can never be eliminated completely. There is always some residual effect, which usually is greater when the variable is randomized than when it is matched. In other words, effects of the irrelevant variables in a dependent-groups design usually are less than those in an independent-groups design. However, three qualifications must be made to this general statement of the superiority of the dependent-groups design.

1 The matched variable (or variables) must actually *have* an effect on the dependent variable. If there is no effect of the matched variable on the dependent variable before matching, there is no effect to be eliminated, and there can be no difference between the two types of design on this ground. From the third qualification (see below), the dependent-groups design will be less satisfactory in this case than the independent-groups design.

2 When subjects are matched with themselves (a procedure sometimes referred to as "subjects used as their own control") the design has a major flaw not found in an independent-groups design: the subject may be changed by experiencing one of the conditions. If this is so, the matched-groups design cannot be used; or, if it is used, the results will be difficult to interpret. It is best that half the subjects first receive condition *A*, then condition *B*, while the other half first receive condition *B*, then condition *A*. This technique, called "counter-balancing", does not eliminate the problem completely, since the subjects are still changed by their experience of the first condition.

3 The next two sections describe the analysis of the results of independent-groups and dependent-groups designs. A *t*-distribution is used in both analyses, but for the same number of subjects, the number of degrees of freedom is greater in the independent-groups than in the dependent-groups design; we say that there is a *loss of degrees of freedom* in the dependent-groups design. A design with more degrees of freedom usually is preferred, since the probability of a Type II error (β) is smaller for a given probability of a Type I error (α). For this reason it might be thought that the independent-groups design would be preferred. The gain achieved by matching, however, is usually much greater than the slight loss related to the decreased degrees of freedom; thus, the dependent-groups design is preferred, even though it has fewer degrees of freedom than the independent-groups design.

11.4 The *t*-test of Two Independent Means

The *t*-test of two independent means is closely related to the z-test of two independent means (Section 10.6). The hypotheses that are tested are the same; the hypotheses are H_0: $\mu_1 = \mu_2$ and H_1: $\mu_1 \neq \mu_2$, where μ_1 and μ_2 are the means of the two populations. (In some cases the alternative hypothesis may be one-sided.) The two samples or groups are independent in the sense explained in Section 11.3.

The test statistic for the *t*-test of two independent means is

$$t_{\text{obs}} = \frac{\bar{y}_1 - \bar{y}_2}{s_p \sqrt{\dfrac{1}{n_1} + \dfrac{1}{n_2}}} \quad \text{(Test of two independent means)}$$

In this formula, $\bar{y}_1$ and $\bar{y}_2$ are the sample means, n_1 and n_2 are the sample sizes, and s_p is the *pooled standard deviation*. The pooled standard deviation is obtained from the standard deviations in the two groups. The formula for s_p is

$$s_p = \sqrt{\frac{(n_1 - 1)s_1^2 + (n_2 - 1)s_2^2}{n_1 + n_2 - 2}}$$

The variances, s_1^2 and s_2^2, in this formula are the *adjusted* variances (Section 11.1) obtained by division by $(n_1 - 1)$ and $(n_2 - 1)$, respectively.[3]

The formula for t_{obs} is quite complicated. The relation of this formula to the formula for z_{obs} is explained in Section 11.6.

---

[3] The pooled standard deviation, s_p, can be called the *adjusted* pooled standard deviation, since it is based on the adjusted variances. In Section 8.5 an *unadjusted* pooled standard deviation, s_p', was defined, based on the unadjusted variances. The numerators of these two formulas are algebraically equal but the denominators are different. In the adjusted pooled standard deviation, s_p, the denominator is $(n_1 + n_2 - 2)$; in the unadjusted pooled standard deviation, s_p', the denominator is $(n_1 + n_2)$.

In the *t*-test of *one* mean the test statistic, t_{obs}, when the null hypothesis is true, has a *t*-distribution with $(n - 1)$ degrees of freedom. Here, in the *t*-test of *two independent* means, the test statistic, t_{obs}, also has a *t*-distribution when the null hypothesis is true; the number of degrees of freedom is $(n_1 + n_2 - 2)$. Note that we "lose" two degrees of freedom—not just one, as before. There are two ways to remember the *df* for two independent means: by considering that one degree of freedom is lost for each sample [$df = (n_1 - 1) + (n_2 - 1)$] or by noting that the expression $(n_1 + n_2 - 2)$ is precisely the denominator of the expression for s_p.

Example 1	In the example given in Section 11.3, two conditions were studied: the application-form condition, in which the interviewer has seen the application form before the interview; and the no-application-form condition, in which the interview is conducted without the application form. Suppose that 18 interviewers are randomly assigned to the two conditions, without matching. The lengths of the interviews are given in Table 11.2a. The *t*-test of two independent means can be used to determine whether the population mean times in the two conditions are equal. We follow the steps of Section 10.4.

1a The hypotheses are H_0: $\mu_1 = \mu_2$ and H_1: $\mu_1 \neq \mu_2$. We choose a two-sided alternative hypothesis because if there is a difference between the means, the difference could be in either direction.

1b There are no additional parameters to specify. It is not necessary to know the population standard deviations, σ_1 and σ_2, in order to carry out this *t*-test. (Recall that these parameters must be specified in the *z*-test of two population means. See Section 10.6.) However, it should be noted at this point that one requirement of the *t*-test of two independent means is that the two population standard deviations must be equal, even though their exact value is not required. (This requirement is discussed further in Section C.3.)

2 Let $\alpha = 0.05$.

3 The sample sizes are $n_1 = 9$ and $n_2 = 9$.

4 The critical values of the test statistic are found as follows. Since H_1 is two-sided, the test is two-tailed. When H_0 is true, t_{obs} has a *t*-distribution with $df = 9 + 9 - 2 = 16$. The upper-tail critical value is $t_{crit} = t_{.025} = 2.120$; the lower-tail critical value is, therefore, -2.120.

5 The acceptance region is $(-2.120 < t_{obs} < 2.120)$ and the rejection region consists of the two intervals $(t_{obs} < -2.120)$ and $(t_{obs} > 2.120)$. These regions can be written more simply by using absolute values: the acceptance region is $(| t_{obs} | < 2.120)$ and the rejection region is $(| t_{obs} | > 2.120)$.

6 As the alternative hypothesis is composite, it is not possible to calculate the probability of a Type II error, β. (See Chapter 13.)

7 The calculation of the test statistic is shown in Table 11.2. The means, variances, and standard deviations are calculated in Table 11.2b. Strictly speaking, the standard deviations are not required, but they can be used to check the calculation of s_p, shown in Table 11.2c. Note that the value of s_p is 3.031, which is between the values of s_1 (2.966) and s_2 (3.095); the pooled standard deviation must always fall between the standard deviations of the two groups. After s_p has been calculated, the calculation of t_{obs} is straightforward. In this example, its value is -0.15, the negative sign indicating that the second group has a greater mean than the first.

8 The test statistic falls in the acceptance region; we are unable to reject the hypothesis that the two population means are the same.

TABLE 11.2 The *t*-test of Two Independent Means

(a) Data

LENGTH OF INTERVIEW (IN MINUTES)

GROUP 1 APPLICATION-FORM CONDITION	GROUP 2 NO-APPLICATION-FORM CONDITION
9.4	15.1
10.3	8.6
6.4	5.2
5.3	12.3
15.4	9.5
12.2	10.6
8.9	8.3
9.7	7.2
10.3	13.0

(b) Basic Calculations

Sum	87.9	89.8
Sum of Squares	928.89	972.64
Sample Size	$n_1 = 9$	$n_2 = 9$
Mean	$\bar{y}_1 = 9.767$	$\bar{y}_2 = 9.978$
Variance	$s_1^2 = 8.800$	$s_2^2 = 9.579$
Standard Deviation	$s_1 = 2.966$	$s_2 = 3.095$

(c) Calculation of Test Statistic

$$s_p^2 = \frac{(n_1 - 1)s_1^2 + (n_2 - 1)s_2^2}{n_1 + n_2 - 2}$$

$$= \frac{8 \times 8.800 + 8 \times 9.579}{16} = \frac{147.032}{16} = 9.1895$$

$$s_p = \sqrt{9.1895} = 3.031$$

$$t_{obs} = \frac{\bar{y}_1 - \bar{y}_2}{s_p \sqrt{\dfrac{1}{n_1} + \dfrac{1}{n_2}}}$$

$$= \frac{9.767 - 9.978}{3.031 \sqrt{\dfrac{1}{9} + \dfrac{1}{9}}} = \frac{-0.211}{1.429} = -0.15$$

Comments. (a) When the two samples have the same size, as in this example, the calculation of s_p can be simplified.

$$s_p = \sqrt{\frac{s_1^2 + s_2^2}{2}} \quad \text{(Equal sample sizes)}$$

This formula gives the same result as the more complex formula, when $n_1 = n_2$. The number of degrees of freedom is still given by the formula $(n_1 + n_2 - 2)$, which is the denominator of the more complex formula. (b) If the p-value procedure is used, the absolute value of the test statistic is compared with the percentiles of the *t*-distribution with 16 *df*. Since 0.15 is less than $t_{.100}$ ($= 1.337$) we can say that the two-tailed *p*-value is greater than $2 \times 0.1 = 0.2$, and report that the observed difference between the two sample means is not significant (*p*-value > 0.20, two-tailed).

Example 2

Consider a study that is carried out with independent groups of unequal size. The data are summarized below.

	GROUP 1	GROUP 2
Sample size	$n_1 = 8$	$n_2 = 11$
Mean	$\bar{y}_1 = 31.3$	$\bar{y}_2 = 23.5$
Standard deviation	$s_1 = 5.63$	$s_2 = 6.86$

The pooled standard deviation must be computed first. Its square is

$$
\begin{aligned}
s_p^2 &= \frac{(n_1 - 1)s_1^2 + (n_2 - 1)s_2^2}{n_1 + n_2 - 2} \\
&= \frac{(8 - 1) \times (5.63)^2 + (11 - 1) \times (6.86)^2}{8 + 11 - 2} \\
&= \frac{7 \times 31.6969 + 10 \times 47.0596}{17} \\
&= 40.734
\end{aligned}
$$

Hence, $s_p = \sqrt{40.734} = 6.38$. This result is reasonable, since it falls between the standard deviations of the two groups.

Now t_{obs} can be calculated.

$$
\begin{aligned}
t_{\text{obs}} &= \frac{\bar{y}_1 - \bar{y}_2}{s_p \sqrt{\dfrac{1}{n_1} + \dfrac{1}{n_2}}} \\
&= \frac{31.3 - 23.5}{6.38 \sqrt{\dfrac{1}{8} + \dfrac{1}{11}}} = \frac{7.8}{6.38 \times 0.4647} = 2.63
\end{aligned}
$$

There are 17 degrees of freedom. For a test at the 0.05 level, two-tailed, $t_{\text{crit}} = t_{.025}$, which is 2.110 (Table A6). Since the observed value exceeds this critical value, the difference between the means is significant at the 0.05 level, two-tailed.

The *p*-value can be determined by comparing 2.63 with the percentiles of the *t*-distribution with 17 *df*. The observed value falls between $t_{.010}$ (2.567) and $t_{.005}$ (2.898). The two-tailed *p*-value is, therefore, less than $2 \times 0.010 = 0.02$ (and greater than $2 \times 0.005 = 0.01$). We can conclude that the two means are significantly different (*p*-value < 0.02, two-tailed).

11.5 The *t*-test of Two Dependent Means

The difference between the means of two populations may be tested by drawing either independent or dependent samples from the two populations. In both cases the hypotheses tested are the same: H_0: $\mu_1 = \mu_2$ and H_1: $\mu_1 \neq \mu_2$, where μ_1 and μ_2 are the means of the two populations. (In certain situations, the alternative hypothesis can be one-sided.) The *t*-test of two dependent means described in this section differs from the *t*-test of two independent means in two ways: the formula for t_{obs} is different and the number of degrees of freedom of the distribution of t_{obs} is different.

The test statistic for the *t*-test of two dependent means is

$$t_{obs} = \frac{\bar{y}_1 - \bar{y}_2}{s_d/\sqrt{n}} \quad \text{(Test of two dependent means)}$$

In this formula, $\bar{y}_1$ and $\bar{y}_2$ are the sample means, n is the number of *pairs* of observations, and s_d is the standard deviation of the *differences* between the paired observations. If a score in the first group is y_1 and the corresponding, matched score in the second group is y_2, the difference between the scores is $d = y_1 - y_2$. This difference is calculated for each pair. The adjusted standard deviation of these differences is calculated using the formula of Section 11.1.

$$s_d = \sqrt{\frac{\Sigma d^2 - (\Sigma d)^2/n}{n - 1}}$$

The subscript d is added to the symbol for the standard deviation simply to remind us that it is a standard deviation of *differences*. Note that n in the formula is the number of pairs, since there are n differences whose standard deviation is calculated. Note also that the *adjusted* standard deviation is calculated.

The test statistic, t_{obs}, has a *t*-distribution when the null hypothesis is true. The number of degrees of freedom is $(n - 1)$. Note that there are $2n$ observations—n in the first group and n in the second—but that there are only n differences. The number of degrees of freedom is based on these differences. Just as in the *t*-test of one mean, there is a loss of one degree of freedom. (Section 11.6 will show that the *t*-test of two dependent means is very closely related to the *t*-test of one mean.)

Example

Consider an experiment with dependent groups that compares the length of an interview when the interviewer has seen the application form and when the interviewer has not seen it. The 18 interviewers in the experiment are matched on the basis of their experience. The two interviewers with longest experience are paired and assigned randomly to the two conditions, then the two interviewers of next-longest experience are paired and assigned randomly to the two conditions, and so on. The data from this experiment are shown in Table 11.3a. The test is carried out by following the steps of Section 10.4.

1a The hypotheses are $H_0: \mu_1 = \mu_2$ and $H_1: \mu_1 \neq \mu_2$.

1b There are no additional parameters to specify.

2 Let $\alpha = 0.05$.

3 The number of pairs is $n = 9$.

4 The critical values of the test statistic are found as follows. Since H_1 is two-sided, the test is two-tailed. When H_0 is true, t_{obs} has a *t*-distribution with $df = 9 - 1 = 8$. The upper-tail critical value is $t_{crit} = t_{.025} = 2.306$. The lower-tail critical value is, therefore, -2.306.

TABLE 11.3 The *t*-test of Two Dependent Means

(a) Data

| PAIR | LENGTH OF INTERVIEW (IN MINUTES) | | DIFFERENCE |
| | GROUP 1 APPLICATION-FORM CONDITION | GROUP 2 NO-APPLICATION-FORM CONDITION | |
	y_1	y_2	$d = y_1 - y_2$
Longest experience	8.6	6.3	2.3
Next longest	10.3	11.4	−1.1
Next longest	9.7	12.3	−2.6
Next longest	11.6	7.4	4.2
Next longest	10.4	12.6	−2.2
Next longest	12.3	11.7	0.6
Next longest	9.8	14.2	−4.4
Next longest	13.6	16.3	−2.7
Shortest experience	12.2	14.0	−1.8

(b) Basic Calculations

Sum	98.5	106.2	$\Sigma d = -7.7$
Sum of Squares			$\Sigma d^2 = 65.99$
Mean	$\bar{y}_1 = 10.94$	$\bar{y}_2 = 11.80$	$\bar{d} = -0.86$

(c) Calculation of Test Statistic

$$s_d^2 = \frac{\Sigma d^2 - (\Sigma d)^2/n}{n - 1}$$

$$= \frac{65.99 - (-7.7)^2/9}{8} = \frac{59.402}{8} = 7.425$$

$$s_d = \sqrt{7.425} = 2.725$$

$$t_{obs} = \frac{\bar{y}_1 - \bar{y}_2}{s_d/\sqrt{n}}$$

$$= \frac{10.94 - 11.80}{2.725/\sqrt{9}} = \frac{-0.86}{0.908} = -0.95$$

5 The acceptance region is ($-2.306 < t_{obs} < 2.306$) and the rejection region consists of the two intervals ($t_{obs} < -2.306$) and ($t_{obs} > 2.306$). These regions can be written more simply by using absolute values: the acceptance region is ($|t_{obs}| < 2.306$) and the rejection region is ($|t_{obs}| > 2.306$).

6 The probability of a Type II error, β, cannot be calculated when there is a composite alternative hypothesis. See Chapter 13.

7 The calculation of the test statistic is shown in Table 11.3. The first step in calculating t_{obs} for two dependent means is to calculate the difference of each pair of scores. These differences are shown at the right of Table 11.3a. It is important to record the *signs* of the differences correctly. The two means, $\bar{y}_1$ and $\bar{y}_2$, can be calculated in the usual way from the sums of the scores in the two groups. The sums and means are shown in Table 11.3b. Also shown are the sum (Σd) and sum of squares (Σd^2) of the differences. These two sums are required for the calculation of s_d shown in Table 11.3c. Finally, t_{obs} is calculated from the formula. Its value is -0.95, the negative sign indicating that the second group has a greater mean than the first group.

8 The test statistic falls in the acceptance region; we are unable to reject the hypothesis that the two conditions have the same population mean.

Comments a) The difference between the means, $\bar{y}_1 - \bar{y}_2$, which forms the numerator of t_{obs}, may be checked by the calculation of the sum of the difference scores, Σd. The mean difference, $\bar{d} = (\Sigma d)/n$ is exactly the same as $(\bar{y}_1 - \bar{y}_2)$. (It can be shown algebraically that the mean of the difference scores is equal to the difference of the group means.) As shown in Table 11.3b, $\bar{d} = -0.86$, which is equal to $(\bar{y}_1 - \bar{y}_2) = 10.94 - 11.80 = -0.86$. (b) The p-value can, of course, be calculated for the t-test of two dependent means. In this example, the observed value of the test statistic, in absolute value, is 0.95, which is less than $t_{.100} (= 1.397)$ for $df = 8$. Hence, the two-tailed p-value is greater than $2 \times 0.1 = 0.2$. The observed difference is not significant (p-value > 0.20, two-tailed).

11.6 Relationships Among the Formulas

Note. This section may be omitted without loss of continuity.

Your understanding of z-tests and t-tests will increase if you learn the relationships among the various formulas. This section outlines first the relationships between the formulas for z_{obs} and for t_{obs}, and then the relationships among the three formulas for t_{obs}.

The Formulas for t_{obs} and z_{obs}

In Section 7.1 the standard score (z-score) for a variable y was defined as

$$z = \frac{y - \text{mean}}{\text{s.d.}}$$

In hypothesis testing, the standard score is called z_{obs}, which has the general form

$$z_{obs} = \frac{\text{Observed statistic} - \left(\begin{array}{c}\text{Mean of the distribution of the}\\ \text{statistic when } H_0 \text{ is true}\end{array}\right)}{\text{Standard deviation of the distribution of the statistic}}$$

(The qualification "when H_0 is true" does not have to be added to the expression in the denominator, since in all tests we have considered, we have assumed that the standard deviation is the same when H_0 is true and when H_1 is true.)

In the *z-test of one mean* (Section 9.3) the null hypothesis is $H_0: \mu = \mu_0$: the observed statistic is $\bar{y}$, the sample mean; the mean of the distribution of $\bar{y}$ when H_0 is true is μ_0; and the standard deviation of the distribution of $\bar{y}$ is $\sigma/\sqrt{n}$, where σ is the population standard deviation and n is the sample size. Hence, the test statistic is

$$z_{obs} = \frac{\bar{y} - \mu_0}{\sigma/\sqrt{n}} \quad \text{(Test of one mean)}$$

In the *z-test of two independent means* (Section 10.6) the null hypothesis is $H_0: \mu_1 = \mu_2$; the observed statistic is $(\bar{y}_1 - \bar{y}_2)$; the mean of the distribution of

$(\bar{y}_1 - \bar{y}_2)$ when H_0 is true is 0 (zero); and the standard deviation of the distribution of $(\bar{y}_1 - \bar{y}_2)$ is

$$\sqrt{\frac{\sigma_1^2}{n_1} + \frac{\sigma_2^2}{n_2}}$$

where σ_1^2 and σ_2^2 are the two population variances and n_1 and n_2 are the two sample sizes. Hence, the test statistic is

$$z_{obs} = \frac{(\bar{y}_1 - \bar{y}_2) - 0}{\sqrt{\frac{\sigma_1^2}{n_1} + \frac{\sigma_2^2}{n_2}}} \text{(Test of two independent means)}$$

The formulas for t_{obs} differ from those for z_{obs} only in the denominator. In general, t_{obs} has the form

$$t_{obs} = \frac{\text{Observed statistic} - \left(\begin{array}{c}\text{Mean of the distribution of the}\\ \text{statistic when } H_0 \text{ is true}\end{array}\right)}{\left(\begin{array}{c}\text{Estimate of the standard deviation of the}\\ \text{distribution of the statistic}\end{array}\right)}$$

In the *t-test of one mean* (Section 11.2) the standard deviation of $\bar{y}$ is estimated by $s/\sqrt{n}$, giving

$$t_{obs} = \frac{\bar{y} - \mu_0}{s/\sqrt{n}} \quad \text{(Test of one mean)}$$

In the *t-test of two independent means* (Section 11.4), we make the additional assumption that $\sigma_1^2 = \sigma_2^2$. If each of these two variances equals σ^2, the formula for the standard deviation of $(\bar{y}_1 - \bar{y}_2)$ can be written

$$\sqrt{\frac{\sigma^2}{n_1} + \frac{\sigma^2}{n_2}} = \sigma\sqrt{\frac{1}{n_1} + \frac{1}{n_2}}$$

If we now estimate the common population standard deviation by the pooled sample standard deviation, s_p, we can obtain the estimate of the standard deviation of $(\bar{y}_1 - \bar{y}_2)$.

$$s_p\sqrt{\frac{1}{n_1} + \frac{1}{n_2}}$$

Then, using the general formula for t_{obs} given above, the statistic is

$$t_{obs} = \frac{(\bar{y}_1 - \bar{y}_2) - 0}{s_p\sqrt{\frac{1}{n_1} + \frac{1}{n_2}}} \quad \text{(Test of two independent means)}$$

The formula for t_{obs} in the *t-test of two dependent means* is also a special case of the general formula, as shown in the following subsection.

The Formulas for t_{obs}

Consider first the formulas for the test of two independent means and for the test of two dependent means. The observed statistic is the same in each case $(\bar{y}_1 - \bar{y}_2)$; the null hypothesis is the same (H_0: $\mu_1 = \mu_2$); and the mean of the distribution of $(\bar{y}_1 - \bar{y}_2)$ when H_0 is true is 0 (zero) in both cases. Hence, the numerator of the test statistic, t_{obs}, is the same for two independent means, as shown above, and for two dependent means.

$$t_{obs} = \frac{(\bar{y}_1 - \bar{y}_2) - 0}{s_d/\sqrt{n}} \quad \text{(Test of two dependent means)}$$

The two formulas differ in their *denominators*.

The *t*-test of two dependent means is closely related to the *t*-test of one mean. The data from a dependent-groups design can be thought of in two ways: (1) there are two sets of scores, y_1 and y_2, and two means; or (2) there is one set of difference scores, d, and one mean. If we think of the data in the first way, we will consider two population means, μ_1 and μ_2, and the null hypothesis H_0: $\mu_1 = \mu_2$. If we think of the data in the second way, we will consider one population mean, μ_d, and the null hypothesis H_0: $\mu_d = 0$. Here μ_d is the mean of the single population from which the difference scores can be considered to be drawn. Clearly, $\mu_d = \mu_1 - \mu_2$. Hence, the two null hypotheses are equivalent. In order to apply the general formula for t_{obs} to the difference scores, we note that the observed statistic is $\bar{d}$; that the mean of the distribution of $\bar{d}$ when H_0 is true is 0 (zero); and that the estimate of the standard deviation of the distribution of $\bar{d}$ is $s_d/\sqrt{n}$, where s_d is the adjusted standard deviation of the difference scores in the sample. The test statistic is, therefore,

$$t_{obs} = \frac{\bar{d} - 0}{s_d/\sqrt{n}}$$

This formula is, of course, the same as the formula that we have used before for two dependent means, since $\bar{d} = \bar{y}_1 - \bar{y}_2$. We have shown, therefore, that the *t*-test of two dependent means is essentially a *t*-test of one mean—namely, the mean of the difference scores.

11.7 Summary

The *t*-distribution is a symmetric continuous distribution. There are many *t*-distributions; each *t*-distribution is labeled by its *degrees of freedom*. When the number of degrees of freedom is large (say, 30 or more), the *t*-distribution is essentially identical to the standard normal distribution.

Three tests were described: the *t-test of one mean*, the *t-test of two independent means*, and the *t-test of two dependent means*. The test statistic for each test has a *t*-distribution when the null hypothesis is true. There are relationships among the formulas for the three test statistics and between the formulas for the *t*-tests and the *z*-tests.

The *t*-tests of two means are used to study the relationship between an *independent* and a *dependent variable*. However, other variables, called *irrelevant variables*, can obscure this relationship. The *basic principle of experimental design* summarizes the techniques that can be used to minimize the effect of irrelevant variables. In the *dependent-groups design*, some of the irrelevant variables are *matched* and the remaining variables are either *randomized* or held constant. In the *independent-groups design*, all the variables are randomized or held constant.

Exercises

1 (Section 11.1) Replace the question mark in each of the following expressions by the appropriate number found from Table A6. The number of degrees of freedom is given

for each expression. (*Hint.* Make a sketch of each distribution and show the given numbers on the sketch.)

	df	EXPRESSION
(a)	6	$t_{.005} = ?$
(b)	10	$P(t > 1.812) = ?$
(c)	10	$P(t < -1.812) = ?$
(d)	10	$P(|t| > 1.812) = ?$
(e)	15	90th percentile $= ?$
(f)	23	$P(t > 2.317) < ?$
(g)	23	$P(t > 2.317) > ?$
(h)	27	$P(t < -2.623) < ?$
(i)	35	$P(t > 2.46) < ?$

2 (Section 11.2) The following data were recorded from a random sample drawn from a population with unknown mean and standard deviation. Test the hypothesis that the unknown mean is 56 by a two-tailed test ($\alpha = 0.01$). The data are 72, 55, 68, 66, 64, 53, 69, 68, 60, 71, 62.

3 (Section 11.3) Find a report in a published paper or in a textbook that describes an experiment. Give the complete reference to the paper or text and write a brief summary of the experiment; then answer the following questions about the experiment.

(a) Identify the manipulated variable, the dependent variable, the observed variable, the independent variable. (*Hint.* Some of these variables are the same variable.)
(b) Identify an irrelevant variable that was held constant in the experiment.
(c) Identify an irrelevant variable that was randomized in the experiment.
(d) Were any variables matched? If so, identify one of them.
(e) Was an independent-groups or a dependent-groups design used in the experiment?
(f) Why do you think the experimenter chose the design in **(e)** over the other type of design?

4 (Section 11.4) Two conditions are studied in a learning experiment. In one condition, the subjects are given *massed* practice: in the other, they are given *distributed* practice. The dependent variable is the number of errors that they make on series of test trials. The results are as follows.

MASSED PRACTICE	DISTRIBUTED PRACTICE
5	4
12	7
12	3
6	6
1	2
7	
3	

Test the hypothesis that the two conditions have equal mean number of errors. Report the result using a *p*-value.

5 (Section 11.4) Test the hypothesis of equal population means ($\alpha = 0.01$) for the following data from two independent groups.

	GROUP 1	GROUP 2
Sample size	14	14
Mean	104.72	108.34
Standard deviation (s)	3.16	4.22

6 (Section 11.5) In a learning experiment, subjects were matched for previous experience with the stimulus materials. The two subjects with the most experience were matched, then one was chosen at random for the massed-practice condition and the other assigned to the distributed-practice condition. Similarly, the two subjects with the next-highest experience were matched and assigned to the two conditions. This procedure was followed for seven matched pairs. The dependent variable is the number of errors in a series of test trials. The results are as follows.

PAIR	MASSED PRACTICE	DISTRIBUTED PRACTICE
1	5	2
2	8	6
3	9	3
4	6	7
5	8	4
6	7	5
7	3	6

Test the hypothesis that the two conditions have equal mean numbers of errors, using a one-tailed test at $\alpha = 0.05$.

12 ESTIMATION

There are two principal inferential techniques in statistics: hypothesis testing and estimation. In hypothesis testing, we decide between two hypotheses about a population parameter; in estimation, we estimate the value of the population parameter.

The theory of estimation applies with little variation to the estimation of many population parameters. The general theory is described in Sections 12.3 and applied to several parameters in Sections 12.4 and 12.5. For a condensed treatment of estimation, skip Sections 12.1 and 12.2 and begin reading the chapter at Section 12.3. You may find, however, that the theory of estimation will be clearer if you begin with detailed examples that illustrate more concretely the new concepts introduced in this chapter. If you prefer this approach, read Sections 12.1 and 12.2 first.

The theory developed in this chapter is very useful for determining the sample size to use in a study; one method of planning the sample size is described in Section 12.6.

12.1 Estimating a Population Mean—an Example

We want to estimate the mean of a certain population from the data in a random sample drawn from that population. We draw a random sample of size two from the population; the two scores in the sample are 5 and 9. The sample mean, $(5 + 9)/2 = 7.0$, is an estimate of the population mean. In order to explain why the sample mean is a good estimate of the population mean, we must consider a large number of samples from the population and show that on the average the sample mean has the properties required for a good estimate. Furthermore, we need to know the value of the population mean in order to show that the sample mean is a good estimate of it. So we begin with a description of the population and the sampling process. Keep in mind, though, that in an actual research study the population mean would not be known and only one sample would be drawn.

The population consists of the 10 digits: 0, 1, 2, . . . , 9. Sampling from this population is done *with* replacement. Each digit has a probability of 1/10 of being selected. The random digits in Table A5 in the Appendix were drawn in this way. Random samples of size two ($n = 2$) are obtained from the table by taking pairs of these digits. Twenty such samples are listed in Table 12.1.[1]

Unbiased Estimate

The mean ($\bar{y}$) of each sample is shown in the second column of Table 12.1. Each sample mean is an estimate of the population mean. The mean of the population is (see Section 6.6)

---

[1] The 40 digits required for these 20 samples were found in lines 58 to 65 of column 14 of Table A5.

TABLE 12.1 Estimates and Errors for 20 Random Samples

SAMPLE	$\bar{y}$	ERROR	SQUARED ERROR
5, 9	7.0	+2.5	6.25
8, 4	6.0	+1.5	2.25
6, 9	7.5	+3.0	9.00
2, 3	2.5	−2.0	4.00
2, 5	3.5	−1.0	1.00
8, 7	7.5	+3.0	9.00
8, 2	5.0	+0.5	0.25
0, 4	2.0	−2.5	6.25
6, 9	7.5	+3.0	9.00
2, 0	1.0	−3.5	12.25
9, 9	9.0	+4.5	20.25
3, 7	5.0	+0.5	0.25
8, 6	7.0	+2.5	6.25
6, 0	3.0	−1.5	2.25
9, 2	5.5	+1.0	1.00
1, 6	3.5	−1.0	1.00
8, 3	5.5	+1.0	1.00
4, 3	3.5	−1.0	1.00
4, 1	2.5	−2.0	4.00
9, 1	5.0	+0.5	0.25
Means	4.95	0.45	4.825

Note. Error $= \bar{y} - \mu = \bar{y} - 4.5$
Squared error $= (\bar{y} - 4.5)^2$

$$\mu = \Sigma y P(y)$$

$$= 0 \times \frac{1}{10} + 1 \times \frac{1}{10} + \ldots + 9 \times \frac{1}{10}$$

$$= 0.0 + 0.1 + \ldots + 0.9$$

$$= 4.5$$

Now consider the sample means. Some are greater than the true mean, 4.5; others are less. None of the sample means is exactly equal to 4.5, although some possible samples would have this mean. How accurately do the sample means estimate the population mean? Let us compute the *mean* of these sample means. As shown at the bottom of the second column of Table 12.1, the mean of the sample means is 4.95—close to but somewhat larger than the true mean, 4.5. If we had studied 1,000 samples of size two, instead of just 20 samples, the mean of the sample means would almost certainly be even closer to the true mean. The term *unbiased* is used to describe the fact that on the average the sample mean equals the population mean.

1 The sample mean, $\bar{y}$, is an unbiased estimate of the population mean, μ, because in a large number of random samples, the mean of the sample means equals the population mean.

Another way to measure the accuracy of the sample mean as an estimate of the population mean is to consider the *error of the estimate*; the error is the difference between the estimate and the true value: $(\bar{y} - \mu)$. These errors, which may be positive or negative, are shown in the third column of Table 12.1; their mean is 0.45. If a larger number of samples were studied, the mean error would almost certainly be even closer to zero. An unbiased estimate can be described alternatively as follows.

2 The sample mean, $\bar{y}$, is an unbiased estimate of the population mean, μ, because in a large number of random samples, the mean error of the estimates is zero.

Standard Error

The fact that on the average the errors are zero does not tell us whether the *absolute values* of the errors are small or large. In statistical work, it is usual to work with the *square* of the errors, rather than with their absolute value. The square root of the mean of the squared errors is called the *standard error of* $\bar{y}$. In this instance of 20 samples, the mean squared error is 4.825 (Table 12.1). Hence, the standard error is $\sqrt{4.825} = 2.197$. In Section 12.3, we will see that if many samples are studied, the standard error of $\bar{y}$ equals the standard deviation of the distribution of sample means, which is $\sigma/\sqrt{n}$, where σ is the standard deviation of the population from which the random samples are drawn and n is the sample size. In our example, $n = 2$ and σ is found by the formula of Section 6.6. We first find the variance, σ^2.

$$\sigma^2 = \Sigma(y - \mu)^2 P(y)$$
$$= (0 - 4.5)^2 \times \frac{1}{10} + (1 - 4.5)^2 \times \frac{1}{10} + \ldots + (9 - 4.5)^2 \times \frac{1}{10}$$
$$= 20.25 \times 0.1 + 12.25 \times 0.1 + \ldots + 20.25 \times 0.1$$
$$= 8.25$$

The standard deviation, σ, is

$$\sigma = \sqrt{8.25} = 2.872$$

The theoretical value of the standard error is, therefore,

$$\sigma/\sqrt{n} = 2.872/\sqrt{2} = 2.031$$

We see that the square root of the mean of the squared errors in the 20 samples, 2.197, is close to this theoretical value. We can summarize the standard error as follows.

3 The standard error of $\bar{y}$ is $\sigma/\sqrt{n}$, because in a large number of random samples, the mean of the squared errors in using $\bar{y}$ as an estimate of μ is σ^2/n and hence the square root of the mean squared error is $\sigma/\sqrt{n}$.

Confidence Interval

Let us return to the first of the random samples of size two: the sample with scores 5 and 9. The estimate of the population mean is 7.0. From the standard error of this estimate (2.031), we can state that the estimated mean is 7.0 ± 2.031. In other words, we are indicating that we do not know how large the error in our estimate is, but the square root of the mean squared error is 2.031; sometimes the error will be greater and sometimes less than 2.031, and it can be either positive or negative. We could say, therefore, that the population mean is estimated to be in the interval extending from 7.0 − 2.031 to 7.0 + 2.031. This interval is (4.969, 9.031). Although we are fairly sure that the population mean is in this interval, we cannot be certain that it is; the error could be greater than 2.031, and if it is, the population mean will be outside the interval.

The basic idea of indicating that the population mean is estimated to be within an interval of values is formalized in statistics by the notion of a *confidence interval*. This section describes the 80% confidence interval. (Section 12.3 defines other confidence intervals, such as the 95% confidence interval, the 99% confidence interval, etc.) The 80% confidence interval is somewhat wider than the interval described in the previous paragraph. The 80% confidence interval for the first random sample extends from a *lower limit* of $7.0 - (1.282 \times 2.031)$ to an *upper limit* of $7.0 + (1.282 \times 2.031)$. The 80% confidence interval is, therefore, (4.4, 9.6), rounded to one decimal place. The number 1.282 that is used to construct this interval is found from the normal table; it is $z_{.10}$. (An explanation[2] of why a percentile of the normal distribution is used is given in Section 12.3. Briefly, note that 80% of the area of a standard normal distribution falls between -1.282 and $+1.282$.)

Why is the 80% confidence interval given that name? It is easiest to understand the name if we construct the 80% confidence intervals for all the 20 random samples from the population of random digits. These intervals are given in the fourth column of Table 12.2. The column is headed "80% confidence interval using known σ" since the upper and lower limits are computed from the known value, 2.872, of the population standard deviation. Recall that the standard error is $\sigma/\sqrt{n} = 2.872/\sqrt{2} = 2.031$. (We will show below what is done if the population standard deviation is not known.) The lower limit of each interval is $\bar{y} - 1.282 \times 2.031 = \bar{y} - 2.604$; the upper limit of each interval is $\bar{y} + 1.282 \times 2.031 = \bar{y} + 2.604$.

Notice that the confidence interval computed from the first sample includes the true population mean, 4.5: that is, the number 4.5 falls in the interval (4.4, 9.6). So a claim that the population mean falls in the confidence interval (4.4, 9.6) is correct. The population mean also falls in the interval computed from the second sample (3.4, 8.6). The third interval does not include the true mean however. The claim that the population mean is in the confidence interval (4.9, 10.1) would be incorrect. In the table, all the intervals that include the true mean are indicated by an asterisk.

There are 15 intervals that include the true mean and 5 that do not. The success rate is 15 out of 20, or 75%, which is close to 80%. The 80% confidence interval is so constructed that if we considered a much larger number of samples, the success rate would be very close to 80%. We can summarize this property of the confidence interval as follows.

4 The interval extending from $\bar{y} - z_{.10} \cdot \sigma/\sqrt{n}$ to $\bar{y} + z_{.10} \cdot \sigma/\sqrt{n}$ is the 80% confidence interval (using known σ), because in a large number of random samples, 80% of such intervals include the true mean μ.

In real studies the population standard deviation usually is not known, but it is still possible to construct the 80% confidence interval for estimating the

---

[2] The explanation depends on the normality of the distribution of $\bar{y}$. For the discrete population and the sample size of two of this example, the distribution of $\bar{y}$ is not normal. But it just so happens that in this case the intervals $(\bar{y} - 2.604, \bar{y} + 2.604)$ include the true mean 80% of the time. Usually, when the distribution of $\bar{y}$ is nonnormal, intervals constructed in this way will not be the correct 80% confidence intervals. (The 80% interval is used so that several intervals, among the 20, do not include the true mean. Many more samples than 20 would be required in the example if the 95% or 99% intervals were used.)

TABLE 12.2 Confidence Intervals for 20 Random Samples

SAMPLE	$\bar{y}$	s	80% C.I. USING KNOWN σ	80% C.I. USING SAMPLE s
5, 9	7.0	2.83	(4.4, 9.6)*	(0.8, 13.2)*
8, 4	6.0	2.83	(3.4, 8.6)*	(−0.2, 12.2)*
6, 9	7.5	2.12	(4.9, 10.1)	(2.9, 12.1)*
2, 3	2.5	0.71	(−0.1, 5.1)*	(1.0, 4.0)
2, 5	3.5	2.12	(0.9, 6.1)*	(−1.1, 8.1)*
8, 7	7.5	0.71	(4.9, 10.1)	(6.0, 9.0)
8, 2	5.0	4.24	(2.4, 7.6)*	(−4.2, 14.2)*
0, 4	2.0	2.83	(−0.6, 4.6)*	(−4.2, 8.2)*
6, 9	7.5	2.12	(4.9, 10.1)	(2.9, 12.1)*
2, 0	1.0	1.41	(−1.6, 3.6)	(−2.1, 4.1)
9, 9	9.0	0.00	(6.4, 11.6)	(9.0, 9.0)
3, 7	5.0	2.83	(2.4, 7.6)*	(−1.2, 11.2)*
8, 6	7.0	1.41	(4.4, 9.6)*	(3.9, 10.1)*
6, 0	3.0	4.24	(0.4, 5.6)*	(−6.2, 12.2)*
9, 2	5.5	4.95	(2.9, 8.1)*	(−5.3, 16.3)*
1, 6	3.5	3.54	(0.9, 6.1)*	(−4.2, 11.2)*
8, 3	5.5	3.54	(2.9, 8.1)*	(−2.2, 13.2)*
4, 3	3.5	0.71	(0.9, 6.1)*	(2.0, 5.0)*
4, 1	2.5	2.12	(−0.1, 5.1)*	(−2.1, 7.1)*
9, 1	5.0	5.66	(2.4, 7.6)*	(−7.3, 17.3)*

Notes. **(a)** The 80% confidence interval using the known σ is

$$\text{Lower limit} = \bar{y} - (z_{.10} \cdot \sigma/\sqrt{n}) = \bar{y} - (1.282 \times 2.872/\sqrt{2})$$
$$= \bar{y} - (1.282 \times 2.031)$$
$$= \bar{y} - 2.604$$
$$\text{Upper limit} = \bar{y} + (z_{.10} \cdot \sigma/\sqrt{n}) = \bar{y} + 2.604$$

(b) The 80% confidence interval using the sample s is

$$\text{Lower limit} = \bar{y} - (t^1_{.10} \cdot s/\sqrt{n}) = \bar{y} - (3.078 \times s/\sqrt{2})$$
$$= \bar{y} - (2.176 \times s)$$
$$\text{Upper limit} = \bar{y} + (t^1_{.10} \cdot s/\sqrt{n}) = \bar{y} + (2.176 \times s)$$

(c) The intervals that include the true μ ($= 4.5$) are indicated by an asterisk.

mean. The standard error is *estimated* by substituting the sample standard deviation for the population standard deviation. The estimated standard error is $s/\sqrt{n}$, which differs for each sample. For the first random sample in Table 12.2 the estimated standard error is $2.83/\sqrt{2} = 2.00$. The lower limit of the 80% confidence interval is $7.0 - (3.078 \times 2.00)$; the upper limit is $7.0 + (3.078 \times 2.00)$. The 80% confidence interval is, therefore, (0.8, 13.2). The number 3.078 that is used to construct this interval is found from the t-distribution; it is $t_{.10}$ for $df = 1$. In general, $df = n - 1$ for one population mean. (The df will henceforth be indicated as a superscript of the percentile; for example, $t^1_{.10}$. Section 12.3 explains why the t-distribution is used to compute this confidence interval.)

Note that the confidence interval just computed, (0.8, 13.2), is wider than the confidence interval computed from the known σ, (4.4, 9.6). However, you can see from the last column of Table 12.2 that for some of the other samples the confidence intervals using the sample s are narrower than the confidence intervals using σ. On the average, 80% of these new confidence intervals include the true mean, μ, just as do 80% of the former confidence intervals. It turns out that 16 of the 20 confidence intervals in the last column of the table include the true mean. In 20 samples, it is fortuitous that *exactly* 80% ($= 16/20$)

of the intervals include the true mean. If a large number of samples were considered, the percentage of confidence intervals including the true mean would be very close to 80%.[3]

5 The interval extending from $\bar{y} - t_{.10}^{n-1} \cdot s/\sqrt{n}$ to $\bar{y} + t_{.10}^{n-1} \cdot s/\sqrt{n}$ is the 80% confidence interval (using sample s), because in a larger number of random samples, 80% of such intervals include the true mean μ.

12.2 Estimating a Population Variance—an Example

Let us continue the study of the sampling example begun in Section 12.1. In this section, we will see how to estimate the population variance and the population standard deviation.

Biased Estimate of σ^2

The sample variance, introduced in Section 4.4, may be used to estimate the population variance. The unadjusted sample variance, s'^2, is

$$s'^2 = \frac{\Sigma(y - \bar{y})^2}{n} = \frac{\Sigma y^2 - (\Sigma y)^2/n}{n}$$

The unadjusted sample variance (for each of the 20 random samples) is shown in the second column of Table 12.3. The mean of these 20 estimates is 4.325.

[3] As pointed out in footnote 2, this statement is not precisely true when the distribution of $\bar{y}$ is not normal. In the present example the distribution of $\bar{y}$ is not normal. However, the general idea of an 80% confidence interval using the sample s should be clear from this example.

TABLE 12.3 Variability Estimates for 20 Random Samples

SAMPLE	s'^2	s^2	s'	s
5, 9	4.00	8.00	2.0	2.828
8, 4	4.00	8.00	2.0	2.828
6, 9	2.25	4.50	1.5	2.121
2, 3	0.25	0.50	0.5	0.707
2, 5	2.25	4.50	1.5	2.121
8, 7	0.25	0.50	0.5	0.707
8, 2	9.00	18.00	3.0	4.243
0, 4	4.00	8.00	2.0	2.828
6, 9	2.25	4.50	1.5	2.121
2, 0	1.00	2.00	1.0	1.414
9, 9	0.00	0.00	0.0	0.000
3, 7	4.00	8.00	2.0	2.828
8, 6	1.00	2.00	1.0	1.414
6, 0	9.00	18.00	3.0	4.243
9, 2	12.25	24.50	3.5	4.950
1, 6	6.25	12.50	2.5	3.536
8, 3	6.25	12.50	2.5	3.536
4, 3	0.25	0.50	0.5	0.707
4, 1	2.25	4.50	1.5	2.121
9, 1	16.00	32.00	4.0	5.657
Means	4.325	8.650	1.800	2.546

Note. $\sigma^2 = 8.25$, $\sigma = 2.872$.

The population variance, σ^2, for the population of random digits is 8.25 (see Section 12.1). We see that, on the average, the estimates are too low. Therefore, the estimate, s'^2, is called a *biased* estimate of σ^2.

6 The unadjusted sample variance, s'^2, is a biased estimate of the population variance, σ^2, because in a large number of random samples, the mean of the unadjusted sample variances does not equal the population variance.

On the average, in fact, the unadjusted sample variance is *less than* the population variance. In other words, the unadjusted sample variance is too small. An unbiased estimate of σ^2 is obtained by adjusting the estimate; the adjustment consists of changing the divisor in the sample variance from n to $(n - 1)$, which increases the value of the sample variance (Section 11.1). The adjusted sample variance is unbiased.

Unbiased Estimate of σ^2

The adjusted sample variance is

$$s^2 = \frac{\Sigma(y - \bar{y})^2}{n - 1} = \frac{\Sigma y^2 - (\Sigma y)^2/n}{n - 1}$$

Its value, for each of the 20 random samples, is shown in the third column of Table 12.3. The mean of these 20 estimates is 8.650, which is close to the population variance ($\sigma^2 = 8.25$). If a larger number of samples were studied, the mean of the adjusted estimates would almost certainly be even closer to the population variance.

7 The adjusted sample variance, s^2, is an unbiased estimate of the population variance, σ^2, because in a large number of random samples, the mean of the adjusted sample variances equals the population variance.

This example illustrates why we prefer the adjusted variance to the unadjusted variance.

Estimates of σ

Since the adjusted sample variance, s^2, is an unbiased estimate of the population variance, σ^2, it might be thought that the sample standard deviation, s, would be an unbiased estimate of the population standard deviation, σ. This is not the case, however. Unfortunately, it is not possible to find an unbiased estimate of the population standard deviation. One reason that statisticians prefer to work with variances rather than with standard deviations is that the adjusted sample variance is an unbiased estimate, whereas there is no unbiased estimate of the population standard deviation. Nevertheless, when an estimate of s is required, s rather than s' usually is preferred.

Look at the two sample standard deviations, s' and s, in the third and fourth columns of Table 12.3. Note that the mean of the 20 unadjusted standard deviations is 1.800—much smaller than the population standard deviation, 2.872. The mean of the 20 adjusted standard deviations is 2.546—much closer to σ—but the discrepancy between the two would remain even in a larger number of random samples.

8 The unadjusted sample standard deviation, s', and the adjusted standard deviation, s, are biased estimates of the population standard deviation, σ, because in a large number of random samples, the means of both estimates are not equal to the population standard deviation. However, the bias in s is usually less than the bias in s'; for this reason, s is preferred to s' as an estimate of σ.

12.3 The Theory of Point and Interval Estimation

An estimate of a population parameter may be reported in two ways. If a single number is given as the estimate, this estimate is called a *point estimate* or a *point estimator*. (The words "estimate" and "estimator" are synonymous. The word "point" is used to indicate that a single value is being reported as the estimate.) The other way to report an estimate is to give an interval of values in which the population parameter is claimed to fall; this estimate is called an *interval estimate* or an *interval estimator*.

Most of this section deals with the population mean, μ, and the estimate $\bar{y}$. The properties of the point and interval estimates of μ depend on the central limit theorem. Recall from Section 7.5 that the central limit theorem states that the distribution of the sample mean, $\bar{y}$ has three characteristics: (1) its mean is μ, the population mean; (2) its standard deviation is $\sigma/\sqrt{n}$, where σ is the population standard deviation and n is the sample size; and (3) its shape is approximately normal.

Point Estimates

The central limit theorem states that the mean of the distribution of the sample mean is equal to the population mean. In some samples the sample mean is greater than the population mean, and in other samples it is less; but on the average, the sample mean is equal to the population mean. The sample mean is, therefore, an *unbiased estimate* of the population mean.

Another way to think of a point estimate is to consider the *error* of the estimate. If the estimate is $\bar{y}$, the error of the estimate is $(\bar{y} - \mu)$. In some samples the error is positive and in other samples the error is negative. The central limit theorem states that on the average the error is zero. [Since the mean of $\bar{y}$ is μ, the mean of $(\bar{y} - \mu)$ must be zero.][4] The sample mean is an unbiased estimate of the population, because the average error in the estimate is zero.

Other parameters also have unbiased estimates. There are two (equivalent) *definitions of an unbiased estimate* of a population parameter: (1) an unbiased estimate of a population parameter is an estimate whose mean, computed from estimates in all possible random samples, is equal to the parameter; or (2) an unbiased estimate of a population parameter is an estimate whose mean error, computed from estimates in all possible random samples, is equal to zero. Note that in these definitions, the expression "on the average", which we have been using, has been replaced by the more precise and technical term "mean". The sample mean, as an estimate of the population mean, satisfies these definitions and is, therefore, an unbiased estimate of the population mean.

Now consider the estimation of another population parameter, the variance σ^2. It can be shown that the adjusted sample variance, s^2, is an unbiased estimate of σ^2. The proof, that the mean of s^2 computed in all possible random

---

[4] This statement is actually a statement of the effect of change of origin on the mean of a distribution. (See Section 7.1.) When the origin of $\bar{y}$ is changed by subtracting μ, the mean is changed from μ to $(\mu - \mu) = 0$.

samples is σ^2, depends on the basic definition of the population variance and also on the central limit theorem. A related proof shows that the mean of the *unadjusted* sample variance, s'^2, equals $[(n - 1)/n]\sigma^2$, which is less than σ^2. Hence, s'^2, is a *biased estimate* of the population variance. The unadjusted sample variance is an *underestimate* of the population variance; however, the bias decreases as n increases, and when, say, $n \geqq 30$, the bias is quite small.

A biased estimate can be defined simply as an estimate that is not unbiased. Or one can use the following two (equivalent) *definitions of a biased estimate* of a population parameter: (1) a biased estimate of a population parameter is an estimate whose mean, computed from estimates in all possible random samples, is *not* equal to the parameter; or (2) a biased estimate of a population parameter is an estimate whose mean error, computed from estimates in all possible random samples, is *not* equal to zero.

It can be shown that both s and s' are biased estimates of the population standard deviation, σ. The adjusted estimate, s, usually is preferred, because it is the square root of the unbiased estimate of σ^2 and because it is usually less biased than s'.

To summarize the results stated above:

PARAMETER	ESTIMATE	BIASED/UNBIASED
μ	$\bar{y}$	Unbiased
σ^2	s^2	Unbiased
σ^2	s'^2	Biased
σ	s	Biased
σ	s'	Biased

The Standard Error

It is not enough that a point estimate be unbiased: that is, that its mean error, computed from estimates in all possible random samples, is zero. It is also desirable that the variability of the errors be as small as possible, since the estimate is more precise if the variability is small than if the variability is large. One measure of variability is the standard deviation. The *standard error* of an estimate is defined as the standard deviation of the errors in the estimates computed in all possible random samples.

Now consider the estimate $\bar{y}$ of the population mean, μ. The standard deviation of the distribution of $\bar{y}$ is $\sigma_{\bar{y}} = \sigma/\sqrt{n}$, as stated by the central limit theorem. The error in the estimate is $(\bar{y} - \mu)$, which is simply a change in origin of the variable $\bar{y}$. As shown in Section 7.1, the standard deviation of a variable does not change when the origin changes. Hence, the standard deviation of the errors—and therefore the standard error of $\bar{y}$—is $\sigma/\sqrt{n}$.

The standard error of $\bar{y}$ is abbreviated $SE_{\bar{y}}$, or simply SE if it is understood that the point estimate is $\bar{y}$. We have

$$SE_{\bar{y}} = \sigma/\sqrt{n}$$

As mentioned above, it is desirable to have the variability of the errors, and hence the standard error, as small as possible. From the formula for $SE_{\bar{y}}$, we can see that the standard error decreases as n increases. Hence, it is desirable to have the sample size as large as possible.

Example 1

The population of random digits used as an example in Sections 12.1 and 12.2 has $\mu = 4.5$ and $\sigma = 2.872$. When the sample size is 2, the standard error of $\bar{y}$ is $2.872/\sqrt{2} = 2.0310$. If the sample size is 8, however, the standard error is $2.872/\sqrt{8} = 1.0155$. From this example, we see that by increasing the sample size by a factor of four (from 2 to 8), the standard error is decreased by a factor of two (from 2.0310 to 1.0155).

The standard error of an estimate is a very useful indicator of the precision of the estimate, but if it is necessary to estimate μ, it is very unlikely that σ will be known. Hence, the standard error of $\bar{y}$, which is $\sigma/\sqrt{n}$, cannot be calculated. However, σ can be estimated by the point estimate s and the standard error can be estimated by $s/\sqrt{n}$. We denote the estimated standard error by "$est(SE_{\bar{y}})$".

$$\boxed{est(SE_{\bar{y}}) = s/\sqrt{n}}$$

When it is clear that the mean, and not some other statistic, is the estimate, its estimated standard error may be abbreviated to $est(SE)$.

Example 2

A sample of 16 persons is studied. Their scores have a mean of 32.837 and a standard deviation of 2.076. Estimate the population mean and the standard error.

The standard error can be estimated from the sample standard deviation and the sample size. The estimated standard error is

$$est(SE) = s/\sqrt{n} = 2.076/\sqrt{16} = 0.519$$

Since the standard error is a measure of the average error in the estimated mean, it is not useful to report the standard error to more than one, or at most two digits. Hence, the standard error should be reported as 0.5 (or 0.52). The number of decimal places that is usefully reported in the estimated mean is the same as that of the standard error. Hence, we report the estimated mean as 32.8, with a standard error of 0.5 (or we could say that the estimated mean is 32.84, with a standard error of 0.52).

Interval Estimates

When we estimate the mean of a population by the mean of a sample, it is very unlikely that the estimate will be exactly equal to the population parameter. We can indicate that the estimate is probably in error by attaching a "plus or minus" to the estimate. The standard error is a measure of the average size of the error, and therefore it is natural to give the estimate as $\bar{y} \pm SE$. The estimate is now an interval estimate extending from a lower limit of $\bar{y} - SE$ to an upper limit of $\bar{y} + SE$. However, even this interval might not include the population mean. The theory of confidence intervals was developed by statisticians to give precise meaning to an interval estimate of this kind.

An interval estimate can only be understood by considering the *probability* that the interval includes the true parameter. The probability of an interval including a particular number is given a special name—"confidence coefficient". The *confidence coefficient* of an interval estimate is the probability (usually expressed as a percentage) that the interval includes the population parameter that is being estimated. For example, the 95% confidence interval for the mean is an interval constructed so that the probability that the interval includes the population mean is 0.95.

Confidence intervals can be constructed for estimating many different parameters. Only confidence intervals for the mean are dealt with in the remainder of this section. Confidence intervals for other parameters are considered in Sections 12.4 and 12.5.

Interpretation A 95% confidence interval is interpreted as follows. The formula for the interval is designed so that if the same formula is used in all possible random samples drawn from the population with mean μ, the confidence interval will include the value μ in 95% of the samples. Note carefully that it is the interval, and not the parameter μ, that varies from sample to sample. The probability of 0.95 refers to the probability of the *interval* (i.e., the probability that the interval includes μ), not to the probability of the parameter. A given population has a fixed mean, which does not vary; the mean cannot have a probability of having a value in a certain interval. Rather, the intervals vary, some containing the population mean and others not containing it—the 95% confidence interval is so constructed that 95% of the intervals contain the population mean.

Formula When σ Is Known To describe a confidence interval with an arbitrary confidence coefficient, we will use the expression $(1 - \alpha)100\%$ for the confidence coefficient. [This complex expression is used because it makes the formulas for the limits of the confidence interval relatively simple. For example, $80 = (1 - 0.20)100$, so that $\alpha = 0.20$.] The lower and upper limits of the $(1 - \alpha)100\%$ confidence interval for the mean, when σ is known, are

$$
\begin{array}{l}
\text{Lower limit} = \bar{y} - z_{\alpha/2} \cdot SE = \bar{y} - z_{\alpha/2} \cdot \sigma/\sqrt{n} \\
\text{Upper limit} = \bar{y} + z_{\alpha/2} \cdot SE = \bar{y} + z_{\alpha/2} \cdot \sigma/\sqrt{n}
\end{array}
\quad (\sigma \text{ known})
$$

Note that percentiles of the *normal* distribution appear in these formulas (for a reason to be given below) and that the area in the upper tail of the normal distribution is $\alpha/2$. For the 80% confidence interval, $\alpha = 0.20$ and $\alpha/2 = 0.10$. Since $z_{.10} = 1.282$, the 80% confidence interval is $(\bar{y} - 1.282 \cdot SE, \bar{y} + 1.282 \cdot SE)$. Think of this in the following way: there is an area of 0.80 in the middle of the distribution and an area of 0.10 in each tail. Another example: the interval $(\bar{y} - SE, \bar{y} + SE)$ is the 68% confidence interval, since $68 = (1 - 0.32)100$, $0.32/2 = 0.16$, and $z_{.16} = 1.00$. Hence, the 68% confidence interval is $(\bar{y} - 1.00 \cdot SE, \bar{y} + 1.00 \cdot SE)$, or $(\bar{y} - SE, \bar{y} + SE)$, as stated.

Justification The central limit theorem states that the distribution of the sample mean, $\bar{y}$, is approximately normal with mean $= \mu$, the population mean, and standard deviation $= \sigma/\sqrt{n}$, where σ is the population standard deviation and n is the sample size. The standard deviation of $\bar{y}$ is, of course, what we now are calling the standard error, SE. The approximation of the distribution of $\bar{y}$ to a normal distribution is quite good when $n > 10$ and the population distribution is continuous. If the population distribution is discrete, larger sample sizes may be necessary for the approximation to be good enough for the formulas of this section to be used.

The distribution of $\bar{y}$ is illustrated in Figure 12.1a. The figure also shows that 95% of the sample means will fall in the interval extending from $\mu - 1.96SE$ to $\mu + 1.96SE$. (Note that $z_{.025} = 1.96$.) (*Warning.* This interval is *not* the 95% confidence interval! The interval in the figure is centered on the unknown

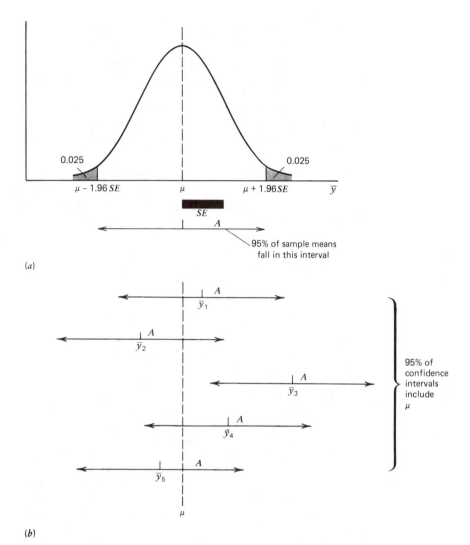

Figure 12.1 Confidence intervals for μ. (a) Distribution of $\bar{y}$. (b) Confidence intervals.

population mean, μ. Confidence intervals are centered on an observed sample mean, $\bar{y}$.) We will use the interval in the figure to show that in 95% of all random samples, the 95% confidence interval includes μ.

Let us denote the *width* of the interval in Figure 12.1a by A. In fact, $A = 1.96SE + 1.96SE = 3.92SE$. The 95% confidence interval has the same width, A, since a 95% confidence interval extends from $\bar{y} - 1.96SE$ to $\bar{y} + 1.96SE$. The 95% confidence intervals are shown for five samples in Figure 12.1b. Each of these intervals is centered on its mean ($\bar{y}_1, \ldots, \bar{y}_5$) and has width A. Notice that μ is included in most of the intervals. Only the interval about $\bar{y}_3$ does not extend as far as μ.

What is the probability that a confidence interval includes μ? The answer is simple: the probability is 0.95. The interval includes μ for precisely those sample means (such as $\bar{y}_1, \bar{y}_2, \bar{y}_4$, and $\bar{y}_5$) that fall in the original interval about the population mean extending from $\mu - 1.96SE$ to $\mu + 1.96SE$. The argument depends on the fact that the confidence intervals have exactly the same width, A, as the interval about the population mean in Figure 12.1a. Since the probability that a sample mean falls in the interval about μ in Figure 12.1a is 0.95, it follows that the probability that a confidence interval of the same width includes μ is also 0.95.

Conversely, if the sample mean (such as $\bar{y}_3$) is outside the interval about the population mean in Figure 12.1a, then the confidence interval drawn about the sample mean will not include μ. The probability of this occurrence is 0.05.

Formula When σ *Is Unknown* In most situations the population standard deviation is not known and therefore the standard error must be estimated by $est(SE) = s/\sqrt{n}$. Furthermore, the percentiles of the t-distribution, instead of the normal distribution, are used. Recall, from Section 11.2, that the number of degrees of freedom is $(n - 1)$ for the t-test of one mean. The same df is used for the confidence interval when σ is unknown. The df is indicated as a superscript of the t in the formulas for the limits of the $(1 - \alpha)100\%$ confidence interval.

$$\text{Lower limit} = \bar{y} - t_{\alpha/2}^{n-1} \cdot est(SE) = \bar{y} - t_{\alpha/2}^{n-1} \cdot s/\sqrt{n}$$
$$\text{Upper limit} = \bar{y} + t_{\alpha/2}^{n-1} \cdot est(SE) = \bar{y} + t_{\alpha/2}^{n-1} \cdot s/\sqrt{n}$$
$(\sigma \text{ unknown})$

Example 3

Consider again the data of Example 2: $\bar{y} = 32.837$, $s = 2.076$, $n = 16$. The estimated standard error is $est(SE) = 2.076/\sqrt{16} = 0.519$. In order to compute the 95% confidence interval for the mean, we need $t_{.025}^{15}$, since we need 2.5% in each tail and $df = 16 - 1 = 15$. The percentile is 2.131. The lower limit is $32.837 - (2.131 \times 0.519) = 32.837 - 1.106 = 31.731$. The upper limit is $32.837 + 1.106 = 33.943$. The 95% confidence interval is, therefore, (31.7, 33.9). (Note that the limits are rounded.)

Let us also find the 99% confidence limits for the mean. We require $t_{.005}^{15} = 2.947$. The lower limit is $32.837 - (2.947 \times 0.519) = 32.837 - 1.529 = 31.308$. The upper limit is $32.837 + 1.529 = 34.366$. The 99% confidence interval is, therefore, (31.3, 34.4). The 99% confidence interval is *wider* than the 95% confidence interval; we are more certain (have greater confidence) that such an interval will include the true mean.

Summary of Estimation Formulas

We have concluded discussion of the theory of estimating one mean. The following two sections present the formulas for estimating several other parameters. These new formulas are similar to the formulas described in this section. The principal formulas are summarized in Table 12.4. The first row of the table gives the formulas for estimating one mean by a point estimate and its standard error and for estimating one mean by a confidence interval. In reading the following sections, you should refer back to the table for a summary of the formulas for estimating other parameters.

12.4 Estimating the Difference Between Two Means

Two basic designs, independent-groups and dependent-groups, were described in Section 11.3. For either design, the population parameters of interest are the two means, μ_1 and μ_2. The t-tests described in Chapter 11 are tests of the equality of the two means, or—what comes to the same thing—tests of whether or not the difference, $\mu_1 - \mu_2$, is zero. In this section, we *estimate*, rather than test, this difference.

TABLE 12.4 Summary of Estimation Formulas

PARAMETER	POINT ESTIMATE	ESTIMATED STANDARD ERROR	CONFIDENCE INTERVAL
One Mean (σ unknown)			
μ	$\bar{y}$	$s/\sqrt{n}$	$\bar{y} \pm t_{\alpha/2}^{n-1} \cdot s/\sqrt{n}$
Two Independent Means ($\sigma_1 = \sigma_2$ unknown)			
$\mu_1 - \mu_2$	$\bar{y}_1 - \bar{y}_2$	$s_p \sqrt{\dfrac{1}{n_1} + \dfrac{1}{n_2}}$	$(\bar{y}_1 - \bar{y}_2) \pm t_{\alpha/2}^{n_1 + n_2 - 2} \cdot s_p \sqrt{\dfrac{1}{n_1} + \dfrac{1}{n_2}}$
Two Dependent Means (σ_d unknown)			
$\mu_1 - \mu_2 = \mu_d$	$\bar{y}_1 - \bar{y}_2 = \bar{d}$	$s_d/\sqrt{n}$	$\bar{d} \pm t_{\alpha/2}^{n-1} \cdot s_d/\sqrt{n}$
One Proportion			
p	$\hat{p} = y/n$	$\sqrt{\dfrac{\hat{p}\hat{q}}{n}}$	$\hat{p} \pm z_{\alpha/2} \cdot \sqrt{\dfrac{\hat{p}\hat{q}}{n}}$
Two Independent Proportions			
$p_1 - p_2$	$\hat{p}_1 - \hat{p}_2$	$\sqrt{\dfrac{\hat{p}_1\hat{q}_1}{n_1} + \dfrac{\hat{p}_2\hat{q}_2}{n_2}}$	$(\hat{p}_1 - \hat{p}_2) \pm z_{\alpha/2} \cdot \sqrt{\dfrac{\hat{p}_1\hat{q}_1}{n_1} + \dfrac{\hat{p}_2\hat{q}_2}{n_2}}$

Since the population standard deviations usually are not known, the formulas given below are for the estimated standard error and for the confidence interval using this estimated standard error.

Two Independent Means

When we wish to estimate the difference between two population means, $\mu_1 - \mu_2$, the point estimate obviously is the difference between the two sample means, $\bar{y}_1 - \bar{y}_2$. The estimated standard error is

$$est(SE_{\bar{y}_1 - \bar{y}_2}) = s_p \sqrt{\frac{1}{n_1} + \frac{1}{n_2}} \quad \text{(Independent means)}$$

where n_1 and n_2 are the two sample sizes and where s_p, the pooled standard deviation (Section 11.4), is

$$s_p = \sqrt{\frac{(n_1 - 1)s_1^2 + (n_2 - 1)s_2^2}{n_1 + n_2 - 2}}$$

(For a justification of the above formula for the estimated standard error, see Section 11.6, where this formula is derived as the estimated standard deviation of the distribution of $\bar{y}_1 - \bar{y}_2$.)

The $(1 - \alpha)100\%$ confidence interval, for the difference between the means, has the following limits.

$$\text{Lower limit} = (\bar{y}_1 - \bar{y}_2) - t_{\alpha/2}^{n_1 + n_2 - 2} \cdot s_p \sqrt{\frac{1}{n_1} + \frac{1}{n_2}}$$
$$\text{(Independent means)}$$
$$\text{Upper Limit} = (\bar{y}_1 - \bar{y}_2) + t_{\alpha/2}^{n_1 + n_2 - 2} \cdot s_p \sqrt{\frac{1}{n_1} + \frac{1}{n_2}}$$

A percentile of the t-distribution is used, since the standard error is estimated. The df is $(n_1 + n_2 - 2)$, which is the same value used in the t-test of two independent means.

Example 1

Consider data from a study with two independent groups.[5]

	GROUP 1	GROUP 2
Sample size	$n_1 = 9$	$n_2 = 9$
Mean	$\bar{y}_1 = 9.797$	$\bar{y}_2 = 9.978$
Standard deviation	$s_1 = 2.966$	$s_2 = 3.095$

The point estimate of the difference between the two population means is

$$\bar{y}_1 - \bar{y}_2 = 9.797 - 9.978 = -0.181 \text{ or } -0.18, \text{ rounded}$$

The negative sign of the difference indicates, of course, that the second mean is estimated to be greater than the first.

The pooled standard deviation, s_p, is calculated using the formula of Section 11.4.

$$s_p = \sqrt{\frac{(n_1 - 1)s_1^2 + (n_2 - 1)s_2^2}{n_1 + n_2 - 2}}$$

$$= \sqrt{\frac{(9 - 1) \times (2.966)^2 + (9 - 1) \times (3.095)^2}{9 + 9 - 2}}$$

$$= 3.031$$

The estimated standard error is, then,

$$\text{est}(\text{SE}_{\bar{y}_1 - \bar{y}_2}) = s_p \sqrt{\frac{1}{n_1} + \frac{1}{n_2}}$$

$$= 3.031 \sqrt{\frac{1}{9} + \frac{1}{9}}$$

$$= 1.429 \text{ or } 1.43, \text{ rounded}$$

Note that the estimated standard error (1.43) is much larger than the estimated difference between the means, in absolute value (0.18). Hence, there does not appear to be a real difference between the population means. If the actual error in the estimated difference was $+1.0$ (less than a standard error), the true mean difference would be $-0.18 + 1.00 = +0.82$, a difference in favor of the first group. But if the actual error was -1.0, the true difference would be $-0.18 - 1.00 = -1.18$, a difference in favor of the second group. The size of the estimated standard error (1.43) shows that errors of ± 1.0 would not be uncommon. Hence, there is no evidence for a real difference between the groups in either direction. (See Section C.1 for further discussion of this argument.)

Consider now a confidence interval estimate of the difference of the means. Let us find the limits of the 90% confidence interval.

$$\text{Lower limit} = (\bar{y}_1 - \bar{y}_2) - t_{\alpha/2}^{n_1 + n_2 - 2} \cdot s_p \sqrt{\frac{1}{n_1} + \frac{1}{n_2}}$$

$$= (9.797 - 9.978) - t_{.05}^{16} \times 3.031 \sqrt{\frac{1}{9} + \frac{1}{9}}$$

$$= -0.181 - (1.746 \times 1.429)$$

$$= -0.181 - 2.495$$

$$= -2.676$$

[5] These data are taken from Example 1 of Section 11.4

$$\text{Upper limit} = -0.181 + 2.495$$
$$= 2.314$$

(Note that two of the numbers in the third line of the calculation of the lower limit have been calculated previously: -0.181 is the point estimate of the difference of the means and 1.429 is the estimated standard error.) The 90% confidence interval for the difference of the population means is $(-2.676, 2.314)$. The interval is so wide that it might well be rounded to $(-2.7, 2.3)$.

The confidence interval indicates that the true difference between the means lies somewhere between -2.7 and $+2.3$. The fact that this confidence interval extends from negative to positive values provides evidence that there is no real difference between the means.[6]

Two Dependent Means

When we have two dependent groups, the point estimate of the difference between the population means ($\mu_1 - \mu_2$ or μ_d) again is $\bar{y}_1 - \bar{y}_2$, or, as shown in Section 11.5, simply $\bar{d}$, the mean of the differences between the pairs of scores. The estimated standard error is

$$\boxed{\,est(SE_{\bar{y}_1 - \bar{y}_2}) = est(SE_{\bar{d}}) = s_d/\sqrt{n}\ \text{ (Dependent means)}\,}$$

where n is the number of pairs and where s_d, the standard deviation of the difference scores (Section 11.5), is

$$s_d = \sqrt{\frac{\Sigma d^2 - (\Sigma d)^2/n}{n - 1}}$$

(The formula for the estimated standard error is justified in Section 11.6.)

The limits of the $(1 - \alpha)100\%$ confidence interval for the difference between the means are

$$\boxed{\begin{aligned}\text{Lower limit} &= \bar{d} - t_{\alpha/2}^{n-1} \cdot s_d/\sqrt{n} \\ \text{Upper limit} &= \bar{d} + t_{\alpha/2}^{n-1} \cdot s_d/\sqrt{n}\end{aligned}\quad \text{(Dependent means)}}$$

Note that a percentile of the t-distribution is used and that the number of degrees of freedom is $(n - 1)$, the same value used in the t-test of two dependent means.

Example 2

Consider data from a study with two dependent groups.[7]

	GROUP 1	GROUP 2	DIFFERENCE
Sample size	$n_1 = 9$	$n_2 = 9$	$n = 9$
Mean	$\bar{y}_1 = 10.94$	$\bar{y}_2 = 11.80$	$\bar{d} = -0.86$
Standard deviation			$s_d = 2.725$

There are nine scores in the first group and nine scores in the second; these scores are matched in pairs, giving $n = 9$ difference scores. The mean and standard deviation of the difference scores are shown in the above table.

[6] Estimation and hypothesis testing have many close relationships. For further discussion, see Section C.1.

[7] These data are taken from the Example of Section 11.5.

The point estimate of the difference between the two population means is

$$\bar{y}_1 - \bar{y}_2 = 10.94 - 11.80 = -0.86$$

which is $\bar{d}$ in the above table. The estimated standard error is

$$est(SE_{\bar{y}_1 - \bar{y}_2}) = est(SE_{\bar{d}}) = est(SE_d) = s_d/\sqrt{n} = 2.725/\sqrt{9}$$
$$= 0.908 \text{ or } 0.91, \text{ rounded}$$

The 90% confidence interval for the difference of the means has the following limits.

$$
\begin{aligned}
\text{Lower limit} \quad &= \bar{d} - t_{\alpha/2}^{n-1} \cdot s_d/\sqrt{n} \\
&= -0.86 - t_{.05}^{8} \times 2.725/\sqrt{9} \\
&= -0.86 - (1.860 \times 0.908) \\
&= -0.86 - 1.69 \\
&= -2.55 \\
\text{Upper limit} \quad &= -0.86 + 1.69 \\
&= +0.83
\end{aligned}
$$

(Note that the number 0.908 in the second line of the calculation of the lower limit has been calculated previously as the estimated standard error.) The 90% confidence interval for the difference of the population means is $(-2.55, 0.83)$ or, rounded further, $(-2.6, 0.8)$.

12.5 Estimates Involving Proportions

The parameter p in the binomial distribution (Section 6.3) may be thought of as the probability of drawing a success from the population or as the proportion of successes in the population. In brief, the binomial parameter p is the *population proportion*.

In this section, the notion of a proportion is extended to samples. This notion certainly is familiar from everyday life. In common expressions such as "75% of the class is female", and "only 38% of the children could add to 20", the percentages are *sample proportions*, since they give the proportion of each sample that has the specified characteristic. The expressions can be changed to "0.75 of the class is female" and "only 0.38 of the children could add to 20". A proportion may be expressed either as a decimal fraction or as a percentage. For a population proportion, however, it is customary to express the proportion as a decimal fraction.

The sample proportion is a *statistic*, since it is calculated in a sample. Previously, the statistic that was computed in a binomial study was y, the number of successes in the sample. Since the word "number" is rather unspecific, we choose to call y the sample *count*, in order to indicate that we count how many successes there are in the sample. In a binomial experiment, then, we can either compute the sample count, y, or the sample proportion, which we will denote by $\hat{p}$ ("p-hat"); the "hat" indicates that the symbol represents a sample estimate of a population parameter. The sample proportion is the count divided by the sample size: $\hat{p} = y/n$.

It is easy to estimate the population proportion, p, from a sample. If there are 23 successes in a sample of 40 observations, the estimate of p is $\hat{p} = 23/40 = 0.58$. We now want to study the properties of this estimate. We therefore need to know the distribution of $\hat{p}$.

The Distribution of the Sample Proportion

The distribution of y is binomial. As $\hat{p}$ is equal to y divided by a constant (n), the change from y to $\hat{p}$ is simply a change of scale. In Section 7.1 it was stated that when the scale is changed by dividing by a constant, the mean and standard deviation also are divided by that constant and the shape of the distribution is unchanged. See Figure 12.2 for an illustration of the distributions for $p = 0.7$ and $n = 5$. The binomial distribution in Figure 12.2a has mean =

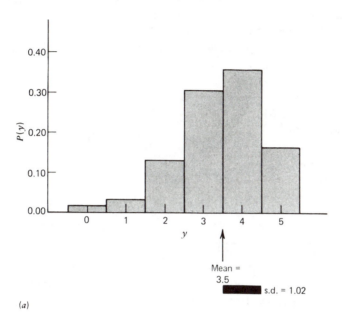

(a)

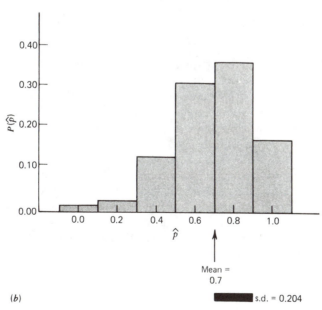

(b)

Figure 12.2 The distribution of the sample proportion.
(a) The distribution of y ($n = 5$, $p = 0.7$).
(b) The distribution of $\hat{p} = y/n$ ($n = 5$, $p = 0.7$).

$np = 5 \times 0.7 = 3.5$ and standard deviation $= \sqrt{npq} = \sqrt{5 \times 0.7 \times 0.3} = 1.02$. (Recall that $q = 1 - p$.) The distribution of $\hat{p}$ is shown in Figure 12.2b. The horizontal scale is now $\hat{p}$, extending from 0.0 to 1.0. The shape of the distribution is unchanged, since the probability of each value of $\hat{p}$ in Figure 12.2b is exactly the same as the probability of the corresponding y in Figure 12.2a. The effect of the change of scale is to divide the mean by n to give a new mean of $3.5/5 = 0.7$, and to divide the standard deviation by n to give a new standard deviation of $1.02/5 = 0.204$.

In general, the mean of the distribution of $\hat{p}$ is $np/n = p$, the population proportion, and the standard deviation of the distribution of $\hat{p}$ is

$$\frac{\sqrt{npq}}{n} = \sqrt{\frac{npq}{n^2}} = \sqrt{\frac{pq}{n}}$$

These two characteristics of the distribution of $\hat{p}$ tell us that (1) $\hat{p}$ is an unbiased estimate of p and (2) the standard error of the estimate is $\sqrt{pq/n}$. In symbols, the standard error may be written

$$\boxed{\text{SE}_{\hat{p}} = \sqrt{\frac{pq}{n}} \quad \text{(One proportion)}}$$

Note that the standard error depends on the population parameter p (and on $q = 1 - p$). If we are estimating p, the value of p is unknown and the standard error is unknown. However, we can estimate the standard error using the sample proportion of successes, $\hat{p}$, and the sample proportion of failures, $\hat{q}$.

$$\boxed{est(\text{SE}_{\hat{p}}) = \sqrt{\frac{\hat{p}\hat{q}}{n}} \quad \text{(One proportion)}}$$

Consider now the shape of the distribution of $\hat{p}$. Figure 12.2 illustrates that the shape is the same as the binomial distribution of y. Since we know that the binomial distribution can be approximated by the normal distribution (Section 9.4), we can also approximate the distribution of $\hat{p}$ by the normal distribution. On the basis of this approximation, confidence intervals for the population proportion can be obtained.

Approximate Confidence Interval for the Proportion

In all the previous examples, the estimated standard error is multiplied by a suitable percentile of the t-distribution to give the limits of the confidence interval. It can be shown that the appropriate multiplier, when estimating a proportion, is a suitable percentile of the *normal* distribution, and not the t-distribution. Hence, the $(1 - \alpha)100\%$ confidence interval for the population proportion is found as follows.

$$\boxed{\begin{aligned} \text{Lower limit} &= \hat{p} - z_{\alpha/2} \cdot \sqrt{\frac{\hat{p}\hat{q}}{n}} \\ &\qquad\qquad\qquad\qquad\qquad \text{(One proportion)} \\ \text{Upper limit} &= \hat{p} + z_{\alpha/2} \cdot \sqrt{\frac{\hat{p}\hat{q}}{n}} \end{aligned}}$$

The most commonly computed confidence intervals are the 95% and the 99% intervals. The above limits are only approximate. The accuracy of these limits depends on the *smaller* of the number of successes, y, and the number of failures, $n - y$. For example, if $n = 50$ and $y = 13$, then the smaller of 13 and 37 (the number of failures) is 13. The accuracy of the confidence limits is as follows.[8]

(a) If the smaller of the number of successes and the number of failures is greater than 20, the limits are accurate to two decimal places.

(b) If the smaller of the number of successes and the number of failures is greater than 5 but not greater than 20, the limits are accurate only to one decimal place.

(c) If the smaller of the number of successes and the number of failures is less than or equal to 5, the formulas should not be used.[9]

(d) Limits that are calculated to be less than 0.05 or greater than 0.95 probably are not accurate. Such limits should be replaced by 0.00, the smallest possible proportion, and by 1.00, the largest possible proportion, respectively.

Example 1

The effectiveness of a patent medicine is studied by finding the proportion of persons obtaining relief from symptoms after taking the medicine. Suppose that 15 out of 25 persons obtain relief. The point estimate of the proportion is $\hat{p} = 15/25 = 0.60$ or 60%. The estimated standard error is

$$est(SE_{\hat{p}}) = \sqrt{\frac{\hat{p}\hat{q}}{n}}$$

$$= \sqrt{\frac{0.60 \times 0.40}{25}} = 0.098 \text{ or } 0.10, \text{ rounded}$$

Let us find the approximate 95% confidence limits for the population proportion.

$$\text{Lower limit} = \hat{p} - z_{\alpha/2} \cdot \sqrt{\frac{\hat{p}\hat{q}}{n}}$$

$$= 0.60 - z_{.025} \cdot \sqrt{\frac{0.60 \times 0.40}{25}}$$

$$= 0.60 - (1.96 \times 0.098)$$

$$= 0.60 - 0.19$$

$$= 0.41$$

$$\text{Upper limit} = 0.60 + 0.19$$

$$= 0.79$$

In determining the accuracy of these limits, we note that the number of successes is 15 and the number of failures is 10. Hence, the smaller of these two quantities is 10. According to the accuracy rules, the limits of the confidence interval are accurate to only one decimal place. Therefore, we should round each limit to one decimal place and report the approximate 95% confidence interval as (0.4, 0.8). The estimated proportion of persons who obtain relief from the medicine is somewhere between 40% and 80% — quite a wide range.

---

[8] These rules were formulated by comparing the approximate limits with the limits given in Chart 1.2 of H. R. Neave, *Statistical Tables* (London: George Allen & Unwin, 1978).

[9] The chart referenced in footnote 8 may be used in this case.

Estimate of the Difference between Two Proportions

We sometimes require an estimate of the difference between the proportions of success in two populations—for example, the difference between the proportion of males who pass a test and the proportion of females who pass it. Only the case of two *independent* samples is considered here.

The unbiased estimate of the difference between the population proportions, $p_1 - p_2$, is the difference between the sample proportions, $\hat{p}_1 - \hat{p}_2$. The standard error of this difference is

$$SE_{\hat{p}_1 - \hat{p}_2} = \sqrt{\frac{p_1 q_1}{n_1} + \frac{p_2 q_2}{n_2}} \quad \text{(Two independent proportions)}$$

where the samples have sizes n_1 and n_2 and the proportions of failure are $q_1 = 1 - p_1$ and $q_2 = 1 - p_2$. The standard error is estimated by

$$\text{est}(SE_{\hat{p}_1 - \hat{p}_2}) = \sqrt{\frac{\hat{p}_1 \hat{q}_1}{n_1} + \frac{\hat{p}_2 \hat{q}_2}{n_2}} \quad \text{(Two independent proportions)}$$

where, now, $\hat{p}_1$, $\hat{q}_1$, $\hat{p}_2$, and $\hat{q}_2$ are the proportions of success and failure in the two samples.

The approximate $(1 - \alpha)100\%$ confidence interval for $p_1 - p_2$ has the following limits.

$$\text{Lower limit} = (\hat{p}_1 - \hat{p}_2) - z_{\alpha/2} \cdot \sqrt{\frac{\hat{p}_1 \hat{q}_1}{n_1} + \frac{\hat{p}_2 \hat{q}_2}{n_2}}$$

(Two independent proportions)

$$\text{Upper limit} = (\hat{p}_1 - \hat{p}_2) + z_{\alpha/2} \cdot \sqrt{\frac{\hat{p}_1 \hat{q}_1}{n_1} + \frac{\hat{p}_2 \hat{q}_2}{n_2}}$$

The accuracy of the 95% and the 99% intervals can be determined from the rules **a**, **b**, and **c** given previously for the approximate confidence interval for one proportion.[10] In these rules, "the smaller of the number of successes and the number of failures" must be replaced by "the smallest of the four numbers: y_1, $(n_1 - y_1)$, y_2, and $(n_2 - y_2)$". For example, if there are 23 successes out of 40 in the first group and 54 successes out of 60 in the second group, the smallest number of successes or failures in any group is 6 (the number of failures in the second group); rule **b** indicates that the confidence limits are accurate to only one decimal place.

Example 2

A patent medicine is compared with a sugar pill to see if the patent medicine is effective in relieving symptoms. Of 30 persons who take the patent medicine, 20 obtain relief. Of an independent sample of 40 persons who take the sugar pill, 10 obtain relief. The proportions of success are $\hat{p}_1 = 20/30 = 0.667$ and $\hat{p}_2 = 10/40 = 0.250$, respectively.

The point estimate of the difference between the proportions is $\hat{p}_1 - \hat{p}_2 = 0.667 - 0.250 = 0.417$ or 0.42, rounded. The estimated standard error is

[10] Rule **d** is no longer relevant, since $p_1 - p_2$ can be less than 0.0.

$$est(SE_{\hat{p}_1 - \hat{p}_2}) = \sqrt{\frac{\hat{p}_1\hat{q}_1}{n_1} + \frac{\hat{p}_2\hat{q}_2}{n_2}}$$

$$= \sqrt{\frac{0.667 \times 0.333}{30} + \frac{0.250 \times 0.750}{40}}$$

$$= \sqrt{0.00740 + 0.00469}$$

$$= \sqrt{0.01209}$$

$$= 0.110$$

The appropriate 99% confidence interval for the difference between the proportions is calculated by using the point estimate of $p_1 - p_2$ and the estimated standard error from above. The limits are

$$
\begin{aligned}
\text{Lower limit} &= (\hat{p}_1 - \hat{p}_2) - z_{\alpha/2} \cdot est(SE_{\hat{p}_1 - \hat{p}_2}) \\
&= 0.417 - z_{.005} \times 0.110 \\
&= 0.417 - (2.576 \times 0.110) \\
&= 0.417 - 0.283 \\
&= 0.134 \\
\text{Upper limit} &= 0.417 + 0.283 \\
&= 0.700
\end{aligned}
$$

The smallest number of successes or failures in either group is 10. The accuracy rules state that the limits are accurate to only one decimal place. The approximate 99% confidence interval is, therefore, (0.1, 0.7); the difference in the proportions of persons who get relief from the patent medicine and from the sugar pill is estimated to be somewhere between 10% and 70%. (Note that these numbers are *differences* in two proportions.) Both limits are positive; there appears to be a real difference between the effectiveness of the two pills.

12.6 The First Method for Planning Sample Size

How do we determine what sample size to use in a study? This section discusses one method for determining the sample size. The method is based on a desired standard error of estimate. (Chapter 13 outlines another method for determining the sample size—one based on hypothesis testing.)

In this chapter we have seen that the calculated standard error of an estimate (or the width of the confidence interval) tells us how precise the estimate is. If the standard error is small, the estimate is very precise. If the standard error is large, the error of the estimate can be large.

In order to plan the sample size, we reverse this process, deciding before doing the study how precise we want the estimate to be. The desired precision dictates the size of the standard error, and we determine the sample size needed to give this standard error.

Example 1— One Mean

We want to determine the mean IQ of a certain population, and we want the error in the estimate to be not more than plus or minus two IQ points. How large must the random sample be?

It should be noted that we cannot guarantee that the error in the estimate does not exceed ± 2 points. We can interpret the given restriction in the sense of confidence interval: let us say that the 95% confidence interval should extend from two IQ points below the estimated mean to two IQ points above it. In other words, the *half-width* of the confidence interval is 2.

The 95% confidence interval extends from $z_{\alpha/2} \cdot SE$ below $\bar{y}$ to $z_{\alpha/2} \cdot SE$ above $\bar{y}$.[11] (See Section 12.3.) As $z_{.025} = 1.96$, we have

$$1.96 SE = 2$$

which shows that the required $SE = 2/1.96 = 1.02$.

Now the standard error of the mean is $SE = \sigma/\sqrt{n}$. We therefore have

$$1.02 = \sigma/\sqrt{n}$$

In order to solve for n, we must know σ! *It is impossible to plan the sample size without some idea of the value of the population standard deviation.* A rough estimate of σ is satisfactory, since we only need a rough value for the sample size. Whether the sample size should be 100 or 120 is not a critical issue in the design of most studies. But it is important to determine whether n should be 100 or 1,000 or even 10,000, in order to have the desired precision; a rough value of σ is sufficient for this purpose.

The standard deviation of IQ is approximately 15; we will use this value here. Substituting this value into the above equation, we obtain

$$1.02 = 15/\sqrt{n}$$

To solve this equation for n, we multiply both sides of the quation by $\sqrt{n}$, giving

$$1.02\sqrt{n} = 15$$

Squaring both sides of the equation,

$$(1.02)^2 \times n = (15)^2$$

and solving for n, we get

$$n = (15)^2/(1.02)^2 = 225/1.0404 = 216.26$$

Since n must be an integer, the required n is 216.

Example 2—Two Independent Means

We want to determine the difference in mean IQ between children in single-child families and children in multichild families. We want the width of the 99% confidence interval to be three IQ points. Independent samples of children from the two populations are to be studied. How large must each random sample be if they are to be of equal size?

The 99% confidence interval extends from $(\bar{y}_1 - \bar{y}_2) - z_{\alpha/2} \cdot SE_{\bar{y}_1 - \bar{y}_2}$ to $(\bar{y}_1 - \bar{y}_2) + z_{\alpha/2} \cdot SE_{\bar{y}_1 - \bar{y}_2}$. The *full-width* of this interval is, therefore, $2z_{\alpha/2} \cdot SE_{\bar{y}_1 - \bar{y}_2}$, which is required to equal 3.0. Since $z_{.005} = 2.576$, we have

$$2 \times 2.576 \times SE_{\bar{y}_1 - \bar{y}_2} = 3.0$$

giving

$$SE_{\bar{y}_1 - \bar{y}_2} = 3.0/(2 \times 2.576) = 0.582$$

The standard error of the difference of two independent means is $\sigma\sqrt{2/n}$, where σ is the standard deviation of each population (assumed equal) and n is the sample size in each group.[12] The equation for n is, therefore,

$$0.582 = \sigma\sqrt{2/n}$$

---

[11] We use $z_{\alpha/2}$ and not $t_{\alpha/2}$ because the sample size is not known and hence the number of degrees of freedom for the t cannot be determined. Furthermore, only an approximate value for the sample size is required; for this purpose the difference between using z and t is unimportant.

[12] This formula can be derived from the formula for the standard deviation of $\bar{y}_1 - \bar{y}_2$, given in Section 11.6.

Let us again assume that $\sigma = 15$. We then have

$$0.582 = 15 \times \sqrt{2/n}$$

The solution of this equation is

$$n = (15)^2 \times 2/(0.582)^2 = 1{,}329$$

So over 1,300 children in each group are required.

This example illustrates the importance of stating the desired precision of an estimate before a study is carried out. If the above study were carried out with, say, 100 children in each group, the 99% confidence interval for the difference in mean IQ would be far wider than desired.

Example 3—One Proportion

We want to estimate the proportion of persons who get relief from a patent medicine with a standard error of 0.1. How large a random sample is required?

Note that the precision of the estimate may be specified directly in terms of the standard error, rather than by using the width of a confidence interval. The standard error of a proportion is $\sqrt{pq/n}$ (Section 12.5). However, p (and q) are unknown; they are the parameters we are trying to estimate. There are two ways to proceed. One way is to use a guessed value for p, just as we used a probable (or likely) value for σ in the two previous examples. An easier, and usually more justifiable, technique is based on the fact that the standard error of a proportion, $\sqrt{pq/n}$, cannot be larger than $0.5/\sqrt{n}$, no matter what the value of p is.[13]

For this example, we have

$$0.1 = 0.5/\sqrt{n}$$

which gives

$$n = (0.5)^2/(0.1)^2 = 25$$

A random sample of 25 persons is needed for the standard error to be 0.1 (or less).

12.7 Summary

There are two ways to estimate a population parameter: a *point estimate* and an *interval estimate*. A point estimate may be *biased* or *unbiased*. The variability of the error of a point estimate is called the *standard error*. The standard error may itself be estimated. A *confidence interval* is an interval estimate; the *confidence coefficient* of the interval is the probability that the interval includes the population parameter. The central limit theorem is basic to the theory of estimation of the population mean.

Point and interval estimates of the following parameters were described: the mean, the difference between two independent means, the difference between two dependent means, the proportion, and the difference between two independent proportions. Point estimates of the variance and the standard deviation also were described.

The sample size required for a study can be determined if the standard error of the estimate is specified. This technique is called the *first method for planning sample size*.

---

[13] If you work out $\sqrt{pq}$ for different values of p (and $q = 1 - p$), you will see that its largest value occurs when $p = 0.5$. Therefore, the standard error is never larger than $\sqrt{0.5 \times 0.5}/\sqrt{n} = 0.5/\sqrt{n}$.

Exercises

1 (Section 12.1) Draw four random samples of three observations each from the population of digits 0, 1, . . . , 9. Use the random number table (Table A5). Obtain the 12 random digits needed by using the *last three digits* of your student number to find the *starting place* in the table. If the last three digits of your student number are ABC, use AB to give the row of the table and C to give the column of the table. Suppose, for example, the last three digits are 072. Then the starting row is row 07 (that is, row 7) and the starting column is column 2, so the first five random digits are 72905. Continue reading digits along the row until you have 12 digits. (If $AB = 00$, use row 100; if $C = 0$, use column 10.)

(a) Record the starting position ABC and the four random samples you have drawn.

(b) What are μ and σ^2 for the population? What is the value of σ^2/n?

(c) For each sample, compute $\bar{y}$, its error as an estimate of μ, and its squared error. Compute the mean, over the four samples, of the sample mean, the error, and the squared error.

(d) What values would you expect the three means, computed in (c), to have? Are the values you obtained close to the expected values?

2 (Section 12.2) Use the four random samples drawn in Question 1 to answer this question.

(a) Compute s^2 and s'^2 for each sample.

(b) Compute the mean, over the four samples, of s^2 and s'^2.

(c) Does s^2 or s'^2 appear to be an unbiased estimate of σ^2? Explain.

3 (Section 12.3) A random sample of 15 scores was drawn from a population. The population is known to have a standard deviation of 12. The scores are 53, 47, 38, 41, 55, 67, 62, 58, 55, 44, 68, 43, 40, 50, 49.

(a) Estimate the population mean by a point estimate and give the standard error of this estimate.

(b) Estimate the population mean by a 90% confidence interval.

(c) State in words the meaning of the confidence interval computed in (b).

(d) Repeat (a) and (b), assuming that the population standard deviation is not known.

4 (Section 12.4)

(a) Estimate the difference between the population means for the data of Question 5 in the Exercises for Chapter 11, using a 99% confidence interval.

(b) Estimate the difference between the population means for the data of Question 6 in the Exercises for Chapter 11, using a 99% confidence interval.

5 (Section 12.5) A study was conducted to determine whether a given inbred strain of mice is susceptible to audiogenic seizure. Of the sample of 1,000 mice tested, 810 responded with seizures.

(a) Estimate the population proportion of mice that had seizures. Use a point estimate and its standard error.

(b) Place a confidence interval on this proportion, using a confidence coefficient of 0.90.

6 (Section 12.5) In a study of the relationship between birth order and college success, an investigator found that 126 in a sample of 180 college graduates were first-born or only children; in a sample of 100 nongraduates of comparable age and socioeconomic background, the number of first-born or only children was 54.

(a) Estimate the difference in proportion of first-born or only children for the two populations from which these samples were drawn. Include the standard error of this estimate.

(b) Construct the 95% confidence interval for the difference between the two population proportions.

7 (Section 12.6) You want to estimate the difference in mean grade-point average between two groups of college students. The estimate must be so accurate that the full width of the 95% confidence interval is 0.4 grade points. If the standard deviation of the grade-point measurements is approximately equal to 0.6, how many students must be included in each group? (Assume that the groups will be of equal size.)

8 (Section 12.6) An experimenter has chosen the dosage level of a certain drug so that it should induce sleep in 60% of all cases treated. He plans to study a sample of cases in order to estimate the true proportion of persons in whom sleep is induced by the drug. The proportion is expected, of course, to be near 0.60, but in any event he wants the estimate to have such an accuracy that the half-width of the 95% confidence interval is 0.10.

(a) Calculate the required sample size, assuming 0.60 as the population proportion.

(b) Calculate the sample size that will ensure the required accuracy regardless of the population proportion.

13 POWER AND EFFECT SIZE

Chapter 9 described how to calculate the probability, β, of a Type II error. A Type II error occurs if the null hypothesis is accepted when the alternative hypothesis is, in fact, true. This chapter deals in detail with a closely related quantity, power $= 1 - \beta$.

The power of a given test is affected by several characteristics of the test. One of these characteristics is called the hypothesized effect size. The effect size was introduced in Chapter 8 as the difference between two sample means. The hypothesized effect size is the absolute value of difference between the value of a parameter when the alternative hypothesis is true and its value when the null hypothesis is true.

Power also is affected by the sample size. Section 13.4 shows how to plan the sample size so that a test has a specified power.

The techniques of this chapter can be used for any statistical test, even though the details are given for only three tests: the test of one mean, the test of two independent means, and the test of one proportion.

13.1 Hypothesized Effect Size and Power

When the alternative hypothesis (H_1) is true, one of two conclusions can be reached on the basis of the observed value of the test statistic: either the null hypothesis (H_0) is accepted (a Type II error) or the alternative hypothesis is accepted (the correct conclusion). The probability of a Type II error is β; its calculation was discussed in Chapter 9. The probability of the correct conclusion is, of course, $1 - \beta$; this probability is called *power*. The power of a statistical test is the probability of accepting the alternative hypothesis when the alternative hypothesis is, in fact, true.

Power and β can be calculated only when both hypotheses are specific. Therefore, we must consider a specific value of the parameter in the interval of values that the parameter has when H_1 is true. We will show that the power of a statistical test depends on the *difference* between this value of the parameter and the value that the parameter has when the null hypothesis is true. The difference between these two values is called the *hypothesized effect size*.

Why is a new term, "power", being introduced, when a closely related probability, β, previously was used? The sum of these two probabilities is 1.0, so that if one is known, the other is readily obtained. Section 10.2 listed several ways to report the result of a significance test. It was stressed that the aim of a significance test is to disprove the null hypothesis (and accept the alternative hypothesis). By emphasizing the rejection of the null hypothesis, we can describe α, the probability of a Type I error, and power in closely related ways.

- α is the probability of rejecting the null hypothesis when the *null* hypothesis is true.

• Power is the probability of rejecting the null hypothesis when the *alternative* hypothesis is true.

In other words, α is the probability of *incorrectly* rejecting the null hypothesis and power is the probability of *correctly* rejecting the null hypothesis. It usually is desirable for the power of a statistical test to be as large as possible, since we usually want to reject H_0. The factors that can increase or decrease the power of a test are described below.

The Hypothesized Effect Size

In general, the hypothesized effect size is defined as the absolute value of the difference between the value of a parameter when the alternative hypothesis is true and its value when the null hypothesis is true. The parameter is μ in a test of one mean and p in a test of one proportion. In a test of two means the parameter is the *difference* between the two population means, $\mu_1 - \mu_2$. The formulas for the hypothesized effect sizes for these three situations are now described.

The hypothesized effect size in a test of two means is simply $|\mu_1 - \mu_2|$. This formula can be thought of in two equivalent ways. First, the hypothesized effect size is the difference, in absolute value, of the two population means, μ_1 and μ_2, when H_1 is true—just as the effect size for two sample means, $\bar{y}_1$ and $\bar{y}_2$, is $\bar{y}_1 - \bar{y}_2$ (Section 8.5). Note that the hypothesized effect size, unlike the effect size in the sample, is always positive, by definition. Second, the hypotheses H_0: $\mu_1 = \mu_2$ and H_1: $\mu_1 \neq \mu_2$ can be rewritten in terms of the difference between the means as H_0: $\mu_1 - \mu_2 = 0$ and H_1: $\mu_1 - \mu_2 \neq 0$. The hypothesized effect size is the *difference* between $|\mu_1 - \mu_2|$ when H_1 is true and $|\mu_1 - \mu_2|$, which is zero, when H_0 is true. Therefore, the hypothesized effect size is $(|\mu_1 - \mu_2| - 0)$ which reduces to $|\mu_1 - \mu_2|$.

Next, consider a test of the mean of one population. The hypotheses about μ are H_0: $\mu = \mu_0$ and H_1: $\mu \neq \mu_0$. Here μ_0 is a specific number. The population mean is hypothesized to be either equal to μ_0 or not equal to μ_0. Let μ_1 be a specific value of μ when H_1 is true. Then the hypothesized effect size (the difference between the values of the parameter when H_1 is true and when H_0 is true) is $|\mu_1 - \mu_0|$.

Finally, consider a test of one population proportion. The hypotheses are H_0: $p = p_0$ and H_1: $p \neq p_0$. If we let p_1 be a specific value of p when H_1 is true, we can define the hypothesized effect size as $|p_1 - p_0|$.

The three formulas for the hypothesized effect size, abbreviated *HES*, can be summarized as follows. (Although the first formula could also apply to two *dependent* means, the case of two dependent means is not considered in this chapter.)

$$HES = |\mu_1 - \mu_2| \qquad \text{(Two independent means)}$$
$$HES = |\mu_1 - \mu_0| \qquad \text{(One mean)}$$
$$HES = |p_1 - p_0| \qquad \text{(One proportion)}$$

The Factors That Affect Power

The power of a statistical test depends on five characteristics: two characteristics of the population, one characteristic of the sample, and two characteristics of the test itself. We shall call these characteristics the *factors* that affect the power of the test. The factors are the hypothesized effect size (*HES*), the standard deviation of the population distribution (σ), the sample size (n), the probability of a Type I error (α), and whether the test is one- or two-tailed.

Consider a test of one mean, with hypotheses $H_0: \mu = 100$ and $H_1: \mu > 100$. Let $\sigma = 10, n = 100$, and $\alpha = 0.05$. The alternative hypothesis is one-sided, so the test is one-tailed. Let us calculate the power of this test if the alternative hypothesis mean is 102.5. We have $\mu_0 = 100$ and $\mu_1 = 102.5$, which give a hypothesized effect size $= HES = |102.5 - 100| = 2.5$. The standard deviation of the sample mean, $\bar{y}$, is the standard error $= SE = \sigma/\sqrt{n} = 10/\sqrt{100} = 1$. The distributions of $\bar{y}$ when H_0 and H_1 are true are shown in Figure 13.1.

The critical value of z_{obs} for a one-tailed test at the 0.05 level is 1.64. Hence, the critical value of $\bar{y}$ is

$$\bar{y}_{crit} = 100 + (1.64 \times 1) = 101.64$$

The rejection region is ($\bar{y} > 101.64$) and the acceptance region is ($\bar{y} < 101.64$). In order to calulate either β or the power of the test, we must convert $\bar{y}_{crit}$ to a standard score in the H_1 distribution. The required standard score is $(101.64 - 102.5)/1 = -0.86$. Let us first calculate β, which is the probability that $\bar{y}$ falls in the acceptance region.

$$\beta = P(z < -0.86) = 0.5 - 0.3051 = 0.19, \text{ approximately}$$

(The value 0.3051 can be found from Table A3 as the area for $z = 0.86$.) The power is easily found from β.

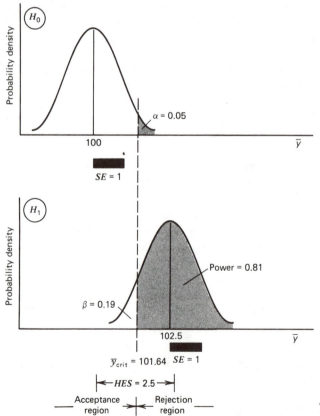

Figure 13.1 The power of a statistical test.

$$\text{Power} = 1 - \beta = 1 - 0.19 = 0.81$$

If we wish to find the power directly, we find the probability that $\bar{y}$ falls in the rejection region.

$$\text{Power} = P(z > -0.86) = 0.5 + 0.3051 = 0.81, \text{approximately}$$

We now consider what happens to the power of the test when the five factors change, one at a time. The changes are illustrated in Figure 13.2 and summarized in Table 13.1. The distributions of Figure 13.1 have been redrawn in the central subfigure of Figure 13.2. Five other subfigures are shown; in each, one of the factors has been changed and the effect on the power is shown.

Increase HES Suppose that the true mean is 103, not 102.5, so that HES = 3. The H_1 distribution is farther to the right than it is when $\mu_1 = 102.5$. The critical

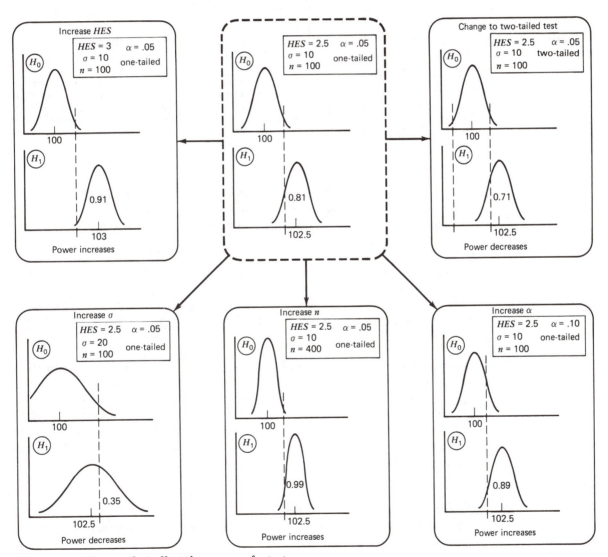

Figure 13.2 Factors that affect the power of a test.

TABLE 13.1 Explanation of Change of Power When a Factor Changes

CHANGE OF FACTOR	CHANGE OF DISTRIBUTIONS AND/OR CRITICAL VALUE	CHANGE OF POWER
Increase *HES*	Distributions are farther apart	Power increases
Increase σ	Distributions are broader and critical value moves to right	Power decreases
Increase *n*	Distributions are narrower and critical value moves to left	Power increases
Increase α	Critical value moves to left	Power increases
Change to two-tailed test (from one-tailed)	Critical value moves to right	Power decreases
Decrease *HES*	Distributions are closer together	Power decreases
Decrease σ	Distributions are narrower and critical value moves to left	Power increases
Decrease *n*	Distributions are broader and critical value moves to right	Power decreases
Decrease α	Critical value moves to right	Power decreases
Change to one-tailed test (from two-tailed)	Critical value moves to left	Power increases

Note. The expressions "critical value moves to left" and "critical value moves to right" assume that the H_1 distribution of the test statistic is to the right of the H_0 distribution. If the H_1 distribution is located to the left of the H_0 distribution, each "left" and "right" must be interchanged above; however, the stated change of power is still correct.

value of $\bar{y}$ is unchanged. Therefore, the area of the H_1 distribution over the rejection region is increased and the power is increased. As *HES* increases, the two distributions move farther apart and the power increases. (See the summary in Table 13.1.)

Increase σ When the population standard deviation increases, the standard error (SE) of $\bar{y}$ also increases. The effect of increasing σ from 10 to 20 is to double *SE*, as shown in Figure 13.2. The two distributions have the same mean but are broader than before. The critical value increases, since 95th percentile has moved to the right. As the figure shows, the power decreases.

Increase n We now consider the effect of increasing the sample size rather than changing any characteristics of the population. When *n* increases, the standard error of $\bar{y}$ decreases. The effect of increasing *n* from 100 to 400 is to halve *SE*, as shown in Figure 13.2. Because the two distributions are narrower than before, the critical value must be decreased in order for α to remain equal to 0.05. The result is that the power increases. Note that the effects of increasing σ and *n* are opposite. The standard error, $\sigma/\sqrt{n}$, increases when σ increases and decreases when *n* increases.

Increase α If α is increased from 0.05 to 0.10, the critical value is decreased, which increases the area of the H_1 distribution over the rejection region. Hence, the power of the test increases.

Change to a Two-tailed Test If the alternative hypothesis is two-sided (H_1: $\mu \neq 100$) instead of one-sided, the test is two-tailed instead of one-tailed. The critical value in the upper tail is increased, since the area in that tail is now 0.025 instead of 0.05. The power of the test for the specific alternative $\mu = 102.5$ is therefore decreased.

Decrease in HES, σ, n, and α For simplicity, Figure 13.2 shows only the effect of increasing these four factors. The effect of decreasing each of these factors, in turn, is summarized in Table 13.1. The effect is, of course, just the opposite of increasing these factors.

Change to a One-tailed Test If a test is changed from a two-tailed to a one-tailed test, the critical value is decreased and the power increases.

How to Use These Results

The relationships between power and the five factors are very important for understanding hypothesis testing. It is essential to consider these relationships when designing any research study. The relationships described above for the test of one mean apply with only minor changes to the test of two independent means and to the test of one proportion—in fact, to all statistical tests.

Sample Size The most obvious application of these results is to the determination of the sample size. We know that the greater the sample size, the greater the power. If the values of the other four factors are given, it is possible to determine the sample size to give any specified power. See Section 13.4 for details.

Standard Deviation In many research situations the population standard deviation is fixed—it cannot be changed. But the standard deviation in effect can be reduced by using certain designs, and the power of the test therefore increased. For example, in the dependent-groups design the relevant standard deviation is not the standard deviation of the populations, but the standard deviation of the difference scores. If matching has been successful, the latter standard deviation will be less than the former and the power of the dependent-groups design is therefore greater than the corresponding independent-groups design.

Hypothesized Effect Size The effect size in the population is fixed. If the difference between males and females on some measure is three points, a statistical test designed to show a significant difference between males and females has a power that can be computed from this actual difference and the other factors of the test. But the experimenter does not know this actual difference, and therefore hypothesizes it. Some ways to hypothesize the unknown effect size are described in Section 13.3.

Probability of a Type I Error Since the power of a test increases as α increases, we could ensure that any statistical test has high enough power simply by making α large enough; however, it usually is not proper to set α greater than 0.05. Conversely, the relationship between power and α shows that a penalty is paid if α is made very small—the power, too, will be very small, which means that it is very unlikely that a significant result will be obtained in the study.

One-tailed versus Two-tailed Tests A one-tailed test is always more powerful than a two-tailed test with the same α. On these grounds, we might select a one-tailed test. However, an argument presented in Section 10.3 showed that in almost all situations it is more defensible to do a two-tailed rather than a one-tailed test. Essentially, the two-tailed test is a more conservative test that is less likely to give a significant result; but since it allows for results in either direction, it is usually preferred.

13.2 Approximate Power Table

The effects of the standard deviation and the sample size on power can be understood in terms of the effect of the standard error, $SE = \sigma/\sqrt{n}$, on power. *Power decreases as the standard error increases.* This summary statement is consistent with the individual effects of σ and n: power decreases as σ increases and as n decreases.

We can take this approach one step further. The effects of the standard error and the hypothesized effect size can be understood in terms of the effect of their ratio, called δ, (the Greek letter "delta").

$$\delta = \frac{\text{Hypothesized effect size}}{\text{Standard error}} = \frac{HES}{SE}$$

Power increases as δ increases. This summary statement is consistent with the individual effects of HES and SE: power increases as HES increases and as SE decreases.

Formulas for δ

Three tests are presented in this chapter: the test of one mean, the test of two independent means, and the test of one proportion. Formulas relating δ to more basic quantities are given in Table 13.2 for each of these tests.

Using the Power Table

It is straightforward but tedious to calculate the power (or β) of a test from first principles (Chapter 9). When designing an experiment, an experimenter must consider several different sample sizes, probabilities of a Type I error, etc., so he or she needs to be able to calculate the power quickly. The approximate power table (Table A7 in the Appendix) is very useful for this purpose.

Section 13.1 pointed out that there are five factors that affect the power of a test. We have seen, however, that three of these factors (the hypothesized effect size, the standard deviation, and the sample size) combine to form δ. Therefore, we need consider only three factors in determining the power: δ, α, and whether the test is one- or two-tailed. These three factors form the borders of Table A7.

The values of δ (from 0.0 to 5.2) label the rows of the table. The four columns are identified by a value of α and whether the test is one- or two-tailed. Each number in the body of the table is the approximate power of the test for the particular values of the factors specified by the row and column labels. Note that except for a few small values of δ, the power values for one- and two-tailed tests have been combined. (Unless δ is small, the power of a one-tailed test and

TABLE 13.2 The Formulas for δ

> **One Mean**
>
> $HES = |\mu_1 - \mu_0|$
>
> $$\delta = \frac{HES}{\sigma} \sqrt{n}$$
>
> $SE = \sigma / \sqrt{n}$
>
> **Two Independent Means**
>
> $HES = |\mu_1 - \mu_2|$
>
> $$\delta = \frac{HES}{\sigma} \sqrt{\frac{n}{2}}$$
>
> $SE = \sigma \sqrt{\frac{2}{n}}$
>
> These formulas assume that both populations have the same standard deviation, σ, and that the samples drawn from the two populations are of the same size, n.
>
> **One Proportion**
>
> $HES = |p_1 - p_0|$
>
> $$\delta = \frac{HES}{\sqrt{p_0 q_0}} \sqrt{n}$$
>
> $SE = \sqrt{p_0 q_0 / n}$
>
> The proportions p_0 and q_0 are the proportions of successes and failures specified by the null hypothesis.

that of a two-tailed test are the same if the value of α for the one-tailed test is one half that of α for the two-tailed test.)

Example 1

Find the power of a two-tailed test at the 0.05 level for $\delta = 0.6$. Look for the row with $\delta = 0.6$ and for the column with two-tailed $\alpha = 0.05$. A power of 0.09 is read from the body of the table. Note that a one-tailed test with $\alpha = 0.025$ would also have a power of 0.09.

Example 2

Find the power of a one-tailed test at the 0.05 level for $\delta = 0.4$. The table has a split column for this value of δ. The power is 0.11. Note that the two-tailed test with twice the α ($\alpha = 0.10$) has slightly greater power (0.13). For values of δ greater than 0.4 these two tests have the same power (approximately).

Example 3

A study is planned to test a population proportion. The null hypothesis is H_0: $p = 0.7$. It is suspected that the true proportion is 0.8. What is the probability that a significant result will be obtained in a study with $n = 50$ if the test is carried out at the 0.01 level, two-tailed?

The "probability of a significant result" (when H_1 is true) is the power of the test. We first must determine δ. The hypothesized effect size is $|0.8 - 0.7| = 0.1$, $p_0 = 0.7$, $q_0 = 0.3$, and $n = 50$. We find

$$\delta = \frac{HES}{\sqrt{p_0 q_0}}\sqrt{n} = \frac{0.1}{\sqrt{0.7 \times 0.3}}\sqrt{50} = 1.54$$

The approximate power of a test with $\delta = 1.54$, $\alpha = 0.01$ two-tailed, is 0.15. The power is very low, which means that it is very unlikely that a significant result will be obtained if the study is carried out as planned. A larger sample size or a less stringent value of α must be used.

Example 4

A test of a population mean is planned. The null hypothesis is H_0: $\mu = 30$. The investigator believes that the true mean is 35. The population standard deviation is believed to be about 5. What is the power of a test at the 0.01 level, one-tailed? A sample size of 16 is planned.

The hypothesized effect size is $|35 - 30| = 5$, $\sigma = 5$, and $n = 16$, which gives

$$\delta = \frac{5}{5}\sqrt{16} = 4.0$$

The power of this one-tailed test at the 0.01 level is 0.95, indicating that it is practically certain that an effect of this size ($HES = 5$) will be detected, even with such a small sample size.

Example 5

A comparison of two conditions is planned. The null hypothesis is H_0: $\mu_1 = \mu_2$. It is expected that the conditions differ in mean by 2 units. The populations have equal standard deviations of about 5 units. What is the power of a two-tailed test at the 0.05 level if there will be 40 subjects in each of the two independent groups?

The hypothesized effect size is 2, $\sigma = 5$, and $n = 40$ per group. We find

$$\delta = \frac{2}{5}\sqrt{40/2} = 1.79$$

[Note that although the total number of subjects is 80, the formula directs us to divide the number of subjects per group (n) by 2, so that the square-root factor is $\sqrt{20}$.] The power of the test is approximately 0.44, a moderate value. Perhaps n should be increased somewhat, in order to make the power greater.

The Accuracy of the Power Table

The power table can be used to find the power of three different tests: the test of one proportion, the test of one mean, and the test of two independent means. It should always be kept in mind, however, that the power values found in the table are approximate. The table has been constructed for the z-test; the accuracy of power values found from the table depends on the degree to which the z-test and the normal distribution are a good approximation for the actual test and distribution.

Another reason that power values obtained from Table A7 should be treated as very approximate is that the values of the hypothesized effect size and of the standard deviation usually are only approximate, or "guessed". (Section 13.3 discusses how these values can be set.) Since HES and σ are approximate, the value of δ that is used to find the power in Table A7 also is approximate. For these reasons it is desirable to round all values of power obtained from Table A7 to the nearest 0.05.

13.3 How to Hypothesize an Unknown Effect Size

It is necessary to hypothesize the effect size and the standard deviation in order to find the power of a test. Hypothesizing values of these two parameters is one of the most difficult aspects of experimental design. Three suggestions are offered in this section.

Hypothesize Directly the Effect Size and the Standard Deviation

In some situations the population standard deviation and the expected effect size can be assessed from previous studies or other sources. For example, many IQ tests are standardized so that $\sigma = 15$. The standard deviation of many populations has been found to be close to this value. On the basis of previous work, the difference in the mean of two populations might be expected to be, say, 2 IQ points. The two values, $HES = 2$ and $\sigma = 15$, and a proposed value of n can then be substituted into the formula for δ; the power can then be determined from Table A7.

Hypothesize the Effect Size as a Proportion of the Standard Deviation

Most measures used by psychologists are not so well standardized as are IQ tests. Before conducting a study, the investigator probably has little idea of the standard deviation of the measure he is going to use, or of the expected effect size.

If you look at the formulas for δ in Table 13.2 for tests of one mean and two independent means, you will see that in each case the ratio HES/σ appears. Since HES and σ appear in the formula as a ratio, it is not necessary to have a separate value for each. It is only necessary to hypothesize a value for the ratio. So if the effect size is hypothesized to be, say, half a standard deviation, the ratio HES/σ is one-half, or 0.5.

It is quite common, and very useful, to describe an effect size as a certain proportion of a standard deviation. For example, a new reading program for elementary school students is expected to increase their reading level, as measured by a certain test, by three-quarters of a standard deviation. The ratio HES/σ is, therefore, 0.75.

Hypothesize the Ratio HES/σ Using Cohen's Scale

It has just been noted that the ratio HES/σ must be set in order for the power of a statistical test to be determined. How can this be done if very little is known about the dependent variable and about the expected difference in means? It is natural to look at published psychological studies to see what ratios typically are found. Very few studies in psychology have ratios greater than 1.0. The psychologist J. Cohen,[1] after looking at many studies, suggested the following scale for the ratio of effect size to standard deviation:

$HES/\sigma = 0.2$—small ratio.

$HES/\sigma = 0.5$—medium ratio.

$HES/\sigma = 0.8$—large ratio.

These values should be used only as a last resort. If the effect size and standard deviation, or their ratio, cannot be guessed from previous research or in other ways, it may still be possible to judge whether the ratio will be "small",

---

[1] J. Cohen, *Statistical Power Analysis for the Behavioral Sciences* (New York: Academic Press, 1969).

"medium", or "large". If so, the appropriate values of HES/σ, suggested by Cohen, may be used in the calculation of power. Remember, however, that this scale is arbitrary.

The relationship between the ratio HES/σ and the overlap of two population distributions is shown in Figure 13.3. When the ratio is small (0.2), the overlap of the two distributions is large. When the ratio is medium (0.5), there is less overlap. And when the ratio is large (0.8), there is still less overlap of the two distributions. Even when the ratio is large on Cohen's scale, there is still some overlap of the distributions, as can be seen in Figure 13.3c. Many scores in the lower distribution are higher than scores in the higher distribution. Of course, if the ratio were still larger—say, 1.5,—the overlap of the distributions would be even less.

Cohen gives a number of examples of small, medium, and large ratios of the effect size to the standard deviation. Consider girls' heights. The difference in mean height between 15- and 16-year-old girls is about one-fifth of a standard deviation—a small ratio. The difference in mean height between 14- and 18-year old girls is about one-half of a standard deviation—a medium ratio. The difference in mean height between 13- and 18-year-old girls is about four-fifths of a standard deviation—a large ratio.

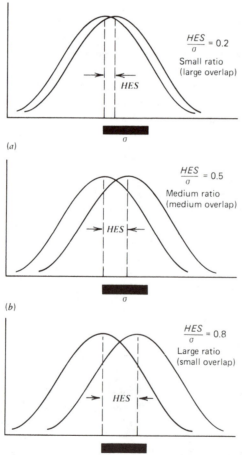

Figure 13.3 The overlap of population distributions.

Differences in IQ provide another example of various ratios of effect size to standard deviation. The difference in the mean IQ of twins and the mean IQ of nontwins is about 0.2σ, a small ratio. (Nontwins have a slightly higher IQ.) The difference in mean IQ between clerical and semiskilled workers is about 0.5σ, a medium ratio. The mean difference between managerial and clerical workers is about 0.8σ, a large difference on Cohen's scale.

13.4 The Second Method for Planning Sample Size

The first method for planning sample size was described in Section 12.6. That method is based on a desired precision in *estimating* a population parameter. The desired precision implies a value of the standard error, SE. The formula relating SE and n is then solved for the value of n that gives the required SE and, hence, the desired precision.

This section describes a method for planning sample size based on *hypothesis testing* rather than on estimation. The general approach is similar to the first method. The desired characteristics of the test imply a value of δ. The formula relating δ and n is then solved for the value of n that gives the required δ and, hence, the desired characteristics of the test.

It is important to keep in mind that the determination of sample size is done while *planning* an experiment. The results of the experiment are not known. In order to find n, however, we must have some idea of what results to expect. In particular, we must know the expected effect size and the population standard deviation. Several ways to hypothesize the effect size and the standard deviation were described in Section 13.3.

From Sections 13.1 and 13.2, we know that the power of a test is related to δ and α, and to whether the test is one- or two-tailed. Given values of δ *and* α, and that the test is one- or two-tailed, we can determine the power of the test from Table A7. In the second method for planning sample size, we use Table A7 in a different way. Given a desired power of the test (and a value for α and that the test is one- or two-tailed) we can find the value of δ by using Table A7 "backwards": that is, we look in the body of the table for the desired power and read off the value of δ from the label on the row in which that power appears.

Having found the value of δ, we then find n by using the appropriate formula relating δ and n. The appropriate formula depends on the type of study that is planned; the three possible formulas are given in Table 13.2.

The method may be better understood by considering the following process, which begins with the answers to a series of questions that a statistical consultant typically would ask a client who came to her inquiring, "How large a sample do I need for my experiment?"

Question 1

"What kind of experiment are you planning?"

The client might say that he is planning a test of the difference between the means of two conditions: a control condition and an experimental condition. He plans to have independent samples in the two conditions and to use an equal number of subjects in each condition. (He wants to know what this number, n, should be.)

Question 2

"What probability of a Type I error are you prepared to tolerate?"

The client usually suggests a value of $\alpha = 0.05$ or 0.01 or some other small, conventional value.

Question 3

"Do you plan a one- or a two-tailed test?"

Usually, a two-tailed test is more appropriate (Section 10.3). There may be a special reason for doing a one-tailed test, however.

The next question relates to the probability of a Type II error but usually is phrased in terms of power.

Question 4

"What power would you like the test to have?"

The client must decide how certain he wants to be that the experiment will give a significant result if the alternative hypothesis is in fact true. Usually, the experimenter will want to have large power. There is no point in carrying out the experiment unless he is quite certain of getting a statistically significant result. He will want the power to be 0.8 or 0.9, or even greater; the probability of a Type II error (β), therefore, will be 0.2 or 0.1, or even smaller.

The answers to Questions 2, 3, and 4 allow the statistician to find the value of δ by using Table A7. For example, suppose that the test is two-tailed with $\alpha = 0.01$ and power = 0.80. In Table A7, we go to the column headed by $\alpha = 0.01$, two-tailed. We look down this column for a power value equal or close to 0.80; we find a power value of exactly 0.80 opposite a value of $\delta = 3.4$. As a second example, suppose that the test is one-tailed with $\alpha = 0.01$ and power = 0.80, as before. In this case, we find power values of 0.78 and 0.81 opposite $\delta = 3.1$ and 3.2, respectively. For most purposes it would be satisfactory to choose that value of δ (here 3.2) that gives a power value closest to 0.80. As we know, Table A7 gives only approximate power values, particularly when the sample size is small.

The statistician and the client have now completed the easiest part of their work. The hardest question to answer is

Question 5

"What effect size do you expect, and what is the population standard deviation?"

The hypothesized effect size (*HES*) and the population standard deviation (σ) have been combined in this question, since we know, from Section 13.3, that it usually is easier to specify their ratio than their individual values. The client, with the help of the statistician, must arrive at a value for the ratio HES/σ.

The answer to Question 1 tells the statistician which of the three formulas relating δ and n (Table 13.2) is appropriate. Having found δ from Table A7 and knowing the ratio HES/σ, she can solve the formula for n. (*Note.* For the test of one proportion, the formula relating δ and n involves *HES* and the proportions of successes and failures if the null hypothesis is true, p_o and q_o.)

Example

Suppose the answers to the five questions are

1 Test of two population means (experimental and control group).

2 $\alpha = 0.01$.

3 Two-tailed test.

4 Power = 0.80.

5 The dependent variable is the time to complete a task, and it is expected that the experimental and control conditions will differ by two seconds. From previous research it is known that the standard deviation of the times in either condition is about five seconds. Hence, the ratio $HES/\sigma = 2/5 = 0.40$.

From the answers to Questions 2, 3, and 4, the required value of δ is 3.4 (Table A7). The answer to Question 1 indicates that the appropriate formula relating δ and n is

$$\delta = \frac{HES}{\sigma}\sqrt{\frac{n}{2}}$$

Substituting the known values, we have

$$3.4 = 0.40\sqrt{\frac{n}{2}}$$

In order to solve this equation for n, square both sides to remove the square root sign.

$$(3.4)^2 = (0.40)^2 \times \frac{n}{2}$$

Isolate n on the right-hand side of the equation.

$$\frac{(3.4)^2 \times 2}{(0.40)^2} = n$$

We now put n on the left and work out the arithmetic expression.

$$n = \frac{(3.4)^2 \times 2}{(0.40)^2} = 144.50$$

This answer should be rounded *up* to the nearest integer (so that the power is greater than specified). In fact, it would be best to round up to the nearest 5, since this method of determining sample size is only used to get a rough idea of what sample size is required for the desired power, α, effect size, and standard deviation. So 145 subjects *per condition* (290 in all) are needed in this experiment. If this number is more than the experimenter can possibly employ in his experiment, he will have to consider a revision of the power of the test (or of α) and repeat the calculations.

13.5 Summary

The *power* of a statistical test is the probability of correctly rejecting the null hypothesis. The *hypothesized effect size (HES)* is the difference between the value of a parameter when the alternative hypothesis is true and its value when the null hypothesis is true. The factors that affect the power of a test include the hypothesized effect size, the population standard deviation, the sample size, the probability of a Type I error, and whether the test is one- or two-tailed.

The power of tests of one mean, of two independent means, and of one proportion can be found by using the *approximate power table*. To use this table, we must know δ, the ratio of the hypothesized effect size to the standard error.

It is difficult to specify the values of *HES* and σ before doing an experiment, but it is impossible to determine the power of the test without them. Three ways were presented for assigning values to them: hypothesizing directly the effect size and the standard deviation, hypothesizing the effect size as a proportion of the standard deviation, and hypothesizing the ratio *HES*/σ using *Cohen's scale*.

An important part of the planning of an experiment is the decision about the number of subjects to use. The *second method for planning sample size* is based on the desired power of the test, the hypothesized effect size, the standard deviation, the probability of a Type I error, and whether the test is one- or two-tailed.

Exercises

1 (Section 13.1) This question is based on Example 1 of Section 10.1. You should read this example before starting the question. The test is the binomial test of one proportion—namely, the proportion of correct judgments. The hypotheses are H_0: $p = 0.5$ and H_1: $p > 0.5$. The sample size used in the example was $n = 20$. (*Note:* use Tables A1 or A2 to find all the probabilities in this question.)

(a) Suppose that the operator's skill is such that in a long series of records, he will make the correct judgment on 70% of them. What is the probability of correctly rejecting the null hypothesis when $n = 20$ if $\alpha \leq 0.01$?

(b) Suppose that the operator's skill is 90% and that all other conditions of the test are the same. What is the probability of correctly rejecting the null hypothesis now?

(c) It should be clear to you that the difference between 70% and 50% (guessing) in (a) is a *hypothesized effect size* of the experiment. [Similarly, the difference between 90% and 50% is the effect size in (b).] The probability of correctly rejecting the null hypothesis is the *power* of the test. Summarize your answers to (a) and (b) by choosing the appropriate words from the following sentence: When the hypothesized effect size (increase/decreases) the power of the test (increases/decreases).

(d) Repeat (a) and (b), but with $n = 25$ instead of 20. Summarize your results in a chart such as this.

		H_1	
POWER		$p = 0.7$	$p = 0.9$
SAMPLE SIZE	20		
	25		

Also, state in words the relationship between power and sample size as found in this chart.

2 (Section 13.1) A psychologist plans to examine the effects of illumination on reading speed. From previous work, he believes that the mean time to read the passage he is using is 230 seconds with a standard deviation of 15 seconds (population values). But he wonders whether in his lab the mean will still be 230 seconds (H_0) or whether it will be larger: for example, 236 seconds (specific value of the alternative hypothesis parameter). Assume that the standard deviation has not changed. Suppose that a sample size of 25 subjects is planned and that the test is made with $\alpha = 0.01$, one-tailed.

(a) Work out the critical value of the statistic $\bar{y}$, β, and the power of this test.

(b) Make two graphs of the distribution of the test statistic under the two hypotheses $\mu = 230$ and $\mu = 236$. You may use Figure 13.1 of the text as a model. Make sure the diagrams are roughly to scale. Show on the diagram the critical value, α, β, and the power.

(c) Repeat **(a)** and **(b)**, but use a sample size of 100 instead of 25. Does the power increase or decrease when the sample size increases? You should be able to see this, and explain it, from your diagrams.

(d) Now repeat **(a)** and **(b)** for the last time, but set up a two-tailed rejection region still using $\alpha = 0.01$ and $n = 25$. Since the critical value will change, β and the power also will change. Does the power increase or decrease when you change to a two-tailed test? You should be able to see this, and explain it, from your diagrams.

3 (Section 13.2) Recalculate all the power values in Question 2 using the power table (Table A7).

4 (Section 13.4) Because of extensive recordkeeping over a long period, it can be taken as known that under standard conditions a given strain of laboratory rats has a mean weight gain of 70 g from birth to 90 days. To test the implications of a developmental theory, an experiment is performed in which a sample of 60 animals is reared from birth in total darkness. The investigator is interested in whether under these experimental conditions the mean weight gain of a population of animals departs even slightly from the standard population mean of 70 g in either direction. Thus, the null hypothesis is H_0: $\mu = 70$. The investigator accepts $HES/\sigma = 0.20$ as a conventional operational definition of a slight departure. He uses the relatively lenient significance criterion of $\alpha = 0.10$. What is the chance that he will get a significant result? (Adapted from Cohen, p. 45; full reference in footnote 1.)

5 (Section 13.4) A political scientist plans to appraise the attitude toward the United Nations of the urban population of a new African republic. He will use an orally administered Thurstone Attitude Scale, which has the property that a neutral response is scaled 6 (on an 11-point scale). His null hypothesis, then, is H_0: $\mu = 6$. Since he wishes to be able to conclude that the average is either "pro" or "anti", he plans a nondirectional test and wishes to use a stringent significance criterion—namely, $\alpha = 0.01$. He also seeks the assurance of relatively high power, 0.90. Furthermore, he is only interested in a "pro" or "anti" attitude if the departure from a scale score of 6 is more than one-fifth of the within-population standard deviation. What sample size should he use? (Adapted from Cohen, p. 59; full reference in footnote 1.)

6 (Section 13.4) An experimental psychologist designs a study to appraise the effect of opportunity to explore a maze without reward on subsequent maze-learning in rats. Random samples of 30 cases each are drawn from the available supply and assigned to an experimental (E) group, which is given an exploratory period, and a control (C) group, which is not. Next, the 60 rats are tested and the number of trials needed to reach a criterion of two successive errorless runs is determined. The (nondirectional) alternative hypothesis is $|\mu_E - \mu_C| \neq 0$. The psychologist anticipates that the effect size will be such that the population means will differ by half a within-population standard deviation.

(a) Using $\alpha = 0.05$, what is the power of the test?

(b) Since the power calculated in **(a)** is not large enough to satisfy him, the psychologist wants to see what sample size he will need to give a power of 0.80. Can you help?

(c) What is the minimum effect size that can be detected with power = 0.9 ($\alpha = 0.05$, two-tailed and $n = 30$ per group)? (Adapted from Cohen, p. 38; full reference in footnote 1.)

REVIEW CHAPTER C

Estimation and hypothesis testing have been treated as separate topics in the last few chapters. In this chapter the connection between these two methods of statistical inference is shown.

A number of statistical procedures (z-tests, t-tests, etc.) already have been described, and many more will be introduced in the remaining chapters of this book. It is helpful to classify these procedures in order to see the relationships among them and to select the appropriate one for a given situation; such a classification is presented here. A brief discussion of the requirements for the validity of each statistical procedure also is given.

One section of the chapter is devoted to a further discussion of the interpretation of statistically significant and nonsignificant results. The interpretation depends on the effect size, a concept described in detail in Chapter 13.

C.1 Estimation and Hypothesis Testing

The two major methods of statistical inference, estimation and hypothesis testing, have so far been treated as quite distinct methods. The purpose of this section is to show the relationship between these methods. The relationship is summarized by the following principle.

> **P11: Estimation and Hypothesis Testing**
> A result stated as a test of a hypothesis can be reexpressed as an estimate, and vice versa. For many purposes, an estimate with its standard error (or a confidence interval estimate) is more useful than a statement of the significance of the result.

How to Test a Hypothesis by a Confidence Interval

An experimenter usually tests a hypothesis by computing a test statistic and observing whether it falls in the acceptance region or in the rejection region. He can also test any hypothesis by computing the appropriate confidence interval and seeing if it includes the null hypothesis value of the parameter: if the test is two-tailed and has a probability of a Type I error $= \alpha$, the $(1 - \alpha)100\%$ confidence interval is the appropriate one. The null hypothesis is rejected if the confidence interval *does not* include the null hypothesis value of the parameter; conversely, the null hypothesis is not rejected if the confidence interval *does* include the null hypothesis value. This procedure for testing a hypothesis is explained and justified in Example 1.

Example 1

Consider the test of two independent means. The null hypothesis is H_0: $\mu_1 = \mu_2$ and the alternative hypothesis is H_1: $\mu_1 \neq \mu_2$. For simplicity, we assume that the standard deviations of the populations are known, so that the critical values can be found from the normal table. The distribution of $(\bar{y}_1 - \bar{y}_2)$, when H_0 is true, is shown in Figure C.1a. If $\alpha = 0.05$, the critical values of z_{obs} are ± 1.96, and therefore the critical values of $(\bar{y}_1 - \bar{y}_2)$ are $\pm 1.96SE$. (We need not be concerned here with the value of the standard error, SE,

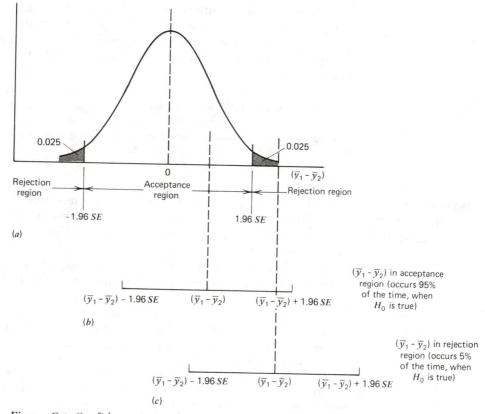

Figure C.1 Confidence interval estimation and hypothesis testing. (*a*) Distribution of $(\bar{y}_1 - \bar{y}_2)$ when H_0 is true. (*b*) 95% confidence interval that includes zero. (*c*) 95% confidence interval that does not include zero.

which depends on the standard deviations of the two populations and the sample sizes.) Note that the acceptance region is centered on zero, since the mean value of $(\bar{y}_1 - \bar{y}_2)$ is zero, if H_0 is true.

The 95% confidence interval is centered on the observed difference of the sample results, $(\bar{y}_1 - \bar{y}_2)$, and extends a distance of $1.96SE$ on each side of this value. The confidence interval is, therefore, exactly as wide as the acceptance region. This means that if the observed difference, $(\bar{y}_1 - \bar{y}_2)$, falls in the acceptance region, the confidence interval will extend beyond the value of zero (Figure C.1*b*). Conversely, if the observed difference falls in the rejection region, the confidence interval will not extend as far as the value of zero (Figure C.1*c*).

We say that the confidence interval either *includes* the value of zero or does not include it. Since zero is the value of $(\mu_1 - \mu_2)$ when the null hypothesis is true, the question is whether or not the confidence interval includes the null-hypothesis value of the parameter. (In this case the null hypothesis specifies that the *difference* of two parameters is zero, but we can think of this difference as a single parameter.) We see from Figure C.1, therefore, that if the 95% confidence interval includes the null-hypothesis value of the parameter, the null hypothesis is not rejected, and that if the 95% confidence interval does not include that value, the null hypothesis is rejected.

Warning. The acceptance region and the confidence interval are not the same, even though they have the same width in the above example. The acceptance region is centered on zero (the null hypothesis value of the parameter); the null hypothesis is accepted if the test statistic, $\bar{y}_1 - \bar{y}_2$, falls in the acceptance region. The *confidence interval* is centered on the value of $\bar{y}_1 - \bar{y}_2$; the null hypothesis is accepted if the confidence interval includes the null hypothesis value.

Example 2

A single population mean is estimated by the 99% confidence interval of (10.8, 14.4). We wish to test the null hypothesis H_0: $\mu = 10$ against the alternative hypothesis H_1: $\mu \neq 10$. The null-hypothesis value of the parameter ($\mu = 10$) is not included in the confidence interval. Therefore, the null hypothesis is rejected by a 0.01-level, two-tailed test.

Making an Estimate after a Hypothesis Test

It is not as easy to go from the result of a hypothesis test to an estimate as it is to go in the reverse order (as above). The mere information that the result of a study is significant (or not significant) at a specified α is not sufficient to make an estimate. However, a report usually also gives a point estimate of the parameter (e.g., a mean or a difference of means), the sizes of the samples, and the value of the test statistic. From this information, a confidence interval can be calculated.

Example 3

A report states that a study of reaction time has been carried out with two independent groups of 10 subjects each; the difference in mean time was 5.0 seconds, which is significantly different from zero ($t_{obs} = 6.2$, p-value < 0.01, two-tailed). Our task is to estimate the difference between the means using a 99% confidence interval.

In order to calculate the confidence limits, we need the difference between the sample means (5.0), the degrees of freedom (10 + 10 − 2 = 18), and the estimated standard error. The estimated standard error can be obtained, by calculation, from the given value of t_{obs}. Recall, from Section 11.6, that

$$t_{obs} = \frac{\bar{y}_1 - \bar{y}_2}{est(SE)}$$

From the given information, we have

$$6.2 = \frac{5.0}{est(SE)}$$

Therefore, $est(SE) = 5.0/6.2 = 0.806$. We find that $t^{18}_{.005} = 2.878$, from Table A6. The limits of the confidence interval are

$$\text{Upper limit} = (\bar{y}_1 - \bar{y}_2) + t^{df}_{\alpha/2} \; est(SE)$$

$$= 5.0 + (2.878 \times 0.806)$$

$$= 5.0 + 2.32$$

$$= 7.3$$

$$\text{Lower limit} = 5.0 - 2.32$$

$$= 2.7$$

The 99% confidence interval is (2.7, 7.3). Note that the confidence interval does not include the value zero, which is consistent with the fact that the null hypothesis was rejected at the 0.01 level.

The report of this study stated that the difference between the means, 5.0, was significant at the 0.01 level. Unless we are careful, we may interpret this significant difference as being *exactly* 5.0. This difference is not exact; it is only an estimate, and we are given no measure of how precise it is. The advantage of computing and reporting the estimated standard error or the confidence interval is that an indication is given of the precision of the estimate, and not simply a statement that it is significant. In the example, the estimated population mean difference is somewhere between 2.7 and 7.3— quite a wide range.

C.2 Classification of Procedures

We have now studied a number of procedures: tests of means, estimates of proportions, z-tests, t-tests, etc. Classification of these procedures helps us to select the appropriate procedure for a given study. A suitable classification is given in Tables C.1 and C.2 (See Chapter D for a different and more extensive classification that includes the procedures introduced in the next four chapters.)

The appropriate procedure can be selected from the classification in Tables C.1 and C.2 by answering the following four questions.

1 Are means or proportions calculated? (So far, these are the only possibilities—see Section B.1 for a discussion of the distinction between means and proportions.) If means are calculated, select Table C.1; if proportions are calculated, select Table C.2.

2 Are one or two samples studied? And if two samples are studied, are they independent or dependent? The appropriate column of the table is selected on the basis of the answers to these questions.

3 If means are calculated, is σ known or unknown? The appropriate row of Table C.1 is selected on the basis of the answer. (The distinction does not apply to proportions.)

TABLE C.1 Tests and Estimates of Means

	ONE MEAN	TWO INDEPENDENT MEANS	TWO DEPENDENT MEANS
σ unknown	t-test of one mean (11.2)[a]	t-test of two independent means (11.4)	t-test of two dependent means (11.5)
	Estimate of one mean (12.3)	Estimate of difference between two independent means (12.4)	Estimate of difference between two dependent means (12.4)
σ known	z-test of one mean (9.3, 10.1)	z-test of two independent means (10.6)	——[b]
	Estimate of one mean (12.3)	——	——

[a] The number in parentheses indicates the section in which the test or estimate is described.
[b] A dash indicates that the test or estimate is not discussed in this text.

TABLE C.2 Tests and Estimates of Proportions

ONE PROPORTION	TWO INDEPENDENT PROPORTIONS
Binomial test —using binomial table (9.1, 10.1)[a] —or using normal approximation (9.4)	(See test in Chapter 14)
Estimate of one proportion (12.5)	Estimate of the difference between two independent proportions (12.5)

[a] The number in parentheses indicates the section in which the test or estimate is described.

4 Which method of statistical inference (hypothesis testing or estimation) is desired? In each cell of the tables, the upper entry is the hypothesis test and the lower entry is the estimation method.

C.3 Requirements for the Validity of Statistical Procedures

When particular tests or estimates were described in previous chapters, the requirements for the validity of the procedures usually were mentioned only briefly. These requirements are summarized in this section.

In many studies it is difficult to determine whether a particular requirement (see below) is satisfied. For this reason these requirements are often called *assumptions* for the validity of statistical procedures. Since in many studies we cannot *prove* that a particular requirement has been met, we must *assume* that it has been met in order to carry out the procedure.

Four requirements are listed below. Some apply to all procedures, others only to certain ones. Some of the requirements may be replaced by other requirements or may be relaxed. These qualifications are indicated for each requirement.

REQUIREMENT 1: RANDOM SAMPLING

The subjects in the sample must be selected at random from the population of interest.

Alternative Requirement

The requirement of random sampling from a population may be replaced in two-group studies by the requirement of random assignment of subjects to the two conditions (Section 11.3).

Application

This requirement, or its alternative, applies to all tests and estimates.

Comment

Statistical inference requires randomness; the randomness may originate in random sampling or in randomization. If a study has neither feature, no inference made in the study is a valid statistical inference.

REQUIREMENT 2: INDEPENDENCE OF OBSERVATIONS

Each observation must be independent of every other observation in the study. This means, roughly, that the probability that an observation has any particular value is unaffected by the values of all other observations. (See Chapter 18 for a technical definition of independence.)

Alternative Requirement

None.

Application

This requirement applies to all designs except those in which dependence of the observations is built-in (dependent-groups designs).

Comment

The requirement that the observations in one-group and independent-groups studies be independent of each other is absolute. Violation of this requirement in such studies definitely makes the analyses invalid. The analysis of the data from a dependent-groups design takes account of the dependence of the observations, and therefore is valid.

REQUIREMENT 3: NORMALITY OF POPULATION DISTRIBUTION

The distribution of scores in the population from which the sample is drawn must be normal. In the case of two populations, both populations must be normally distributed.

Application

This requirement applies to all tests of hypotheses about means and to confidence-interval estimates of means. Normality is not required in order that point estimates and standard errors be valid, however. The requirement does not apply to tests and estimates of proportions.

Relaxation of Requirement

The requirement of normality can be relaxed considerably. If the sample size exceeds 10, the population can have almost any shape and the distribution of the statistic will still be close to its theoretical shape (normal or t, as the case may be). If the sample size is 10 or less, the population distribution must be unimodal if the distribution of the statistic is to be close to its theoretical shape.

Comment

The normality requirement can be relaxed, because the central limit theorem (Section 7.5) states that the distribution of the sample mean is approximately normal even if the population distribution is not normal. The proofs of this theorem and related theorems ensure that certain other statistics (e.g., $\bar{y}_1 - \bar{y}_2$ and t_{obs}) have approximately the normal distribution, or the t-distribution, even if the population distributions are not normal.

REQUIREMENT 4: EQUALITY OF VARIANCE

The variances of the two population distributions must be equal.

Application

Of the procedures described so far in this text, this requirement applies only to the test and estimate of two independent means (Sections 11.4 and 12.4).

Relaxation of Requirement

The test and estimate of two independent means can still be used if the population variances are different, provided that they are not *too* different. It is possible to evaluate, roughly, whether this requirement is met by comparing the *sample* variances: if the ratio of the larger sample variance to the smaller does not exceed 4, it is safe to use the t-test. (For example, sample variances of 45.3 and 17.2 are alright, but variances of 100.3 and 7.6 are too different to allow the t-test to be used.)

Comment

This requirement often is stated as a requirement of *homogeneity* of variance; the term "homogeneity" is used in statistics to mean "equality", particularly of variances. The requirement could also be stated in terms of the population standard deviations: the standard deviations must be equal, or homogeneous.

C.4 Statistical Significance and Effect Size

A test of a hypothesis can have two possible results: it can be statistically significant or not statistically significant. The null hypothesis is rejected if the result is statistically significant, and the null hypothesis cannot be rejected if the result is not statistically significant. (See Section 10.2 for a review of these and related terms.) In this section the interpretation of significant and nonsignificant results is explored further.

Interpretation of a Statistically Significant Result

A statistically significant result cannot be interpreted properly if the effect size is not known. The effect size in a two-group study is the absolute value of difference between the two means: the difference between the two *population* means is the effect size in the population; the difference between the two *sample* means is the effect size in the sample. The former is unknown, but the latter is known after the study has been completed.

It was shown in Chapter 13 that the power of a statistical test is related to the hypothesized effect size in the population and to the sizes of the two samples. The smaller the effect size in the population, the smaller the power of the test. On the other hand, the larger the sample sizes, the larger the power. Even if the effect size in the population is very small, the test can have large power if the sample sizes are large enough.

If a test has very large power, the probability is high that a statistically significant result will be obtained, assuming that the effect size in the population is not zero. This means that a statistically significant result can be obtained even if the effect size is very small. But a small effect size often indicates that the effect of one variable on the other is trivial. It is not sufficient to show that the result is statistically significant; it is also necessary to show that the effect is not trivial.

The preceding argument can be summarized in a principle.

P12: Statistical Significance and Effect Size
A statistically significant result shows that there is an effect, but the effect may be large or small. In evaluating a statistical study, it is important to consider both the effect size and whether the result is statistically significant.

Example 1

The difference between two independent means is found to be significant at the 0.05 level, two-tailed. The value of t_{obs} is 2.5. If the pooled standard deviation, s_p, is 5.0, determine the effect size.

The given information is not sufficient to determine the effect size. We must know the sizes of the two samples. Consider two possibilities: **(a)** $n = 20$ in both groups; and **(b)** $n = 1,000$ in both groups.

(a) The estimated standard error for two independent groups of equal size, n, is

$$\text{est(SE)} = s_p \sqrt{\frac{1}{n} + \frac{1}{n}} = s_p \sqrt{\frac{2}{n}}$$

(See Sections 11.6 and 12.4 for an explanation of this formula.) For $s_p = 5.0$ and $n = 20$, the formula gives $est(SE) = 5.0\sqrt{2/20} = 1.58$. We also know, from Section 11.6, that

$$t_{obs} = \frac{\bar{y}_1 - \bar{y}_2}{est(SE)} = \frac{\text{Effect size}}{\text{est}(SE)}$$

Therefore,

$$\text{Effect size} = t_{obs} \times est(SE)$$

The effect size in the study is $2.5 \times 1.58 = 3.95$.

Lacking further information about the variable measured in the study, we cannot say whether an effect size of 3.95 is large or small. However, we can use Cohen's scale, described in Section 13.3, to evaluate the ratio of the effect size to the standard deviation (s_p). Here this ratio is $3.95/5.0 = 0.79$, which is a large effect size according to Cohen's scale.

(b) Let us repeat these calculations, assuming that there are 1,000 subjects in each group. The estimated standard error is $est(SE) = 5.0\sqrt{2/1{,}000} = 0.224$. The effect size is $2.5 \times 0.224 = 0.56$. In Cohen's scale, the ratio of this effect size to s_p is $0.56/5.0 = 0.11$—a very small effect size.

Comment A report that the result of a study is statistically significant by itself does not indicate that the effect size is large or small. If the sample size is small, the effect size must be quite large for the result to be statistically significant. If the sample size is very large, on the other hand, the effect size may be very small and still be significant.

Accepting the Null Hypothesis

When a result is nonsignificant, the conclusion usually is that the null hypothesis cannot be rejected. Considerations of effect size allow us, in certain circumstances, to conclude that the null hypothesis can be accepted, rather than just that it cannot be rejected.[1]

Suppose that an experiment is designed in the following way. A certain effect size is hypothesized to exist in the population. Given this value of *HES* (and the desired power, the standard deviation, a small value of α, and that the test is one- or two-tailed), the required sample size is determined by the method described in Chapter 13. Suppose that the power is specified to be very large—say, 0.99. This means that the probability of a Type II error, β, is very small (0.01). With the experiment designed in this way, the study is conducted. Suppose that it turns out that the test statistic falls in the acceptance region.

Normally, we would conclude that the null hypothesis cannot be rejected—however, in this experiment we can justifiably accept the null hypothesis. Essentially, we have set up two *specific* hypotheses: H_0 is the hypothesis of zero effect, and H_1 is the hypothesis that the effect size is *HES*. The values of α and β are both small. Just as in Chapter 9, we accept H_1 if the test statistic falls in the acceptance region for H_1, and we accept H_0 if the test statistic falls in the acceptance region for H_0. The two hypotheses are on an equal footing: we accept H_1 if the test statistic has certain values, and we accept H_0 if the test statistic has other values.

This argument for accepting the null hypothesis can be made even stronger if *HES* is defined to be the smallest population effect size that the investigator would consider as nonzero; presumably, there are effect sizes so

---

[1] For an extensive discussion, see A. G. Greenwald, "Consequences of prejudice against the null hypothesis", *Psychological Bulletin* 82 (1975), 1–20.

small that the investigator treats them, for all intents and purposes, as zero. *HES* then is set to the smallest value that would be treated as *nonzero*. If the test has high power and the result is not significant, the investigator can claim that the population effect size is not as large as the hypothesized effect size and may therefore be treated as zero: that is, the null hypothesis is accepted.

Example 2

Consider two hypotheses about a population mean: H_0: $\mu = 50$ and H_1: $\mu \neq 50$. The standard deviation is known to be 10. The investigator decides that if the true population mean were between 48 and 52, the effect would be trivial and the null hypothesis would be essentially true. A sample of 400 is studied and the test carried out at the 0.05 level. A sample mean of 50.82 is obtained. What conclusion should be drawn?

The smallest nonzero effect size (*HES*) is 2 (that is, $52 - 50$). The value of δ is $(HES/\sigma)\sqrt{n} = (2/10)\sqrt{400} = 4.0$. The power of the two-tailed test with $\alpha = 0.05$ is 0.98 (Table A7). Hence, $\beta = 0.02$, a small value. The upper critical value of the test is

$$\bar{y}_{crit} = 50 + 1.96 \times 10/\sqrt{400} = 50.98$$

The observed mean, 50.82, does not exceed this value; therefore, it falls in the acceptance region for H_0. Since β is so small, we are justified in accepting H_0 and concluding that the population mean is 50.

C.5 Summary

The result of any study may be expressed as a test of a hypothesis or as an estimate of a parameter. A result expressed in one way can be reexpressed in the other way.

The appropriate statistical procedure may be selected from those discussed so far in this text on the basis of four questions: (1) Are means or proportions calculated? (2) Are one or two samples studied? (3) If means are calculated, is σ known or unknown? (4) Which method of statistical inference (hypothesis testing or estimation) is desired?

The validity of any statistical procedure depends on a number of requirements (or assumptions). The following requirements were described: random sampling, independence of observations, normality of population distribution, and equality of variance. Some of these requirements can be relaxed or replaced by other requirements.

It is important to consider effect size when we interpret the significance of the result of a test of a hypothesis. Trivial effect sizes can be statistically significant. The null hypothesis can be accepted if a test has high power for a hypothesized effect size and the test statistic falls in the acceptance region.

Exercises

1 (Section C.1) A study of the IQs of a random sample of 16 teenage boys found their mean to be 105.3 and the standard deviation to be 14.3.
(a) Find the 99% confidence interval for the mean IQ of the population of teenage boys from which the sample was drawn.
(b) *Without calculation*, state whether the null hypothesis that the population mean IQ is 100 would be rejected at the 0.01 level, two-tailed.
(c) Confirm your answer to (b) by computing the test-statistic and carrying out the test.

2 (Section C.1) A control and an experimental condition were compared by running each of 20 subjects first in one condition and then, the following day, in the other condition. The response variable was the time taken to complete a task. The mean time for completion was 52.0 seconds in the control condition and 78.3 seconds in the experimental condition. The standard deviation of the *difference* between each subject's control and experimental condition times was 31.2 seconds.

(a) Test the hypothesis that the mean times are the same for the two conditions, using a two-tailed test at the 0.10 level.

(b) *Without calculation*, state whether the 90% confidence interval for the difference in mean times would include zero.

(c) Confirm your answer to **(b)** by computing the 90% confidence interval.

3 (Section C.2) A binomial experiment was conducted to test $H_o: p = 0.5$ against $H_1: p > 0.5$. Of the 20 subjects run, 15 showed the "successful" response. (p is the probability of the "successful" response.) Choose the correct answers to **(a)** and **(b)**, below.

(a) This experiment is a test of a
   (i) Proportion.
   (ii) Mean.

(b) The p-value of the result of the experiment (15 successes) is
   (i) 0.5.
   (ii) > 0.5.
   (iii) 0.021.
   (iv) 0.042.
   (v) Not calculable from the given information.

4 (Section C.2) This question is an exercise in identification. For each of the following examples, state which statistical procedure is appropriate. In order to do so, first answer the four questions described in Section C.2 of the text. (*Note.* Do not perform any calculations nor carry out the procedure; this question simply asks you to identify the procedure that would be used in each situation.)

(a) It is hypothesized that one child in 10 between the ages of five and six has acquired the concept of volume, as evidenced by their choice of the "larger" but shorter of 2 three-dimensional figures. To test this hypothesis, 25 children are studied; five of them acquired the concept of volume.

(b) In a study of language acquisition, the investigator wished to compare boys and girls for the age at which the first recognizable word is spoken. From parents' reports of the age at which the child's first word was spoken, the average age for 20 boys was found to be 10.7 months with a standard deviation of 3.1 months, and the average age for 18 girls was found to be 9.4 months with a standard deviation of 2.6 months.

(c) In order to validate a measure of social dominance, an experiment was conducted comparing two groups, extroverts and introverts. On the basis of prior tests, nine subjects were identified as extroverted and nine subjects were identified as introverted. All 18 subjects were administered the social-dominance scale, and the results were used to test if the mean social-dominance scores were different for the two populations.

(d) A psychiatrist claims that over 80% of all people who visit doctors have problems of a psychosomatic nature. He selects 100 patients at random to test this theory. He finds that 88 of these patients have psychosomatic problems.

(e) From each of 14 litters of rats, all of the same strain, two male rats of approximately equal weight were selected. One of these was placed on feeding schedule A and the other on feeding schedule B. You wish to test the hypothesis that the two feeding schedules produce equal gains or losses in weight.

5 (Section C.2) Although it was emphasized in Section C.2 that we can determine what statistical procedure is appropriate simply by answering four questions, this is not always true. Sometimes there are several appropriate procedures. Consider the following study to determine if vocabulary-test performance improves on a second administration of a test. Ten examinees are given the same test twice, on two consecutive days. The resulting scores are

EXAMINEE	FIRST ADMINISTRATION	SECOND ADMINISTRATION
1	52	61
2	38	44
3	65	63
4	56	59
5	44	47
6	72	89
7	33	29
8	62	67
9	55	60
10	77	80

The research question (does vocabulary-test performance improve on the second administration?) can be approached in two ways. Perform the statistical tests suggested in **(a)** and **(b)** in order to see this.

(a) First, work with proportions. The null hypothesis is that the probability of a higher score on the second administration is 0.5, against the alternative hypothesis that it is greater than 0.5. The test statistic is y, the number of examinees obtaining higher scores on the second administration.

(b) Next, work with the actual difference between the second and first scores. The null hypothesis is that the mean difference score is zero, against the alternative hypothesis that it is greater than zero. The test statistic is t_{obs}.

(c) Do the conclusions of the two tests agree? Do you see any reasons why one way of testing the research question may be better than the other?

6 (Section C.3) If you were reading the report of a study which used a t-test of two independent groups, which of the following statements, if true about the experiment, would make you doubt the conclusions of the study? (Identify all such statements.)

 (i) The population distributions are normal.

 (ii) The population distributions are distinctly bimodal; subjects get either very low scores or very high scores.

(iii) The two populations have almost the same standard deviations.

 (iv) The two populations have almost the same variances.

 (v) The variance of scores in one condition is 10 times the variance of scores in the other condition.

 (vi) There were 20 subjects in each group, but three of the subjects in the first group were also used in the second group.

7 (Section C.4) This question has been designed to ensure that you can interpret the reports of statistical results found in research papers.

(a) Find a research paper that reports the result of a statistical test. The test should be a binomial test, a z-test, or a t-test. Give a complete reference to the paper, including page numbers.

(b) Identify the statistical test that has been made. What hypotheses were tested? Was a p-value reported? If so, what was its value? If not, can you determine approximately what it must have been? Was the test one- or two-tailed?

(c) Sometimes a statistical test will conclude with the letters n.s., rather than a p-value. This means that the result was *not* statistically *significant*. Are there any results reported like that in your paper? In any case, what would the p-value be for such a result?

(d) What conclusion was drawn from the statistical test? Was this conclusion based on a statistically significant or a statistically nonsignificant result? If the result was *nonsignificant*, do the authors use an expression equivalent to "accept H_0", or do they use expression equivalent to "fail to reject H_0"? Try to comment on which of these expressions would be appropriate in this study if the result was nonsignificant.

8 (Section C.4) An investigator used a certain maze to study the learning process in rats. She collected extensive norms, from which she estimated the population parameters: $\mu = 120$ errors and $\sigma = 15$ errors. These are the mean and standard deviation of the number of errors until the rat has learned the maze.

Recently, a new strain of rats was developed. According to the advertisements, this new strain learns significantly faster. To test this claim, the investigator examined 36 rats of the new strain. Test the hypothesis that the new population mean is 120 errors (H_0) against the hypothesis that it is 116 errors (H_1). Assume that the standard deviation remains the same.

Let the test statistic be the mean number of errors, $\bar{y}$, in the sample of 36 rats. Answer all questions so that the probability of a Type I error is 0.10.

(a) What is the region for rejecting the hypothesis that the mean is 120 errors? You may express this region in terms of either $\bar{y}$ or z_{obs}.

(b) If in the sample the mean was 117 errors, what would you conclude? Why?

(c) What is the value of β for this test?

(d) Would it be appropriate in this test to accept the null hypothesis if the test statistic falls in the acceptance region?

14 TESTS FOR DICHOTOMOUS VARIABLES

The tests described in preceding chapters were tests of means and tests of proportions. In Section 14.1 we introduce a distinction between variables that are numerical and variables that are dichotomous. Tests of means (such as the t-tests) and tests of medians are tests of numerical variables. Tests of proportions (such as the binomial test, the sign test, and the chi-square tests) are tests of dichotomous variables. The median tests, the sign test, and some chi-square tests are described in this chapter.[1]

14.1 Dichotomous Variables

Many of the variables described in previous chapters were numerical variables. A *numerical variable* may be defined as a variable whose possible values are numbers. This chapter deals with dichotomous variables. A *dichotomous variable* is a variable that has two values. There are three types of dichotomous variables: a variable can be *naturally* dichotomous; a dichotomous variable can result from a *comparison*; or a numerical variable can be *reduced* to a dichotomous one.

Examples of naturally dichotomous variables are: sex (male/female); result from tossing a coin (heads/tails); outcome on a task (success/failure); and attitude toward a statement (agree/disagree).

Examples of dichotomous variables arising from a comparison are: result of a series of tennis games played between members of two clubs (in each game the winner is either a player from club A or one from club B); spouse earning higher income (wife or husband); and member of a matched pair, in a dependent-groups design, who is rated by an observer, as better, on some characteristic (first or second member is better).

Examples of numerical variables reduced to dichotomous variables are: score on an exam (0 to 100) reduced to pass or fail; attitude measured on a nine-point scale reduced to favorable or not favorable; and years of education reduced to high school graduate versus not completed high school.

Tests of dichotomous variables are essentially tests of proportions. The binomial test of one proportion has already been described in detail. Several other tests of dichotomous variables are discussed below.

---

[1] The tests in this chapter, along with some of the tests introduced in Chapters 15 and 16, are often called "nonparametric" tests. The relation of these tests to tests such as the t-test is described in Section D.5.

14.2 The Sign Test

The sign test is simply the binomial test of a null hypothesis that $p = 0.5$. The sign test can be used to analyze the results of a dependent-groups (matched-groups) design. The two members of each matched pair are compared on the dependent variable. If the first member is higher than the second, the pair is marked plus (+); if the second member is higher than the first, the pair is marked minus (−). The analysis is based on these signs, one for each pair —hence the name "sign test".

Hypotheses

Let the two groups be labeled 1 and 2. The scores of the two members of a pair are x_1 and x_2. If there is no difference between the population distributions from which the samples are drawn, the probability that x_1 is greater than x_2 is 0.5 and the probability that x_1 is less than x_2 is also 0.5. We assume, for the moment, that it is impossible that $x_1 = x_2$; therefore,

$$P(x_1 < x_2) + P(x_1 > x_2) = 1$$

If one probability is known, the other can be determined from the above formula. Therefore, we can use just one probability—say, $P(x_1 > x_2)$. The null hypothesis of no difference between the population distributions is then

$$H_0: P(x_1 > x_2) = 0.5$$

The alternative hypothesis can be one- or two-sided. If the alternative to the null hypothesis is simply that the two distributions differ, then

$$H_1: P(x_1 > x_2) \neq 0.5$$

If the alternative is that the first distribution is located at higher values than the second—that is, that the first has a greater value of its central tendency than the second—then

$$H_1: P(x_1 > x_2) > 0.5$$

Conversely, if it is more likely that x_1 will be less than x_2, the alternative hypothesis is

$$H_1: P(x_1 > x_2) < 0.5$$

It is not necessary that the members of each pair be *measured* on the variable x; it is only necessary that the two members be compared and a decision made whether $x_1 < x_2$ or $x_1 > x_2$. In some studies the two members of each pair are directly compared by means of a contest or game in which they compete: if the first member wins, then $x_1 > x_2$; if the second member wins, then $x_1 < x_2$.

Test Statistic

The test statistic (y) is the number of successes: that is, the number of pairs in which $x_1 > x_2$. If each pair is marked with a sign, + or −, the test statistic is the number of plus signs. When the null hypothesis is true, the number of plus signs has a binomial distribution with $p = 0.5$, and $n =$ number of pairs.

Ties

In some studies it may be impossible to discriminate between the members of some pairs. Such a pair is tied. When the members of the pair are actually measured, a tie occurs when the same score is obtained by both members. A tied pair can be assigned neither a plus nor a minus sign. The simplest procedure is to drop all tied pairs from the analysis and to decrease n accordingly.[2]

Example 1

Two methods of therapy are compared. Thirty clients are matched in pairs; one member of each pair is randomly assigned to therapy A and the other member to therapy B. At the conclusion of the therapy the members of each pair are compared with each other: either the person in therapy A or the person in therapy B is rated as most improved. The results of the 15 comparisons are shown in Table 14.1.

The null hypothesis of no difference between the therapies is expressed statistically as $H_0: P(A > B) = 0.5$. The two-sided alternative hypothesis is $H_1: P(A > B) \neq 0.5$. The test statistic is y = number of pairs in which $A > B$. In this study, $y = 9$. The critical values of y for $\alpha = 0.05$ are 3.5 and 11.5 (Table A2, $n = 15$). The observed result is not significant and the null hypothesis cannot be rejected.

Example 2

Two methods of therapy again are compared. Forty clients are matched in pairs, and the members of each pair are assigned at random to therapy A or to therapy B. At the conclusion of therapy, each client is rated on a scale extending from 1 to 9: the low end of the scale indicates no improvement and the high end indicates great improvement. (Note the difference between Examples 1 and 2: in Example 1 the members of the pair are directly compared, whereas in Example 2 the members are independently rated.)

The ratings are shown in Table 14.2. The ratings for each pair now are compared. If the rating for A is greater than for B, a *plus* is recorded; conversely, if A is less than B, a *minus* is recorded. There are three *ties*.

The null hypothesis is that there is no difference between the population distributions of ratings for the two therapies. This null hypothesis may be expressed, as in Example 1, as $H_0: P(A > B) = 0.5$. The two-sided alternative hypothesis is $H_1: P(A > B) \neq 0.5$.

The ties are dropped from the analysis. The test statistic is the number of pairs in which $A > B$. Here $y = 4$. Since $n = 17$ after the tied pairs are dropped, we cannot use

[2] A more conservative procedure is to count the number of plus and minus nontied pairs and give signs to the tied pairs in such a way that the *difference* between the number of plus and minus pairs is reduced.

TABLE 14.1 The Sign Test—Example 1

PAIR	MOST IMPROVED MEMBER
1	A
2	B
3	B
4	A
5	B
6	A
7	A
8	A
9	B
10	A
11	A
12	A
13	B
14	B
15	A

TABLE 14.2 The Sign Test—Example 2

PAIR	THERAPY A	THERAPY B	PLUS IF $A > B$
1	7	2	+
2	3	7	−
3	4	6	−
4	5	6	−
5	3	3	Tie
6	1	3	−
7	4	3	+
8	3	6	−
9	7	8	−
10	6	6	Tie
11	4	7	−
12	3	5	−
13	4	9	−
14	8	2	+
15	3	4	−
16	5	9	−
17	6	6	Tie
18	5	4	+
19	5	8	−
20	2	3	−

Table A2 to find the critical value. Let us use Table A1 and compute the two-tailed p-value.

$$
\begin{aligned}
p\text{-value} &= 2 \times P(y \leqq 4) \\
&= 2 \times [P(y = 0) + P(y = 1) + P(y = 2) + P(y = 3) + P(y = 4)] \\
&= 2 \times (0.000 + 0.000 + 0.001 + 0.005 + 0.018) \\
&= 2 \times 0.024 \\
&= 0.048
\end{aligned}
$$

The p-value is less than 0.05. Hence, the result is significant at the 0.05 level, two-tailed. There appears to be a difference between the therapies in favor of therapy B.

14.3 The Chi-square Test of Two Independent Proportions

The sign test, like the binomial test on which it is based, is a test of one proportion. We now turn to the test of two independent proportions, called the chi-square test. (The Greek letter is χ, which is pronounced "kye", although written in English as *chi*.) The test is based on the chi-square distribution described in this section. Further details about this test, including an alternative test based on the normal distribution, are given in Section 14.4.

Hypotheses

Two independent groups are selected from two populations. The dependent variable is a dichotomous variable; it usually is a naturally dichotomous variable, but occasionally is a numerical variable reduced to a dichotomous one. Each subject in the study, therefore, is assigned to one of two categories: success or failure. The null hypothesis states that the proportions of successes in the two populations are the same; expressed in terms of probability, the null hypothesis is

$$H_0\colon p_1 = p_2$$

where p_1 is the probability of success in the first population and p_2 is the probability of success in the second. The alternative hypothesis is

$$H_1: p_1 \neq p_2$$

The Data

The observations can be summarized in a 2×2 table of frequencies as shown in Table 14.3a. These frequencies are called *observed frequencies* to distinguish them from *expected frequencies* (see below). In the 2×2 table of observed frequencies the columns represent the two groups. The rows represent failure and success; the rows are also labeled 1 and 2, respectively, in order to be consistent with notation that will be used in Chapter 15.

Each observation in Group 1 is either a failure or a success. The number of failures is denoted O_{11}, the number of successes O_{21}. (The letter O is a symbol for "observed".) Note that the first subscript of these symbols denotes failure ("1") or success ("2"), and that the second subscript denotes the group number ("1" for the first group). Similarly, the numbers of failures and successes in the second group are O_{12} and O_{22}, respectively. The sum of the observed frequencies in each column is the sample size, n_1 or n_2, shown under each column. (The total sample size is $n = n_1 + n_2$.)

Test Statistic

If the null hypothesis is true, then the proportions of success are the same ($p_1 = p_2$), and therefore the proportions of failure are also the same ($q_1 = q_2$). Since the proportions (or probabilities) are the same in the two populations, we can let the *overall* proportion of success = p (= $p_1 = p_2$) and let the overall proportion of failure = q (= $q_1 = q_2$). These overall proportions can be estimated by pooling the results from the two groups. The number of failures is $O_{11} + O_{12}$; the estimated overall proportion, called the *pooled proportion* of failure, is

$$\hat{q} = \frac{O_{11} + O_{12}}{n_1 + n_2}$$

Similarly, the pooled proportion of success is

$$\hat{p} = \frac{O_{21} + O_{22}}{n_1 + n_2}$$

TABLE 14.3 Observed and Expected Frequencies

(a) Observed Frequencies (O_{ij})

	GROUP 1	GROUP 2	POOLED PROPORTIONS
1:FAILURE	O_{11}	O_{12}	$\hat{q} = \frac{O_{11} + O_{12}}{n_1 + n_2}$
2:SUCCESS	O_{21}	O_{22}	$\hat{p} = \frac{O_{21} + O_{22}}{n_1 + n_2}$
	n_1	n_2	

(b) Expected Frequencies (E_{ij})

	GROUP 1	GROUP 2
1:FAILURE	$E_{11} = \hat{q}n_1$	$E_{12} = \hat{q}n_2$
2:SUCCESS	$E_{21} = \hat{p}n_1$	$E_{22} = \hat{p}n_2$

We can now compute how many successes and failures we would expect to have in each group, assuming that the null hypothesis is true. There are n_1 subjects in Group 1; the overall proportion of failure is estimated to be $\hat{q}$. Hence, the expected number of failures is $E_{11} = \hat{q}n_1$. The other three expected frequencies are computed in the same way and are shown in Table 14.3b.

If the *null* hypothesis is in fact true, the observed frequencies should be close to these expected frequencies. If the *alternative* hypothesis is true, on the other hand, the observed frequencies usually will differ from the expected frequencies. The test statistic, called χ^2_{obs} ("chi-square observed"), is based on the *differences* between each observed frequency and its corresponding expected frequency. Each difference is squared and then divided by the expected frequency itself. The sum of the results, for each cell in the 2×2 table, is the test statistic. In symbols, the test statistic is

$$\chi^2_{obs} = \sum_i \sum_j \frac{(O_{ij} - E_{ij})^2}{E_{ij}} \quad \text{(Definitional formula)}$$

In this formula the rows of the table are denoted by i ($= 1$ and 2) and the columns by j ($= 1$ and 2).

Chi-square Distribution

The test statistic, χ^2_{obs}, has a distribution that is discrete since the observed frequencies, O_{ij}, have integral values. However, the distribution of χ^2_{obs} can be approximated by a continuous distribution called the χ^2-distribution. (The Greek letter is χ, "chi"; hence the "chi-square" distribution.) There are many χ^2-distributions, each labeled by its number of degrees of freedom (df). The number of degrees of freedom depends on the dimensions of the frequency table, not on the sample size. For the 2×2 table of this section, $df = 1$. [See the example that follows for an explanation of why the number of degrees of freedom in a 2×2 table is 1. In Section 14.5 the number of columns of the table will be increased, and in Chapter 15 both the number of columns and the number of rows will be increased. In general, for an $r \times c$ table (that is, a table with r rows and c columns), the number of degrees of freedom is $(r - 1) \times (c - 1)$.]

The χ^2-distribution is nonsymmetric; its values extend from zero to plus-infinity. Percentiles of the χ^2-distribution are given in Table A8 in the Appendix. In this section we need only the first row of the table —the row for $df = 1$. Note that percentiles are given for both the lower and the upper tails of the distribution (since the distribution is nonsymmetric). The χ^2-test, based on the statistic χ^2_{obs}, requires only the upper-tail percentiles, however. The reason for this is that the observed frequencies, on the whole, differ more from the expected frequencies when H_1 is true than when H_0 is true, so that χ^2_{obs} tends to be larger when H_1 is true than when H_0 is true. Accordingly, the rejection region for a given α is [$\chi^2_{obs} > \chi^2_\alpha(1)$]. Note that the df (here, $df = 1$) is indicated in parentheses after the notation for the percentile.

Example

Two methods of instruction, lecture and tutorial, were compared by forming two independent groups of students. There were 30 students in the lecture group and 40 in the tutorial group. At the end of the period of instruction, each student took a common final exam on which the student either passes or fails. The number of students in each category, for each group, is shown in the observed frequency table in Table 14.4a.

TABLE 14.4 Chi-square Test of Two Independent Proportions

(a) Observed Frequencies (O_{ij})

		METHOD OF INSTRUCTION		
		LECTURE	TUTORIAL	
FINAL EXAM	FAIL	12	8	20
	PASS	18	32	50
		30	40	70

(b) Expected Frequencies (E_{ij})

$\hat{p} = 50/70 = 0.7143$ $\hat{q} = 20/70 = 0.2857$

$E_{11} = 0.2857 \times 30 = 8.57$ $E_{12} = 0.2857 \times 40 = 11.43$

$E_{21} = 0.7143 \times 30 = 21.43$ $E_{22} = 0.7143 \times 40 = 28.57$

EXPECTED FREQUENCIES	8.57	11.43	20.00
	21.43	28.57	50.00
	30.00	40.00	70.00

(c) Computation of χ^2_{obs}

$$\chi^2_{obs} = \sum_i \sum_j \frac{(O_{ij} - E_{ij})^2}{E_{ij}}$$

$$= \frac{(12 - 8.57)^2}{8.57} + \frac{(8 - 11.43)^2}{11.43} + \frac{(18 - 21.43)^2}{21.43} + \frac{(32 - 28.57)^2}{28.57}$$

$$= \frac{(+3.43)^2}{8.57} + \frac{(-3.43)^2}{11.43} + \frac{(-3.43)^2}{21.43} + \frac{(+3.43)^2}{28.57}$$

$$= 1.373 + 1.029 + 0.549 + 0.412$$

$$= 3.36$$

The null hypothesis states that there is no difference between the success rates of the two groups: that is, that $p_1 = p_2$, where p_1 is the probability of passing the exam in the first group and p_2 is the probability of passing the exam in the second group. The test of this hypothesis, against the alternative that the probabilities are different, is carried out by computing χ^2_{obs}.

Table 14.4b shows the computation of the expected frequencies. The pooled proportions of students passing ($\hat{p}$) and failing ($\hat{q}$) are computed first. Then the expected frequency in each of the four cells of the table is computed. These expected frequencies can be checked by calculating the sum of each row and column of the expected frequency table. These sums should be identical to the corresponding sums of the rows and columns of the observed-frequency table. (Note that the frequencies in the margins of the expected-frequency table are the same as the frequencies in the margins of the observed-frequency table.)

The test statistic χ^2_{obs} is then computed (Table 14.4c). Note that the absolute value of each of the four differences, $O_{ij} - E_{ij}$, is the same: two differences are $+3.43$ and two are -3.43. It can be shown algebraically that the absolute value of these differences must always be the same. There is, therefore, only one difference, not four different differences, between the observed and expected frequencies. Hence, if we know one difference, we know all the others. The number of degrees of freedom in a χ^2-test is the minimum number of differences needed in order to determine all the others. The number of degrees of freedom is, therefore, *one* for a test of the frequencies in a 2×2 table.

The value of χ^2_{obs} is 3.36. The critical value of χ^2 is $\chi^2_{.05}(1) = 3.841$ (Table A8). Hence, the observed result is not significant at the 0.05 level. There does not appear to be a difference in the probability of passing the final exam between the two groups.

14.4 Further Details About the Chi-square Test

Computational Shortcuts

It is easy to show that the expected frequencies can be calculated by the following formula.

$$ E_{ij} = \frac{O_{i.} \times O_{.j}}{n} \quad \text{(Expected frequencies)} $$

where $O_{i.}$ (read "oh sub i dot") is the sum of the *i*th row, $O_{.j}$ (read "oh sub dot j") is the sum of the *j*th column, and *n* is the total number of observations. In other words, the expected frequency in a given cell of the table can be found by multiplying the row sum for that cell by the column sum for that cell and dividing by the total frequency.

The following (computational) formula is algebraically equivalent to the definitional formula for χ^2_{obs}, and usually is more convenient to use.

$$ \chi^2_{\text{obs}} = \sum_i \sum_j \frac{O^2_{ij}}{E_{ij}} - n \quad \text{(Computational formula)} $$

Note that the *n* in this formula represents the total number of observations.

Example 1

The two shortcuts described above are illustrated in Table 14.5. The observed frequencies in this table are the same as those in the Example of Section 14.3 (Table 14.4). However, the expected frequencies and χ^2_{obs} are calculated differently. The value of χ^2_{obs} is, of course, the same as that found in Table 14.4.

The *z*-test of Two Independent Proportions

The test statistic for this alternative test is based on the confidence interval for two independent proportions (Section 12.5). (Recall, from Section C.1, the close relation between confidence-interval estimation and hypothesis testing.) The test statistic is

$$ z_{\text{obs}} = \frac{\hat{p}_1 - \hat{p}_2}{\sqrt{\dfrac{\hat{p}_1 \hat{q}_1}{n_1} + \dfrac{\hat{p}_2 \hat{q}_2}{n_2}}} \quad \begin{array}{l}\text{(z-test of two independent}\\ \text{proportions)}\end{array} $$

In this formula, $\hat{p}_1$ and $\hat{p}_2$ are the estimated proportions of successes in the two populations, $\hat{q}_1$ and $\hat{q}_2$ are estimated proportions of failures, and n_1 and n_2 are

TABLE 14.5 Computational Short Cuts

(a) Observed Frequencies (O_{ij})

METHOD OF INSTRUCTION

		LECTURE	TUTORIAL	
FINAL EXAM	FAIL	12	8	$O_{1.} = 20$
	PASS	18	32	$O_{2.} = 50$
		$O_{.1} = 30$	$O_{.2} = 40$	$n = 70$

(b) Expected Frequencies (E_{ij})

$E_{11} = O_{1.} \times O_{.1}/n$ $= 20 \times 30/70 = 8.57$	$E_{12} = O_{1.} \times O_{.2}/n$ $= 20 \times 40/70 = 11.43$
$E_{21} = O_{2.} \times O_{.1}/n$ $= 50 \times 30/70 = 21.43$	$E_{22} = O_{2.} \times O_{.2}/n$ $= 50 \times 40/70 = 28.57$

(c) Computation of χ^2_{obs}

$$\chi^2_{obs} = \sum_i \sum_j \frac{O^2_{ij}}{E_{ij}} - n$$

$$= \frac{(12)^2}{8.57} + \frac{(8)^2}{11.43} + \frac{(18)^2}{21.43} + \frac{(32)^2}{28.57} - 70$$

$$= 3.36$$

the two sample sizes. This statistic has, approximately, the standard normal distribution when H_0 is true. The rejection region for a given α consists of the two tails of this distribution: ($z_{obs} < -z_{\alpha/2}$) and ($z_{obs} > z_{\alpha/2}$).[3]

Example 2

The z-test is illustrated in Table 14.6. The observed frequencies in this table are the same as those in the Example of Section 14.3 (Table 14.4). Table 14.6b shows the computation of z_{obs}. We first compute the proportions of passes ($\hat{p}_1$ and $\hat{p}_2$) and failures ($\hat{q}_1$ and $\hat{q}_2$) in the two groups. These values are then put in the formula for z_{obs}, given above. The result is $z_{obs} = -1.83$, the negative sign indicating that the proportion of passes is greater in the second (tutorial) group than in the first (lecture) group.

The rejection region for $\alpha = 0.05$ has two parts: ($z_{obs} < -1.96$) and ($z_{obs} > +1.96$). Since the observed result is not in this rejection region, the proportions of students passing in the two groups are not significantly different. (The null hypothesis cannot be rejected.)

Relationship between the Two Tests Because the z-test and the χ^2-test almost always produce the same decision, either test may be used. The *critical values* are exactly related as follows: $(z_{\alpha/2})^2 = \chi^2_\alpha(1)$. For example, when $\alpha = 0.05$, $z_{.025} = 1.96$ and $\chi^2_{.05}(1) = 3.841$. These two critical values are related: $(1.96)^2 = 3.841$. The two *test statistics* are not exactly related in this way (i.e.,

---

[3] Occasionally, a one-sided alternative hypothesis may be justified. If the hypothesis is $H_1: p_1 > p_2$, the rejection region for the z-test is ($z_{obs} > z_\alpha$) and the rejection region for the χ^2-test is [$\chi^2_{obs} > \chi^2_{2\alpha}(1)$]: that is, the critical value for χ^2_{obs} is listed in the χ^2-table as having twice the desired α. However, the data should first be inspected to be sure that $\hat{p}_1 > \hat{p}_2$. Only in that case can H_1 be accepted if χ^2_{obs} is in the rejection region. The probability that $\hat{p}_1 > \hat{p}_2$, if H_0 is true, is 1/2; therefore, the probability of a significant result, when H_0 is true, is $\frac{1}{2}(2\alpha) = \alpha$, as required.

TABLE 14.6 The z-test of Two Independent Proportions

(a) Observed Frequencies (O_{ij})

		METHOD OF INSTRUCTION		
		LECTURE	TUTORIAL	
FINAL EXAM	FAIL	12	8	20
	PASS	18	32	50
		30	40	70

(b) Computation of z_{obs}

$$\hat{p}_1 = 18/30 = 0.60 \qquad \hat{q}_1 = 12/30 = 0.40$$
$$\hat{p}_2 = 32/40 = 0.80 \qquad \hat{q}_2 = 8/40 = 0.20$$

$$z_{obs} = \frac{\hat{p}_1 - \hat{p}_2}{\sqrt{\dfrac{\hat{p}_1 \hat{q}_1}{n_1} + \dfrac{\hat{p}_2 \hat{q}_2}{n_2}}}$$

$$= \frac{0.60 - 0.80}{\sqrt{\dfrac{0.60 \times 0.40}{30} + \dfrac{0.80 \times 0.20}{40}}}$$

$$= \frac{-0.20}{\sqrt{0.012}} = -1.83$$

z^2_{obs} is not exactly equal to χ^2_{obs}), but the relationship is close enough that the two tests may be used interchangeably.[4]

Condition for Satisfactory Approximation The two test statistics, z_{obs} and χ^2_{obs}, are only *approximately* distributed as standard normal or chi-square with one degree of freedom, respectively. Both approximations are usually satisfactory if *all* the four *expected* frequencies are equal to five or more. Note that this requirement applies to both tests, even though the expected frequencies of the four cells are explicitly needed only for the χ^2-test.[5]

Test of Independence of Two Dichotomous Variables

The chi-square test thus far has been viewed as a test of the equality of the proportions of successes in two independent groups. The two groups are distinguished by an experimentally-defined variable: for example, method of instruction (lecture or tutorial), type of practice (massed or distributed), or color of stimulus (red or green). The subjects in the experiment are assigned at

---

[4] It is possible to define z_{obs} in a slightly different way so that z^2_{obs} is exactly equal to χ^2_{obs}. The products $\hat{p}_1\hat{q}_1$ and $\hat{p}_2\hat{q}_2$ in the denominator are both replaced by $\hat{p}\hat{q}$, the product of the pooled proportions. There are some advantages to this latter definition. See K. R. Eberhardt and M. A. Fligner, "A Comparison of Two Tests for Equality of Two Proportions", *American Statistician* 31 (1977), 151–55. The formula given in the text is preferred however, since it corresponds exactly to the formula for the confidence interval of two independent proportions (Section 12.5)

[5] Some statisticians claim that the approximation of χ^2_{obs} to the χ^2-distribution can be improved by subtracting 1/2 from the absolute value of each difference, $O_{ij} - E_{ij}$, before squaring. This claim is disputed by other statisticians, however, and this correction, often called the Yates correction for continuity, is not recommended here. [See, for example, G. Camilli and K. D. Hopkins, "Applicability of Chi-Square to 2×2 Contingency Tables with Small Expected Cell Frequencies", *Psychological Bulletin* 85 (1978), 163–67.]

random to one of the two conditions. (Refer to Section 8.2 for further discussion of experimentally-defined variables.)

In many studies, on the other hand, there is not an experimentally-defined variable. Rather, the subjects are distinguished by a naturally-occurring variable such as sex (male or female), age (young or old), or political orientation (liberal or conservative). A chi-square test can then be used to test whether the proportions of successes are equal in the two groups distinguished by the naturally-occurring variable.

In both types of studies, there are two dichotomous variables: one variable distinguishes the groups and the other variable is the dependent variable (success or failure). It is often convenient (particularly when both variables are naturally-occurring) to view the study as a study of the *relationship* between the two variables. The two variables may be related (or associated) or they may be unrelated (or unassociated). Instead of these terms, however, we often will use the terms "dependence" and "independence". All these terms are ways to describe the relationship between two variables when the null and alternative hypotheses are true.

H_0: the two variables are independent (i.e., they are not related, or not associated).

H_1: the two variables are dependent (i.e., they are related, or associated).

[The use here of the terms "independence" and "dependence" to refer to the relationship between two variables should be distinguished very carefully from two other uses of these words. The first use comes into play when a variable is described as the independent variable or the dependent variable (Section 8.2). The second use comes into play when an experimental design is described as an independent-groups or a dependent-groups design (Section 11.3).]

The above hypotheses about two dichotomous variables are essentially equivalent to the hypotheses, stated in Section 14.3, about two independent proportions.[6] Let us call the two variables A and B, and denote the two values of A by A_1 and A_2, and of B by B_1 and B_2. The 2×2 table can then be represented as follows.

One population (or one sample) is represented by this table. But we can think of the population as divided into two populations: one population whose value of variable B is B_1, and another population whose value of variable B is B_2. The two derived populations are represented by the two columns of the table. We can then consider the proportion of persons in each of the two populations who have value A_2, say, of variable A. *The independence of the two variables, A and*

---

[6] Independent proportions come from an independent-groups design, mentioned in the preceding paragraph.

B, is equivalent to the equality of these two proportions. Informally, the variables A and B are independent if the proportion of B_1s that are A_2 equals the proportion of B_2s that are A_2.[7] (See Chapter 18 for further discussion of independence.)

The test of the independence of two dichotomous variables is carried out in exactly the same way as the test of equality of two independent proportions. We begin with the observed frequency table, compute the expected frequencies from the marginal frequencies, compute the test statistic χ^2_{obs}, and compare the value of χ^2_{obs} with the critical value of χ^2 with 1 df. The following example shows these steps in more detail.

Example 3

In a study conducted to determine the relationship between political orientation (liberal versus conservative) and party preference (Democratic versus Republican), a random sample of persons is interviewed. Several questions are asked to determine whether the interviewed person is liberal or conservative. Then the person is asked his or her party preference. Hence, each person in the single sample is scored on two dichotomous variables.

The data for this study can be summarized in an observed-frequency table (Table 14.7a). Each of the 87 subjects falls in one of the cells of the 2×2 table—for example, there are 16 subjects who are classified as conservative and whose party preference is Democratic.

The question being asked about these data is whether or not party preference is related to (or dependent on) political orientation. Recall, from Section 8.2, that two variables are related if the distributions of one variable, for the different values of the other, are different; the two variables are unrelated (or independent) if the distributions are the same.

The proportion of liberals who prefer the Republicans is 9/34 and the proportion who prefer the Democrats is 25/34. The corresponding proportions of conservatives preferring each party are 37/53 and 16/53. Since each pair of proportions adds to 1.0, we can work with just one proportion from each pair—say, the proportion preferring the Republicans. The proportion of liberals preferring the Republicans is $9/34 = 0.26$, and the proportion of conservatives preferring the Republicans is $37/53 = 0.70$. The question to be answered is whether these two proportions are *significantly* different. If they are, party preference is related to political orientation and the variables are dependent; if the proportions are not significantly different, the null hypothesis of no relationship between (or independence of) the two variables cannot be rejected.

A χ^2-test is carried out to see whether or not these two variables are independent. In Table 14.7b the expected frequencies are calculated from the marginal frequencies. The expected frequency table can be checked by computing its marginal frequencies, which must be the same as the marginal frequencies of the observed-frequency table. The test statistic, χ^2_{obs}, can be calculated in two ways (Table 14.7c): (*i*) from the difference of the observed and expected frequencies (each difference is the same, except for sign, confirming that the table has 1 degree of freedom); or (*ii*) from the computational formula.

The value of χ^2_{obs} is quite large: $\chi^2_{obs} = 15.6143$. This value exceeds $\chi^2_{.005}(1)$ (which is 7.879), so the null hypothesis of independence can be rejected at the 0.005 level. The data support the conclusion that political orientation and party preference are related, or dependent.

---

[7] Alternatively, the population can be divided into two parts on the basis of the values of variable A. Then we can define the independence of the two variables as the equality of the proportion of A_1s that are B_2 and the proportion of A_2s that are B_2.

TABLE 14.7 Test of Independence of Two Dichotomous Variables

(a) Observed Frequencies (O_{ij})

		PARTY PREFERENCE		
		DEMOCRATIC	REPUBLICAN	
	LIBERAL	25	9	34
POLITICAL ORIENTATION	CONSERVATIVE	16	37	53
		41	46	87

(b) Expected Frequencies (E_{ij})

$\dfrac{34 \times 41}{87} = 16.0230$	$\dfrac{34 \times 46}{87} = 17.9770$	34.0000
$\dfrac{53 \times 41}{87} = 24.9770$	$\dfrac{53 \times 46}{87} = 28.0230$	53.0000
41.0000	46.0000	87.0000

(c) Two Calculations of χ^2_{obs}

(i) $\chi^2_{obs} = \dfrac{(25 - 16.0230)^2}{16.0230} + \dfrac{(9 - 17.9770)^2}{17.9770} + \dfrac{(16 - 24.9770)^2}{24.9770} + \dfrac{(37 - 28.0230)^2}{28.0230}$

$= 15.6143$

(ii) $\chi^2_{obs} = \dfrac{(25)^2}{16.0230} + \dfrac{(9)^2}{17.9770} + \dfrac{(16)^2}{24.9770} + \dfrac{(37)^2}{28.0230} - 87$

$= 15.6143$

14.5 More Than Two Independent Proportions

The test of the equality of two independent proportions can be extended to more than two independent proportions. The steps for J independent proportions are:

- Form the $2 \times J$ table of observed frequencies.
- Compute the expected frequencies (by either the method of Section 14.3 or that of Section 14.4).
- Compute χ^2_{obs} by either the definitional or the computational formula (Sections 14.3 or 14.4).
- Find the critical value of χ^2, based on a chosen α and the number of degrees of freedom $(J - 1)$; the critical value is, therefore, $\chi^2_\alpha(J - 1)$.
- Make the appropriate decision by comparing χ^2_{obs} with the critical value. The null hypothesis of equal proportions is rejected if χ^2_{obs} exceeds the critical value; otherwise, the null hypothesis cannot be rejected.

Example

Four methods of instruction are compared by forming four independent groups of students. The methods of instruction are lecture, tutorial, self-study with opportunity for help from an instructor, and self-study with no opportunity for help. The dependent variable is the result (fail or pass) on a common final exam. (Here, $J = 4$).

The four groups are of unequal size. The sample sizes and the frequency of failing and passing in each group are shown in the observed-frequency table (Table 14.8a).

TABLE 14.8 Test of Four Independent Proportions

(a) Observed Frequencies (O_{ij})

METHOD OF INSTRUCTION

		LECTURE	TUTORIAL	SELF-STUDY (HELP)	SELF-STUDY (NO HELP)	
FINAL EXAM	FAIL	20	23	24	28	95
	PASS	30	37	16	22	105
		$n_1 = 50$	$n_2 = 60$	$n_3 = 40$	$n_4 = 50$	200

(b) Expected Frequencies (E_{ij})

23.75	28.50	19.00	23.75	95.00
26.25	31.50	21.00	26.25	105.00
50.00	60.00	40.00	50.00	200.00

(c) Observed Minus Expected Frequencies ($O_{ij} - E_{ij}$)

-3.75	-5.50	5.00	4.25	0.00
3.75	5.50	-5.00	-4.25	0.00
0.00	0.00	0.00	0.00	0.00

The null hypothesis is

$$H_0: p_1 = p_2 = p_3 = p_4$$

where p_j is the probability of passing in the jth population ($j = 1, 2, 3, 4$). The alternative hypothesis states that not all the probabilities are equal. The easiest way to state H_1 is

$$H_1: H_0 \text{ is false}$$

The row sums of the observed frequency table are 95 and 105. Hence, the pooled proportions of failing and passing are $\hat{q} = 95/200 = 0.475$ and $\hat{p} = 105/200 = 0.525$, respectively. The expected frequencies in the first group are $E_{11} = n_1\hat{q} = 50 \times 0.475 = 23.75$ and $E_{21} = n_1\hat{p} = 50 \times 0.525 = 26.25$. The other expected frequencies are calculated in a similar way. (Alternatively, each expected frequency can be calculated as the product of the row sum and the column sum, divided by the total sample size. For example, $E_{23} = 105 \times 40/200 = 21.00$.) The expected frequencies can be checked by computing the row and column sums of the expected frequency table. These sums should match exactly the corresponding row and column sums of the observed frequency table.

The test statistic, χ^2_{obs}, is computed from the differences between the observed and expected frequencies.

$$\chi^2_{obs} = \sum_i \sum_j \frac{(O_{ij} - E_{ij})^2}{E_{ij}}$$

$$= \frac{(20 - 23.75)^2}{23.75} + \frac{(23 - 28.50)^2}{28.50} + \ldots + \frac{(22 - 26.25)^2}{26.25}$$

$$= \frac{(-3.75)^2}{23.75} + \frac{(-5.50)^2}{28.50} + \ldots + \frac{(-4.25)^2}{26.25}$$

$$= 0.5921 + 1.0614 + \ldots + 0.6881$$

$$= 7.10$$

All eight differences between the observed and expected frequencies are shown in Table 14.8c. Note that each *column* sum is 0.00. This means that knowing one difference in each column is sufficient to determine the other. Hence, at most, four differences (one for each column) are needed to determine all the others. However, the sum of the four differences in the first *row* is also zero. This means that only three of them are needed to determine all other differences. Hence, the number of degrees of freedom in a 2×4 table is 3 $(= 4 - 1)$. In general, for J groups, $df = J - 1$.

(An alternative calculation of χ^2_{obs} is

$$\chi^2_{obs} = \sum_i \sum_j \frac{O^2_{ij}}{E_{ij}} - n$$

$$= \frac{20^2}{23.75} + \frac{23^2}{28.50} + \ldots + \frac{22^2}{26.25} - 200$$

$$= 16.8421 + 18.5614 + \ldots + 18.4381 - 200$$

$$= 7.10)$$

The critical value for this χ^2-test, for $\alpha = 0.05$, is $\chi^2_{.05}(3) = 7.815$. The observed value of χ^2_{obs} does not exceed the critical value, and hence the null hypothesis cannot be rejected. The four sample proportions ($\hat{p}_1 = 30/50 = 0.60$, $\hat{p}_2 = 37/60 = 0.62$, $\hat{p}_3 = 16/40 = 0.40$, and $\hat{p}_4 = 22/50 = 0.44$) are not sufficiently different to permit the conclusion that the population proportions are different. In other words, the sample proportions are not significantly different ($\alpha = 0.05$).

14.6 Tests of the Median

Two tests are described in this section. The test of one median is closely related to the sign test. The test of two or more independent medians is closely related to the test of equality of independent proportions. In both these new tests the dependent variable is numerical. The sample data are divided into scores above and below the median; in this way, a dichotomous variable is created from the basic numerical variable.

The median tests will be compared with other tests of central tendency in Chapter D. In some data sets, a few scores are indeterminate but lie at the extremes of the distribution of the scores. In such a data set the median (but not the mean) can be calculated and a median test can be carried out. Some examples are given below.

Median of One Population

The test of one *mean* is a test of the null hypothesis that the population mean has a certain specified value (i.e., H_0: $\mu = \mu_0$, where μ_0 is a given, known number). Similarly, the test of one *median*, described here, is a test of the null hypothesis that the population median has a certain specified value. We use the symbol M for the population median and M_0 for a specific value. The null hypothesis is, therefore, H_0: $M = M_0$, where M_0 is a given, known number.

The test of one median is very easy to carry out. Draw a random sample from the population and record each of the n scores. Each score above M_θ is called "+" and each score below M_0 is called "−". (Scores that are exactly equal to the hypothesized median are dropped and n decreased accordingly.) The test then reduces to a test of H_0: $p = 0.5$, where p is the probability of "+": that is, the probability of a score greater than M_0. This test is simply the sign test (the binomial test of $p = 0.5$).

Example 1

A university course is arranged in such a way that a student passes the course once a certain number of criteria have been satisfied. Assuming that students are allowed unlimited time in which to satisfy these criteria, we wish to test whether the median time for completion is 100 days. The data for 20 students who start the course at the same time are given in Table 14.9. The data are collected after 200 days, by which time, as noted in the table, two students have not completed the course. Even though we do not have a number or score for these students, their score is certainly greater than the hypothesized median of 100 days, and they can each be recorded as "+".

The number of pluses is 15, out of 20. The critical values of a two-tailed test, for $\alpha = 0.05$, are 5.5 and 14.5. Hence, the observed result is significant. The data support the conclusion that the median length of time to pass the course is greater than 100 days.

Medians of Two or More Independent Populations

This test is a test of the equality of the medians of J independent populations. If we label the medians $M_1, M_2, \ldots, M_J$, the hypotheses are

$$H_0: M_1 = M_2 = \ldots = M_J$$

and

$$H_1: H_0 \text{ is false}$$

The population medians are unknown but are equal if H_0 is true. Let this common median be M. If H_0 is true, we would expect that about half the scores in each group would be greater than M, and half less than M. This overall *population* median, M, is not known but can be estimated by the overall *sample* median, $\hat{M}$. We still expect, if H_0 is true, that about half the scores in each group will be greater than $\hat{M}$ and about half will be less than $\hat{M}$.

TABLE 14.9 Median Test—One Population

STUDENT	NUMBER OF DAYS TO COMPLETE COURSE	ABOVE (+) OR BELOW (−) 100 DAYS
A	125	+
B	142	+
C	91	−
D	122	+
E	117	+
F	184	+
G	170	+
H	(Not yet completed)	+
I	73	−
J	82	−
K	(Not yet completed)	+
L	144	+
M	191	+
N	82	−
O	109	+
P	134	+
Q	88	−
R	153	+
S	159	+
T	127	+

The test is carried out as follows. First, compute the overall sample median, $\hat{M}$. Second, form a $2 \times J$ observed-frequency table in which the first row contains the frequencies, in each group, of scores below the overall sample median and the second row contains the frequencies of scores above it. Third, test whether the proportions above the overall median are equal. This test is simply the χ^2-test of the equality of J independent proportions (Section 14.5).

Example 2

(This example is an extension of Example 1.) Suppose that there are three methods of instruction and that students are assigned randomly to each method, forming groups of 10, 12, and 15 students. All students must satisfy the same criteria in order to complete the course. Table 14.10a shows the number of days taken by each student. Students who have not completed the course in 200 days are recorded as (>200).

The overall median of the 37 scores is the 19th score in the ranking of the scores, which is the score 151 in Group 3. Since this score is exactly equal to the overall median, it must be dropped from the data. The remaining scores are categorized as either below or above the median, and the frequencies of such scores, in each group, are shown in Table 14.10b. The expected frequencies, assuming the null hypothesis is true, are computed in the usual way; they are shown in Table 14.10c.

The test statistic, χ^2_{obs}, is 8.30. Because the critical value is $\chi^2_{0.05}(2) = 5.991$, the result is significant. The proportions of scores, in the three groups, below and above the overall sample median are significantly different; hence, we can conclude that the medians of the three populations differ.

14.7 Summary

This chapter described a number of tests of *dichotomous variables*. Dichotomous variables can be *naturally* dichotomous, can result from a *comparison*, or can result from a numerical variable that has been *reduced* to a dichotomous one.

TABLE 14.10 Median Test—Three Populations

(a) Data

Group 1: (n = 10) 110, 142, 163, 127, 128, 182, (>200), 162, 164, 157

Group 2: (n = 12) 108, 93, 142, 163, 141, 86, 112, 149, 121, 133, 102, 153

Group 3: (n = 15) 129, 148, (>200), (>200), 182, 173, 187, 146, 183, 151, 191, (>200), 108, 186, 162

(b) Observed Frequencies (O_{ij})

(*Note.* The score of 151 in Group 3, which is equal to the overall median, has been dropped.)

	GROUP 1	GROUP 2	GROUP 3	
BELOW MEDIAN	4	10	4	18
ABOVE MEDIAN	6	2	10	18
	10	12	14	36

(c) Expected Frequencies (E_{ij})

	5	6	7	18
	5	6	7	18
	10	12	14	36

The *sign test* is a binomial test of $p = 0.5$. The sign test can be used in a number of different studies. It can be used in a *dependent-groups design* in which the members of each pair are compared on the dependent variable and one member is designated as higher than the other. The sign test also can be used to test the *median of one population*.

In this chapter the χ^2-*test* was applied only to 2×2 and $2 \times J$ frequency tables. (In Chapter 15 the χ^2-test will be extended to a general $r \times c$ table.) The χ^2-test can be used to test the equality of *two or more independent proportions* and the equality of *two or more independent medians*. The χ^2-test is also used to test the *independence of two dichotomous variables*.

The z-*test of two independent proportions* is essentially equivalent to the χ^2-test of two independent proportions.

Exercises

1 (Section 14.2) A psychologist is studying a new method for developing visual-motor coordination in brain-damaged children. He plans a matched-groups study with two conditions: in the *experimental* condition, the children are given a total of 20 hours of training in the new method; in the *control* condition, the children spend 20 hours in supervised play that does not include specific visual-motor training. Twenty-four children are matched, before the study begins, on visual-motor ability. One member of each pair of matched children is assigned, at random, to the experimental condition and the other to the control condition. After the 20 hours of training or play, each child is tested on a puzzle task and the time, in seconds, taken to complete the task is recorded. Analyze the data, given below, using a two-tailed sign test, at the 0.05 level.

PAIR	EXPERIMENTAL CONDITION	CONTROL CONDITION
1	37	26
2	47	49
3	16	72
4	43	25
5	62	27
6	15	26
7	18	18
8	26	53
9	31	81
10	19	39
11	15	27
12	21	29

2 (Section 14.3)
   (a) What are the values of the following percentiles of the χ^2-distribution?

df	PERCENTILE
1	95th
2	99th
2	90th
2	5th

   (b) What is the probability that a variable, distributed as χ^2 with one degree of freedom, exceeds 7.3? (State the probability as a double inequality—for example, $0.05 < \text{prob} < 0.10$.)

(c) What is the probability that a variable, distributed as χ^2 with one degree of freedom, exceeds 4.2?

3 (Section 14.3) Male and female subjects are asked whether they believe in abortion. In the male group of 53 subjects, 29 (=54.7%) believe in abortion. In the female group of 83 subjects, 69 (=83.1%) believe in abortion. Are these proportions significantly different? Use the χ^2-test.

4 (Section 14.4) Repeat Question 3 using the z-test. Are the p-values the same for the two tests?

5 (Section 14.4) The term "transposition" is used in discrimination studies to indicate that a subject, trained to choose the larger of two small stimuli, chooses the larger of two large test stimuli, rather than choosing the one that is closer in size to the original small stimuli. Subjects who choose the correct stimulus in the test situation are said to exhibit transposition. An investigator studies the relationship between transposition and age in a group of 60 young children. The number of subjects falling into each of four age-response categories are given below.

	AGE	
	5 YEARS OR LESS	OVER 5 YEARS
FAILS TO TRANSPOSE	26	7
TRANSPOSES	7	20

Is there a significant relationship between age and transposition?

6 (Section 14.5) The attitude toward abortion in three religious groups is studied. There are 70 subjects in each of the three groups. The number of subjects who believe in abortion is 30 in group A, 37 in group B, and 42 in group C. Test whether the proportions favoring abortion differ among the three groups. Use $\alpha = 0.01$.

7 (Section 14.6) A group of female subjects was studied to determine whether their median resting heart-rate differs from 80 beats per minute. The heart-rates for the subjects were 79, 64, 99, 95, 104, 83, 85, 67, 69, 88, 100, 86, 74, 89, 86, 77, 81, 83, 90, 70, 78, 81, 85, 83, 82. Make the test at the 0.01 level of significance.

8 (Section 14.6) A study was conducted to determine whether there is a relationship between size of audience and speaker's heart-rate. Forty males were assigned at random to one of four conditions. The conditions were audiences of sizes 3, 10, 25, and 50. Ten subjects were assigned to each condition. Each speaker gave a standard speech to an audience of the assigned size. His heart-rate was measured one minute after the beginning of his speech. Test whether there is a significant difference between the median heart-rates in the four conditions. The heart-rates of the 10 subjects in each condition are given below.

CONDITION			
SIZE 3	SIZE 10	SIZE 25	SIZE 50
60	64	83	75
62	87	77	88
83	82	69	101
59	73	92	95
67	72	75	96
66	66	73	79
67	69	81	90
72	71	71	88
74	72	70	85
74	81	75	91

9 (All sections) The purpose of this question is to increase your skill in identifying which test applies to a given study. For each of the studies described below, state which test would be used to analyze the data from the study. Do *not* carry out the analysis.

 Hint: As an aid in identifying the correct test, make up a sample data table. This table could be either a raw-data table (like Tables 14.1, 14.2, 14.9 and 14.10a) or an observed-frequency table (like many of the other tables in the chapter). Put some fictitious numbers in the table. The table will help you to identify the correct test to apply to the data, and hence to the study.

(a) The number of Jewish students at two universities is to be compared by drawing random samples of 100 students at each university and comparing the number of Jewish students in each sample.

(b) The average age of students at five universities is to be compared by drawing random samples of 50 students at each university and recording each student's age.

(c) A study is designed to compare the athletic skill of children brought up in the city with children brought up on a farm. Twenty children from each population are selected and matched for age. Each age-matched pair (one child from a city, one from a farm) runs a race to see which member of the pair is faster.

(d) In order to determine whether convenient access to a public swimming pool is related to regularity of swimming, a sample of 500 persons is selected. Each person is scored as either "swimming occasionally or regularly" or "never swimming." Each person's place of residence is considered either "close to a pool" (less than 1 km away) or "not close to a pool" (1 km or more away).

15 TESTS FOR CATEGORICAL AND RANK VARIABLES

In this chapter the tests described in Chapter 14 are extended to variables that are not dichotomous. These variables are called categorical variables and rank variables. The tests of categorical variables are straightforward extensions of the χ^2-tests of dichotomous variables (Chapter 14). The tests of rank variables are called rank-sum tests; these tests are often used as alternatives to the t-tests of two independent or of two dependent groups (Chapter 11).

15.1 Types of Variables

In classifying statistical procedures, it is useful to distinguish three types of variables: a *numerical* variable, whose possible values are numbers; a *categorical* variable, which has only a few values that usually are stated in words rather than in numbers; and a *rank* variable, whose values are ranks (that is, integers from 1 up to the number of objects being ranked). The distinctions among these three types are not absolute—a given variable can sometimes be considered to be of two types. For example, a variable with the values 1, 2, 3, and 4 can be considered to be a numerical or a categorical variable (and even, rarely, as a rank variable). However, the general distinctions among these three types are useful.

Most of the preceding chapters have described analyses of numerical variables. The exceptions have been the analyses of dichotomous variables (in Chapter 14 and the binomial test of earlier chapters). A dichotomous variable is simply a categorical variable with only two values. Categorical variables with more than two values are sometimes called *polytomous* variables. Since the techniques described in the next few sections are natural extensions of the techniques for dichotomous variables, they shall be described as techniques for categorical variables, rather than restricted to polytomous variables.

Numerical Variables

A numerical variable is a variable whose possible values are numbers. Two types of numerical variables have been distinguished previously: *discrete* and *continuous* (Section 2.4). A discrete variable usually has only a few possible values. As noted above, a discrete numerical variable with only a few values may be considered to be a categorical variable, and can be analyzed as such.

The values of continuous numerical variables are numbers that can take any possible value within a certain range. Continuous variables are important in theoretical mathematical work (see, for example, Chapter 7) and as an idealized form of a discrete variable (Section 2.7). A measured variable must be discrete, because no measuring instrument is infinitely precise. Such a variable is often analyzed as if it were continuous, however.

Categorical Variables

A categorical variable can arise in two ways. A variable can be *naturally* categorical; or a numerical variable can be *reduced* to a categorical one.

Examples of naturally categorical variables are: psychiatric diagnosis (phobic neurosis, anxiety neurosis, obsessive-compulsive neurosis, hysterical neurosis); brain hemisphere dominance (left, right, mixed); and attitude toward a statement (strongly agree, agree, neutral, disagree, strongly disagree).

Examples of numerical variables reduced to categorical variables are: final exam score (0 to 100) reduced to A, B, C, D, fail; and number of years of schooling reduced to five categories (some elementary school, elementary school graduation, high school graduation, bachelor's degree, doctoral degree).

Rank Variables

Rank variables can be either variables that are naturally ranked or numerical variables reduced to rank variables. In the two rank-sum tests described in this chapter, a numerical variable is reduced to a rank variable during the calculations required by each test. The ranking of a variable was described in Section 3.3. The method is modified slightly for the rank-sum tests.

Examples of naturally ranked variables are: 10 ordinary members of a committee ranked from 1 to 10, by the chairperson of the committee, on their degree of participation in the committee's meetings (the person who participated the most is ranked 1, the next-most-active person is ranked 2, and so on); and five stimuli ranked by a subject in terms of their degree of attractiveness to him (the most attractive is ranked 1, the next-most attractive is ranked 2, and so on).

Examples of numerical variables reduced to rank variables are: the lengths of time (in seconds) 30 subjects in an experiment took to complete a task reduced to ranks from 1 to 30, with rank 1 assigned to the shortest time, rank 2 to the next shortest time, and so on to rank 30; and achievement test scores of 45 subjects reduced to ranks from 1 to 45.

15.2 The Chi-square Test in an $r \times c$ Table

The chi-square test in a 2×2 table (Sections 14.3 and 14.4) can easily be extended to larger tables. In general, the table can have r rows (corresponding to a categorical variable with r categories) and c columns (corresponding to a categorical variable with c categories). This $r \times c$ table is sometimes called a *contingency table*.

The chi-square test in a 2×2 table was viewed in two ways: as a test of two independent proportions (Section 14.3) and as a test of the independence of two dichotomous variables (Section 14.4). Similarly, the chi-square test in an $r \times c$ table may be viewed in two ways (both of which are described below): as a test of the homogeneity of J independent distributions and as a test of the independence of two categorical variables.

The calculations for the chi-square test in the larger table are carried out in the same way as those for the 2×2 table. The expected frequencies may be calculated either from pooled proportions (as in Section 14.3) or from the row and column sums of the observed frequency table (as in Section 14.4). The test statistic, χ^2_{obs}, may also be calculated in one of two ways: either by the definitional formula (Section 14.3) or by the computational formula (Section 14.4). The condition for the distribution of χ^2_{obs} to be approximately χ^2 is the same condition described in Section 14.4—namely, that all the expected frequencies should be five or more. The number of degrees of freedom for the χ^2-distribution is $(r - 1) \times (c - 1)$. An informal justification of this formula is given in Example 1, below.

Test of Homogeneity of J Independent Distributions

Tests of the equality of two or more independent proportions were described in Sections 14.3 and 14.5. These tests can be looked at in another, equivalent way: there are J (= 2, or 3, or 4, etc.) populations, and each population contains a certain proportion of successes and a certain proportion of failures, where "success" and "failure" are the two values of a dichotomous variable. In other words, there is a *distribution* in each population over the two values of the dichotomous variable. The tests in Sections 14.3 and 14.5 are tests of the equality of these distributions in the J populations. The term "homogeneity" often is used for this equality; the tests are therefore known as tests of the *homogeneity* of J independent population distributions.

The distributions of a *dichotomous* variable in $J = 3$ populations are shown in Table 15.1a. The distribution in the first population has two proportions, q_1 and p_1; the distribution in the second population has proportions q_2 and p_2; and the distribution in the third population has proportions q_3 and p_3. If these distributions are homogeneous (equal), the three proportions of failure will be equal $(q_1 = q_2 = q_3)$ and the three proportions of success will be equal $(p_1 = p_2 = p_3)$. Note that the homogeneity of the distributions implies that the proportions of successes are equal. Conversely, the equality of the proportions of successes implies that the distributions are homogeneous. (There are only two values of the variable; if the proportions of successes are equal, the proportions of failures must also be equal, and hence, the distributions are equal or homogeneous.)

This example is extended to a *categorical* variable (with four categories) in Table 15.1b, in which each distribution has four proportions. In the first popu-

TABLE 15.1 Population Distributions

(a) Dichotomous Variable			
	$J = 3$ POPULATIONS		
	1	2	3
Failure	q_1	q_2	q_3
Success	p_1	p_2	p_3

(b) Categorical Variable (Four Categories)			
	$J = 3$ POPULATIONS		
	1	2	3
a	p_{a1}	p_{a2}	p_{a3}
b	p_{b1}	p_{b2}	p_{b3}
c	p_{c1}	p_{c2}	p_{c3}
d	p_{d1}	p_{d2}	p_{d3}

lation the proportions of the four possible values of the categorical variable are called p_{a1}, p_{b1}, p_{c1}, and p_{d1}. The corresponding proportions in the second population are p_{a2}, p_{b2}, p_{c2}, p_{d2}, and in the third population the proportions are p_{a3}, p_{b3}, p_{c3}, p_{d3}. If these three distributions are homogeneous, the three proportions of category a will be equal ($p_{a1} = p_{a2} = p_{a3}$), the three proportions of category b will be equal ($p_{b1} = p_{b2} = p_{b3}$), and so on.

The test of the (null) hypothesis, that the J distributions are homogeneous, is carried out by the drawing independent random samples from the J populations and recording the frequencies of each category in each sample. The calculations are carried out as they are in any other χ^2-test: the expected frequencies are calculated, the test statistic χ^2_{obs} is calculated from the observed and expected frequencies, and χ^2_{obs} is compared to a percentile of the χ^2-distribution. The number of degrees of freedom for a general table with r rows and c columns is

$$df = (r - 1) \times (c - 1)$$

so the rejection region for the test is $\{\chi^2_{obs} > \chi^2_\alpha [(r - 1) \times (c - 1)]\}$. Note that the number of columns, c, is the number of populations, J, and that the number of rows is the number of categories of the categorical variable. The following example explains why $df = (r - 1) \times (c - 1)$.

Example 1

Consider a study of people's attitudes toward abortion. The people live in three different environments (rural, suburban, and inner city). The attitude toward abortion is measured on a four-point scale (strongly oppose, oppose, in favor, and strongly in favor). Independent samples of persons living in the three areas are drawn: 60 persons from a rural environment, 70 persons from the suburbs, and 80 persons from the inner city. The observed frequencies—that is, the number of persons in each group holding each of the four attitudes—are shown in Table 15.2a.

TABLE 15.2 Test of the Homogeneity of Three Independent Distributions

(a) Observed Frequencies (O_{ij})

		RURAL	SUBURBAN	INNER CITY		POOLED PROPORTIONS
ATTITUDE	STRONGLY OPPOSE	7	10	18	35	0.16667
TOWARD	OPPOSE	12	15	23	50	0.23810
ABORTION	IN FAVOR	18	16	25	59	0.28095
	STRONGLY IN FAVOR	23	29	14	66	0.31429
		60	70	80	210	1.00001

(b) Expected Frequencies (E_{ij})

RURAL	SUBURBAN	INNER CITY	
10.000	11.667	13.334	35.001
14.286	16.667	19.048	50.001
16.857	19.667	22.476	59.000
18.857	22.000	25.143	66.000
60.000	70.001	80.001	210.002

The null hypothesis may be stated in two, equivalent, ways.

H_0: the three population distributions are equal

or

$$H_0: p_{a1} = p_{a2} = p_{a3} \text{ and } p_{b1} = p_{b2} = p_{b3} \text{ and}$$
$$p_{c1} = p_{c2} = p_{c3} \text{ and } p_{d1} = p_{d2} = p_{d3}$$

The alternative hypothesis may be stated as:

H_1: the three population distributions are not all equal

or

$$H_1: H_0 \text{ is false}$$

(It is incorrect to state H_1 in a manner similar to the second statement of H_0—replacing all the equality signs by inequality signs—since only *some* of these equalities need to be false in order for H_0 to be false.)

The (overall) pooled proportions of persons who respond with each of the four possible attitudes are

$$\hat{p}_a = 35/210 = 0.16667; \hat{p}_b = 50/210 = 0.23810$$
$$\hat{p}_c = 59/210 = 0.28095; \hat{p}_d = 66/210 = 0.31429$$

The expected frequencies in each cell are calculated as the product of the pooled proportion for the cell's row and the number of subjects for the cell's column. (For example, the first expected frequency is $0.16667 \times 60 = 10.000$.) The expected frequencies are shown in Table 15.2b. (Alternatively, each expected frequency can be calculated as the product of the row sum and the column sum, divided by the total sample size. For example, the first expected frequency is $35 \times 60/210 = 10.000$.) The expected frequencies should be checked by computation of the row and column sums of the expected-frequency table. These sums, in Table 15.2b, match the corresponding sums of the observed frequencies (Table 15.2a), within rounding error.

Explanation of the Formula for df The fact that the row sums of the expected-frequency table must equal the corresponding row sums of the observed-frequency table means that the expected frequencies in the third column can be determined from the expected frequencies in the first two columns. We say that only the frequencies of the first two cells in each row are "free to vary". Furthermore, not all the frequencies in the first two columns are free to vary, since the column sums must equal the column sums of the observed-frequency table. Only the frequencies of the top three cells of each column are free to vary; the last cell in each column can be determined from them. Only $3 \times 2 = 6$ frequencies, therefore, are free to vary. Hence, the number of degrees of freedom for the table is 6. A similar argument shows that the number of degrees for an r by c table is $(r - 1) \times (c - 1)$.

Test Statistic and Test The test statistic, χ^2_{obs}, can be computed by either of the following formulas.

$$\chi^2_{obs} = \sum_i \sum_j \frac{(O_{ij} - E_{ij})^2}{E_{ij}}$$

or

$$\chi^2_{obs} = \sum_i \sum_j \frac{O^2_{ij}}{E_{ij}} - n$$

The value of χ^2_{obs} is 13.24. The number of degrees of freedom (as explained above) is $(4 - 1) \times (3 - 1) = 3 \times 2 = 6$. The critical value for a test at the 0.05 level is $\chi^2_{.05}(6) = 12.59$. Since the observed value exceeds the critical value, we can reject the null hypothesis and conclude that the distributions of responses to the abortion question are different for the different environments.

Test of Independence of Two Categorical Variables

The test of the independence of two *dichotomous* variables was described in Section 14.4. The test is a χ^2-test based on a 2 × 2 table. The test of the independence of two *categorical* variables is a χ^2-test based on an *r* × *c* table, where one of the variables has *r* categories and its values form the rows of the table and the other variable has *c* categories and its values form the columns of the table.

The hypotheses to be tested are

H_0: the two variables are independent.
H_1: the two variables are dependent.

The test of independence is based on *one* sample, each member of which has a value on each of the *two* categorical variables. The test of homogeneity, described above, is based on *J* samples, each member of which has a value on just *one* categorical variable. Despite this fundamental difference between the two tests, the hypotheses for the two tests are essentially equivalent and the procedures for carrying out the two tests are identical. The equivalence was briefly explained for dichotomous variables in Section 14.4. For categorical variables, the argument is similar: the *J* samples, in the test of homogeneity, are distinguished by what is, in effect, a categorical variable, with *J* categories; the test of homogeneity is, therefore, a test of two categorical variables: one that distinguishes the groups, and the other as measured for each subject.

The calculation of χ^2_{obs} proceeds in the usual way for χ^2-tests. The critical value is found from the χ^2-distribution with $df = (r - 1) \times (c - 1)$.

Example 2

A study is conducted to determine if there is a relationship between political orientation and income level. A random sample of 140 persons is surveyed. Political orientation is measured on a four-point scale extending from 1 (left orientation) to 4 (right orientation). Income is defined in a suitable way and reduced to three levels: low, medium, and high.

Each person in the sample has a political orientation and an income level and, hence, belongs to one of the cells in the 3 × 4 table of observed frequencies (Table 15.3*a*). From these observed frequencies we compute expected frequencies (Table 15.3*b*) as the product of the row and column sums divided by the total sample size. (For example, the expected frequency for high-income level and political orientation = 2 is $40 \times 54/140 = 15.429$.)

The null hypothesis is

H_0: political orientation and income level are independent

The alternative hypothesis is

H_1: political orientation and income level are dependent

The test statistic, χ^2_{obs}, is calculated, as usual, by either of the following formulas.

TABLE 15.3 Test of the Independence of Two Categorical Variables

(a) Observed Frequencies (O_{ij})

POLITICAL ORIENTATION

		LEFT			RIGHT	
		1	2	3	4	
	LOW	17	28	6	2	53
INCOME LEVEL	MEDIUM	15	14	11	7	47
	HIGH	8	12	17	3	40
		40	54	34	12	140

(b) Expected Frequencies (E_{ij})

15.143	20.443	12.871	4.543	53.000
13.429	18.129	11.414	4.029	47.001
11.429	15.429	9.714	3.429	40.001
40.001	54.001	33.999	12.001	140.002

$$\chi^2_{obs} = \sum_i \sum_j \frac{(O_{ij} - E_{ij})^2}{E_{ij}}$$

or

$$\chi^2_{obs} = \sum_i \sum_j \frac{O_{ij}^2}{E_{ij}} - n$$

The value, for these data, is $\chi^2_{obs} = 18.75$. Suppose that the test is to be carried out with $\alpha = 0.01$. The number of degrees of freedom is, as usual, $(r - 1) \times (c - 1)$, where r and c are the number of rows and columns, respectively. Here $df = (3 - 1) \times (4 - 1) = 2 \times 3 = 6$. The critical value is $\chi^2_{.01}(6) = 16.81$. Since the observed value exceeds the critical value, the null hypothesis is rejected and we can conclude that political orientation and income level are related (or dependent).

Comment Some of the expected frequencies in Table 15.3b are quite small. All the expected frequencies in the last column are less than 5. One of the requirements for the validity of the χ^2-test, discussed in Section 15.4 below, is that the expected frequencies not be too small. One solution to this problem is to combine two or more categories so that all the expected frequencies exceed 5. In this example, this can be done by combining categories 3 and 4 of political orientation and repeating all the calculations.[1] The observed and expected frequencies of this reduced, or collapsed, table are shown in Table 15.4. The value of χ^2_{obs} for the reduced table is 15.02. The critical value also is changed, since df now is $(3 - 1) \times (3 - 1) = 2 \times 2 = 4$. The critical value is $\chi^2_{.01}(4) = 13.28$. The result is still significant and the test is now valid, since all expected frequencies are considerably greater than 5.

---

[1] It is not legitimate to try different combinations of categories in order to find a particular combination that gives a significant result. The categories that are combined should go together sensibly. In this example, categories 3 and 4 of political orientation go together because they are adjacent categories. It would not make sense and would not be legitimate to combine, for example, categories 2 and 4.

TABLE 15.4 A Reduced Table

(a) Observed Frequencies (O_{ij})

		POLITICAL ORIENTATION			
		1	2	3 & 4	
	LOW	17	28	8	53
INCOME LEVEL	MEDIUM	15	14	18	47
	HIGH	8	12	20	40
		40	54	46	140

(b) Expected Frequencies (E_{ij})

15.143	20.443	17.414	53.000
13.429	18.129	15.443	47.001
11.429	15.429	13.143	40.001
40.001	54.001	46.000	140.002

15.3 Test of Goodness-of-fit

All the χ^2-tests described in Chapters 14 and 15 are, in a sense, tests of goodness-of-fit. Each test compares a set of observed frequencies to a set of expected frequencies (calculated on the basis of a null hypothesis). The test statistic is small when the two sets of frequencies are similar (good fit) and large when the two sets of frequencies are quite different (bad fit). If the fit is good, the null hypothesis is supported (strictly: H_0 cannot be rejected); if the fit is bad, the alternative hypothesis is supported.

Despite this way of looking at all χ^2-tests, the term "goodness-of-fit" usually is applied only to χ^2-tests of a certain kind, described in this section. A *goodness-of-fit test* is a test of *one* categorical variable measured in *one* population. The null hypothesis specifies the probability (proportion) of each possible value of the categorical variable. The proportions of each value of the categorical variable are then observed in a sample. The goodness-of-fit test determines whether the observed proportions are close to the specified probabilities (good fit, H_0 cannot be rejected) or the observed proportions are discrepant from the specified probabilities (bad fit, H_0 is rejected). The decision between a good fit and a bad fit is made on the basis of a test statistic, χ^2_{obs}, which is computed from the observed and expected frequencies. The test statistic is compared to a critical value found from the χ^2-distribution with $df = r - 1$, where r is the number of categories of the categorical variable.

Dichotomous Variable—Hypotheses

The goodness-of-fit test of a categorical variable with two categories (that is, a dichotomous variable) has already been described. The test is the binomial test. The null hypothesis usually is written H_0: $p = 0.5$, or H_0: $p = 0.7$, etc. In general, the null hypothesis is

$$H_0: p = p_0$$

where p_0 is a specified (known) number. As $q = 1 - p$, the null hypothesis could also be written as

$$H_0: p = p_0 \text{ and } q = q_0$$

where $q_0 = 1 - p_0$. For the examples of $p_0 = 0.5$ and $p_0 = 0.7$, the null hypotheses would be

$$H_0: p = 0.5 \text{ and } q = 0.5$$
$$H_0: p = 0.7 \text{ and } q = 0.3$$

Null hypotheses of this kind are tested by the binomial test; they can also be tested by the χ^2-test, as shown in Example 3, below.

Categorical Variable— Hypotheses

When there are more than two categories, the null hypothesis is extended in a natural way. If there are four categories, called a, b, c, and d, the null hypothesis is

$$H_0: p_a = p_{a0}, p_b = p_{b0}, p_c = p_{c0}, \text{ and } p_d = p_{d0}$$

This hypothesis states that the probabilities of the categories (p_a, p_b, etc.) have specific values (p_{a0}, p_{b0}, etc.). The alternative hypothesis states that the probabilities do not (all) have these values.

$$H_1: H_0 \text{ is false}$$

A particular example of a null hypothesis is

$$H_0: p_a = 9/16, p_b = 3/16, p_c = 3/16, p_d = 1/16$$

(This hypothesis is tested in Example 2, below.)

The category probabilities in a goodness-of-fit test often are hypothesized to have *equal* values—if there are r categories, then each probability is hypothesized to be $1/r$. For four categories, for example, the null hypothesis would be

$$H_0: p_a = 1/4, p_b = 1/4, p_c = 1/4, p_d = 1/4$$

This hypothesis can also be written as

$$H_0: p_a = p_b = p_c = p_d$$

Of course, these two ways of writing the null hypothesis are equivalent, but there is a real danger in *misinterpreting* the second way as an expression of the equality of four *independent* proportions (Section 14.5). The four proportions in the goodness-of-fit test are *not* independent—they are the proportions of four *categories* of *one* variable measured in *one* sample. (To make the distinction clear, the null hypothesis in Section 14.5 was written as

$$H_0: p_1 = p_2 = p_3 = p_4$$

where the numbers 1, 2, 3, 4 represent four populations, whereas in the goodness-of-fit test the letters a, b, c, d represent four categories.)

The Test Statistic

A sample of size n is drawn from the population and the number of elements in each category is called the observed frequency, O_i, of the category. There are r categories. If the null hypothesis is true, the probabilities of the categories are p_{a0}, p_{b0}, etc.; hence, the expected frequencies, E_i, in the categories are np_{a0}, np_{b0}, etc. The test statistic is

$$\chi^2_{\text{obs}} = \sum_i \frac{(O_i - E_i)^2}{E_i}$$

which can also be computed by the formula

$$\chi^2_{obs} = \sum_i \frac{O_i^2}{E_i} - n$$

As there are r categories, there are r expected frequencies and these frequencies must sum to n. This means that only $(r - 1)$ of them are free to vary—the last expected frequency can be determined from the other $(r - 1)$ of them. Hence, $df = r - 1$. The critical value for a test with probability of a Type I error $= \alpha$ is $\chi^2_\alpha(r - 1)$.

Example 1

A die is tossed 100 times in order to test whether it is fair or not. The observed frequencies (O_i) of each of the six possible results are shown in Table 15.5a. The null hypothesis that the die is fair is

$$H_0: p_1 = 1/6, p_2 = 1/6, \ldots, p_6 = 1/6$$

(Here we label the categories, $1, 2, \ldots, 6$ instead of a, b, and so on, since the faces of the die shows digits, not letters.)

In 100 tosses the expected frequency, E_i, of each category, is $100 \times 1/6 = 16.667$. The differences between the observed and expected frequencies, $O_i - E_i$, and the computation of the test statistic also are shown in the table. (Alternatively, the test statistic can be calculated by

$$\chi^2_{obs} = \sum_i \frac{O_i^2}{E_i} - n = \frac{18^2}{16.667} + \frac{14^2}{16.667} + \ldots + \frac{23^2}{16.667} - 100 = 3.92)$$

The number of degrees of freedom is $(6 - 1) = 5$. The critical value for a test at the 0.05 level is $\chi^2_{.05}(5) = 11.07$. Since the observed value of χ^2_{obs} does not exceed the critical value, the null hypothesis cannot be rejected. There is no evidence from these data that the die is unfair. In other words, there is a "good fit" between the observed frequencies (or proportions) and the expected frequencies (or probabilities).

Example 2

A genetic theory predicts that the offspring in a breeding experiment will be of types a, b, c, and d in proportions 9/16, 3/16, 3/16, and 1/16, respectively. This theory is, essentially, a statement of a null hypothesis.

TABLE 15.5 Test of Goodness-of-fit—Example 1

(a) Frequencies

CATEGORY	O_i	E_i	$(O_i - E_i)$
1(a)	18	16.667	1.333
2(b)	14	16.667	-2.667
3(c)	15	16.667	-1.667
4(d)	13	16.667	-3.667
5(e)	17	16.667	0.333
6(f)	23	16.667	6.333
	100	100.002	-0.002

(b) Test Statistic

$$\chi^2_{obs} = \sum_i \frac{(O_i - E_i)^2}{E_i} = \frac{(1.333)^2}{16.667} + \frac{(-2.667)^2}{16.667} + \ldots + \frac{(6.333)^2}{16.667} = 3.92$$

TABLE 15.6 Test of Goodness-of-fit—Example 2

(a) Frequencies

CATEGORY	O_i	E_i	$(O_i - E_i)$
a	131	120.375	10.625
b	52	40.125	11.875
c	18	40.125	−22.125
d	13	13.375	−0.375
	214	214.000	0.000

(b) Test Statistic

$$\chi^2_{obs} = \sum_i \frac{(O_i - E_i)^2}{E_i} = \frac{(10.625)^2}{120.375} + \frac{(11.875)^2}{40.125} + \frac{(-22.125)^2}{40.125} + \frac{(-0.375)^2}{13.375} = 16.66$$

$$H_0: p_a = 9/16, \ p_b = 3/16, \ p_c = 3/16, \ p_d = 1/16$$

In an experiment, 214 offspring are observed. The frequency, O_i, of each type (category) is shown in Table 15.6a. The expected frequencies, E_i, are the product of n (214) and the hypothesized proportions. (For example, the expected frequency of type a is $214 \times 9/16$ = 120.375.) The table shows the differences between the observed and expected frequencies, $O_i - E_i$, and the computation of the test statistic, χ^2_{obs}. The value is 16.66, which exceeds the 99.5th percentile of the χ^2-distribution with $r - 1 = 4 - 1 = 3$ degrees of freedom [$\chi^2_{.005}(3) = 12.84$]. Hence, these data indicate that the genetic theory does not apply to the experiment that was run.

Example 3

As shown earlier in this section, the binomial test is a test of goodness-of-fit for two categories. Hence, the test can be carried out by the χ^2-test in a similar way to the previous examples. The critical value is found from the χ^2-distribution with $df = r - 1 = 2 - 1 = 1$. It is convenient to use the χ^2-test when $n > 25$ (the largest value in the binomial tables) or when p_{a0}, the hypothesized proportion of successes, is not one of the values given in the binomial tables. The χ^2-distribution also provides an alternative to the normal approximation to the binomial distribution (Section 9.4). In fact, the critical values of the normal distribution and the χ^2-distribution with the $df = 1$ are exactly related as follows: $(z_{\alpha/2})^2 = \chi^2_\alpha(1)$. (This relationship was noted in Section 14.4.) Furthermore, the test statistic χ^2_{obs} is approximately equal to the square of the test statistic (z^2_{obs}) that is used in the normal approximation to the binomial.[2]

15.4 Requirements for the Validity of Chi-square Tests

Four requirements for the validity of the statistical procedures described in the first 13 chapters of this book were summarized in Section C.3. The first two of these requirements (random sampling and independence of observations) also apply to all χ^2-tests. These two requirements, together with an additional requirement that applies only to χ^2-tests, are described below. (The new requirement is called Requirement 5, to be consistent with the numbering in

---

[2] The two test statistics, χ^2_{obs} and z^2_{obs} can be made exactly equal by the following change to the formula for χ^2_{obs}.

$$\chi^2_{obs} = \frac{(|O_a - E_a| - \frac{1}{2})^2}{E_a} + \frac{(|O_b - E_b| - \frac{1}{2})^2}{E_b}$$

Section C.3.) These three requirements apply to all the χ^2-tests of Chapters 14 and 15.

REQUIREMENT 1: RANDOM SAMPLING

The subjects in the sample (or samples) must be selected at random from the population(s) of interest.

REQUIREMENT 2: INDEPENDENCE OF OBSERVATIONS

Each observation must be independent of every other observation in the study.

Application

This requirement applies to all χ^2-tests and to all experimental designs except those in which dependence of the observations is built-in (dependent-groups designs).

Comment

This requirement is sometimes overlooked even in published studies, whose conclusions are invalidated accordingly. Consider a study of the homogeneity of the distributions in three conditions of an experiment. Suppose the variable has five categories, giving a 5×3 table. If three independent groups of subjects are used, one group in each condition, the observations are independent and the χ^2-test is valid. However, if each subject is run in all three conditions—that is, if three observations are made on each subject—the observations are not independent. It is easy to determine whether the independence requirement is met: each subject must appear in one and only one cell of the table.[3]

REQUIREMENT 5: LARGE EXPECTED FREQUENCIES

Each expected frequency must be 5 or more in order for the χ^2-distribution to be a satisfactory approximation to the distribution of χ^2_{obs}.

Application

This requirement applies to all χ^2-tests.

Comment

The stated requirement is a rough "rule of thumb". Some statisticians recommend that all expected frequencies exceed 10. Others say that some expected frequencies can be as small as 1, provided that no more than 10% of the cells have expected frequencies below 5. The requirement, as given above, provides a reasonable compromise between these views. Note that the requirement applies to the *expected*, not the *observed*, frequencies.

15.5 Rank-sum Test for Two Independent Groups

The rank-sum tests described in this section and in Section 15.6 are alternatives to the *t*-tests for two groups (Chapter 11). The rank-sum test for two *independent* groups is described in this section, and the rank-sum test for two *dependent* groups is described in Section 15.6.

---

[3] This is a necessary but not sufficient condition for independence.

Like the t-tests, rank-sum tests are tests of a numerical variable. In a t-test the data are used directly in the computation of the group means, the standard deviations, and the test statistic, t_{obs}. In a rank-sum test, on the other hand, the data are first converted to *ranks*, and the computation of the test statistic is based solely on these ranks.

Rank-sum tests and t-tests are tests of the central tendency of two populations. In a t-test, the central tendency is measured by the mean. For the t-test of two independent means to be strictly valid, two requirements must be met: the population distributions must be normal and must have the same standard deviation (Section C.3). Because of these two requirements, the two population distributions are *identical* when the two means are equal: that is, when the null hypothesis is true. The two population distributions are *different* when the two means are unequal; that is, when the alternative hypothesis is true. The t-test can, therefore, be considered to be a test of the identity (or equality) of two population distributions, rather than a test of the equality of the two means.

The rank-sum tests make no requirements about the shape of the population distributions. Nor do they single out a particular measure of central tendency, such as the mean. The rank-sum tests are tests of the identity (or equality) of two population distributions and are sensitive to differences in the central tendency in these distributions.

Hypotheses

The hypotheses for the rank-sum test for two independent groups[4] can be stated as follows.

H_0: the two population distributions are the same.

H_1: the two population distributions are different.

The test is carried out by drawing independent samples from the two populations. Let the number of subjects in the *smaller* sample be n_S and the number in the *larger* sample be n_L. (We will use S and L to label the two groups and the two populations, even when the two samples are of equal size.) The $(n_S + n_L)$ scores are ranked as a single group—the ranks are $1, 2, \ldots, (n_S + n_L)$. We then consider the ranks assigned to the scores in each group. If the null hypothesis were true, we would expect the average ranks in the two groups to be about the same; if the alternative hypothesis were true, we would expect the average ranks to be different. The hypotheses can, therefore, be written in a different way.

H_0: average (rank_S) = average (rank_L).

H_1: average (rank_S) $\neq$ average (rank_L).

Here H_1 is two-sided, allowing for the possibilities that the central tendency of one population is greater or less than the central tendency of the other. If we wished to test a one-sided hypothesis, H_1 would be written as

H_1: average (rank_S) < average (rank_L)

---

[4] This test is also called the Wilcoxon test or the Mann-Whitney U test.

In this case the population from which the smaller sample is drawn is hypothesized to have a lesser central tendency. On the other hand, H_1 would be written as

$$H_1: \text{average (rank}_S) > \text{average (rank}_L)$$

if the population from which the smaller sample is drawn is hypothesized to have a greater central tendency.

Test Statistic

The test statistic is W_S.

$$\boxed{W_S = \text{sum of the ranks of the smaller group}}$$

Note that it is the sum of the ranks of the *smaller*, not the *larger*, group that is calculated.[5] The subscript S on W_S is a reminder of this definition of the test statistic. If the two groups are of equal size, W_S can be computed from either group.

Distribution of the Test Statistic

If the null hypothesis is true, all possible assignments of ranks to the subjects are equally likely. The distribution of W_S is derived by listing all these possible assignments, computing W_S for each, and computing the probability of each value of W_S.

Consider a study with seven subjects, of which two are in the smaller group ($n_S = 2$) and five are in the larger group ($n_L = 5$). The ranks are the numbers 1, 2, 3, 4, 5, 6, 7. The two subjects in the smaller group could have the ranks 1, 2 or 1, 3, and so on. The 21 possible assignments of ranks to the two subjects are listed in Table 15.7a. The table also shows W_S for each of these 21 assignments. If the null hypothesis were true, each of these assignments would be equally likely (probability = 1/21). The probability of each value W_S is, therefore, 1/21 times the number of assignments that have that value. These probabilities, shown in Table 15.7b, form the distribution of W_S. (Note that the distribution is symmetric.)

Rejection Region

When the alternative hypothesis is *two-sided*, the null hypothesis is rejected if W_S is either very large or very small. The smallest value of W_S in the distribution in Table 15.7 is $W_S = 3$; its probability is 0.048. The largest value, $W_S = 13$, also has probability = 0.048. The sum of these two probabilities is 0.096, which is less than 0.10. Hence, if we use $\alpha = 0.10$, we would reject H_0 for values of $W_S = 3$ and 13. For the purpose of another illustration, consider $\alpha = 0.20$ (a rather large value). The sum of the probabilities for $W_S = 3, 4, 12, 13$ is 0.192. Hence, for $\alpha = 0.20$, we would reject H_0 for values of $W_S = 4$ or *smaller*, or $W_S = 12$ or *larger*.

The rejection region for a *one-sided* alternative hypothesis is determined in a similar way. The rejection region will be one-tailed. The particular tail of

---

[5] The tabled distribution (Table A9) is the distribution of W_S, not W_L. Hence, it is incorrect to use W_L, the sum of the ranks of the larger group, as the test statistic (unless the groups are of equal size, in which case either sum may be used).

TABLE 15.7 Distribution of W_S for $n_S = 2, n_L = 5$

(a) The Possible Ranks Assigned to Smallest Group

RANKS	W_S = SUM OF RANKS	RANKS	W_S = SUM OF RANKS
1, 2	3	3, 4	7
1, 3	4	3, 5	8
1, 4	5	3, 6	9
1, 5	6	3, 7	10
1, 6	7	4, 5	9
1, 7	8	4, 6	10
2, 3	5	4, 7	11
2, 4	6	5, 6	11
2, 5	7	5, 7	12
2, 6	8	6, 7	13
2, 7	9		

(b) Distribution

W_S	PROBABILITY
3	1/21 = .048
4	1/21 = .048
5	2/21 = .095
6	2/21 = .095
7	3/21 = .143
8	3/21 = .143
9	3/21 = .143
10	2/21 = .095
11	2/21 = .095
12	1/21 = .048
13	1/21 = .048

the W_S distribution depends on the direction of the inequality in the statement of H_1. If H_1 is

$$H_1: \text{average (rank}_S) < \text{average (rank}_L)$$

the sum of ranks in the smaller group will be *small* if H_1 is true, and hence the lower tail of the W_S distribution is used. Conversely, if H_1 is

$$H_1: \text{average (rank}_S) > \text{average (rank}_L)$$

the sum of ranks in the smaller group will be *large* if H_1 is true, and hence the upper tail of the W_S distribution is used.

As Table 15.7b shows, the lower-tail rejection region for a one-tailed test with $\alpha = 0.05$, for $n_S = 2$ and $n_L = 5$, consists of the one value, $W_S = 3$; the probability of that value, when H_0 is true, is 0.048, which is less than 0.05. For these sample sizes, the upper-tail rejection region for a one-tailed test with $\alpha = 0.10$ consists of the values $W_S = 12$ or greater.

Rejection regions for this test are given in Table A9 in the Appendix. The table gives the rejection regions for both one- and two-tailed tests, for all combinations of n_S and n_L up to $n_S = 10$ and $n_L = 10$. For a one-tailed test we must decide, on the basis of the direction in which H_1 is stated, which one of the two tails given in the table is the correct tail for the test.

Ties

In some sets of data, two or more scores may have the same value. What ranks should be assigned to these scores? The best procedure to follow is to assign to each score the average (mean) of the ranks that the scores would have if they were not tied. For example, if nine scores are 8, 10, 10, 12, 13, 14, 14, 14, 15, the rank of 8 is 1, the ranks of the two scores of 10 are each 2.5 [= (2 + 3)/2], the rank of 12 is 4, the rank of 13 is 5, the ranks of the three scores of 14 are each 7 [= (6 + 7 + 8)/3], and the rank of 15 is 9.

The rejection regions in Table A9 were calculated assuming that there are no ties in the data. The distribution of W_S changes when there are ties. Table A9 is still satisfactory, however, provided that there are not too many ties (say, not more than 20% of the scores are tied).

Example 1

Eight subjects are tested in a learning experiment, three of the subjects in one condition and five in the other. The conditions are called S and L, respectively, corresponding to the sizes of the groups. The number of errors the subjects make are

Condition S: 15, 2, 8
Condition L: 22, 20, 5, 27, 16

Do the distributions of scores in the two conditions differ significantly?

The scores, in increasing order, are

2, 5, 8, 15, 16, 20, 22, 27

These numbers have the ranks

1, 2, 3, 4, 5, 6, 7, 8

Hence, the ranks of the scores in the two groups are

Condition S: 4, 1, 3
Condition L: 7, 6, 2, 8, 5

The test statistic (the sum of the ranks in the smaller group) is $W_S = 4 + 1 + 3 = 8$. The rejection region for a two-tailed test at $\alpha = 0.05$ is $W_S = 6$ or smaller, or $W_S = 21$ or larger (Table A9). The observed value of W_S does not fall in the rejection region, and hence the null hypothesis cannot be rejected. There is not a significant difference between the two conditions.

Example 2

Consider an example with ties. The scores in two conditions are

Condition S: 15, 3, 8, 7, 3
Condition L: 22, 6, 8, 8, 19

Here the two groups are of equal size and the labeling of one as S and the other as L is arbitrary. The scores, in increasing order, are

3, 3, 6, 7, 8, 8, 8, 15, 19, 22

These numbers have the ranks

1.5, 1.5, 3, 4, 6, 6, 6, 8, 9, 10

Note that the tied scores are each given the average of the ranks they would have had if they were not tied. The ranks of the scores in the two groups are

Condition S: 8, 1.5, 6, 4, 1.5
Condition L: 10, 3, 6, 6, 9

The test statistic is $W_S = 8 + 1.5 + 6 + 4 + 1.5 = 21$. Suppose that H_1 is one-sided, with condition S hypothesized to have a lower average rank than condition L. The rejection region for the one-tailed test is then $W_S = 19$ or smaller ($\alpha = 0.05$, Table A9). The observed result is not significant. Note, however, that 5 of the 10 scores are tied. The distribution in Table A9 is probably not correct when 50% of the scores are tied.

15.6 Rank-sum Test for Two Dependent Groups

In a dependent-groups design, the subjects are matched in pairs, and the subjects of each pair are assigned at random to the two experimental conditions. In the t-test of two dependent means the analysis is based on the differences, d_i, between the scores of the n pairs of subjects. In the rank-sum test[6] these n differences are ranked, in a manner described below, and the test statistic is based on the ranks. (The sign test could also be used to analyze the data from such a study, but the rank-sum test or the t-test usually is preferred.)

Hypotheses

The rank-sum test for two dependent groups (like the test for two independent groups) is a test of the identity of the two population distributions. The hypothesis may be written as follows.

H_0: the two population distributions are the same.

H_1: the two population distributions are different.

The test is sensitive to differences in the central tendencies of the two distributions. A one-sided alternative hypothesis may be expressed (using the term "center" as short for "measure of central tendency") as

H_1: the center of distribution one is greater than the center of distribution two.

Of course, the alternative hypothesis may be stated in the other direction, instead.

H_1: the center of distribution one is less than the center of distribution two.

Test Statistic

To compute the test statistic, called T_+,

- Compute the n differences, $d_i = y_{1i} - y_{2i}$.
- Rank the *absolute value* of these differences, $|d_i|$, so that the smallest absolute difference is ranked 1 and the largest absolute difference is ranked n.[7]
- Affix a plus or minus sign to each rank, so that the sign of the rank agrees with the sign of the original difference, d_i.
- Compute the test statistic, T_+.

$$T_+ = \text{sum of the } positive \text{ signed ranks}$$

[6] This test is also called the Wilcoxon signed-rank test.

[7] It is important that the ranks be assigned in this way, and not in the reverse order (smallest ranked n and largest ranked 1). This rank-sum test is the only statistical procedure described in this text in which the order in which the data are ranked is important. Either order of ranking can be used in every other procedure.

Distribution of the Test Statistic

If the null hypothesis is true, it is just as likely that a given difference, d_i, is positive or negative. This means that all possible assignments of $+$ and $-$ to the n ranks are equally likely. The distribution of T_+ is derived by listing all these possible assignments, computing T_+ for each, and computing the probability of each value of T_+.

Consider a study with $n = 5$ pairs of subjects. One possible assignment of signs to the five ranks is $(+1, -2, -3, +4, -5)$. In this assignment there are two positive differences and the two positive signed ranks are $+1$ and $+4$. The sum of the positive ranks is $T_+ = 5$. Another possible assignment of signs to the ranks is $(-1, -2, +3, +4, +5)$. Here there are three positive differences and $T_+ = 3 + 4 + 5 = 12$. There are, in fact, 32 possible assignments of signs to ranks, when $n = 5$, as shown in Table 15.8a. Each of these assignments is equally likely (probability $= 1/32$) when the null hypotheses is true. The probability of each value of T_+ is, therefore, $1/32$ times the number of assignments that have that value. These probabilities form the distribution of T_+, shown in Table 15.8b.

Rejection Region

If the alternative hypothesis is *two-sided* the null hypothesis will be rejected if T_+ is either very large or very small. The smallest value of T_+ in the distribution in Table 15.8 is $T_+ = 0$ (that is, all the differences, d_i, are negative); its probability is 0.031. The largest value is $T_+ = 15$ (that is, all the differences, d_i, are positive); its probability is also 0.031. The sum of these probabilities is 0.062, which is less than 0.10. Hence, if we use $\alpha = 0.10$, we would reject H_0 for values of $T_+ = 0$ and 15. Consider, instead, $\alpha = 0.20$ (a very large value). The sum of the probabilities for $T_+ = 0, 1, 2, 13, 14, 15$ is 0.186. Hence, for $\alpha = 0.20$, we would reject H_0 for values of $T_+ = 2$ or *smaller* or $T_+ = 13$ or *larger*.

The rejection region for a *one-sided* alternative hypothesis is determined in a similar way. The particular one-tailed rejection region depends on the direction of the inequality in the statement of H_1. If H_1 is

H_1: the center of distribution one is *greater* than the center of distribution two

most of the differences, $d_i = y_{1i} - y_{2i}$, will be *positive* if H_1 is true, and T_+ will tend to be large. The *upper* tail of the T_+ distribution is used in this case. Conversely, if H_1 is

H_1: the center of distribution one is *less* than the center of distribution two

most of the differences will be *negative* if H_1 is true and T_+ will tend to be small. The *lower* tail of the T_+ distribution is used.

Table 15.8b shows that the upper-tail rejection region for a one-tailed test with $\alpha = 0.05$, for $n = 5$, consists of the one value $T_+ = 15$, since when H_0 is true the probability of that value is 0.031, which is less than 0.05. The lower-tail rejection region for a one-tailed test with $\alpha = 0.05$, for $n = 5$, consists of the one value $T_+ = 0$.

Rejection regions for the rank-sum test for two dependent groups are given in Table A10 in the Appendix. The table gives the rejection regions for both one- and two-tailed tests, for all values of n up to 20. For a one-tailed test, we must decide, on the basis of the direction in which H_1 is stated, which one of the two tails given in the table is the correct tail for the test.

TABLE 15.8 Distribution of T_+ for $n = 5$

(a) The Possible Positive-Signed Ranks

NUMBER OF POSITIVE DIFFERENCES	POSITIVE-SIGNED RANKS	T_+	NUMBER OF POSITIVE DIFFERENCES	POSITIVE-SIGNED RANKS	T_+
None	None	0	Three	1, 2, 3	6
One	1	1		1, 2, 4	7
	2	2		1, 2, 5	8
	3	3		1, 3, 4	8
	4	4		1, 3, 5	9
	5	5		1, 4, 5	10
Two	1, 2	3		2, 3, 4	9
	1, 3	4		2, 3, 5	10
	1, 4	5		2, 4, 5	11
	1, 5	6		3, 4, 5	12
	2, 3	5	Four	1, 2, 3, 4	10
	2, 4	6		1, 2, 3, 5	11
	2, 5	7		1, 2, 4, 5	12
	3, 4	7		1, 3, 4, 5	13
	3, 5	8		2, 3, 4, 5	14
	4, 5	9	Five	1, 2, 3, 4, 5	15

(b) Distribution

T_+	PROBABILITY
0	$1/32 = .031$
1	$1/32 = .031$
2	$1/32 = .031$
3	$2/32 = .062$
4	$2/32 = .062$
5	$3/32 = .094$
6	$3/32 = .094$
7	$3/32 = .094$
8	$3/32 = .094$
9	$3/32 = .094$
10	$3/32 = .094$
11	$2/32 = .062$
12	$2/32 = .062$
13	$1/32 = .031$
14	$1/32 = .031$
15	$1/32 = .031$

Ties

There are two distinct ways in which ties can occur in data from a dependent-groups design.

Zero Difference For a matched pair, the two scores may be the same, giving a difference, $d_i = 0$, for that pair. It is impossible to give a sign to the rank of this zero difference if this sign is to represent the direction of the difference. The simplest procedure is to drop each such tied pair (a zero difference) from the analysis and to decrease n accordingly.[8]

[8] This recommendation is in accord with the recommendation for tied data in the sign test (Section 14.2). A more conservative procedure is to rank all zero differences as tied differences: if there are r such differences, each would receive rank $= (r + 1)/2$, the mean of the ranks $1, 2, \ldots, r$. Give half these ranks a positive sign and half a negative sign.

Tied Difference Two or more differences (in absolute value) may be tied. The best procedure (also used in the rank-sum test for independent groups) is to assign to each such difference the average (mean) of the ranks that the absolute differences would have if they were not tied. For example, if nine absolute differences are 2, 5, 6, 6, 6, 9, 10, 10, 15, the rank of 2 is 1, the rank of 5 is 2, the rank of each 6 is 4 $[= (3 + 4 + 5)/3]$, the rank of 9 is 6, the rank of each 10 is 7.5 $[= (7 + 8)/2]$, and the rank of 15 is 9.

The rejection regions in Table A10 were calculated assuming that there are no ties in the data. The distribution of T_+ changes when there are ties. Table A10 is still satisfactory, however, provided that there are not too many ties (say, not more than 20% of the differences are tied).

Example 1

Two conditions of learning are studied by selecting 10 subjects, matching them in pairs on their previous experience in similar learning experiments, and assigning the members of each pair at random to the two conditions. The number of errors made by the subjects are shown in Table 15.9a.

The table shows the calculation of the signed ranks. First, the differences, d_i, between the paired scores are calculated. (Note that a plus sign indicates that the group 1 score is greater than the group 2 score and that a negative sign indicates the reverse.) Second, the absolute values, $|d_i|$, of these differences are listed and then ranked with the smallest absolute difference (2) ranked 1 and the largest absolute difference (7) ranked 5. Third, the signs of the differences are transferred to the ranks, giving the signed ranks. Four of the signed ranks are positive and one is negative. Fourth, the sum of the positive signed ranks is calculated: $T_+ = 10$.

The significance of this value is determined by consulting Table A10 for $n = 5$. We

TABLE 15.9 Rank-sum Test for Two Dependent Groups

(a) Example 1

PAIR	GROUP 1	GROUP 2	d_i	$|d_i|$	RANK	SIGNED RANK
1	15	22	−7	7	5	−5
2	16	12	+4	4	3	+3
3	8	2	+6	6	4	+4
4	7	5	+2	2	1	+1
5	13	10	+3	3	2	+2
						$T_+ = 10$

(b) Example 2

PAIR	GROUP 1	GROUP 2	d_i	$|d_i|$	RANK	SIGNED RANK
1	72	69	+3	3	2.5	+2.5
2	46	83	−37	37	9	−9
3	72	75	−3	3	2.5	−2.5
4	68	68	0	Drop	Drop	Drop
5	73	83	−10	10	7.5	−7.5
6	47	55	−8	8	6	−6
7	62	68	−6	6	5	−5
8	69	79	−10	10	7.5	−7.5
9	59	57	+2	2	1	+1
10	57	62	−5	5	4	−4
						$T_+ = 3.5$

see that $T_+ = 10$ is not in the rejection region for any tabled value of α; we also know that this is true from the complete distribution of T_+ for $n = 5$ given in Table 15.8. Hence, the difference between the two conditions in the learning experiment is not significant.

Example 2

The data for an example with ties is shown in Table 15.9b. Here there are 10 pairs of scores. One of the pairs is tied, giving a difference, d_i, equal to zero. This pair is dropped from further analysis, reducing n to 9. The nine absolute differences are ranked. Note that two of the absolute differences are 3 (each is ranked 2.5) and that two other absolute differences are 10 (each is ranked 7.5). Each rank is given the sign of the original difference. Here there are only two positive signed ranks and $T_+ = 3.5$. The small value of T_+ indicates that the difference is in favor of group 2 (most of the differences are negative). The rejection region for $n = 9$, $\alpha = 0.05$, two-tailed, is $T_+ = 5$ or smaller, or $T_+ = 40$ or larger. The observed value, 3.5, is in this rejection region. Hence, we conclude that the two experimental conditions differ significantly. Note, however, that four of the nine differences are tied. The distribution in Table A10 is probably not correct when 44% of the scores are tied.

15.7 Summary

In this chapter the χ^2-tests of dichotomous variables (Chapter 14) were extended to χ^2-tests of *categorical* variables. A variable can be naturally categorical, or a *numerical* variable can be reduced to a categorical one.

The χ^2-tests included the test of the *homogeneity of J independent distributions*, the test of *independence of two categorical variables*, and the test of *goodness-of-fit*. The requirements for the validity of all χ^2-tests also were discussed.

Two rank-sum tests were described. These tests are applied to numerical variables that are reduced to rank variables during the course of the calculations of each test. The two tests are the *rank-sum test for independent groups* and the *rank-sum test for dependent groups*.

Exercises

1 (Section 15.2) A study is conducted to determine if the proportions of three types of cancer are different in three cities. In each city a sample of 100 cancer patients is selected and each patient is classified as having one of three types of cancer (A, B, or C). The number of patients in each category is found to be

	CITY 1	CITY 2	CITY 3
CANCER A	21	47	38
CANCER B	43	26	28
CANCER C	36	27	34
	100	100	100

Test the hypothesis that the distribution of cancers is different in the three cities. Make the test at the 0.01 level.

2 (Section 15.2) A study of transposition in size discrimination such as described in Question 5 in the Exercises for Chapter 14 can be extended to consider the degree and type of verbalization associated with age (and with transposition). Suppose that the four categories of verbalization are

A: no verbalization during experiment.

B: size mentioned spontaneously but with no mention of how it related to the correct or incorrect response.

C: the principle relating size and correct response was verbalized when the subject was questioned after the experiment.

D: the principle was verbalized spontaneously during the experiment.

Sixty subjects in the experiment are categorized by age and by type of verbalization, giving the following frequency table:

	4 YEARS	5 YEARS	6 YEARS
CATEGORY A	15	8	2
CATEGORY B	2	7	3
CATEGORY C	2	4	8
CATEGORY D	1	1	7

(a) Is type-of-verbalization related to age in this 4×3 table? (Carry out the chi-square test.)

(b) Note that many of the expected frequencies calculated in **(a)** are less than 5. Why is this a problem?

(c) In order to make all the expected frequencies greater than 5, reduce the number of verbalization categories to two: combine A and B and combine C and D. Reanalyze the data. Why are categories combined in this particular way, rather than, say, by combining A and C and combining B and D?[9]

3 (Section 15.3) A speciality butcher plans to take a two-week vacation in either July or August. He wants to determine which of the 4 two-week blocks during those two months would be least disruptive to his customers. He selects a random sample of 40 customers and asks each of them to identify one of the 4 two-week blocks as the preferred block for the vacation. The number of customers who select each block is shown below.

BLOCK	NUMBER
July 4–17	6
July 18–31	10
August 1–14	15
August 15–28	9

The butcher must first determine whether the proportions choosing each time block are significantly different. What conclusion would the butcher make if he performs the test at the 0.05 level?

4 (Section 15.5) The reaction times, in milliseconds, in two conditions (experimental and control) are compared. Eighteen subjects are assigned at random to the two conditions. Analyze the data using the appropriate rank-sum test.

EXPERIMENTAL	CONTROL
430	843
221	416
645	604
317	733
548	490
374	668
478	452
315	520
504	661

[9] Adapted from Margaret R. Kuenne, "Experimental investigation of the relation of language to transposition behavior in young children", *Journal of Experimental Psychology* 36 (1946), 471–90.

5 (Section 15.6) Each of the 15 subjects in an experiment views two figures simultaneously projected on a screen. One of these figures is complex in structure; the other is simple in structure. The display is visible for 10 seconds. The time during which each subject fixates each figure is recorded. (Because each subject spends time fixated on neither figure, the times given below add to less than 10 seconds in each case.) Test the hypothesis that the average time fixated on the more complex figure is greater than the average time fixated on the simpler figure. Use the appropriate rank-sum test.[10]

SUBJECT	COMPLEX FIGURE	SIMPLE FIGURE
1	4.7	3.7
2	3.2	3.6
3	3.7	4.2
4	5.5	2.7
5	4.6	3.2
6	6.3	1.8
7	7.2	2.0
8	4.6	3.7
9	5.3	2.6
10	4.8	3.2
11	5.6	2.8
12	3.1	5.2
13	2.8	4.1
14	4.1	2.1
15	5.1	3.0

6 (All sections) The purpose of this question is to increase your skill in identifying which test (from *either* Chapters 14 or 15) applies to a given study. For each of the studies described below, state which test (or tests) would be used to analyze the data from the study. Do *not* carry out the analysis.

Hint: As an aid in identifying the correct test, make up a sample data table (either a raw-data table or an observed-frequency table). Put some fictitious numbers in the table. The table will help you to identify the correct test to apply to the data and hence to the study.

(a) The afterwork activities of residents of four districts in a city are compared. The activities are classified as "education", "music", "sport", "other".

(b) A class of 12-year-olds begins ballet lessons. Each student is classified either as having skated previously or as not having skated. The students' scores on a dance ability test are to be compared at the end of six months of ballet lessons.

(c) After a nuclear accident, the proportion of leukemia victims in the area near the accident and in a control area are compared by studying samples of medical records in each of the two areas.

(d) A group of young children are matched on age. One member of each pair is trained on color names; the other is given a control task not involving color-naming. After the training or control task, each child is asked to name the colors of 10 objects, and the number of correct responses for each child is recorded.

(e) A study is made of the relationship between women's ages and their principal activity. The ages are divided into five intervals (20–29, 30–39, etc.), and there are four principal activities (full-time job, homemaker, part-time job, volunteer work).

[10] Adapted from D. E. Berlyne, "The influence of complexity and novelty in visual figures on orienting responses", *Journal of Experimental Psychology* 55 (1958), 289–96.

16 CORRELATION AND OTHER MEASURES OF ASSOCIATION

Statistics was defined in Section 8.1 as a set of techniques for making inferences about a single variable or about the relationship between two or more variables. This chapter focuses on several tests and measures of the relationship or association between two variables.

Two *measures* of relationship were introduced in Chapter 8: the effect size (difference of means) in a two-group study and the Pearson correlation in a correlational study. Chapter 11 dealt with how to *test* the difference of two means. In Chapters 14 and 15, several *tests* of the relationship between two dichotomous or two categorical variables were described. These measures and tests differ in the types of variables to which they apply, and therefore in the type of association that they measure and test. In this chapter, several new measures and tests are added to those introduced previously.

The terms "relationship" and "association" are treated synonymously in this book. The term "association" will be used here, because it is less likely to be confused with "correlation", which is a specific measure of association or relationship, and because "measure of association" is used more frequently than "measure of relationship" by statisticians studying categorical variables. The two words or the two phrases may be used interchangeably, however.

16.1 Types of Association

The association between two variables can be measured in several ways. The method that should be used depends on the types of the variables (Section 15.1) whose association is being measured and the type of association in which the investigator is interested. Many types of association have been considered by statisticians. Four types that have been discussed previously in this text or are described in this chapter are summarized in Table 16.1.

Recall, from Section 8.2, that there is an association (or relationship) between two variables if the distribution of the second variable is different for the different values of the first variable. Conversely, there is no association (or no relationship) between the two variables if these distributions are the same. We test whether or not there is an association by computing a (test) *statistic*. We estimate the size of the association by computing a *measure* of the association. Table 16.1 gives a statistic and a measure for each of the four types of association. (For some types, the statistic and the measure are the same.)

Caution. The various types of association are discussed in this section only in a general way, and a strict distinction between associations in samples and associations in populations is not maintained. The examples and graphs given here are those of small sets of data: that is, small samples. However, the various

TABLE 16.1 Types of Association

TYPE OF ASSOCIATION	TYPES OF THE TWO VARIABLES	TEST STATISTIC	MEASURE
Difference of two means	One dichotomous, other numerical	t_{obs} (11.4)[a]	Effect size or standardized effect size (8.5)
Linear association	Both numerical	r or Z (8.6, 16.2)	r or r^2 (8.6, 16.3)
Monotonic association	Both numerical (reduced to rank variables)	r_S (16.4)	r_S (16.4)
General association	Both categorical (one or both may be dichotomous)	χ^2_{obs} (14.3, 15.2)	$\hat{\phi}'$ or $\hat{\phi}'^2$ (16.5)

[a] The number in parentheses identifies the section in which the statistic or measure is described.

terms introduced in this section also apply to populations. The following sections show how to infer an association in a population from the data in a sample.

Association between a Dichotomous and a Numerical Variable

In a two-group study the independent variable is a dichotomous variable. If the dependent variable is a numerical variable, the most natural measure of association is the difference between the two group means: that is, the *effect size* (Section 8.5). The effect size usually is standardized by dividing by the pooled standard deviation.

The presence or absence of an association between the two variables is tested by a *t*-test (Section 11.4). There is no association if the null hypothesis of equal population means is true. There is an association if the alternative hypothesis of different population means is true. The test is made by computing the test statistic t_{obs} and comparing its value to the appropriate *t*-distribution. (See Figure 16.1 for two graphs illustrating, respectively, no association and association between a dichotomous and a numerical variable. Similar graphs are discussed in detail in Chapter 8.)

This type of association between two variables may be called simply a *difference of two means*. Although it will not be discussed further in this chapter, keep in mind that this type is probably the most commonly studied of all types of association. Many researchers who use the *t*-test are not aware that they are making a test of the association of two variables.

Association between Two Numerical Variables

In a correlational study the two variables are numerical variables. The association between the two variables usually is studied by computing the *regression function* of one variable on the other (Section 8.4). The regression function is the curve or line connecting the means of the distribution of one variable for each different value of the other variable. Several examples of regression functions are shown in Figure 16.2.

If the regression function is a horizontal straight line, there is *no association* between the variables (Figure 16.2a). If the regression function

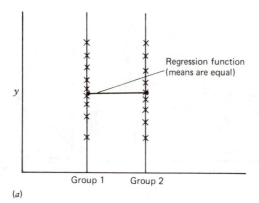

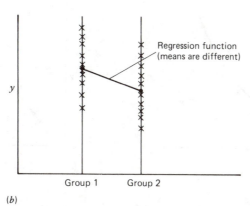

Figure 16.1 Association between a dichotomous and a numerical variable. (*a*) No association. (*b*) Association.

always increases—that is, the means of variable y increase as the values of variable x increase—the association is described as a *positively monotonic association* (Figure 16.2*b*). On the other hand, if the regression function decreases—that is, the means of y decrease as the values of x increase—the association is described as a *negatively monotonic association* (Figure 16.2*c*).

If the regression function is a straight line that is not horizontal, there is a *linear association* between the variables (Figure 16.2*d*). The linear association can be positive, as in Figure 16.2*d*; or negative, in which case the straight line would have negative slope. If the regression function is relatively straight, it is often practical to approximate it by a straight line and to describe the association as linear.

Two other types of association are shown in Figures 16.2*e* and 16.2*f*. The associations shown in these two figures are certainly not linear, nor even monotonic. They are described as *nonlinear* or *curvilinear* associations. If a best-fitting straight line *is* drawn in Figure 16.2*e*, it will be a horizontal straight line, showing that there is *no linear* association between the two variables. We say that the association has a nonlinear component but does not have a linear component. On the other hand, the nonlinear association in Figure 16.2*f* does also have a linear component, since the best-fitting straight line to the regression function in that figure does have a nonzero (positive) slope. (Nonlinear association is not discussed further in this text.)

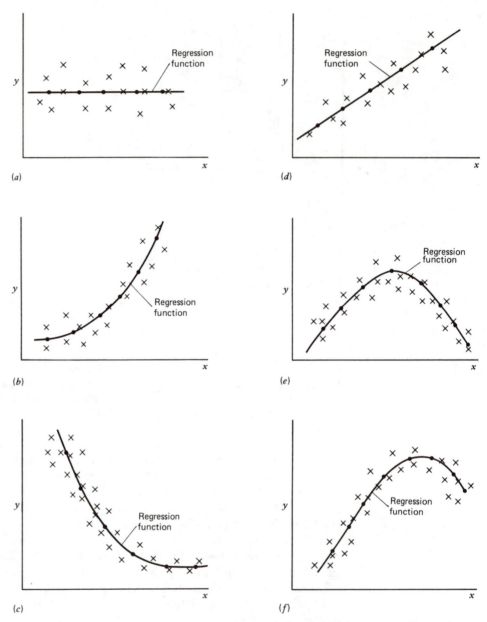

Figure 16.2 Association between two numerical variables. (*a*) No association. (*b*) Monotonic association (positive). (*c*) Monotonic association (negative). (*d*) Linear association. (*e*) Nonlinear association (no linear component). (*f*) Nonlinear association (also has a linear component).

The degree of linear association between two variables is measured by the Pearson correlation, r (Section 8.6). Ways to test whether or not this correlation has a value of zero (or some other specified value) in the population are described in Section 16.2.

Another correlation measure, called the Spearman correlation, r_S, is used to measure the degree of monotonic association between two numerical variables. The Spearman correlation is computed by reducing the numerical variables to rank variables (Section 16.4).

Association between Two Categorical Variables

The association between two *numerical* variables can be described in terms of the regression function computed from the means. The association between two *categorical* variables, on the other hand, cannot be described in this way, since means cannot be computed for categorical variables. To describe this association, we must return to the general definition of association given at the beginning of this section: there is an association between two variables if the distribution of the second variable is different for the different values of the first variable. An association of this kind is referred to as a *general association*, in contrast to the more specific types of association such as linear association and monotonic association.

The chi-square tests described in Chapters 14 and 15 are tests of the general association between two categorical variables. A measure of general association, called Cramér's measure, is described in Section 16.5. The chi-square tests and Cramér's measure may also be used when one or both of the variables are dichotomous (that is, have only two categories).

Examples of no association and association in frequency tables are shown in Table 16.2. In Table 16.2*a*, the column proportions are all equal, which indicates that the distribution of the row variable is the same for all values of the column variable, and hence that there is no association between the row and column variables. This lack of association also appears in the row proportions; the distribution of the column variable is the same for all values of the row variable. In Table 16.2*b*, on the other hand, the column proportions are not equal, which indicates that there is association between the row and column variables. (The row proportions are also not equal.) Recall, however, the *Caution* given at the beginning of this section. The apparent association in the frequency table must be tested (by a chi-square test) in order to see if an association can be inferred in the population.

TABLE 16.2 Association between Two Categorical Variables

(a) No Association

FREQUENCIES				COLUMN PROPORTIONS				ROW PROPORTIONS			
12	10	18	40	.20	.20	.20	.20	.30	.25	.45	1.00
6	5	9	20	.10	.10	.10	.10	.30	.25	.45	1.00
24	20	36	80	.40	.40	.40	.40	.30	.25	.45	1.00
18	15	27	60	.30	.30	.30	.30	.30	.25	.45	1.00
60	50	90	200	1.00	1.00	1.00	1.00	.30	.25	.45	1.00

(b) Association

FREQUENCIES				COLUMN PROPORTIONS				ROW PROPORTIONS			
12	5	23	40	.20	.10	.26	.20	.30	.12	.58	1.00
8	8	4	20	.13	.16	.04	.10	.40	.40	.20	1.00
28	25	27	80	.47	.50	.30	.40	.35	.31	.34	1.00
12	12	36	60	.20	.24	.40	.30	.20	.20	.60	1.00
60	50	90	200	1.00	1.00	1.00	1.00	.30	.25	.45	1.00

16.2 Tests and Estimates of the Pearson Correlation

When two numerical variables have a regression function that is linear (or nearly so), the association between the variables can be measured by the Pearson correlation.[1] Three tests of the value of the population correlation are described in this section. The population correlation has the symbol ρ—the Greek letter r, pronounced "rho". Estimates of ρ also are described in this section.

Recall, from Section 8.6, the two equivalent formulas for the Pearson correlation of n pairs of scores (x, y).[2]

$$r = \frac{\Sigma(x - \bar{x})(y - \bar{y})}{n \, s'_x \, s'_y} \quad \text{(Definitional formula)}$$

$$r = \frac{\Sigma xy - (\Sigma x)(\Sigma y)/n}{n \, s'_x \, s'_y} \quad \text{(Computational formula)}$$

See Example 2, below, for an illustration of the calculations, using the computational formula.

The Test of Zero Correlation

Suppose we want to see which of the following two hypotheses about the association between two variables is true: the null hypothesis (H_0: $\rho = 0$), which states that there is no association between the two variables; or the alternative hypothesis (H_1: $\rho \neq 0$), which states that there is a nonzero *linear* association between the variables. Note that these hypotheses assume that if there is an association between the variables, it is a linear association. The test now described is sensitive primarily to a linear association, and not to a curvilinear, or a more general, association. If we know that the correlation must have a certain sign (say, positive), the alternative hypothesis can be one-sided (say, H_1: $\rho > 0$). However, the usual arguments in favor of two-sided alternatives, and two-tailed tests, apply to this test, just as they apply to tests of means (see the discussion in Section 10.3).

The statistical test is constructed by considering the distribution of the test statistic (here, r, the sample correlation)[3] when H_0 is true and when H_1 is true. The distribution of r when H_0 is true is a symmetric, unimodal distribution whose mean is zero. The distribution when $n = 20$ is shown in Figure 16.3a.

---

[1] Strictly speaking, there is an additional requirement—that the two variables have a *bivariate normal* distribution. See, for example, J. Neter and W. Wasserman, *Applied Linear Statistical Models* (Homewood, Ill.: Irwin, 1974). Roughly speaking, the requirement combines the requirements of normality and equality of variance with the requirement that the regression function be linear.

[2] In the formulas for r, we revert to the use of the unadjusted sample deviation s', rather than using the adjusted sample deviation s as used in Chapters 11 and 12. The formulas for r could be written in terms of s, but then n would have to be replaced by $(n - 1)$. The value of r would be unchanged.

[3] The test of zero correlation can also be based on a test statistic that has a t-distribution with $(n - 2)$ degrees of freedom. The test statistic is

$$t_{obs} = \frac{r\sqrt{n - 2}}{\sqrt{1 - r^2}}$$

The test based on t_{obs} is equivalent to the test using r and Table A11. The latter test is preferred, because of its simplicity.

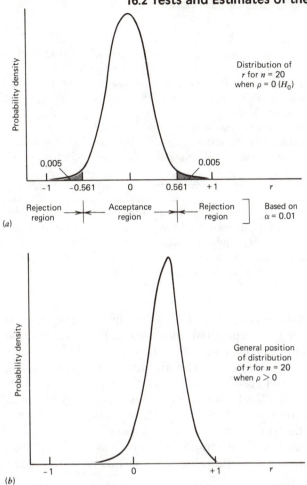

Figure 16.3 Distributions for the test of H_0: $\rho = 0$.

The variability of the distribution is different for different sample sizes, n. It is important to note that the distribution is *not* a normal distribution (such as shown in Figure 7.6 and tabulated in Table A3). However, the distribution of r has been calculated and certain percentiles of the distribution are given in Table A11 in the Appendix. The percentiles are upper-tail. Since the distribution of r is symmetric about zero, lower-tail percentiles are the negative of the corresponding upper-tail percentiles.

The distribution of r when H_1 is true, is complex. Its "general position", however, is centered either on values of r greater than zero (if $\rho > 0$; see Figure 16.3b) or on values of r less than zero (if $\rho < 0$). A value of r sufficiently discrepant from zero is evidence for the alternative hypothesis. If we want the probability of a Type I error (α) to be less than 0.01, we choose the critical values of r so as to cut off areas of $0.01/2 = 0.005$ in each tail of the H_0 distribution. The upper-tail critical value (for $n = 20$) is found (from Table A11) to be 0.561; the lower-tail critical value is, therefore, -0.561. The null hypothesis is rejected if the sample r is less than -0.561 or greater than $+0.561$. Absolute values of r smaller than 0.561 lead to the conclusion that the null hypothesis of zero correlation cannot be rejected.

Example 1

The Pearson correlation between two variables is -0.37 in a random sample of 100 subjects. Is this correlation significantly different from zero?

The critical values of r for a two-tailed test at $\alpha = 0.05$ are -0.197 and $+0.197$ (Table A11). Hence, the observed correlation of -0.37 is significantly different from zero at the 0.05 level, two-tailed.

Alternatively, we could note from Table A11 that $|r| = 0.37$ exceeds the greatest percentile shown in the table $r_{.001}$); thus, the sample correlation of -0.37 is significant (p-value < 0.002, two-tailed). Note that the α given in the table has been doubled for a two-tailed p-value.

Example 2

Consider the sample of $n = 6$ pairs of scores (x, y) shown in Table 16.3. Is the correlation between x and y significantly different from zero, at the 0.01 level?

The calculation of the correlation is shown in the table. The correlation is $r = 0.960$, exceeding the critical value, 0.917, which cuts off an area of $0.01/2 = 0.005$ in the upper tail (Table A11). Hence, the sample correlation is significantly different from zero.

The Test of a Specified (Nonzero) Correlation

This section describes how to test whether a sample correlation is significantly different from a specified, nonzero population correlation. The specified correlation might be 0.40, 0.70, -0.60, etc. We will use the symbol ρ_0 to denote this specified correlation. The null hypothesis is H_0: $\rho = \rho_0$ and the alternative hypothesis is H_1: $\rho \neq \rho_0$. If ρ_0 is 0.50, the hypotheses are H_0: $\rho = 0.50$ and H_1: $\rho \neq 0.50$. (One-sided alternative hypotheses also may be used.)

You might expect these hypotheses to be tested in a way similar to the way the hypothesis of *zero* correlation is tested, as previously described. However, the distribution of r when ρ is not zero is skewed and has a different shape for every different possible value of ρ. For this reason, it is not convenient to use r

TABLE 16.3 Calculation of the Pearson Correlation

x	y
2	6
3	7
4	8
5	8
6	9
7	9

$\Sigma x = 27$ $\Sigma y = 47$
$\Sigma x^2 = 139$ $\Sigma y^2 = 375$
$\Sigma xy = 222$

$\bar{x} = (\Sigma x)/n = 27/6 = 4.500$
$\bar{y} = (\Sigma x)/n = 47/6 = 7.833$

$s'_x = \sqrt{\dfrac{\Sigma x^2 - (\Sigma x)^2/n}{n}} = \sqrt{\dfrac{139 - (27)^2/6}{6}} = 1.708$

$s'_y = \sqrt{\dfrac{\Sigma y^2 - (\Sigma y)^2/n}{n}} = \sqrt{\dfrac{375 - (47)^2/6}{6}} = 1.067$

$r = \dfrac{\Sigma xy - (\Sigma x)(\Sigma y)/n}{n\, s'_x s'_y} = \dfrac{222 - (27)(47)/6}{6 \times 1.708 \times 1.067} = 0.960$

as a test statistic. The statistician R. A. Fisher showed that a statistic Z (called Fisher's Z) has a distribution that is approximately normal regardless of the value of ρ, and whose standard deviation does not depend on ρ, but only on the sample size.

Figure 16.4 shows the distributions of r and Z for $\rho = 0.50$ and for $n = 20$. The distribution of r (Figure 16.4a), is skewed to the left. If ρ were larger, the distribution would be less variable and even more skewed. The distribution of Z, on the other hand, is approximately normal (Figure 16.4b).

Fisher's Z is computed from r by

$$Z = Z(r)$$

where $Z(r)$ represents a complicated formula.[4] The formula is rarely needed, however, since there is a table giving the value of Z for various values of r

---

[4] The formula is

$$Z = \tfrac{1}{2} \log_e\left(\frac{1 + r}{1 - r}\right)$$

In this formula, $\log_e$ is the natural logarithm (or the logarithm to the base e, where $e = 2.71828$, approximately).

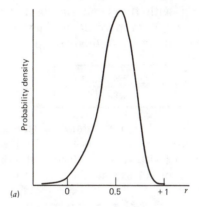

(a)

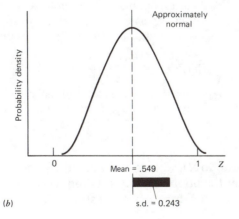

(b)

Figure 16.4 Fisher's Z. (a) Distribution of r for $\rho = 0.50$ ($n = 20$). (b) Distribution of Z for $\rho = 0.50$ ($n = 20$).

(Table A12 in the Appendix). The parameters of the (approximately) normal distribution of Z are

$$\text{mean}(Z) = Z(\rho)$$

$$\text{s.d.}(Z) = \frac{1}{\sqrt{n-3}}$$

where the population correlation is ρ, the sample size is n, and $Z(\rho)$ represents, again, the complicated formula. Note that the mean of the distribution depends only on ρ and the standard deviation depends only on n. The formula for the mean is rarely needed, since the value of Z, for the given ρ, can be found in Table A12. For example, when $\rho = 0.50$ the corresponding value of Z is 0.549 (Table A12). Hence, the mean of the normal distribution of Z in Figure 16.4b is 0.549, and its standard deviation is $1/\sqrt{n-3} = 1/\sqrt{20-3} = 1/\sqrt{17} = 0.243$, for $n = 20$.

The statistical test is constructed by considering the distribution of the test statistic, Z, when H_0 is true and when H_1 is true. When $H_0: \rho = \rho_0$ is true, the distribution of Z is approximately normal, with mean $= Z(\rho_0)$ and standard deviation $= 1/\sqrt{n-3}$. When H_1 is true, the distribution of Z is also approximately normal, with mean either greater than or less than the mean of the null hypothesis distribution and with the same standard deviation as the null hypothesis distribution (since the standard deviation depends only on n). The test of H_0 is most easily carried out by standardizing Fisher's Z as follows.

$$z_{obs} = \frac{Z - \text{mean}(Z)}{\text{s.d.}(Z)}$$

[Note carefully the distinction between capital Z (Fisher's Z) and lowercase z (standardized normal variable).] When H_0 is true, the test statistic z_{obs} has a *standard* normal distribution (with mean $= 0$ and standard deviation $= 1$); hence, we can use Tables A3 or A4 to find the critical values.

Example 3

Is a correlation of 0.62 obtained in a sample of 20 subjects significantly different from 0.50? Use a two-tailed test at the 0.05 level.

The value of Fisher's Z corresponding to the sample r of 0.62 is $Z = 0.725$ (found in Table A12). Fisher's Z has mean $= 0.549$ (the value of Z corresponding to $\rho = 0.50$) and standard deviation $= 1/\sqrt{20-3} = 0.243$. Hence, the test statistic, z_{obs}, has the value

$$z_{obs} = \frac{Z - \text{mean}(Z)}{\text{s.d.}(Z)} = \frac{0.725 - 0.549}{0.243} = \frac{0.176}{0.243} = 0.724$$

This value does not exceed the critical value of the normal distribution, 1.960, for a 0.05-level, two-tailed test; the sample correlation is not significantly different from 0.50.

Estimating ρ By a Confidence Interval

It is often more useful to *estimate* the value of the population correlation rather than to *test* whether the population correlation has a specific value. (Recall the discussion of the relationship between estimation and hypothesis testing in Section C.1.)

The confidence interval is most readily determined by using the charts in Table A13 in the Appendix. There are two charts, one for the 95% confidence interval and the other for the 99% confidence interval. To use the charts, you need the sample correlation (r) and the sample size (n). A transparent ruler is placed vertically on the appropriate chart so that it intersects the abscissa at the given value of the sample correlation. The ruler then intersects the two curves for the given sample size at two points whose positions on the ordinate are the two required limits of the confidence interval.

Example 4

Estimate the population correlation if the sample correlation is 0.62 in a sample of 20 subjects.

On the 95% chart, the ruler placed vertically at $r = 0.62$ intersects the two curves for $n = 20$ at the points with vertical positions 0.22 and 0.82 (approximately). Hence, the 95% confidence interval for ρ is (0.22, 0.82). Note that this interval extends farther below the sample $r = 0.62$ than above it. The sample correlation is not in the center of the confidence interval.

Example 5

Estimate the population correlation if the sample correlation is 0.40 in a sample of nine subjects.

There are no curves for $n = 9$ in the charts in Table A13. Obviously, the curve for $n = 9$ would fall about halfway between the curves for $n = 8$ and $n = 10$. So we imagine a curve drawn in that position and proceed as usual. The 99% confidence interval is $(-0.55, +0.89)$, a very wide interval. Confidence intervals are very wide when n is small.

The Test of the Equality of Two Independent Correlations

In some studies the correlation between two variables is computed in two independent samples, and the investigator wishes to test whether or not the two correlations are significantly different. The test is based on Fisher's Z.

The two sample correlations, r_1 and r_2, are each converted to Fisher's Z using Table A12. The values are called Z_1 and Z_2, respectively. It can be shown that the distribution of the difference of these two values, $Z_1 - Z_2$ is approximately normal, with

$$\text{Mean} = Z_1(\rho_1) - Z_2(\rho_2)$$

$$\text{Standard deviation} = \sqrt{\frac{1}{n_1 - 3} + \frac{1}{n_2 - 3}}$$

where the two population correlations are ρ_1 and ρ_2, the two sample sizes are n_1 and n_2, the value $Z_1(\rho_1)$ is the value of Fisher's Z corresponding to ρ_1, and the value $Z_2(\rho_2)$ is the value of Fisher's Z corresponding to ρ_2.

The two hypotheses are $H_0: \rho_1 = \rho_2$ and $H_1: \rho_1 \neq \rho_2$ (or H_1 may be one-sided). When H_0 is true, $\rho_1 = \rho_2$ and, therefore, $Z_1(\rho_1) = Z_2(\rho_2)$. Hence, the mean of the distribution of $Z_1 - Z_2$ is zero. The test statistic, which has a standard normal distribution when H_0 is true, is

$$z_{\text{obs}} = \frac{Z_1 - Z_2}{\sqrt{\frac{1}{n_1 - 3} + \frac{1}{n_2 - 3}}}$$

To decide between the two hypotheses, the investigator compares the value of this test statistic in the usual way with percentiles of the standard normal distribution.

Example 6

The correlation between verbal and performance IQ is compared in the male and female samples. The correlation between the two IQ measures is 0.57 for males ($n_1 = 75$) and 0.71 for females ($n_2 = 100$). Are these correlations signficantly different?

The values of Fisher's Z for the two sample correlations are 0.648 (males) and 0.887 (females). The test statistic is

$$z_{obs} = \frac{Z_1 - Z_2}{\sqrt{\dfrac{1}{n_1 - 3} + \dfrac{1}{n_2 - 3}}} = \frac{0.648 - 0.887}{\sqrt{\dfrac{1}{75 - 3} + \dfrac{1}{100 - 3}}}$$

$$= \frac{-0.239}{\sqrt{0.01389 + 0.01031}} = \frac{-0.239}{0.1556} = -1.54$$

The probability of this or a more-extreme value (two-tailed) is found from Table A3 to be $2 \times 0.0618 = 0.1236$. Hence, the hypothesis of the equality of the correlation for males and females cannot be rejected (p-value > 0.10, two-tailed.)

16.3 Correlation and Prediction

In this section a new interpretation of the Pearson correlation is introduced. The new interpretation is based on the fact that the correlation represents the degree to which one variable—say, y—can be *predicted* from another variable—say, x.

Consider the scatterplot of six pairs of scores (x, y) in Figure 16.5. (See Table 16.3 for the numerical values of these six pairs.) Chapter 8 stated that the regression function of y on x is the curve of the mean value of y for each value (or interval of values) of x. The regression function may be either curved or straight, but in either case, the straight line that best fits the regression function is called the linear regression function. This line is shown in Figure 16.5. (The linear regression function is computed directly from the six points. Its exact equation is given later in this section.)

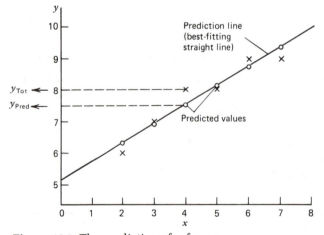

Figure 16.5 The prediction of y from x.

Section 8.6 noted that the Pearson correlation has two properties that facilitate interpretation:

1 The value of the Pearson correlation must lie between -1 and $+1$; values near -1 or $+1$ indicate a high degree of correlation; values near zero indicate a low degree of correlation.

2 The absolute value of the correlation measures the spread of the points of the scatterplot about the regression line—the greater the spread, the closer the correlation to zero.

We now encounter a third property of the correlation:

3 The *square* of the correlation between two variables is the proportion of the variance of one variable that can be predicted from the other variable.

Prediction

The regression function is more than just the curve of means of y for each value of x. In some situations the regression function is known but the scatterplot itself is not known. The regression function can then be used to predict the value of y for any value of x by reading the ordinate (height) of the regression function for that value of x. The regression function, therefore, can serve as a *prediction function*. If the regression function is linear (or has been fitted by a straight line) the prediction function usually is called the *prediction line*.

It will be shown below that the prediction line for the data in Figure 16.5 has the equation

$$y_{Pred} = 5.133 + 0.600x$$

The equation is different for other sets of data. The equation means that to find the predicted value of y (called y_{Pred}) for a particular value of x, we multiply the value of x by 0.600 and add 5.133 to the result. The values obtained in this way for $x = 2, 3, \ldots, 7$ are shown in Table 16.4 and indicated by circles on the prediction line in Figure 16.5.

It is clear from the figure and the table that some of the predicted values are less than the actual values (for example, when $x = 4$ the predicted value is $y_{Pred} = 7.533$ and the actual value is $y = 8$) and other predicted values are greater than the actual values (for example, when $x = 7$ the predicted value is $y_{Pred} = 9.333$ and the actual value is $y = 9$). It is a general characteristic of prediction that predictions rarely are perfect. The imperfection is reflected in the scatter of

TABLE 16.4 The Prediction of y from x

Prediction equation: $y_{Pred} = 5.133 + 0.600x$	
x	y_{Pred}
2	6.333
3	6.933
4	7.533
5	8.133
6	8.733
7	9.333

the data points about the regression function: that is, about the prediction line. We will see shortly that the square of the Pearson correlation is a measure of the accuracy of the prediction.

Equation of the Prediction Line

Recall, from analytic geometry, that the equation of any straight line may be written in the following way.

$$y = a + bx$$

where x and y represent the coordinates of points on the line and a and b are fixed constants. The constants a and b have the following interpretation.

a = y-intercept of the straight line.
b = slope of the straight line.

The y-intercept is found from the point at which the line cuts the y-axis. The y-intercept is the distance between this point and the x-axis. The *slope* of a line is the ratio of the rise to the run, for any run (Section 8.6). The steeper the line, the greater the slope.

In order to emphasize that we are *predicting* y from x, we replace y by y_{Pred} in the equation of the prediction line and write

$$y_{Pred} = a + bx$$

where a and b are numbers calculated from the data. The coefficient (b) of x can be found from the following formula.

$$b = r \left(\frac{s'_y}{s'_x} \right)$$

where r is the Pearson correlation between x and y, and s'_x and s'_y are the standard deviations of x and y, respectively. (The coefficient, b, is the slope of the prediction line.) The constant term (a) in the prediction formula can be found from the following formula.

$$a = \bar{y} - b\bar{x}$$

where b is the number given by the preceding formula and $\bar{x}$ and $\bar{y}$ are the means of x and y, respectively. (The constant term, a, is the y-intercept of the prediction line.) A proof of the above formulas can be found in Section 16.6.

Example 1

The six pairs (x, y) shown in Figure 16.5 are listed in Table 16.3. The table also contains the calculations of the means and standard deviations of x and y and the Pearson correlation, r, of x and y. These values may be substituted into the above formulas for the slope and y-intercept to get the equation of the prediction line. The slope is

$$b = r \times s'_y/s'_x = 0.960 \times 1.067/1.708 = 0.600$$

The y-intercept is

$$a = \bar{y} - b \times \bar{x} = 7.833 - 0.600 \times 4.500 = 7.833 - 2.700 = 5.133$$

The equation of the prediction line is, therefore,

$$y_{\text{Pred}} = 5.133 + 0.600x$$

You should compare these values for the slope and y-intercept with the position of the straight line in Figure 16.5. Note that the straight line increases 0.600 in the y direction for each increase of 1.0 in the x direction, confirming that the slope is 0.600. The line intersects the y-axis at $y = 5.133$, confirming that the y-intercept is 5.133.

Predicted and Unpredicted Variances

As noted earlier, the prediction of a score, y, by the prediction line usually is not perfectly accurate. The predicted value, y_{Pred}, differs from the actual value, y, by an amount called y_{Unpred} (the unpredicted part of y). Replacing y by the symbol y_{Tot} (the totality of y, or all of y), we can see that

$$y_{\text{Unpred}} = y_{\text{Tot}} - y_{\text{Pred}}$$

This equation says simply that the unpredicted part of y equals the difference between y itself (y_{Tot}) and the predicted part of y. This relationship can be seen even more clearly if the terms in the above equation are transposed.

$$\boxed{y_{\text{Tot}} = y_{\text{Pred}} + y_{\text{Unpred}}}$$

This equation says that the original score y is the sum of two parts: a predicted part and an unpredicted part.

The values of y_{Tot}, y_{Pred}, and y_{Unpred} for each of the six points in Figure 16.5 are listed in Table 16.5. The variances of these three variables also are shown in the table. It is remarkable that there is a very simple relationship among the three variances. The relationship is described by the following principle.[5]

P13: The Principle of Prediction

Consider the prediction of a dependent variable from one (or more than one) independent variable. The total variance of the dependent variable equals the sum of the following two variances: the variance of that part of the dependent variable that can be predicted from the independent variable(s) (the predicted variance) and the variance of that part of the dependent variable that cannot be predicted from the independent variable(s) (the unpredicted variance). In brief,

$$\boxed{s_{\text{Tot}}'^2 = s_{\text{Pred}}'^2 + s_{\text{Unpred}}'^2}$$

---

[5] The principle applies to linear or nonlinear prediction, to prediction from one or from more than one independent variable, and to prediction in samples as well as in populations. Since the three variances in the formula given in the principle have the same denominators, the numerator of the total variance equals the sum of the numerators of the other two variances: this relationship appears throughout the analysis of variance (see Chapter 17). For the principle of prediction to hold, the prediction must be done in such a way that the predicted part of the dependent variable is uncorrelated with the unpredicted part. This condition is satisfied by most prediction methods used in statistical work.

TABLE 16.5 Predicted and Unpredicted Variances

x	y_{Tot}	y_{Pred}	y_{Unpred}
2	6	6.333	−0.333
3	7	6.933	+0.067
4	8	7.533	+0.467
5	8	8.133	−0.133
6	9	8.733	+0.267
7	9	9.333	−0.333
	$\Sigma y_{Tot} = 47$	$\Sigma y_{Pred} = 46.998$	$\Sigma y_{Unpred} = +0.002$
	$\Sigma y_{Tot}^2 = 375$	$\Sigma y_{Pred}^2 = 374.435$	$\Sigma y_{Unpred}^2 = 0.533$
	$s_{Tot}^{\prime 2} = 1.139$	$s_{Pred}^{\prime 2} = 1.050$	$s_{Unpred}^{\prime 2} = 0.089$

Details of the Variance Calculations

$$s_{Tot}^{\prime 2} = \frac{375 - (47)^2/6}{6} = \frac{6.833}{6} = 1.139$$

$$s_{Pred}^{\prime 2} = \frac{374.435 - (46.998)^2/6}{6} = \frac{6.300}{6} = 1.050$$

$$s_{Unpred}^{\prime 2} = \frac{0.533 - (0.002)^2/6}{6} = \frac{0.533}{6} = 0.089$$

In the example of six points in Figure 16.5, the variable y is the dependent variable and the variable x is the independent variable. (The roles of the two variables also can be reversed; the principle of prediction will still apply.) The dependent variable is, in fact, the *predicted* variable, whereas the independent variable is the variable used to make the prediction.

In the example, the total variance of y is $s_{Tot}^{\prime 2} = 1.139$. The predicted variance is $s_{Pred}^{\prime 2} = 1.050$ and the unpredicted variance is $s_{Unpred}^{\prime 2} = 0.089$. We see that $s_{Tot}^{\prime 2} = s_{Pred}^{\prime 2} + s_{Unpred}^{\prime 2}$: that is, $1.139 = 1.050 + 0.089$, as stated in the principle.

The Inter-pretation of r^2

We have seen that the total variance of y can be divided into two variances: the predicted variance and the unpredicted variance. This relationship can be represented diagrammatically, as in Figure 16.6a; the total variance is represented as a circle and the two component variances are represented as sectors of this circle. The two sectors together, of course, comprise the whole circle. An even more useful representation can be made if each variance is considered as a *proportion* of the total variance. Then the predicted variance is a proportion, $s_{Pred}^{\prime 2}/s_{Tot}^{\prime 2}$, of the total variance and the unpredicted variance is a proportion, $s_{Unpred}^{\prime 2}/s_{Tot}^{\prime 2}$, of the total variance. As shown in Figure 16.6b, we have the total variance (which is a proportion 1.0 of itself) represented as the sum of the two proportions.

$$1 = \frac{s_{Pred}^{\prime 2}}{s_{Tot}^{\prime 2}} + \frac{s_{Unpred}^{\prime 2}}{s_{Tot}^{\prime 2}}$$

So the proportion of the total variance of y that can be predicted from x plus the proportion of the total variance that cannot be predicted from x, equals 1.0.

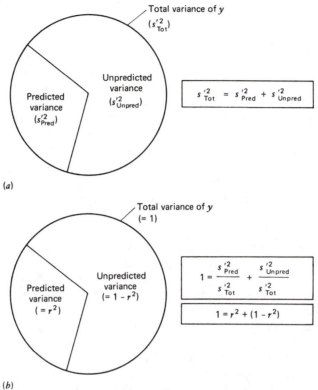

Figure 16.6 The principle of prediction. (a) Variances.
(b) Variances as proportions of total variance.

It can be shown (Section 16.6) that the first of these two proportions is precisely equal to the square of the Pearson correlation between x and y: r^2—that is,

$$r^2 = \frac{s'^2_{Pred}}{s'^2_{Tot}}$$

The second of these proportions must, therefore, be $1 - r^2$, since the two proportions sum to 1.0—that is,

$$1 - r^2 = \frac{s'^2_{Unpred}}{s'^2_{Tot}}$$

The relationship of the variances, expressed as proportions of the total variance of y, can be written as

$$1 = r^2 + (1 - r^2)$$

This important result, that the square of the correlation is the proportion of the variance of y that can be predicted linearly from x, is the basis of a very important way to interpret any given correlation. For example, the six data points in Figure 16.5 have a correlation of 0.960. The square of this correlation is 0.922, so we can say that about 92% of the variance of y can be predicted from

x. This result can be confirmed from Table 16.5, which shows that $s_{Tot}'^2 = 1.139$ and $s_{Pred}'^2 = 1.050$. The ratio $s_{Pred}'^2/s_{Tot}'^2$ is equal to $1.050/1.139 = 0.922$, which is r^2—as it should be.

The ratio $s_{Pred}'^2/s_{Tot}'^2$ is given a special name: *the coefficient of determination*. This name derives from the notion that the independent variable (here, x) causes or determines a certain proportion (r^2) of the dependent variable (here, y). However, the correlation between two variables does not imply a causal relationship between them; the term "coefficient of determination" is really a misnomer.

Note. The easiest way to calculate the coefficient of determination is to calculate r and then square it, giving r^2. The variances $s_{Pred}'^2$ and $s_{Tot}'^2$ have been calculated here solely to show that their ratio is, in fact, equal to r^2.

When x is predicted linearly from y (rather than y being predicted from x, as above), the proportion of the total variance of x that can be predicted from y is also r^2. Hence, the coefficient of determination is the proportion of the total variance of either variable that can be predicted from the other. Loosely speaking, we can say that r^2 is the proportion of the variance of the two variables that is shared, or held in common. It often is said that r^2 is the proportion of shared variance. The coefficient of determination is always the proportion of the total variance of one variable that can be predicted from the other. The denominator of this proportion is $s_{Tot}'^2$, which is $s_y'^2$ if y is being predicted, or $s_x'^2$ if x is being predicted.

We have seen that the proportion of variance that can be predicted is r^2. Similarly, the proportion of variance that cannot be predicted is $(1 - r^2)$. Occasionally, this latter quantity is called the *coefficient of alienation*.[6]

Example 2

Table 16.6 shows the values of r^2 and $1 - r^2$ for a number of values of r. Note that the absolute value of the correlation must be greater than 0.7 (or 0.707, to be precise) in order that the predicted variance be greater than one-half the total variance.

16.4 The Spearman Correlation

The Spearman correlation is a measure of degree to which two numerical variables are *monotonically* related or associated (Section 16.1). The Spearman correlation (defined below) is used both as a measure of the degree of monotonic association and as a test statistic to test the hypothesis of no monotonic association. Since variables that are linearly associated are necessarily also monotonically related, the Spearman correlation may be used in situations in which the Pearson correlation (a measure of linear association) is used. The values and interpretations of the two correlations are essentially the same when the actual relationship is linear.

The Spearman correlation between two variables, x and y, is calculated by first reducing the sample values of each variable, separately, to ranks. Hence, the variables become rank variables, with values ranging from 1 to n, the number of pairs of observations. *The Spearman correlation, r_S, is defined as the Pearson correlation of the rank variables.* (The Spearman correlation is sometimes called the Spearman *rank* correlation.)

---

[6] In most, but not all, textbooks of statistics the coefficient of alienation is defined as the *square root* of $(1 - r^2)$. This definition spoils the symmetry between the definitions of the two coefficients. The square root of $(1 - r^2)$ cannot be interpreted as a proportion, whereas $(1 - r^2)$ can be so interpreted.

TABLE 16.6 Typical Values of r, r^2, and $(1 - r^2)$

r	r^2	$1 - r^2$
-1.0	1.00	0.00
-0.9	0.81	0.19
-0.8	0.64	0.36
-0.7	0.49	0.51
-0.6	0.36	0.64
-0.5	0.25	0.75
-0.4	0.16	0.84
-0.3	0.09	0.91
-0.2	0.04	0.96
-0.1	0.01	0.99
0.0	0.00	1.00
$+0.1$	0.01	0.99
$+0.2$	0.04	0.96
$+0.3$	0.09	0.91
$+0.4$	0.16	0.84
$+0.5$	0.25	0.75
$+0.6$	0.36	0.64
$+0.7$	0.49	0.51
$+0.8$	0.64	0.36
$+0.9$	0.81	0.19
$+1.0$	1.00	0.00

Example 1

Consider the seven pairs of scores, x and y, given in Table 16.7. The calculation of the Spearman correlation is straightforward. First, the x scores are reduced to ranks (the score of 11 is the smallest score and is assigned rank 1, the score of 13 is the next smallest and is assigned rank 2, etc.). Second, the y scores are reduced to ranks. Third, the Pearson correlation between these two sets of ranks is calculated. This calculation, shown in Table 16.7b, is based on the sums, the sums of squares, and the sum of products of the ranks of x and y. The result is $r_S = 0.857$.

An Alternative Formula

The ranks of x are the integers from 1 to n. The ranks of y, too, are the integers from 1 to n. This means that the standard deviation of the ranks of x must be the same as the standard deviation of the ranks of y. (This is the case in Table 16.7b.) Furthermore, for a given n the standard deviation is the same regardless of the original data, since it is the standard deviation of the first n integers. Using facts such as these, it is possible greatly to simplify the formula for the Spearman correlation. The formula depends only on n and the sum of the squared *differences* of the ranks. If these differences are called d, their sum of squares is Σd^2, and the Spearman correlation is

$$r_S = 1 - \frac{6\Sigma d^2}{n(n^2 - 1)}$$

Although this formula no longer looks like the formula for a correlation, it is important to remember that this formula is mathematically equivalent to the preceding definition of the Spearman correlation as the Pearson correlation

TABLE 16.7 The Spearman Correlation

(a) Data, Ranks, and Difference of Ranks

x	y	RANK x	RANK y	DIFFERENCE OF RANKS (d)
13	15	2	4	−2
14	12	3	2	1
15	13	4	3	1
16	19	5	6	−1
19	20	7	7	0
17	16	6	5	1
11	10	1	1	0

(b) Spearman Correlation from the Pearson Formula

$\Sigma(\text{rank}x) = 28$, $\Sigma(\text{rank}x)^2 = 140$, $s'^2_{\text{rank}x} = 4.000$, $s'_{\text{rank}x} = 2.000$

$\Sigma(\text{rank}y) = 28$, $\Sigma(\text{rank}y)^2 = 140$, $s'^2_{\text{rank}y} = 4.000$, $s'_{\text{rank}y} = 2.000$

$\Sigma(\text{rank}x)(\text{rank}y) = 136$

$$r_S = \frac{\Sigma(\text{rank}x)(\text{rank}y) - \Sigma(\text{rank}x)\Sigma(\text{rank}y)/n}{n\, s'_{\text{rank}x}\, s'_{\text{rank}y}} = \frac{136 - (28)(28)/7}{7 \times 2.000 \times 2.000} = 0.857$$

(c) Spearman Correlation from Alternative Formula

$\Sigma d^2 = (-2)^2 + 1^2 + 1^2 + (-1)^2 + 0^2 + 1^2 + 0^2 = 8$

$$r_S = 1 - \frac{6\Sigma d^2}{n(n^2 - 1)} = 1 - \frac{6 \times 8}{7(7^2 - 1)} = 1 - \frac{48}{7 \times 48} = 0.857$$

between the ranks derived from the original scores. (*Note.* The number 6 in this formula is a constant; it does not change, regardless of the value of n.)

Example 2

The calculation of r_S by the alternative formula is illustrated in Table 16.7. As in Example 1, the original scores, x and y, are converted to ranks, and then the difference of each pair of ranks is computed (Table 16.7a). The sum of the squares of these differences is computed in Table 16.7c. This sum is inserted into the alternative formula and r_S is computed. The result is exactly the same as in Table 16.7b.

Ties

In many data sets, two or more scores may be tied. The procedure described in Section 15.5 should be followed: assign to each of the tied scores the average (mean) of the ranks that the scores would have if they were not tied.

An additional complication ensues when there are tied scores: the two formulas for r_S are no longer equivalent. Furthermore, the statistical test and confidence interval described below are no longer exact. If there are only a few ties (say, 20% or less), however, the two formulas for r_S will give very similar results and the tests and intervals may still be used.

Test of the Hypothesis of No Association

Suppose that the seven pairs of scores in Table 16.7a are a random sample drawn from a population and that we want to determine whether or not there is a monotonic association between x and y. The two hypotheses are

$$H_0: \text{no association between } x \text{ and } y$$

and

$$H_1: \text{a monotonic association between } x \text{ and } y$$

If we define ρ_S as the *population* Spearman correlation, the hypotheses can be expressed as

$$H_0: \rho_S = 0$$

and

$$H_1: \rho_S \neq 0$$

(As usual, the alternative hypothesis in certain situations may be one-sided.)

The distribution of the test statistic, r_S, when the null hypothesis is true has been calculated, and percentiles of this distribution are given in Table A14 in the Appendix. These percentiles are used as critical values for the test of the two hypotheses. The percentiles depend on the sample size. As noted in the table, when the sample size is greater than 60 the percentiles for the Pearson correlation (Table A11) may be used.

Example 3

The data in Table 16.7 have a Spearman correlation of 0.857. The critical value for a test of association for $\alpha = 0.05$, two-tailed, is 0.786 (Table A14, $n = 7$, $\alpha = 0.05/2 = 0.025$ one-tailed). Hence, the sample Spearman correlation is statistically significant at the 0.05 level, two-tailed, and we conclude that there is evidence for a monotonic relationship between x and y.

Confidence Interval Estimate of the Population Spearman Correlation

The confidence interval estimate can be derived from the fact that the mean and standard deviation of the distribution of r_S have the following values.

$$\boxed{\text{mean } (r_S) = \rho_S}$$

where ρ_S is the population value, and

$$\boxed{\text{s.d. } (r_S) = \frac{1}{\sqrt{n-1}}}$$

where n is the sample size. When n is large, the distribution of r_S is approximately normal. Hence, the $100(1 - \alpha)\%$ confidence interval for ρ_S, based on a sample value r_S from n observations, has the approximate limits

$$\boxed{\begin{aligned} \text{Lower limit} &= r_S - (z_{\alpha/2} \cdot 1/\sqrt{n-1}) \\ \text{Upper limit} &= r_S + (z_{\alpha/2} \cdot 1/\sqrt{n-1}) \end{aligned}}$$

It is difficult to state how large n must be if these limits are to have a specified accuracy. Certainly, though, the limits calculated by these formulas should not exceed ± 1.0. If they do, these formulas should not be used.

Example 4

Suppose a Spearman correlation of 0.46 is obtained in a sample of 50 pairs of scores. The 99% confidence interval for the population Spearman correlation has the following limits.

$$
\begin{aligned}
\text{Lower limit} &= r_S - (z_{\alpha/2} \cdot 1/\sqrt{n-1}\,) \\
&= 0.46 - (2.576 \times 1/\sqrt{49}) \\
&= 0.46 - 0.368 \\
&= 0.09 \\
\text{Upper limit} &= 0.46 + 0.368 \\
&= 0.83
\end{aligned}
$$

The population Spearman correlation is estimated to fall between 0.09 and 0.83.

16.5 Association Measures for Categorical Variables

We return now to the study of the association between two *categorical* variables, which formed the main topic of Chapters 14 and 15. In many cases, the association between two categorical variables is tested by χ^2_{obs}, which has approximately a χ^2-distribution when the null hypothesis of no association is true.

The statistic χ^2_{obs} is an appropriate *test statistic* but is not a useful *measure of the association* between two variables. Measures of association (Table 16.1), which most frequently are correlations or squared correlations, have values in the range from -1 to $+1$ or from 0 to $+1$. (The standardized effect size is an exception to this statement, as its values can sometimes exceed 1.0.) The values of χ^2_{obs}, on the other hand, are rarely less than 1.0 and get larger and larger as the sample size (or number of degrees of freedom) increases. A value of $\chi^2_{\text{obs}} = 21.5$, say, between two variables gives no idea of the degree of association between the two variables. The value of χ^2_{obs} is useful only to test the statistical significance between the variables. (To review the distinction between statistical significance and effect size or size of the association, see Section C.4.)

Many measures of the association of two categorical variables have been developed.[7] Here we deal with only one such measure, called *Cramér's measure*. The symbol for the measure is $\hat{\phi}'^2$ which is read "phi-hat prime squared". [The Greek letter is "phi"; the "hat" indicates that the measure is an estimate of a population value; the "prime" indicates that this measure is related but not identical to another measure, $\hat{\phi}^2$ (not defined here); and the "squared" indicates that the measure is a squared measure, like the squared correlation, r^2 (Section 16.3).]

Cramér's measure depends on four quantities: χ^2_{obs}, the sample size n, and the number of categories of each of the two variables. Suppose that there are r

---

[7] For a description of many of these measures, see, for example, L. A. Marascuilo and M. McSweeney, *Nonparametric and Distribution-Free Methods for the Social Sciences* (Monterey, Cal.: Wadsworth, 1977).

categories of the variable represented by the rows of the frequency table and c categories of the variable represented by the columns.[8] Let q be the smaller (minimum) of $(r - 1)$ and $(c - 1)$: that is,

$$q = \min (r - 1, c - 1)$$

Then, Cramér's measure is

$$\hat{\phi}'^2 = \frac{\chi^2_{\text{obs}}}{nq}$$

It can be shown that $\hat{\phi}'^2$ must have a value between 0 and 1. Clearly, the larger the value of χ^2_{obs}, the larger the value of $\hat{\phi}'^2$ (closer to 1) and the greater the association between the two variables. On the other hand, the smaller the value of χ^2_{obs}, the smaller the value of $\hat{\phi}'^2$ (closer to 0) and the smaller the association between the two variables. Cramér's measure can be interpreted, *very roughly*, as the proportion of the variance of one variable that can be predicted from or that is associated with the other variable. In this way $\hat{\phi}'^2$ is similar to the square of the Pearson correlation, r^2. This interpretation of $\hat{\phi}'^2$ is not very exact, however, and should therefore be used with caution.

Example 1

Example 2 in Section 15.2 illustrated a study of the association between political orientation and income level. There were three levels of income (the rows of the table) and four levels of political orientation (the columns of the table). Hence, $r = 3$ and $c = 4$ and the value of q, which is the smaller of $r - 1 = 2$ and $c - 1 = 3$, is 2. The sample size is 140 and the value of χ^2_{obs} is 18.75. Hence, Cramér's measure of association is

$$\hat{\phi}'^2 = \frac{\chi^2_{\text{obs}}}{nq} = \frac{18.75}{140 \times 2} = 0.067$$

It was shown, in Section 15.2, that this value of χ^2_{obs} is statistically significant (at the 0.01 level). However, we see from the value of Cramér's measure that the degree of association between the two variables, political orientation and income level, is very small.

The square root of Cramér's measure sometimes is reported. This measure, $\hat{\phi}'$, is also called Cramér's measure. When Cramér's measure is reported, it is important to know whether it is the squared or the "unsquared" measure. The unsquared measure, $\hat{\phi}'$, is roughly analogous[9] to r, just as $\hat{\phi}'^2$ is roughly analogous to r^2. (For the preceding example, $\hat{\phi}' = \sqrt{0.067} = 0.26$.)

---

[8] Do not confuse this use of the symbol r with its use to represent a (Pearson) correlation.

[9] The unsquared measure, $\hat{\phi}'$, can be shown to be *exactly* equal to the Pearson correlation, r, when both variables are dichotomous: that is, when each variable has two categories, and numerical values are assigned to each category. (For example, the two categories of x could have the values 0 and 1, and the two categories of y could also have the values 0 and 1.)

Confidence Interval Estimate of Cramér's Measure

The distribution of Cramér's unsquared measure, $\hat{\phi}'$, is approximately normal (if n is large enough), and its mean and standard deviation have the following values.

$$\text{mean } (\hat{\phi}') = \phi'$$

where ϕ' is the population value, and

$$\text{s.d. } (\hat{\phi}') = \frac{1}{\sqrt{nq}}$$

where n is the sample size and q is as defined above. The $100(1 - \alpha)\%$ confidence interval for ϕ', based on a sample value $\hat{\phi}'$, has the following limits.

$$\text{Lower limit} = \hat{\phi}' - (z_{\alpha/2} \cdot 1/\sqrt{nq})$$
$$\text{Upper limit} = \hat{\phi}' + (z_{\alpha/2} \cdot 1/\sqrt{nq})$$

The confidence limits for the squared measure, ϕ'^2, may be found by obtaining and then squaring the limits for ϕ'. (The sample size, n, should be large enough that these limits do not exceed ± 1.)

Example 2

In Example 1 the sample values of Cramér's measure were $\hat{\phi}'^2 = 0.067$ and $\hat{\phi}' = 0.26$. The 95% confidence limits for ϕ' are

$$
\begin{aligned}
\text{Lower limit} &= \hat{\phi}' - (z_{\alpha/2} \cdot 1/\sqrt{nq}) \\
&= 0.26 - (1.960 \times 1/\sqrt{140 \times 2}) \\
&= 0.26 - 0.117 \\
&= 0.14 \\
\text{Upper limit} &= 0.26 + 0.117 \\
&= 0.38
\end{aligned}
$$

Hence, the population value of Cramér's measure (unsquared) is estimated to fall between 0.14 and 0.38. The squares of these limits are $(0.14)^2 = 0.02$ and $(0.38)^2 = 0.14$. Hence, the population value of ϕ'^2 is estimated to fall between 0.02 and 0.14.

16.6 Theory of Least-squares Prediction

Note. This section may be omitted with no loss of continuity.

The prediction line was defined in Section 16.3 as that straight line which best fits the regression function (the set of means of the dependent variable for each value of the independent variable). The prediction line equivalently can be defined as that straight line which best fits the set of data points themselves. In this definition, "best fit" means that the sum of squares of the *vertical* distances of the points from the line is the smallest possible. In this section, this definition is used to derive the equation of the prediction line and to prove the principle of prediction.

Preliminaries

Consider the set of seven pairs of scores (x, y) shown in Figure 16.7a. (The actual numbers are in Table 16.8.) We want to predict the y scores from the x

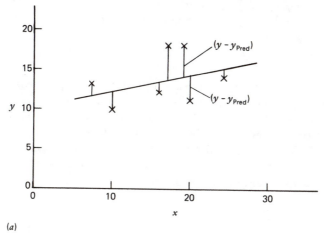

(a)

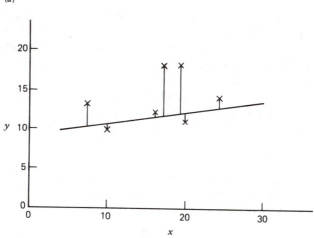

(b)

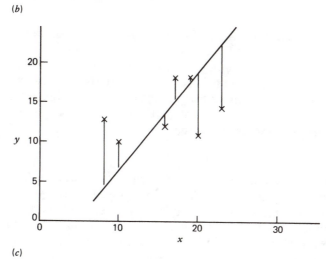

(c)

Figure 16.7 Various prediction lines for the same data. (a) Small SSE. (b) Large SSE. (c) Large SSE.

scores by using a straight line. An example of a possible prediction line is shown in the figure. The mathematical equation for such a straight line is, in general,

$$y_{Pred} = a + b \cdot x$$

where a and b are the y-intercept and the slope, respectively, of the straight line. (See Section 16.3 for further explanation.)

We want to find the line that most accurately predicts the y scores. For any trial line, such as the line in Figure 16.7a, we can measure the accuracy of prediction by comparing the predicted scores (y_{Pred}) with the actual scores (y). In each case the difference between y and y_{Pred} is the error in prediction (this error was called y_{Unpred} in Section 16.3).

$$\text{Error} = y_{Unpred}$$
$$= y - y_{Pred}$$

Note that in the figure these errors are the vertical distances between the points and the prediction line. The errors are *vertical* distances, because we are predicting y from x. (If, instead, we wanted to predict x from y, the errors would be the horizontal distances from the prediction line. A different prediction line is obtained—see later in this section.)

As already stated, we want to find the line that has the smallest errors. The required line is the line that minimizes the sum of squares of these errors, or SSE.

$$SSE = \Sigma(y - y_{Pred})^2$$

Hence, the required line is known as the *least-squares prediction line*. This line can be found more readily if we divide SSE by n, where n is the number of pairs (x, y). The problem, then, is to minimize

$$\frac{SSE}{n} = \frac{\Sigma(y - y_{Pred})^2}{n}$$

The straight line in Figure 16.7a is close to the optimal line—the sum of squared errors (SSE), and hence SSE/n, is small. Consider two other straight lines as possible prediction lines. The line in Figure 16.7b has a slope similar to that of the line in Figure 16.7a but a smaller y-intercept; the errors are greater, on the average, and SSE is larger than it is in Figure 16.7a. The line in Figure 16.7c has a much greater slope than the other lines do, and it is clear from the figure that SSE is also large.

Prediction in Standard Scores

The problem of finding the least-squares prediction line—that is, the values of a and b that minimize SSE—is a straightforward problem in calculus. By first transforming the scores to standard scores (using the sample mean and standard deviation), however, we can to derive the values of a and b by a simple algebraic technique. Consider Figure 16.8a, in which the seven data points are shown again. The least-squares prediction line, whose equation we will derive, also is shown. In Figure 16.8b the seven points have been drawn with axes in standard score units. (The details of the calculations of means and standard deviations and the actual standard score values are shown in Table 16.8.) It should be intuitively clear that if we find the least-squares prediction

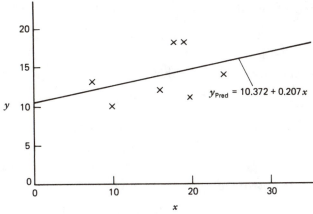

(a)

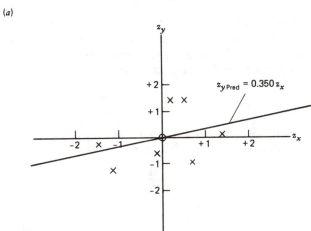

(b)

Figure 16.8 Least-squares prediction line. (a) Original scores. (b) Standard scores.

line for the standard scores, z_x and z_y, we can find the least-squares prediction line for the original scores, x and y, by the usual transformation: multiplying the standard score by the standard deviation and adding the mean (Section 7.1).

We therefore set ourselves the task of finding the least-squares prediction line for the set of scores (z_x, z_y). Recall that standard scores have the property that their mean is zero and their variance (and standard deviation) is 1.

$$\overline{z}_x = \frac{\Sigma z_x}{n} = 0$$

and

$$s_{z_x}'^2 = \frac{\Sigma z_x^2 - (\Sigma z_x)^2/n}{n} = 1$$

The first equation implies that

$$\Sigma z_x = 0$$

and because of this, the second reduces to

$$\frac{\Sigma z_x^2}{n} = 1$$

In the same way, it is clear that

$$\Sigma z_y = 0$$

and

$$\frac{\Sigma z_y^2}{n} = 1$$

Let the prediction line for predicting z_y from z_x be

$$z_{y\text{Pred}} = c + d \cdot z_x$$

where c and d are constants we want to determine. (Here we use symbols c and d for the y-intercept and slope, in order to avoid confusion with the y-intercept and slope, a and b, of the prediction line of original scores.) The error in predicting z_y using this prediction line is

$$\begin{aligned} \text{error} &= z_y - z_{y\text{Pred}} \\ &= z_y - c - d \cdot z_x \end{aligned}$$

The quantity we wish to minimize is the sum of squares of these errors (called SSE_z) divided by n. Its value can be written as follows.

$$\begin{aligned} \frac{SSE_z}{n} &= \frac{\Sigma(z_y - z_{y\text{Pred}})^2}{n} \\[2mm] &= \frac{\Sigma(z_y - c - d \cdot z_x)^2}{n} \\[2mm] &= \frac{\Sigma(z_y^2 + c^2 + d^2 z_x^2 - 2cz_y - 2dz_x z_y + 2cdz_x)}{n} \\[2mm] &= \frac{\Sigma z_y^2}{n} + \frac{\Sigma c^2}{n} + \frac{\Sigma d^2 z_x^2}{n} - \frac{\Sigma 2cz_y}{n} - \frac{\Sigma 2dz_x z_y}{n} + \frac{\Sigma 2dcz_x}{n} \end{aligned}$$

Each of the six terms in the above expression can be simplified. The *first term* is 1, as shown above, from the fact that the variance of z_y is 1. The *second term* is $nc^2/n = c^2$, since the sum of a constant (c^2) is simply n times that constant. The *third term* is simply d^2, since d^2, being a constant, can be taken out of the summation and the remaining sum is 1, the variance of z_x. The *fourth term* is zero, since the constant $2c$ can be taken out of the summation, leaving Σz_y, which is zero (see above). The *fifth term* is $-2dr$, where r is the Pearson correlation. To understand this, note that $2d$ can be taken out of the summation, leaving $\Sigma z_x z_y/n$. Writing the standard scores in terms of original scores, we have

$$\begin{aligned} \frac{\Sigma z_x z_y}{n} &= \frac{\Sigma\{(x - \bar{x})/s_x'\}\{(y - \bar{y})/s_y'\}}{n} \\[2mm] &= \frac{\Sigma(x - \bar{x})(y - \bar{y})}{n\, s_x'\, s_y'} \end{aligned}$$

which is the definitional formula for r (Section 16.2). Finally, the *sixth term* is zero, since the constant $2cd$ can be taken out of the summation leaving Σz_x, which is zero (see above).

We have, therefore, shown that

$$\frac{SSE_z}{n} = 1 + c^2 + d^2 - 0 - 2dr + 0$$

$$= 1 + c^2 + d^2 - 2dr$$

We now require that this expression be made as small as possible by suitable choice of the constants c and d, which represent the y-intercept and the slope, respectively, of the prediction line. Consider, first, various choices for c, which will affect only the value of the second term. Of all the possible values for c (both positive and negative) that could be considered, the second term will be smallest when $c = 0$: the second term is zero when $c = 0$ and is greater than zero for any other value of c. Clearly, the least-square prediction line must have $c = 0$.

We are now left with the following expression to be minimized.

$$\frac{SSE_z}{n} = 1 + d^2 - 2dr$$

This expression is algebraically equivalent to

$$(d - r)^2 + (1 - r^2)$$

For what value of d is *this* expression a minimum? The first term varies as d varies, but the second term does not change. Since the first term is a square, its minimum value is zero, which occurs when $d - r = 0$: that is, when $d = r$. Hence, for the least-squares prediction line, the value of d must be equal to the correlation, r.

We have shown that the least-squares prediction line, in standard scores, is

$$\boxed{z_{y\text{Pred}} = r \cdot z_x \quad \text{(Prediction line in standard scores)}}$$

The y-intercept of this line is zero; in other words, this line goes through the origin. The slope of this line is r, the Pearson correlation. As a by-product of the above development, we also have shown that

$$\frac{SSE_z}{n} = 1 - r^2$$

Prediction in Original Scores

The equation for the least-squares prediction line in original scores easily can be found from the standard-score equation derived above, by using the formula for converting standard scores to original scores.

$$\text{Standard score} = \frac{\text{Original score} - \text{Mean}}{\text{Standard deviation}}$$

We get

$$\frac{y_{\text{Pred}} - \bar{y}}{s'_y} = r \cdot \frac{x - \bar{x}}{s'_x}$$

This equation can be rearranged to give

$$y_{\text{Pred}} = \left(\bar{y} - r\,\frac{s'_y}{s'_x}\,\bar{x}\right) + r\,\frac{s'_y}{s'_x}\,x \qquad \text{(Prediction line in original scores)}$$

The slope of this line is

$$b = r\frac{s'_y}{s'_x}$$

and the y-intercept of the line is

$$a = \bar{y} - r\frac{s'_y}{s'_x}\,\bar{x}$$

which can also be written

$$a = \bar{y} - b\,\bar{x}$$

Example 1

Let us return to the data of Figure 16.8; the scores and the calculations are given in Table 16.8. First, working with the original scores (Table 16.8a), we calculate the means, standard deviations, and correlation of the two variables. These values are then inserted in the above formulas for the slope and y-intercept to give the final equation for the least-squares prediction line.

$$y_{\text{Pred}} = 10.372 + 0.207x$$

This line is shown in Figure 16.8a, where its slope (0.207) and y-intercept (10.372) may be checked. This prediction line is that line which minimizes the sum of squares of the vertical distances from the line.

The prediction line is usually calculated in this way from the original scores. However, as an illustration of the least-squares prediction line for standard scores, some calculations are given in Table 16.8b. Here the original scores, x and y, have been converted to standard scores, z_x and z_y, using their respective means and standard deviations. As a check on the standard scores, their means and standard deviations have been calculated to make sure that they are 0 and 1, respectively, within rounding error. The correlation, r, is the sum of the products of the standard scores, divided by $n = 7$.

$$r = \frac{z_x z_y}{n} = \frac{2.452}{7} = 0.350$$

the same value, of course, that was obtained from the original scores. The prediction line for the standard scores,

$$z_{y\text{Pred}} = 0.350z_x$$

is plotted in Figure 16.8b. Its y-intercept is zero and its slope is 0.350; it is the same as the line for the original scores except for changes of origin and scale of both the variables x and y.

Prediction of x from y

Recall, from Section 8.4, that the regression of y on x is not the same as the regression of x on y. In the same way, the least-squares prediction line for predicting y from x, which we have described above, is *not* the prediction line that would be used to predict x from y. In order to predict x from y, we must

TABLE 16.8 Calculation of the Least-squares Prediction Line

(a) Prediction in Original Scores

x	y
19	18
16	12
20	11
8	13
10	10
17	18
23	14

$\Sigma x = 113$ $\Sigma y = 96$

$\Sigma x^2 = 1999$ $\Sigma y^2 = 1378$

$\Sigma xy = 1586$

$\bar{x} = 113/7 = 16.143$ $\bar{y} = 96/7 = 13.714$

$$s_x' = \sqrt{\frac{1999 - (113)^2/7}{7}} \qquad s_y' = \sqrt{\frac{1378 - (96)^2/7}{7}}$$

$= 4.998$ $= 2.962$

$$r = \frac{1586 - (113 \times 96)/7}{7 \times 4.998 \times 2.962} = 0.350$$

$b = r \cdot s_y'/s_x' = 0.350 \times 2.962/4.998 = 0.207$

$a = \bar{y} - b \cdot \bar{x} = 13.714 - 0.207 \times 16.143 = 10.372$

Prediction line: $y_{Pred} = 10.372 + 0.207x$

(b) Prediction in Standard Scores

x	y	z_x	z_y
19	18	+0.572	+1.447
16	12	−0.029	−0.579
20	11	+0.772	−0.916
8	13	−1.629	−0.241
10	10	−1.229	−1.254
17	18	+0.171	+1.447
23	14	+1.372	+0.097

$\bar{z}_x = 0.000$ $\bar{z}_y = 0.000$

$s_{z_x}' = 1.000$ $s_{z_y}' = 1.000$

$$r = \frac{\Sigma z_x z_y}{7} = \frac{2.452}{7} = 0.350$$

Prediction line: $z_{yPred} = 0.350 z_x$

find that line which minimizes the differences $x - x_{Pred}$—that is, the *horizontal* distances of the points from the line. Of course, the theory and formulas are exactly the same as before, except that x and y are interchanged throughout. The prediction line in standard scores is

$$z_{xPred} = r \cdot z_y$$

and the prediction line in original scores is

$$x_{Pred} = \left(\bar{x} - r\frac{s_x'}{s_y'}\bar{y}\right) + r\frac{s_x'}{s_y'}y$$

Example 2

The least-squares prediction line for the data of Example 1 (Table 16.8) can be found from the above formulas. The slope of the line in original scores is

$$b = r \frac{s'_x}{s'_y} = 0.350 \times \frac{4.998}{2.962} = 0.591$$

The x-intercept is

$$a = \bar{x} - r \frac{s'_x}{s'_y} \bar{y}$$
$$= \bar{x} - b \cdot \bar{y} = 16.143 - 0.591 \times 13.714 = 8.038$$

This line is drawn in Figure 16.9a. The line minimizes the horizontal distances of the points from itself. This line is quite different from the line in Figure 16.8a (which minimizes the vertical distances).

The least-squares prediction line for predicting z_x from z_y has the equation $z_{xPred} = r \cdot z_y$; for the data of Example 1, the equation is

$$z_{xPred} = 0.350 z_y$$

This line is drawn in Figure 16.9b. Again, this line is different from the line in Figure 16.8b, which is

$$z_{yPred} = 0.350 z_x$$

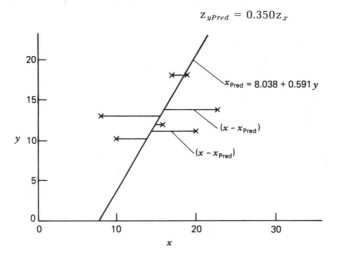

(a)

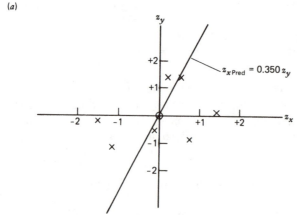

(b)

Figure 16.9 Least-squares prediction of x from y. (a) Original scores. (b) Standard scores.

Note that the positions of x and y in these two equations are reversed. In both cases the coefficient is 0.350 (the correlation between x and y), but the lines are different since in one case x predicted from y and in the other case y is predicted from x.

The Principle of Prediction in Standard Scores

The principle of prediction (Section 16.3) states that the total variance of a dependent variable equals the sum of the predicted variance and the unpredicted variance: that is,

$$s_{Tot}'^{2} = s_{Pred}'^{2} + s_{Unpred}'^{2}$$

This equation is true for one variable linearly predicted by least-squares from a second variable, as the following proof shows. The equation is true whether the variables are treated in standard scores or in original scores (see below, "The Principle of Prediction in Original Scores").

A by-product of the proof, given above, of the equation of the prediction line in standard scores was the following formula for the sum of squared errors divided by n.

$$\frac{SSE_z}{n} = 1 - r^2$$

The errors are the unpredicted scores, which are

$$z_{y\text{Unpred}} = z_y - z_{y\text{Pred}}$$
$$= z_y - r \cdot z_x$$

using the least-squares prediction formula. The mean of the unpredicted scores is

$$\frac{\Sigma z_{y\text{Unpred}}}{n} = \frac{\Sigma(z_y - r \cdot z_x)}{n}$$

$$= \frac{\Sigma z_y}{n} - r \cdot \frac{\Sigma z_x}{n}$$

$$= 0 - r \cdot 0 = 0$$

since the means of the standard scores, z_y and z_x, are zero. Since the mean of the unpredicted scores is zero, the variance of these scores is just SSE_z/n, which, as pointed out above, is simply $(1 - r^2)$. We have shown that, for standard scores,

$$s_{Unpred}'^{2} = 1 - r^2$$

This result, along with others shown below, is displayed in Table 16.9.

The mean of the predicted scores, $z_{y\text{Pred}}$, is zero, since the mean is

$$\frac{\Sigma z_{y\text{Pred}}}{n} = \frac{\Sigma(r \cdot z_x)}{n}$$

$$= r \cdot \frac{\Sigma z_x}{n} = 0$$

TABLE 16.9 Variances in Prediction

	VARIANCES OF VARIABLES EXPRESSED IN	
	STANDARD SCORES	ORIGINAL SCORES
Total variance $(s_{\text{Tot}}'^2)$	1	$s_y'^2$
Predicted variance $(s_{\text{Pred}}'^2)$	r^2	$s_y'^2 \cdot r^2$
Unpredicted variance $(s_{\text{Unpred}}'^2)$	$1 - r^2$	$s_y'^2 \cdot (1 - r^2)$
$\dfrac{s_{\text{Pred}}'^2}{s_{\text{Tot}}'^2}$	$\dfrac{r^2}{1} = r^2$	$\dfrac{s_y'^2 \cdot r^2}{s_y'^2} = r^2$
$\dfrac{s_{\text{Unpred}}'^2}{s_{\text{Tot}}'^2}$	$\dfrac{1 - r^2}{1} = 1 - r^2$	$\dfrac{s_y'^2 \cdot (1 - r^2)}{s_y'^2} = 1 - r^2$

The variance of these scores is, therefore,

$$s_{\text{Pred}}'^2 = \frac{\Sigma z_{y\,\text{Pred}}^2}{n} = \frac{\Sigma(r \cdot z_x)^2}{n}$$

$$= r^2 \cdot \frac{\Sigma z_x^2}{n} = r^2 \cdot 1 = r^2$$

since the variance of the standard scores, z_x, is 1.

Finally, the variance of z_y is denoted by $s_{\text{Tot}}'^2$ and is, of course, 1. It is clear, then, that

$$s_{\text{Tot}}'^2 = s_{\text{Pred}}'^2 + s_{\text{Unpred}}'^2$$

since

$$1 = r^2 + (1 - r^2)$$

This confirms the principle of prediction for prediction in standard scores.

Also note that, as shown in Table 16.9, the ratio of the predicted variance to the total variance (the coefficient of determination) is r^2 and that the ratio of the unpredicted variance to the total variance is $(1 - r^2)$.

The Principle of Prediction in Original Scores

The formulas for the variances of original scores easily can be found from the standard-score formulas, since the two scores are related by a change of scale and a change of origin. Since a change of origin does not affect the variance of a score (Section 7.1), we need only consider the change of scale. The scale of the original scores is s_y' times the scale of the standard scores. This is true for the total scores (y), the predicted scores (y_{Pred}), and the unpredicted scores (y_{Unpred}). Hence, the variance of each of these scores is $s_y'^2$ times the variance of the scores is standard-score form. These results are shown in the last column of Table 16.9.

As each of the three variances has been multiplied by $s_y'^2$, it is still true that

$$s_{\text{Tot}}'^2 = s_{\text{Pred}}'^2 + s_{\text{Unpred}}'^2$$

the principle of prediction.

The table also shows that the ratio of the predicted variance to the total variance, expressed in original scores, reduces, by cancellation of $s_y'^2$ in numerator and denominator, to r^2, the same value that the ratio has in standard-score form. Similarly, the ratio of the unpredicted variance to the total variance is $(1 - r^2)$.

The standard deviation of the unpredicted scores (s'_{Unpred}) is sometimes called the *standard error of estimate*.

16.7 Summary

There are many ways to measure the association between two variables. The measure chosen depends on the types of the two variables: that is, on whether they are numerical, dichotomous, categorical, or rank variables. One must also distinguish between the *measure* of association and the *statistic* used to test whether or not there is any association. (Sometimes, as shown in Table 16.1, the measure is, itself, used as the statistic.)

The *Pearson correlation* is a measure of the linear relationship between the two numerical variables. Several tests and estimates of the Pearson correlation were described. The square of the Pearson correlation can be interpreted as the proportion of the total variance of one of the variables that can be (linearly) predicted from the other variable. This interpretation is based on the *principle of prediction*.

Two other measures of association are the *Spearman correlation* and *Cramér's measure*. A test of the hypothesis of no association, based on the Spearman correlation, was described. Confidence interval estimates for both the Spearman correlation and Cramér's measure also were described.

Exercises

1 (Section 16.1) Consider a study of the association between attendance at school and score on a test given at the end of 20 days of instruction. Suppose that the class has 25 students.

(a) First, assume that both measures are numeric: attendance ranges from 0 to 20 days; the test scores range from 0 (the lowest possible score) to 10 (the highest possible score). Draw four scatterplots using fictitious data of your own to illustrate four possible relationships between the two variables. Also, show on these graphs the regression functions. The four relationships to illustrate are

(i) No association.

(ii) Positive monotonic association.

(iii) Negative linear association.

(iv) Nonlinear association.

Be sure to label each graph and its axes appropriately.

(b) Second, assume that attendance is just recorded in three categories (poor, fair, good) and that the test score is recorded as pass or fail. Make up two fictitious frequency tables, of 99 subjects each, to illustrate the following relationship between the two nonnumerical variables.

(i) No association.

(ii) General association.

2 (Section 16.2) The following scores were obtained in a study of 20 subjects who were measured on two scales: a measure of self-reported loneliness (L) and a measure of dissatisfaction with existing social relationships (D). Test whether the Pearson correlation between these measures is significantly different from zero.

SUBJECT	L	D
1	2	6
2	3	6
3	7	4
4	7	6
5	9	5
6	9	3
7	4	2
8	5	3
9	7	7
10	8	9
11	8	7
12	6	8
13	4	9
14	1	2
15	2	3
16	1	3
17	2	2
18	4	4
19	6	5
20	6	4

3 (Section 16.2) A correlation of 0.63 between loneliness and dissatisfaction is found in a sample of 25 males.

(a) Test whether this correlation is significantly different from 0.40, using a two-tailed test at the 0.01 level.

(b) Find the 99% confidence interval for the population correlation between loneliness and dissatisfaction in males.

(c) Are the answers to (a) and (b) consistent? (Review Section C.1 for the relationship between hypothesis tests and confidence intervals.)

4 (Section 16.2) The correlation between loneliness and dissatisfaction is 0.63 in a sample of 25 males and 0.48 in a sample of 40 females. Is there a significant difference between these correlations?

5 (Section 16.3) The following data for a measure of anxiety and a measure of performance on a certain task were obtained in a sample of 10 subjects.

SUBJECT	ANXIETY	PERFORMANCE
1	1	5
2	3	6
3	5	5
4	4	3
5	6	4
6	5	2
7	7	2
8	9	1
9	2	3
10	9	4

(a) Calculate the Pearson correlation between anxiety and performance.

(b) Calculate the slope and y-intercept of the prediction line for predicting performance from anxiety.

(c) Plot the 10 subjects' scores on the two variables, and also plot the prediction line. [The easiest way to plot the line is to compute the values of y for two different values of x and to plot the two pairs of (x, y) values found; these two points are then joined, forming the prediction line. Do not choose values of x that are too close together.]

6 (Section 16.3) Use the prediction equation found in Question 5 to answer the following questions.
(a) For each subject, compute the predicted performance using the prediction line. (Use the computed prediction equation, not the graph, in order to ensure sufficient accuracy.)
(b) Compute the variance of the observed performance scores and the variance of the predicted performance scores.
(c) Compute the ratio of the predicted variance to the total variance of the performance scores. Compare this ratio to r^2.
(d) Interpret the numerical value of r^2.

7 (Section 16.3) Consider once more the correlation of 0.63 between loneliness and dissatisfaction in a sample of 25 males (Question 3).
(a) What proportion of the variance of the dissatisfaction scores can be predicted from the loneliness scores? What proportion of the variance of the dissatisfaction scores *cannot* be predicted from the loneliness scores?
(b) What proportion of the variance of the loneliness scores can be predicted from the dissatisfaction scores? What proportion of the variance of the loneliness scores *cannot* be predicted from the dissatisfaction scores?

8 (Section 16.4) The following data for two variables, x and y, were obtained for a sample of 10 subjects.

x	y
44	28
37	27
32	38
31	30
49	26
42	29
50	23
39	24
34	31
36	36

(a) Compute the Spearman correlation, r_S, by either method shown in the text.
(b) Make a scatterplot of the data and show that your computed value is reasonable.

9 (Section 16.4) The Spearman correlation in a sample of 100 pairs of scores is 0.371.
(a) Test the hypothesis of no association between x and y at the 0.05 level (two-tailed).
(b) Compute the 95% confidence interval for the population Spearman correlation.
(c) Show that the results from **(a)** and **(b)** are consistent. (*Hint:* look at Section C.1.)

10 (Section 16.5) Calculate Cramér's measure (squared) for the data in Questions 1 and 2a in the Exercises for Chapter 15. In each study, interpret the values in words. Compute the 95% confidence interval for the population value in each study.

11 (Section 16.6) Use the data in Question 8 to perform the following tasks.
(a) Compute the least-squares prediction line for predicting y from x.
(b) Compute the least-squares prediction line for predicting x from y.
(c) Plot the data and the two prediction lines on graph paper.
(d) Explain why the two lines are different and show that the lines are reasonable.
(e) Compute r^2 and $(1 - r^2)$ for these data. Interpret these values, both when y is predicted from x and when x is predicted from y.

17 ANALYSIS OF VARIANCE

The analysis of variance is used extensively in psychological research. Like most of the techniques that have already been described, the analysis of variance is a technique for testing and measuring the relationship between an independent and a dependent variable.

The analysis of variance, as described in this chapter, is closely related to the t-test of two independent means. Both the t-test and the analysis of variance apply to independent-groups designs (Section 11.3). The t-test is appropriate when there are *two* independent groups; the analysis of variance is appropriate when there are *more than two* independent groups. (In fact, the analysis of variance can also be used when there are just two groups—the conclusions are exactly the same as those from the t-test of two independent means.)

Two types of analysis of variance are described in this chapter: the one-factor and the two-factor analyses of variance. (See Section 17.1 for a definition of these types.) The one-factor analysis of variance is described in Sections 17.2 through 17.7. The two-factor analysis of variance is described in Sections 17.8 and 17.9. Both these analyses apply to independent groups. (The analysis of variance also can be extended to *dependent* groups, but such analyses are not described in this text.)

17.1 Factors and Levels

The *analysis of variance* is a technique for studying the relationship between a dependent variable and one or more independent variables. The independent variables in the analysis of variance are usually referred to as *factors*. In the *one-factor analysis of variance*, there is one independent variable; in the *two-factor analysis of variance*, there are two independent variables.

The *dependent variable* in an analysis of variance is a *numerical* variable. (See the discussion of types of variables in Section 15.1). For example, the dependent variable might be the number of errors on a task, the time taken to complete the task, or weight gain in two weeks. Each *factor* in an analysis of variance is a *categorical* variable. The factor is usually an experimentally-defined variable: that is, an experimenter has set up several conditions and each subject in the experiment is studied in just one condition. (See Section 8.2 for a previous discussion of experimentally-defined variables.)

Each of the conditions, or values of the independent variable, is referred to as a *level* of that variable or factor. For example, if there are four experimental conditions in a one-factor analysis of variance, the factor has four levels.

17.2 The Relationship Between an Independent and a Dependent Variable

As many preceding chapters have pointed out, there are two ways to study the relationship between an independent and a dependent variable. The first way is to test whether there is or is not a relationship between the two variables. This is a test of the *regression function*, which is the set of means of the dependent variable, one mean for each value (or interval of values) of the independent variable. If the regression function in the population is horizontal, there is no relationship between the two variables; if it is not horizontal, there is a relationship. The second way to study the relationship between an independent and a dependent variable is to compute the proportion of the variance of the dependent variable that can be predicted from the independent variable. This proportion is called the *coefficient of determination*. As stated in Section 16.3, the coefficient of determination equals the square of the Pearson correlation (r^2) when the prediction is made by the best-fitting straight line. In the analysis of variance, prediction is not made from a straight line, nevertheless, a coefficient of determination can be defined and calculated.

The contrast between the two ways to study the relationship of two variables is a contrast between a *test* whether or not there is a relationship and a *measure* of the size of the relationship. Both approaches are essential to a complete analysis of the relationships between two variables. (See Section 16.1 for further discussion of these two approaches.)

Let us now consider a set of data from an experimental study and see how these two approaches would be applied. To study the retention of a list of nonsense syllables that were learned perfectly by 20 subjects, a psychologist divided the subjects into four groups, representing the four conditions in the study. The groups differed in whether the subjects were warned that they would be tested later and in how much time elapsed before they were tested. The four conditions[1] were

1 Warned that will be tested later; tested one hour later.
2 Warned that will be tested later; tested one day later.
3 Not warned; tested one hour later.
4 Not warned; tested one day later.

For simplicity, we can refer to these four conditions or levels as

1 Warn/short delay.
2 Warn/long delay.
3 No warn/short delay.
4 No warn/long delay.

There were five subjects in each group. The number of syllables correctly recalled (after one hour or one day) by each subject is shown in Table 17.1. The subjects are indicated by letters in the first column of the table. Each subject has a "score" on two variables, x and y. The first variable, x, is the condition or group to which the subject was assigned. The first five subjects were in group 1, the warn/short-delay group; the next five were in group 2, the warn/long-

---

[1] At present, these four groups or conditions are treated as one factor. In Section 17.8 these same conditions will be treated as two factors (warning and delay). It is legitimate, however, to analyze the experiment as a one-factor study, as done in this and following sections.

TABLE 17.1 Results from a Learning Study

SUBJECT	CONDITION x	SYLLABLES RECALLED y	MEAN OF CONDITION
A	1	9	
B	1	11	
C	1	8	8.40
D	1	7	
E	1	7	
F	2	5	
G	2	7	
H	2	6	6.00
I	2	3	
J	2	9	
K	3	5	
L	3	9	
M	3	7	6.60
N	3	4	
O	3	8	
P	4	3	
Q	4	4	
R	4	5	3.40
S	4	1	
T	4	4	

delay group, etc. The condition or group number is shown in the second column of the table. The third column of the table shows the values of the variable y, the number of syllables recalled. Note that x is the independent (experimentally-defined) variable and that y is the dependent (measured) variable.

Regression Function

The mean number of syllables recalled in each condition can be calculated. These means are shown in the last column of Table 17.1. The means can also be added to the scatterplot of the 20 data pairs in Figure 17.1. The four means form the *regression function* of y on x for these data.

Since the four means are not equal, the regression function is not horizontal. This statement is a statement about the *sample* means. Section 17.5 will show how to test the hypothesis that the regression function (in the *population*) is horizontal.

Coefficient of Determination

The dependent variable, y, can be predicted from the independent variable, x. In the absence of any other information, the best prediction of the number of syllables which any subject in the first condition will recall is the mean of that condition: that is, 8.40. Similarly, the best prediction for subjects in the second condition is the mean of the second condition, 6.00. And so on. These predicted values for each of the 20 subjects are shown in Table 17.2, under y_{Pred}.

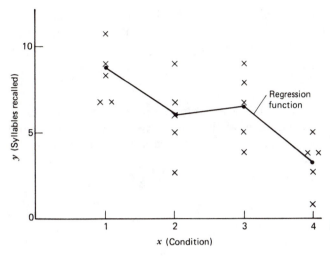

Figure 17.1 Results from a learning study.

TABLE 17.2 Predicted and Unpredicted Variables in the Learning Study

SUBJECT	x	y_{Tot}	y_{Pred}	y_{Unpred}
A	1	9	8.40	0.60
B	1	11	8.40	2.60
C	1	8	8.40	−0.40
D	1	7	8.40	−1.40
E	1	7	8.40	−1.40
F	2	5	6.00	−1.00
G	2	7	6.00	1.00
H	2	6	6.00	0.00
I	2	3	6.00	−3.00
J	2	9	6.00	3.00
K	3	5	6.60	−1.60
L	3	9	6.60	2.40
M	3	7	6.60	0.40
N	3	4	6.60	−2.60
O	3	8	6.60	1.40
P	4	3	3.40	−0.40
Q	4	4	3.40	0.60
R	4	5	3.40	1.60
S	4	1	3.40	−2.40
T	4	4	3.40	0.60
		$\Sigma y_{Tot} = 122$	$\Sigma y_{Pred} = 122.00$	$\Sigma y_{Unpred} = 0.00$
		$\Sigma y^2_{Tot} = 866$	$\Sigma y^2_{Pred} = 808.40$	$\Sigma y^2_{Unpred} = 57.60$
Variance numerators:		121.80	64.20	57.60
		$s'^2_{Tot} = 6.090$	$s'^2_{Pred} = 3.210$	$s'^2_{Unpred} = 2.880$

For consistency with the terminology in Section 16.3, we will refer to the variable y as y_{Tot}. Each "total" variable (y_{Tot}) equals the sum of a predicted variable (y_{Pred}), the condition mean, plus a remainder, the unpredicted variable (y_{Unpred}).

$$y_{Tot} = y_{Pred} + y_{Unpred}$$

The values of these variables are shown in Table 17.2 for each subject.[2]

The variances of these three variables can be calculated in the usual way. Some details of the calculations are shown at the bottom of Table 17.2. The variance of y_{Tot} is

$$s'^2_{Tot} = \frac{\Sigma y^2_{Tot} - (\Sigma y_{Tot})^2/20}{20}$$

$$= \frac{866 - (122)^2/20}{20}$$

$$= \frac{121.80}{20} = 6.090$$

The variance of y_{Pred} is

$$s'^2_{Pred} = \frac{\Sigma y^2_{Pred} - (\Sigma y_{Pred})^2/20}{20}$$

$$= \frac{808.40 - (122.00)^2/20}{20}$$

$$= \frac{64.20}{20} = 3.210$$

The coefficient of determination is the ratio of these two variances.

$$R^2 = \frac{s'^2_{Pred}}{s'^2_{Tot}} = \frac{3.210}{6.090} = 0.53$$

This result means that 53% of the variability in the dependent variable can be predicted from the independent variable: that is, 53% of the variability in the number of syllables learned can be accounted for by the condition in which the subject was run.

Note that a capital R is now used in the symbol for the coefficient of determination.[3] The letter has been changed to a capital to remind us that the coefficient of determination is no longer the square of a Pearson correlation (r). The regression function used to calculate R^2 is the set of condition means, whereas the regression function used to calculate r^2 is the best-fitting straight line.

[2] Note that this prediction is not linear prediction, as used in Section 16.3. The prediction is simply based on the means of each condition. A linear regression function cannot be used, since the values of x are just labels for the conditions; these labels could have been letters (A, B, C, D), or the numbers 1, 2, 3, and 4 could have been assigned in another order to the four conditions.

[3] The symbol R^2 has been chosen because it is also the symbol for the squared *multiple* correlation, the generalization of the Pearson correlation to more than two variables. There is a very close relationship between the analysis of variance and prediction from two or more independent variables. See, for example, J. Cohen and P. Cohen, *Applied Multiple Regression/Correlation Analysis for the Behavioral Sciences*, (Hillsdale, N.J.: Lawrence Erlbaum, 1975).

17.3 Sums of Squares in the Analysis of Variance

The preceding section outlined the two approaches to the study of the relationship of two variables: the regression function and the coefficient of determination. The regression function is based on the means of each condition. (These means will henceforth be called the *group means*.) The coefficient of determination is based on the predicted and total variances. This section concentrates on the variances rather than the means, since, as will be pointed out in Section 17.5, the test of whether or not the regression function is horizontal is based on the variances (hence the origin of the term "analysis of variance").

The principle of prediction (Section 16.3) states that the total variance of a dependent variable equals the sum of the predicted variance and the unpredicted variance.

$$s_{Tot}'^2 = s_{Pred}'^2 + s_{Unpred}'^2$$

Table 17.2 shows that this equation is satisfied by the data of the learning study.

$$6.090 = 3.210 + 2.880$$

Each of the variances has the same denominator, 20—the total number of observations. Hence, the numerators of the variances must also satisfy an equality.

$$\begin{pmatrix} \text{Numerator of} \\ \text{total variance} \end{pmatrix} = \begin{pmatrix} \text{Numerator of} \\ \text{predicted variance} \end{pmatrix} + \begin{pmatrix} \text{Numerator of} \\ \text{unpredicted variance} \end{pmatrix}$$

These numerators are used so frequently in the analysis of variance that they are given special symbols. Each numerator is a sum of squared deviations from a mean and is called, for short, a *sum of squares (SS)*. The numerator of the total variance is called SS_{Tot} (*total sum of squares*). The numerator of the predicted variance is called SS_{BG} (*sum of squares between groups*) and the numerator of the unpredicted variance is called SS_{WG} (*sum of squares within groups*). The three sums of squares are related by the following formula (which, as shown above, is derived from the prediction principle).

$$\boxed{SS_{Tot} = SS_{BG} + SS_{WG}}$$

(This formula is proven directly in Section 17.7.)

There are two formulas for the sample variance: the definitional formula and the computational formula (Section 4.4). Similarly, there are two formulas for the numerator of the variance, which is the sum of squares. These formulas are given below for each of the three sums of squares.

The Total Sum of Squares

The total sum of squares is the sum of squares of the deviation of each score, y, from the mean of all the scores, $\bar{y}$ (the *grand mean*).

$$\boxed{SS_{Tot} = \Sigma(y - \bar{y})^2 \quad \text{(Definitional formula)}}$$

The summation (Σ) includes all the scores in all the conditions or groups. This formula is equivalent to the following computational formula.

$$SS_{Tot} = \Sigma y^2 - (\Sigma y)^2/N$$

where N is the total number of scores. In order to reduce the number of summation signs in this formula, and in the formulas to be given below, we define T as the sum of all the scores: that is, the *grand total*.

$$\boxed{T = \Sigma y \quad \text{(Grand total)}}$$

The computational formula for the total sum of squares is, then,

$$\boxed{SS_{Tot} = \Sigma y^2 - T^2/N \quad \text{(Computational formula)}}$$

The Sum of Squares between Groups

The sum of squares between groups is the predicted sum of squares. It is the sum of squares of the deviation of each predicted score from the mean of all the predicted scores. It can be shown that when the predicted scores are the group means (as they are in the analysis of variance), the mean of the predicted scores is simply the grand mean, $\bar{y}$. The formula for SS_{BG} is, therefore,

$$SS_{BG} = \Sigma(\bar{y}_j - \bar{y})^2$$

The group means are denoted by $\bar{y}_j$: that is, the mean of the first group is $\bar{y}_1$, the mean of the second group is $\bar{y}_2$, etc. It is important to note that there is a predicted score, $\bar{y}_j$, for *each* of the subjects in the study, so that the summation (Σ) extends over all the subjects. All the subjects in the first group have the same predicted score, $\bar{y}_1$. If there are n_1 subjects in the first group, their contribution to the summation in SS_{BG} is $n_1(\bar{y}_1 - \bar{y})^2$. Similarly, the contribution of the n_2 subjects in the second group is $n_2(\bar{y}_2 - \bar{y})^2$. And so on. If there are three groups, SS_{BG} can be written

$$SS_{BG} = n_1(\bar{y}_1 - \bar{y})^2 + n_2(\bar{y}_2 - \bar{y})^2 + n_3(\bar{y}_3 - \bar{y})^2$$

In general, the sum of squares between groups can be written as

$$\boxed{SS_{BG} = \sum_j n_j (\bar{y}_j - \bar{y})^2 \quad \text{(Definitional formula)}}$$

where the summation ($\sum_j$) indicates that the sum is now just over the groups 1, 2, . . . , J. There are J groups in all.

The origin of the term "between groups" should now be clear. The predicted sum of squares is essentially a sum of squares of deviations of the group means from the grand mean. The sum of squares between groups measures how far apart the group means are from the grand mean, and hence from each other.

Each group mean, $\bar{y}_j$, is calculated from the ratio of the sum of the scores in the group to the size of the group. The symbol T_j will represent the sum of scores in the jth group: that is, the *group total* in the jth group.

$$\boxed{T_j = \text{sum of scores in the } j\text{th group}}$$

(We cannot give a formula for T_j using summation notation without greatly complicating the notation we are using.)

The sum of squares between groups can be computed from the group and grand totals by the following computational formula.

$$SS_{BG} = \sum_j (T_j^2/n_j) - T^2/N \quad \text{(Computational formula)}$$

The Sum of Squares within Groups

The sum of squares within groups is the unpredicted sum of squares. (The sum of squares within groups is also called the *error sum of squares*.) It is the sum of squares of the difference between each total score, y_{Tot} or y, and the corresponding predicted score, y_{Pred}. Since the predicted score is the mean of the group in which that score falls, the sum of squares within groups can be written

$$SS_{WG} = \Sigma(y - \bar{y}_j)^2 \quad \text{(Definitional formula)}$$

This sum consists of a sum in each group j of the squares of the deviations of the scores in that group from the group mean, $\bar{y}_j$. Therefore, this sum of squares is called a sum of squares within groups—it measures the variability within the groups rather than the variability between the groups.

The sum of squares within groups can be calculated from the sum of squares of the scores, Σy^2, and the group totals by the following formula.

$$SS_{WG} = \Sigma y^2 - \sum_j (T_j^2/n_j) \quad \text{(Computational formula)}$$

Example

It is helpful to carry out the calculations in the analysis of variance in a structured way such as shown in Table 17.3. In Table 17.3a the data from the learning study (Table 17.1) are presented in four columns, one column of scores (y) for each condition or group. The group and grand totals are computed in Table 17.3b. Only three different terms in the computational formulas must be calculated, and these are shown in Table 17.3c. Finally, in Table 17.3d the three sums of squares are calculated from these three terms.

The sum of SS_{BG} and SS_{WG} must equal SS_{Tot}. However, this relationship can be true even if the calculations in the analysis of variance are incorrect. The relationship will be true no matter what values of the terms are calculated in Table 17.3c. Therefore, the calculation of these three terms should always be checked.

The coefficient of determination can be calculated from the sums of squares. Recall that the coefficient was defined as the ratio of the predicted variance to the total variance. Since the denominator of each of these variances is the same, the coefficient of determination is simply the ratio of the numerators of these variances. These numerators are simply SS_{BG} and SS_{Tot}, respectively. We therefore have

$$R^2 = \frac{SS_{BG}}{SS_{Tot}} = \frac{64.2}{121.8} = 0.53$$

the same value as found in Section 17.2.

17.4 The F-distribution

The test of whether or not the regression function is horizontal is made by computing a test statistic (described in Section 17.5). This statistic has a

TABLE 17.3 Sums of Squares in the Learning Study

(a) Data

CONDITION

1	2	3	4
9	5	5	3
11	7	9	4
8	6	7	5
7	3	4	1
7	9	8	4

(b) Group and Grand Totals

$T_1 = 9 + 11 + 8 + 7 + 7 = 42$
$T_2 = 5 + 7 + 6 + 3 + 9 = 30$
$T_3 = 5 + 9 + 7 + 4 + 8 = 33$
$T_4 = 3 + 4 + 5 + 1 + 4 = 17$
$T = T_1 + T_2 + T_3 + T_4 = 42 + 30 + 33 + 17 = 122$

(c) Terms in the SS

$\Sigma y^2 = (9)^2 + (11)^2 + \ldots + (4)^2 = 866$
$\sum_j (T_j^2/n_j) = (42)^2/5 + (30)^2/5 + (33)^2/5 + (17)^2/5$

$\qquad = 352.8 + 180.0 + 217.8 + 57.8 = 808.4$
$T^2/N = (122)^2/20 = 744.2$

(d) Sums of Squares

$SS_{Tot} = \Sigma y^2 - T^2/N = 866 - 744.2 = 121.8$
$SS_{BG} = \sum_j (T_j^2/n_j) - T^2/N = 808.4 - 744.2 = 64.2$
$SS_{WG} = \Sigma y^2 - \sum_j (T_j^2/n_j) = 866 - 808.4 = 57.6$

distribution, when the null hypothesis is true, called the F-distribution. As with the χ^2-distribution and the t-distribution, there are many F-distributions, depending on the number of degrees of freedom. The F-distribution differs from the χ^2- and t-distributions, however, since *two* numbers, called df_1 and df_2, are required to specify a particular F-distribution.[4]

Three F-distributions are shown in Figure 17.2. The distributions are nonsymmetric and the values of F extend from zero to infinity. The rejection region for all tests in the analysis of variance consists of large values of F: that is, the rejection region consists of the upper tail of the F-distribution. In order to make these tests, upper-tail percentiles of the F-distribution have been tabulated. Table A15 in the Appendix gives the 95th and 99th percentiles of F-distributions with various numbers of degrees of freedom, df_1 and df_2. These percentiles are denoted $F_{.05}(df_1, df_2)$ and $F_{.01}(df_1, df_2)$, respectively—the subscripts 0.05 and 0.01 denoting the area in the upper tail (α).

Example

The 95th percentile of the F-distribution with 4 and 25 degrees of freedom can be found in Table A15. Its value is $F_{.05}(4, 25) = 2.76$. The 99th percentile is $F_{.01}(4, 25) = 4.18$. These values are consistent with the distribution shown in Figure 17.2, since they cut off

[4] The two numbers are sometimes called ν_1 and ν_2, instead. The Greek letter, pronounced "nu", corresponds to the Latin letter n.

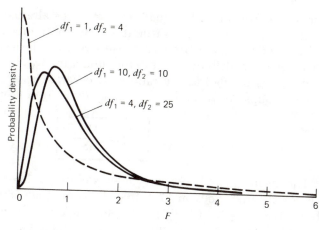

Figure 17.2 The F-distribution.

5% and 1%, respectively, of the upper-tail area of the F-distribution with $df_1 = 4$ and $df_2 = 25$.

You should also check the following values in Table A15 and confirm that they are consistent with Figure 17.2.

$$F_{.05}(1, 4) = 7.71 \qquad F_{.01}(1, 4) = 21.20$$
$$F_{.05}(10, 10) = 2.98 \qquad F_{.01}(10, 10) = 4.85$$

The distribution of $F(1, 4)$ has a very long tail and hence its upper-tail percentiles are beyond the bounds of the figure.

Note that the order in which the two degrees of freedom are written is important. The first is called df_1 and the second is called df_2. The F-distribution with $df_1 = 5$ and $df_2 = 20$ is a different distribution from the F-distribution with the degrees of freedom reversed ($df_1 = 20$ and $df_2 = 5$). For example, the 99th percentiles of these two distributions are

$$F_{.01}(5, 20) = 4.10 \qquad F_{.01}(20, 5) = 9.55$$

17.5 The F-test

This section shows how to test the hypothesis that in the population the regression function is horizontal. The regression function is the set of means of the dependent variable, one mean for each value of the independent variable. The independent variable is the group, or condition, to which the subject belongs. If the regression function (in the population) is horizontal, all the group, or condition, means of the dependent variable are equal. We denote the condition means, in the population, by $\mu_1, \mu_2, \ldots, \mu_J$, so that the population mean in the first condition is μ_1, the population mean in the second condition is μ_2, and so forth to the population mean in the last (Jth) condition, which is μ_J. The null hypothesis, then, is

$$H_0: \mu_1 = \mu_2 = \ldots = \mu_J$$

The alternative hypothesis is the hypothesis that the regression function is not horizontal: that is, that the population means are not all equal.

H_1: the population means are not all equal (H_0 is false)

(Note that it is incorrect to state the hypothesis as $H_1: \mu_1 \neq \mu_2 \neq \ldots \neq \mu_J$, since some of the means may be equal even when the alternative hypothesis is true. The alternative hypothesis is true if any two of the means are unequal.)

In order to decide between these two hypotheses, we must find a test statistic whose distribution is different under the two hypotheses. The statistic, called F_{obs}, is defined as the ratio of two variances, called the *mean square between groups*, MS_{BG}, and the *mean square within groups*, MS_{WG}.

$$F_{obs} = \frac{MS_{BG}}{MS_{WG}}$$

The mean square between groups is closely related to the predicted variance. Recall, from Section 17.3, that the sum of squares between groups was defined as $SS_{BG} = \sum_j n_j(\bar{y}_j - \bar{y})^2$. This sum involves the deviations of the group means from the grand mean. There are only J such deviations, one deviation for each group. These deviations must sum to zero, since they are deviations from their overall mean.[5] Hence, we need know only $(J - 1)$ deviations in order to determine the remaining deviation and, therefore, to calculate SS_{BG}. This means that the number of *degrees of freedom between groups* is just $(J - 1)$, where J is the number of groups. In symbols,

$$df_{BG} = J - 1$$

The mean square between groups is the ratio of SS_{BG} and df_{BG}.

$$MS_{BG} = \frac{SS_{BG}}{df_{BG}} = \frac{SS_{BG}}{J - 1}$$

(*Note.* See Section 17.3 for the computational formula for SS_{BG}.)

The mean square within groups is closely related to the unpredicted variance. Recall, from Section 17.3, that the sum of squares within groups was defined as $SS_{WG} = \sum(y - \bar{y}_j)^2$. This sum is a sum, in each group, of the deviations of the scores in that group from the group mean. The group sums are themselves summed to give SS_{WG}. In group j there are n_j deviations from the group mean. These deviations must sum to zero. Hence, only $(n_j - 1)$ of them need to be known in order for the group's contribution to SS_{WG} to be calculated. There are, therefore, $(n_j - 1)$ degrees of freedom associated with the sum of squares of the deviations in the jth group. The number of *degrees of freedom within groups* is the sum of $(n_j - 1)$ over all groups.

$$df_{WG} = \sum_j (n_j - 1)$$

that is,

$$df_{WG} = N - J$$

---

[5] When the sample sizes are unequal, it is the weighted sum of the deviations of the sample means from the grand mean that is zero.

$$\sum_j n_j (\bar{y}_j - \bar{y}) = 0$$

The weights are the sample sizes. There are still $(J - 1)$ degrees of freedom.

Recall that N is the total number of scores; therefore, $N = \Sigma n_j$. Also, J is the number of groups; therefore, $J = \sum_j 1$. The mean square within groups is the ratio of SS_{WG} and df_{WG}.

$$MS_{WG} = \frac{SS_{WG}}{df_{WG}} = \frac{SS_{WG}}{N - J}$$

(Note. See Section 17.3 for the computational formula for SS_{WG}.)

It can be shown that $F_{obs} = MS_{BG}/MS_{WG}$ has the F-distribution (Section 17.4) when the null hypothesis is true. Recall that the null hypothesis states that all the population means are equal. The degrees of freedom, df_1 and df_2, of this F-distribution are df_{BG} and df_{WG}, respectively. In brief, when H_0 is true, F_{obs} has the F-distribution with $(J - 1)$ and $(N - J)$ degrees of freedom. Since they are associated with the variances in the numerator and denominator of F_{obs}, these two degrees of freedom $(J - 1$ and $N - J)$ are often called the *numerator degrees of freedom* and the *denominator degrees of freedom*, respectively.

The rejection region for the F-test consists of the upper tail of the F-distribution. The reason is as follows. The F-test is a test of equality of the means of J populations. These means are either all equal (H_0) or not all equal (H_1). Let us call all the J populations together an *aggregate*. Consider two possible aggregates: in one possible aggregate, H_0 is true (all the means are equal); in the other possible aggregate, H_1 is true (not all the means are equal). Let us suppose that the variabilities of the J populations are the same in each of the two aggregates and that samples of the same size are drawn from each aggregate. Let us then compare the probable values of the test statistic F_{obs} calculated from the sample data in each of the two aggregates. The mean square within groups, MS_{WG}, should be approximately the same in each aggregate, since MS_{WG} depends solely on the variability *within* groups and reflects the variability of the populations, which are assumed to be the same. On the other hand, the mean square between groups, MS_{BG}, depends on the differences *between* the groups and reflects the differences in the J population means. The mean square between groups should be greater in the aggregate in which H_1 is true than in the aggregate in which H_0 is true. Hence, F_{obs}, which is the ratio of MS_{BG} to MS_{WG}, should be greater in the aggregate in which H_1 is true, since the numerator is greater in that aggregate and the two denominators are about the same. Hence, large values of F_{obs} tend to occur when H_1 is true. Accordingly, the rejection region should consist of large values of F_{obs}. For this reason, the rejection region consists of values in the upper tail of the F-distribution.

Example 1

Let us now return to the data of the learning study (Table 17.3). The sums of squares, as shown in Table 17.3, are

$$SS_{BG} = 64.2 \text{ and } SS_{WG} = 57.6$$

The number of degrees of freedom between groups is one less than the number of groups.

$$df_{BG} = J - 1 = 4 - 1 = 3$$

The number of degrees of freedom within groups is the total number of scores minus one for each group (i.e., minus J, the number of groups).

$$df_{WG} = N - J = 20 - 4 = 16$$

The mean squares can now be calculated. The mean square between groups is

$$MS_{BG} = \frac{SS_{BG}}{df_{BG}} = \frac{64.2}{3} = 21.40$$

The mean square within groups is

$$MS_{WG} = \frac{SS_{WG}}{df_{WG}} = \frac{57.6}{16} = 3.60$$

The ratio of these two mean squares is

$$F_{obs} = \frac{MS_{BG}}{MS_{WG}} = \frac{21.40}{3.60} = 5.94$$

If the null hypothesis of equal population means were true, the distribution of F_{obs} would be the F-distribution with 3 and 16 degrees of freedom. The 95th percentile of this F-distribution (Table A15) is 3.24 (and the 99th percentile is 5.29). This means that when H_0 is true, 95% of all possible experiments with five observations in each of four conditions will have values of F_{obs} less than 3.24. Since our experiment has a value of $F_{obs} = 5.94$ (greater than the 95th percentile), we reasonably can conclude that the null hypothesis is false.

More formally, if we set $\alpha = 0.05$ so that the probability of a Type I error is 0.05, we reject the null hypothesis on the basis of the observed value of the test statistic. If we prefer to use the p-value procedure, we see that $F_{obs} = 5.94$ exceeds the 99th percentile of the F-distribution with 3 and 16 degrees of freedom. Hence, we reject the null hypothesis and conclude that the population means are not all the same (p-value < 0.01).

Summary of Steps in the Analysis of Variance

The eight steps listed below are modeled on the steps for the binomial test and the z- and t-tests, given in Chapters 9, 10, and 11.

1 State the two hypotheses: H_0 and H_1. In the analysis of variance that has been described in this section (the one-factor analysis of variance), the hypotheses are

$$H_0: \mu_1 = \mu_2 = \ldots = \mu_J$$

and

$$H_1: H_0 \text{ is false}$$

where $\mu_1, \mu_2, \ldots, \mu_J$ are the population means in the J conditions or groups. (Example: $H_0: \mu_1 = \mu_2 = \mu_3 = \mu_4$; $H_1: H_0$ is false.)

2 Choose a value for α. (Example: $\alpha = 0.05$.)

3 Choose the sizes of the samples in each condition. (The sample sizes do not have to be equal, although they often are.) Calculate the two degrees of freedom, df_{BG} and df_{WG}. (Example: $n_1 = 5, n_2 = 5, n_3 = 5, n_4 = 5$; $df_{BG} = J - 1 = 4 - 1 = 3$; $df_{WG} = N - J = 20 - 4 = 16$.)

4 Find the critical value of the test statistic, F_{obs}. The analysis of variance is always a one-tailed test, and the rejection region consists of values in the upper tail of the F-distribution. Hence, the critical value is an upper-tail percentile. [Example: critical value $= F_{.05}(3, 16) = 3.24$.]

5 Locate the acceptance and rejection regions. [Example: The acceptance region is ($F_{obs} < 3.24$) and the rejection region is ($F_{obs} > 3.24$).]

6 There is no Step 6! Recall that, whenever possible, Step 6 is the calculation of the probability of a Type II error, β, or, equivalently, the calculation of the power of the test. (The calculation of β and power in the analysis of variance lie beyond the scope of this text.[6])

7 Carry out the study and calculate the value of the test statistic. (*Example:* $SS_{BG} = 64.2$, $SS_{WG} = 57.6$, $MS_{BG} = 64.2/3 = 21.40$, $MS_{WG} = 57.6/16 = 3.60$, $F_{obs} = 21.40/3.60 = 5.94$.)

8 Make the decision, based on whether the test statistic falls into the acceptance or the rejection region. There are various ways to state the conclusion—see Section 10.2. (*Example:* H_0 can be rejected; the sample means are significantly different.)

The Analysis of Variance Summary Table

It is convenient to summarize the results of an analysis of variance in a standard form, called a summary table. The table for Example 1 is shown in Table 17.4. All the terms in the first two lines of the table should be self-explanatory. The third line shows SS_{Tot} and df_{Tot}. The definitional and calculational formulas for SS_{Tot} were given in Section 17.3. It is easy to see from the summary table that $SS_{Tot} = SS_{BG} + SS_{WG}$. The *total degrees of freedom* is

$$\boxed{df_{Tot} = N - 1}$$

where N is the total number of scores. The total degrees of freedom equals the sum of the other two degrees of freedom.

$$\begin{aligned} df_{Tot} &= N - 1 \\ &= (J - 1) + (N - J) \\ &= df_{BG} + df_{WG} \end{aligned}$$

(It is not customary to compute MS_{Tot}—the ratio of SS_{Tot} to df_{Tot}—because this mean square has no significance in the analysis of variance.)

Example 2

As a second example, consider an experiment with three conditions and an unequal number of subjects in each condition. The data and the calculations through the sums of squares are shown in Table 17.5. The three sample means also should be calculated soon

[6] The techniques of Chapter 13 can be extended to cover the analysis of variance. See, for example, J. Cohen, *Statistical Power Analysis for the Behavioral Sciences* (New York: Academic Press, 1969).

TABLE 17.4 Summary Table for Example 1

SOURCE OF VARIABILITY	SS	df	MS	F
Between groups (BG)	64.2	3	21.40	5.94[a]
Within groups (WG)	57.6	16	3.60	
Total	121.8	19		

[a] p-value < 0.01.

TABLE 17.5 Sums of Squares in Example 2

(a) Data

	CONDITION	
1	2	3
5	8	8
7	12	6
6	8	9
3	6	5
9	7	7
7		4
4		4
2		
$(n_1 = 8)$	$(n_2 = 5)$	$(n_3 = 7)$

(b) Group and Grand Totals

$$T_1 = 5 + 7 + \ldots + 2 = 43$$
$$T_2 = 8 + 12 + \ldots + 7 = 41$$
$$T_3 = 8 + 6 + \ldots + 4 = 43$$
$$T = T_1 + T_2 + T_3 = 43 + 41 + 43 = 127$$

(c) Terms in the SS

$$\Sigma y^2 = (5)^2 + (7)^2 + \ldots + (4)^2 = 913$$
$$\sum_j (T_j^2/n_j) = (43)^2/8 + (41)^2/5 + (43)^2/7$$
$$= 231.125 + 336.200 + 264.143 = 831.468$$
$$T^2/N = (127)^2/20 = 806.450$$

(d) Sums of Squares

$$SS_{Tot} = \Sigma y^2 - T^2/N = 913 - 806.450 = 106.550$$
$$SS_{BG} = \sum_j (T_j^2/n_j) - T^2/N = 831.468 - 806.450 = 25.018$$
$$SS_{WG} = \Sigma y^2 - \sum_j (T_j^2/n_j) = 913 - 831.468 = 81.532$$

after the data have been collected, since it is the equality of the three corresponding population means which is being tested by the F-test. These sample means are

$$\bar{y}_1 = T_1/n_1 = 43/8 = 5.38$$
$$\bar{y}_2 = T_2/n_2 = 41/5 = 8.20$$
$$\bar{y}_3 = T_3/n_3 = 43/7 = 6.14$$

The sample means obviously are different. However, the question to be answered is whether they are *significantly* different: that is, whether or not it is reasonable to believe that they came from populations whose (population) means are different.

The number of groups is $J = 3$ and the total sample size is $N = 20$. From these numbers, the two degrees of freedom are calculated.

$$df_{BG} = J - 1 = 3 - 1 = 2$$
$$df_{WG} = N - J = 20 - 3 = 17$$

As a check, we should calculate $df_{Tot} = N - 1 = 20 - 1 = 19$, which equals the sum of $df_{BG} = 2$ and $df_{WG} = 17$. For $\alpha = 0.05$, the critical value of the test statistic is $F_{.05}(2, 17)$, which is found, in Table A15, to be 3.59. The null hypothesis will be rejected if F_{obs} exceeds 3.59 and will be accepted otherwise.

The test statistic, F_{obs}, can now be calculated.

$$MS_{BG} = \frac{SS_{BG}}{df_{BG}} = \frac{25.018}{2} = 12.509$$

$$MS_{WG} = \frac{SS_{WG}}{df_{WG}} = \frac{81.532}{17} = 4.796$$

$$F_{obs} = \frac{MS_{BG}}{MS_{WG}} = \frac{12.509}{4.796} = 2.61$$

We see that $F_{obs} = 2.61$ does not exceed the critical value of 3.59. Therefore, we cannot reject the null hypothesis that the population means are equal. Another way of stating this conclusion is to say that the three sample means (5.38, 8.20, 6.14) are not significantly different. (The summary table is shown in Table 17.6. The value of F_{obs} is marked "*ns*" to indicate that it is not significant.)

17.6 Measures of Relationship in the Analysis of Variance

The coefficient of determination, as defined in Sections 17.2 and 17.3, is the ratio of the predicted variance to the total variance—which, using the same denominators for the two variances, is equivalent to the ratio of the sum of squares between groups (SS_{BG}) to the total sum of squares (SS_{Tot}).

$$R^2 = \frac{SS_{BG}}{SS_{Tot}}$$

The coefficient of determination[7] is defined and computed in the sample. It is possible to define, in the population, a related measure of the relationship between the independent and dependent variables. This related measure is called ω^2 ("omega-squared").[8] An estimate of ω^2, made from the data in a sample, is

$$\hat{\omega}^2 = \frac{(J-1)(F_{obs}-1)}{(J-1)(F_{obs}-1)+N}$$

[7] The ratio SS_{BG}/SS_{Tot} is given various names in the statistical and psychological literature. The two most common names are *correlation ratio* and η^2 ("eta-squared"). The former name is misleading, since the ratio is not a ratio of correlations. In fact, as mentioned in footnote 3, the ratio is related to the squared multiple correlation. The other name, η^2, is widely misunderstood to stand only for a measure of *nonlinear* relationship; in fact, SS_{BG}/SS_{Tot} measures the degree of relationship, whether linear or nonlinear. It is also unfortunate that a sample quantity is represented by a Greek letter, since Greek letters are usually reserved for population parameters.

[8] For a definition of ω^2, see W. L. Hays, *Statistics for Psychologists* (New York: Holt, Rinehart, & Winston, 1963).

TABLE 17.6 Summary Table for Example 2

SOURCE OF VARIABILITY	SS	df	MS	F
Between groups (BG)	25.018	2	12.509	2.61 (ns)
Within groups (WG)	81.532	17	4.796	
Total	106.550	19		

This formula requires J, the number of groups; N, the total number of subjects; and F_{obs}, the value of the test statistic.

The value of R^2 tends to overestimate the true degree of relationship in the population. The value of $\hat{\omega}^2$ usually is smaller than R^2 and is a better (but not, in fact, unbiased) estimate of the degree of relationship.

Example 1

The coefficient of determination for the learning study of Tables 17.1 and 17.3 previously was worked out to be

$$R^2 = \frac{SS_{BG}}{SS_{Tot}} = \frac{64.2}{121.8} = 0.53$$

This result can be interpreted to mean that in the sample, 53% of the variance of the dependent variable can be predicted from the independent variable.

The estimate of the relationship in the population is

$$\hat{\omega}^2 = \frac{(J-1)(F_{obs}-1)}{(J-1)(F_{obs}-1)+N}$$

$$= \frac{(4-1) \times (5.94-1)}{(4-1) \times (5.94-1)+20}$$

$$= \frac{3 \times 4.94}{3 \times 4.94 + 20} = \frac{14.82}{14.82 + 20} = \frac{14.82}{34.82}$$

$$= 0.43$$

(The value of F_{obs} was calculated in Example 1 of Section 17.5.) The value of $\hat{\omega}^2$ is somewhat less than the value of R^2. The estimated proportion of the total variance of the dependent variable that can be predicted, in the population, from the independent variable, is 43%.

Example 2

This example continues Example 2 of Section 17.5. The data are in Table 17.5. The coefficient of determination is

$$R^2 = \frac{SS_{BG}}{SS_{Tot}} = \frac{25.018}{106.550} = 0.23$$

The estimate of $\hat{\omega}^2$ is

$$\hat{\omega}^2 = \frac{(J-1)(F_{obs}-1)}{(J-1)(F_{obs}-1)+N}$$

$$= \frac{(3-1) \times (2.61-1)}{(3-1) \times (2.61-1)+20}$$

$$= \frac{2 \times 1.61}{2 \times 1.61 + 20} = \frac{3.22}{3.22 + 20} = \frac{3.22}{23.22} = 0.14$$

Here the proportion of the total variance that can be predicted from the independent variable (the conditions) is much less than in Example 1. Also note that we found in Section 17.5 that the difference between the means was not statistically significant. (However, small values of $\hat{\omega}^2$ can be statistically significant if the sample size is large enough—see Section C.4.)

17.7 Proof of the Principle of Prediction

Note: This section may be omitted with no loss of continuity.

The principle of prediction (Section 16.3) states that the total variance of a dependent variable equals the sum of the predicted and unpredicted variances when the dependent variable is predicted from an independent variable. From this principle, we can derive the relationship between the three sums of squares in a one-factor analysis of variance (Section 17.3).

$$SS_{Tot} = SS_{BG} + SS_{WG}$$

This equation is proven directly in this section—showing once again that the principle of prediction holds. (See Section 16.6 for a proof that the principle holds in a regression situation.)

The definitional formulas for the three sums of squares were given in Section 17.3 as

$$SS_{Tot} = \Sigma(y - \bar{y})^2$$
$$SS_{BG} = \sum_j n_j (\bar{y}_j - \bar{y})^2$$
$$SS_{WG} = \Sigma(y - \bar{y}_j)^2$$

In the formulas for SS_{Tot} and SS_{WG} the individual scores are represented by y. It is convenient to change the notation to y_{ij}, where j represents the group number and i represents the subject number within the group. In the jth group the subject numbers i extend from 1 to n_j, the number of subjects in the group. The summations in the formulas for SS_{Tot} and SS_{WG} are then changed to summations over both i and j, as follows.

$$SS_{Tot} = \sum_{j=1}^{J} \sum_{i=1}^{n_j} (y_{ij} - \bar{y})^2$$

$$SS_{WG} = \sum_{j=1}^{J} \sum_{i=1}^{n_j} (y_{ij} - \bar{y}_j)^2$$

(In the proof below, the initial and final values of i and j are dropped from the summations.)

We now expand the formula for SS_{Tot} and show that it equals the sum of SS_{BG} and SS_{WG}:

1 $SS_{Tot} = \sum_j \sum_i (y_{ij} - \bar{y})^2$

2 $\qquad = \sum_j \sum_i (y_{ij} - \bar{y}_j + \bar{y}_j - \bar{y})^2$

3 $\qquad = \sum_j \sum_i [(y_{ij} - \bar{y}_j) + (\bar{y}_j - \bar{y})]^2$

4 $\qquad = \sum_j \sum_i (y_{ij} - \bar{y}_j)^2 + \sum_j \sum_i (\bar{y}_j - \bar{y})^2 + 2\sum_j \sum_i (y_{ij} - \bar{y}_j)(\bar{y}_j - \bar{y})$

5 $\qquad = SS_{WG} + \sum_j n_j(\bar{y}_j - \bar{y})^2 + 2\sum_j (\bar{y}_j - \bar{y})\sum_i (y_{ij} - \bar{y}_j)$

6 $\qquad = SS_{WG} + SS_{BG} + 2\sum_j (\bar{y}_j - \bar{y}) \times 0$

7 $\qquad = SS_{WG} + SS_{BG} + 0$

8 $\qquad = SS_{WG} + SS_{BG}$

Explanation In line 2, $\bar{y}_j$ is added and subtracted within each squared term, which makes no change to the terms. In line 3, each term in the double sum is written as the square of the sum of two terms, each enclosed in parentheses. In line 4, this square is expanded into two squares and a cross-product term. The first sum of squares is exactly the definitional formula for SS_{WG} (line 5). The squares in the second sum do not depend on the index i; the summation over i, for fixed j, is, therefore, just n_j times the constant square, as shown in the line 5.

The sum of products in the third term in line 4 can be rearranged, since $(\bar{y}_j - \bar{y})$ does not depend on the index i; the rearrangement is shown in line 5. In the transition from line 5 to line 6, the second term is exactly the definitional formula for SS_{BG}. The sum over i in the last term of line 5 is zero, since it is a sum of deviations of all the scores in group j from the group mean $\bar{y}_j$. Hence, the cross-product term vanishes, leaving $SS_{Tot} = SS_{WG} + SS_{BG}$.

17.8 Two Factors—the Concept of Interaction

This section extends the analysis of variance to two factors (independent variables) by introducing the concept of interaction. This section is only introductory, however: no tests are described nor is the distinction between sample and population relationships stressed. The goal here is simply to explain how the relationship between a dependent variable and two independent variables is described. The explanation is based on the regression function, just as the relationship in a one-factor analysis of variance is based on the regression function. (Statistical tests are described in Section 17.9.)

The two-factor analysis of variance will be illustrated by the same learning study first introduced in Section 17.2. That study featured four conditions, described in abbreviated form as follows.

1 warn/short delay.

2 warn/long delay.

3 no warn/short delay.

4 no warn/long delay.

These four conditions were treated as four values or *levels* of one *factor* in all the analyses described so far in this chapter. The basic analysis showed that the mean numbers of syllables correctly recalled in the four conditions were significantly different ($F_{obs} = 5.94$, p-value < 0.01; see Example 1 of Section 17.5).

We can consider the four conditions in a different way: there are two factors (or independent variables) that define the four conditions. These factors are

Factor A: *Delay*, with levels "short" and "long".

Factor B: *Warning*, with levels "warn" and "no warn".

The design of the study can then be described as a 2×2 design, indicating that there are two levels of Factor A and each of these levels is combined with each of the two levels of Factor B, making $2 \times 2 = 4$ levels in all.

Five subjects were observed in each of the four conditions. The mean number of syllables correctly recalled in each of the four conditions was computed in Table 17.1. The four conditions and their means can be reported in a 2×2 table.

		FACTOR A: DELAY		
		SHORT	LONG	
FACTOR B:	WARN	8.4	6.0	7.2
WARNING	NO WARN	6.6	3.4	5.0
		7.5	4.7	6.1

The number in each cell is the mean for one condition in the experiment. For example, 6.6 is the mean number of syllables recalled in the condition in which the delay was short and no warning was given. The numbers of the *borders* of the table are the means of the two means in the corresponding row or column. For example, 5.0 is the mean of 6.6 and 3.4, so that the mean of all subjects in "no warn" conditions, whatever the delay, is 5.0. The two row means are the means for the two levels of Factor B; the two column means are the means for the two levels of Factor A. (The number in the bottom-right corner is the overall mean.)

The fact that the two column means differ (7.5 versus 4.7) indicates that Factor A (Delay) has an effect on syllables recalled. The difference between these two means is referred to as the *main effect of Factor A* or the *main effect of Delay*. (Refer to Section 8.5 for a discussion of this use of the word "effect".) The main effect of Factor A is illustrated in Figure 17.3a by the regression function of y (syllables recalled) on Factor A (Delay). The two means in this graph are the two column means in the preceding table.

Similarly, the row means (7.2 and 5.0) are different, which shows that Factor B (Warning) has an effect on syllables recalled. This difference is called the *main effect of Factor B* or the *main effect of Warning*. This effect is illustrated in Figure 17.3b by the regression function of y (syllables recalled) on Factor B (Warning). The two means in this graph are the two row means in the preceding table.

Can the effect of the two factors on the dependent variable be accounted for completely by these two main effects? The answer in this example is no—there is an effect beyond the main effects. Look at Figure 17.3c. The four cell means are plotted in this graph as two regression functions: the regression of y on Factor A for the "warn" level of Factor B (the upper curve) and the regression of y on Factor A for the "no warn" level of Factor B (the lower curve). Consider the two means for the "short" level of Factor A. The means are 8.4 and 6.6, showing that the difference between "warn" and "no warn" is 1.8 when the delay is short. On the other hand, the two means for the "long" level of Factor A are 6.0 and 3.4, showing that the difference between "warn" and "no warn" is 2.6 when the delay is long. Hence, the effect of Factor B is different, depending on the level of Factor A. The (overall) main effect of Factor B in Figure 17.3b is not the same as the separate effects of Factor B for each level of Factor A. We describe this nonconstancy of the effect of one factor at various levels of the other factor as an *interaction effect*, or simply as an *interaction*.

The two regressions in Figure 17.3c are not parallel; this fact goes hand in hand with the fact that the effect of Factor B is not the same for the different levels of Factor A. There are, therefore, several ways to describe or define an interaction. The definition of interaction can be summarized as a principle.

P14: Definition of Interaction

The interaction effect of two factors (independent variables) on a third (dependent) variable is the joint effect of the two factors that cannot be accounted for by the main effects of the two factors separately. There is an interaction if the effect of one factor is not the same for all levels of the other factor. Interaction may be defined in terms of the regression functions of the dependent variable, y, on one factor—say, A—for different levels of the

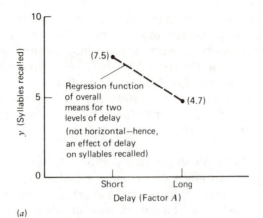

(a)

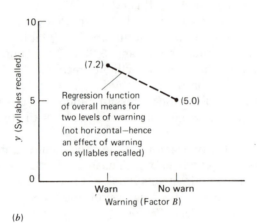

(b)

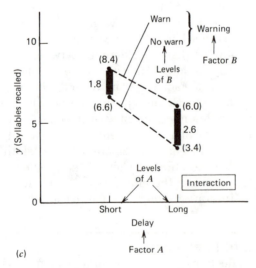

(c)

Figure 17.3 The learning study as a two-factor study. (a) Main effect of Delay (Factor A). (b) Main effect of Warning (Factor B). (c) Interaction of the two factors.

other factor—say, *B*. *There is no interaction* if all the regression functions of *y* on *A* are parallel: that is, they are the same except for the addition or subtraction of a constant. *There is an interaction* if one pair (or more) of the regression functions of *y* on *A* are not parallel: that is, if the differences between the pair of regression functions are not constant.

Caution: As noted at the beginning of this section, the above definition does not distinguish interaction effects in *samples* from interaction effects in *populations*. Strictly speaking, we want to *infer* the presence or absence of interaction effects in the population from data in a sample. For the present, however, we are ignoring the distinction between population and sample; we look at any set of means, whether sample means or population means, and determine whether or not there is an interaction by using the definition of interaction given in the above principle. Techniques of inference in the two-factor analysis of variance are described in Section 17.9.

Data with No Interaction

Consider now some different results from the learning study. Suppose the means for the four conditions were as follows.

		FACTOR *A*: DELAY		
		SHORT	LONG	
FACTOR *B*:	WARN	8.7	7.3	8.0
WARNING	NO WARN	5.7	4.3	5.0
		7.2	5.8	6.5

Here, as in the previous data, there is a main effect of Delay (7.2 versus 5.8, a difference of 1.4) and also a main effect of Warning (8.0 versus 5.0, a difference of 3.0). These data differ, however, from the previous set, in that there is no interaction between the two factors. Look at the effects of Warning at each of the two levels of Delay: the difference between "Warn" and "No Warn" is 8.7 − 5.7 = 3.0 when the delay is short and the difference is 7.3 − 4.3 = 3.0 when the delay is long. These differences are the same (and equal to the overall main effect of Warning), and hence there is no interaction effect between the two factors in their effect on the dependent variable in this study.

The absence of any interaction can also be seen from a graph of the two regression functions in Figure 17.4. The effects of Warning at the two levels of Delay are shown as the distances between the two regression functions at the two levels. Since these differences are equal, there is no interaction. Note also that the two regressions are parallel, which is another way to determine that there is no interaction, according to the definition of interaction given previously.

Interchanging the Two Factors

The two examples discussed so far (one of which showed interaction and the other of which showed no interaction) were both analyzed by considering the regression functions of syllables recalled on Factor *A* (Delay) for the two levels of Factor *B* (Warning). The analysis may also be done by interchanging the roles of the two factors. In Figure 17.5 the regression function of syllables

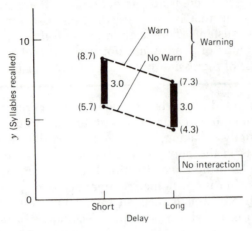

Figure 17.4 A 2 × 2 design with no interaction. *Note:* Differences (indicated by vertical bars) are equal; hence, there is no interaction.

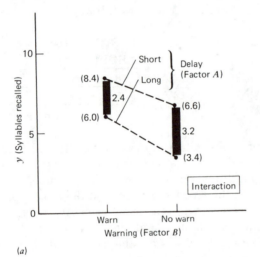

(a)

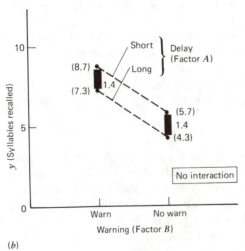

(b)

Figure 17.5 Interchanging the two factors. (*a*) Replot of Figure 17.3*c*. (*b*) Replot of Figure 17.4.

recalled on Factor B (Warning) is plotted for the two levels of Factor A (Delay). In Figure 17.5a the two regressions are not parallel (the effect of Delay is different for the two levels of Warning), and hence there is interaction. In Figure 17.5b, on the other hand, the two regressions are parallel (the effect of Delay is the same for the two levels of Warning), and hence there is no interaction.

Note that it does not matter which factor plays which role in the plot of the two regressions. It can be shown algebraically and geometrically that if there is interaction, the regressions will not be parallel whether the regressions are on Factor A or on Factor B. It is therefore unnecessary to plot the regression in both ways; the choice between the two ways can be based on other considerations such as the number of levels of each of the factors, as shown in Example 1 of the next subsection.

Factors with More Than Two Levels

Example 1

Let us suppose that the learning study is redesigned so that there are three levels of Factor A (Delay): a one-hour delay, a one-day delay, and a two-day delay. The two levels of Factor B (Warning) are unchanged. Suppose that the means in the six conditions of this 3×2 design are as follows.

		FACTOR A: DELAY			
		ONE HOUR	ONE DAY	TWO DAYS	
FACTOR B:	WARN	8.4	6.8	7.2	7.47
WARNING	NO WARN	7.2	3.6	6.6	5.80
		7.80	5.20	6.90	6.63

The column means are different, indicating that there is a main effect of Factor A; similarly, the row means are different, indicating that there is a main effect of Factor B. In order to see whether there is an interaction between the two factors, consider the regression functions in Figure 17.6a. Note that the regressions on Factor A (for the two levels of Factor B) are drawn. The choice of these regressions means that there are just two regressions in the graph. If the regressions on Factor B (for the three levels of Factor A) had been drawn, there would have been three regressions in the graph, making the graph more difficult to interpret.

It is clear from the figure that the two regressions are not parallel, and hence that there is an interaction between the factors. The effect of Warning is different at the different delays: the effect is 1.2 at a one-hour delay, 3.2 at a one-day delay, and 0.6 at a two-day delay.

Consider now another set of means for the 3×2 design.

		FACTOR A: DELAY			
		ONE HOUR	ONE DAY	TWO DAYS	
FACTOR B:	WARN	8.4	6.8	7.2	7.47
WARNING	NO WARN	6.6	5.0	5.4	5.67
		7.50	5.90	6.30	6.57

These means are graphed in Figure 17.6b. Here the regressions are parallel (the differences between the two regressions are the same for all three levels of Delay). Hence, there

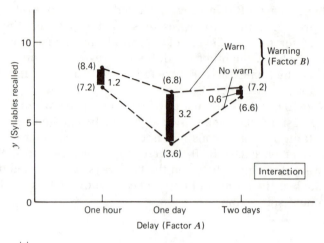

(a)

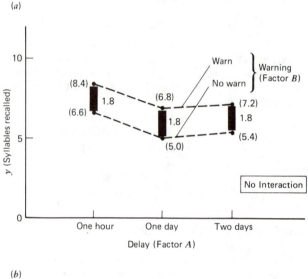

(b)

Figure 17.6 A 3 × 2 design: two possible results. (a) Interaction between Factors A and B. (b) No interaction between Factors A and B.

is no interaction between Delay and Warning in these data. However, there are still main effects of each factor, as shown by the fact that the row means differ and the column means differ.

Example 2

The learning study could be changed into a 3 × 3 design by adding another level to Factor B (Warning). Suppose that we have two kinds of "warn" condition: in one condition the subject is warned that he or she will be tested subsequently and is told when that testing will take place (the "warn/specified" condition); in another condition the subject is warned about the subsequent testing but it is not told when this testing will take place (the "warn/unspecified" condition). The "no warn" condition is retained as before.

Sample data (cell means) are given in Table 17.7. In Table 17.7a the data exhibit interaction; in Table 17.7b there is no interaction. The presence or absence of interaction can be determined in two ways. First, compute the difference between the numbers in each pair of rows in the 3 × 3 table. These differences are shown in Tables 17.7a and b. In Table 17.7a the three differences between rows 1 and 2 are different, showing that

TABLE 17.7 Data for a 3 × 3 Design

(a) Interaction between Factors A and B

		FACTOR A: DELAY			
		ONE HOUR	ONE DAY	TWO DAYS	
FACTOR B: WARNING	WARN/SPECIFIED	8.6	8.2	7.0	7.93
	WARN/UNSPECIFIED	8.4	6.8	7.2	7.47
	NO WARN	7.2	3.6	6.6	5.80
		8.07	6.20	6.93	7.07

DIFFERENCE BETWEEN ROWS 1 AND 2

	ONE HOUR	ONE DAY	TWO DAYS
	0.2	1.4	−0.2

DIFFERENCE BETWEEN ROWS 2 AND 3

	ONE HOUR	ONE DAY	TWO DAYS
	1.2	3.2	0.6

DIFFERENCE BETWEEN ROWS 1 AND 3

	ONE HOUR	ONE DAY	TWO DAYS
	1.4	4.6	0.4

(b) No Interaction Between Factors A and B

		FACTOR A: DELAY			
		ONE HOUR	ONE DAY	TWO DAYS	
FACTOR B: WARNING	WARN/SPECIFIED	8.9	7.3	7.7	7.97
	WARN/UNSPECIFIED	8.4	6.8	7.2	7.47
	NO WARN	6.6	5.0	5.4	5.67
		7.97	6.37	6.77	7.03

DIFFERENCE BETWEEN ROWS 1 AND 2

	ONE HOUR	ONE DAY	TWO DAYS
	0.5	0.5	0.5

DIFFERENCE BETWEEN ROWS 2 AND 3

	ONE HOUR	ONE DAY	TWO DAYS
	1.8	1.8	1.8

DIFFERENCE BETWEEN ROWS 1 AND 3

	ONE HOUR	ONE DAY	TWO DAYS
	2.3	2.3	2.3

there is interaction. It is not necessary to look at any more differences (since one pair of differences that differ indicates interaction), but the differences between rows 2 and 3 and between rows 1 and 3 are shown for illustration. In Table 17.7b, on the other hand, the differences between rows 1 and 2 are all the same (all 0.5). We must also check the differences between the other pairs of rows. The differences between rows 2 and 3 are all 1.8, and the differences between rows 1 and 3 are all 2.3. Hence, there is no interaction between the two factors.

The second way to determine whether there is interaction between the two factors is to look at a graph of the regressions of the dependent variable on one of the factors, for each level of the other factor. The regressions of syllables recalled on Factor A, for the three levels of Factor B, are shown in Figure 17.7. For the first set of data (Figure 17.7a), the three regressions are not parallel, indicating that there is interaction between Factors A and B. For the second set of data (Figure 17.7b), the three regressions are parallel, indicating that there is no interaction between the factors.

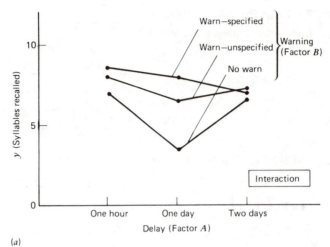

(a)

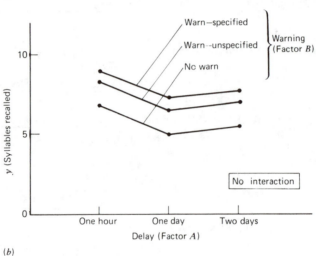

(b)

Figure 17.7 A 3 × 3 design: Two possible results. (*a*) Interaction between Factors *A* and *B*. (*b*) No interaction between Factors *A* and *B*.

17.9 Tests in the Two-factor Analysis of Variance

In a two-factor analysis of variance, three different effects on the dependent variable can be studied: the *main effects* of each of the two factors and the *interaction effect* of these factors. The interaction effect is the joint effect of the two factors on the dependent variable that cannot be accounted for by the main effects separately.

In Section 17.8, no distinction was made between effects found in a sample of observations and the effects in the population. In this section, it is shown how to *infer* a main effect or an interaction effect in the population from the data in a sample. The tests for significant main effects or for a significant interaction are carried out in a manner that resembles closely the test of a significant difference between the means in a one-factor analysis of variance (Section 17.5). For each effect, we compute the sum of squares, the number of degrees of freedom, and the mean square. We also obtain MS_{WG}, the mean square within groups. Then the test statistic for each effect is the ratio of the mean square for that effect to MS_{WG}. Each test statistic is statistically significant if it exceeds the upper-tail critical value of the F-distribution with appropriate numbers of degrees of freedom.

The details of these calculations are shown below. Two examples follow the details. *Note.* All the formulas assume that there are equal numbers of observations (n) in each of the conditions of the analysis of variance study.[9]

Means and Totals

The two factors will be called Factor A (the *column factor*) and Factor B (the *row factor*). Each factor has a certain number of *levels*: there are J levels of Factor A and K levels of Factor B. There are $J \times K$ conditions in the experiment, as every level of Factor A is combined with every level of Factor B. In each condition (or *cell*), n observations are made. These observations are denoted by y (without subscripts).

A 4×3 design is shown in Figure 17.8. The mean of the cell in the kth row and jth column is denoted by $\bar{y}_{kj}$. If we denote the cell total (sum of the n scores in the cell) by T_{kj}, the cell mean is

$$\bar{y}_{kj} = T_{kj}/n \quad \text{(Cell mean)}$$

The column means ($\bar{y}_{.j}$) are computed from the column totals ($T_{.j}$).

$$\bar{y}_{.j} = T_{.j}/(nK) \quad \text{(Column mean)}$$

[9] When the sample sizes are unequal, consult, for example, B. J. Winer, *Statistical Principles in Experimental Design*, 2nd ed. (New York: McGraw-Hill, 1971).

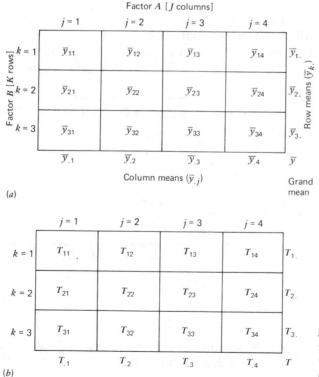

Figure 17.8 Symbols in the two-factor analysis of variance. (a) Means. (b) Totals.

There are nK observations in each column, so that the column total is divided by nK to give the column mean. Note the use of "dot notation". The dot (".") replaces a subscript, in this case the first, over which the sum ($T_{.j}$) or mean ($\bar{y}_{.j}$) has been taken. Similarly, the row means ($\bar{y}_{k.}$) are computed from the row totals ($T_{k.}$).

$$\bar{y}_{k.} = T_{k.}/(nJ) \quad \text{(Row mean)}$$

(There are nJ observations in each row.) Finally, the grand mean ($\bar{y}$) is computed from the grand total (T).

$$\bar{y} = T/(nJK) = T/N \quad \text{(Grand mean)}$$

where $N = nJK$ is the total number of observations in the study.

Sums of Squares

There are five sums of squares in a two-factor analysis of variance. The *sum of squares within groups* measures the variability within the cells. Its formula is

$$SS_{WG} = \Sigma(y - \bar{y}_{kj})^2 \quad \text{(Definitional formula)}$$
$$= \Sigma y^2 - \frac{\Sigma\Sigma T_{kj}^2}{n} \quad \text{(Computational formula)}$$

(The summation, Σy^2, without explicit indices of summation, is a summation over all observations in all cells.)

The *sum of squares for Factor A* (also called the column sum of squares) measures the variability of the column means. Its formula is

$$SS_A = nK \sum_j (\bar{y}_{.j} - \bar{y})^2 \quad \text{(Definitional formula)}$$
$$= \frac{\sum_j T_{.j}^2}{nK} - \frac{T^2}{N} \quad \text{(Computational formula)}$$

The *sum of squares for Factor B* (also called the row sum of squares) measures the variability of the row means. Its formula is

$$SS_B = nJ \sum_k (\bar{y}_{k.} - \bar{y})^2 \quad \text{(Definitional formula)}$$
$$= \frac{\sum_k T_{k.}^2}{nJ} - \frac{T^2}{N} \quad \text{(Computational formula)}$$

The *sum of squares for the interaction of Factors A and B* (also called the interaction sum of squares) measures the variability of the cell means after the variabilities of the row and column means have been taken into account. Its formula is

$$SS_{AB} = n \sum_k \sum_j (\bar{y}_{kj} - \bar{y}_{.j} - \bar{y}_{k.} + \bar{y})^2 \quad \text{(Definitional formula)}$$

$$= \frac{\sum_k \sum_j T_{kj}^2}{n} - \frac{\sum_j T_{.j}^2}{nK} - \frac{\sum_k T_{k.}^2}{nJ} + \frac{T^2}{N} \quad \text{(Computational formula)}$$

The *total sum of squares* is the total of the other four sums of squares.

$$SS_{Tot} = SS_A + SS_B + SS_{AB} + SS_{WG}$$

The total sum of squares can also be defined and computed directly.

$$SS_{Tot} = \Sigma(y - \bar{y})^2 \quad \text{(Definitional formula)}$$

$$= \Sigma y^2 - \frac{T^2}{N} \quad \text{(Computational formula)}$$

The total sum of squares usually is computed by the latter formula and then checked by computation of the sum of the other four sums of squares.

Degrees of Freedom

The degrees of freedom associated with the sums of squares are

$$df_{WG} = N - JK$$
$$df_A = J - 1$$
$$df_B = K - 1$$
$$df_{AB} = (J - 1)(K - 1)$$
$$df_{Tot} = N - 1$$

The first four of these numbers are used as denominators in computing mean squares (see below). The total degrees of freedom (df_{Tot}) usually is computed only as a check on the other four.

$$df_{Tot} = df_A + df_B + df_{AB} + df_{WG}$$

Mean Squares

Four mean squares are computed in a two-factor analysis of variance. Each is the ratio of a sum of squares to a number of degrees of freedom.

$$MS_{WG} = SS_{WG}/df_{WG}$$
$$MS_A = SS_A/df_A$$
$$MS_B = SS_B/df_B$$
$$MS_{AB} = SS_{AB}/df_{AB}$$

Test of the Main Effect of Factor A

The null hypothesis about the main effect of Factor A states that the population means of the J levels of Factor A are equal. The alternative hypothesis states that these means are not all equal.

H_0: column population means are equal (no main effect of Factor A).

H_1: column population means are not all equal (main effect of Factor A).

The test statistic for deciding between these two hypotheses is

$$F_A = \frac{MS_A}{MS_{WG}}$$

When H_0 is true, this statistic is distributed as an F-distribution with $df_A = J - 1$ and $df_{WG} = N - JK$ degrees of freedom. When H_1 is true, F_A tends to have larger values than it has when H_0 is true. Hence, the null hypothesis is rejected for large values of F_A: that is, for values exceeding $F_\alpha(J - 1, N - JK)$ if the test is to have the probability of a Type I error $= \alpha$.

Test of the Main Effect of Factor B

The null hypothesis about the main effect of Factor B states that the population means of the K levels of Factor B are equal. The alternative hypothesis states that these means are not all equal.

H_0: row population means are equal (no main effect of Factor B).

H_1: row population means are not all equal (main effect of Factor B).

The test statistic for deciding between these two hypotheses is

$$F_B = \frac{MS_B}{MS_{WG}}$$

When H_0 is true, this statistic is distributed as an F-distribution with $df_B = K - 1$ and $df_{WG} = N - JK$ degrees of freedom. When H_1 is true, F_B tends to have larger values than it has when H_0 is true. Hence, the null hypothesis is rejected for large values of F_B: that is, for values exceeding $F_\alpha(K - 1, N - JK)$ if the test is to have the probability of a Type I error $= \alpha$.

Test of the Interaction Effect

The null hypothesis about the interaction effect states that the population regression functions of y on Factor A are parallel: that is, that the regression functions are the same except for the addition or subtraction of a constant. (Refer to Section 17.8 for further explanation of the parallelism of regression functions.) The alternative hypothesis states that the population regression functions are not parallel.

H_0: population regression functions are parallel (no interaction effect of Factors A and B).

H_1: population regression functions are not parallel (interaction effect of Factors A and B).

The test statistic for deciding between these two hypotheses is

$$F_{AB} = \frac{MS_{AB}}{MS_{WG}}$$

When H_0 is true, this statistic is distributed as an F-distribution with $df_{AB} = (J - 1)(K - 1)$ and $df_{WG} = N - JK$ degrees of freedom. When H_1 is true, F_{AB} tends to have larger values than it has when H_0 is true. Hence, the null hypothesis is rejected for large values of F_{AB}: that is, for values exceeding $F_\alpha[(J - 1)(K - 1), N - JK)]$ if the test is to have the probability of a Type I error $= \alpha$.

Example 1

Stomach ulcers in rats can be produced by certain brain lesions. A study is carried out to determine the effect of two independent variables on lesioned rats. The variables are two aspects of food intake: the *consistency* of the diet—water or solid (Factor A)—and the *nutrition* of the diet—none or sucrose (Factor B). The rats in the water-none condition are given access only to water, the rats in the water-sucrose condition are given access to a sucrose solution; the rats in the solid-none condition are given access to nonnutritive bulk; and the rats in the solid-sucrose condition are given bulk containing sucrose.[10] The dependent variable in the study is the number of stomach ulcers found in the rats when they were sacrificed.

Thirty-two rats are used in the experiment. They are assigned at random to the four conditions. The number of ulcers for each of the eight rats in each condition is given in Table 17.8a. The total number of ulcers for the eight rats in each condition is T_{kj}, where $j = 1$ or 2 identifies the level of Factor A (consistency) and $k = 1$ or 2 identifies the level of Factor B (nutrition). These totals, together with column, row, and grand totals, are given in Table 17.8b. The corresponding means are in Table 17.8c. The cell means are plotted in Figure 17.9. The regression functions are almost parallel; there appear to be many more ulcers in the two water groups ($j = 1$) than in the two solid groups ($j = 2$); and there appears to be little difference between the nutrition groups ($k = 1$ versus $k = 2$). The statements in the preceding sentence are statements about the sample data. In the analysis of variance an inference is made about the corresponding statements in the population.

The calculations for the analysis of variance are shown in the remaining sections of Table 17.8. The results usually are represented in a summary table such as Table 17.8g. Three F-tests are summarized in this table and are elaborated briefly here.

Main Effect of Factor A The test statistic is $F_A = 4.94$. The critical value (Table A15) for $\alpha = 0.05$ is $F_{.05}(1, 28) = 4.20$. The observed value of F exceeds the critical value; hence, the null hypothesis that the effects of water and solid diets are the same can be rejected. There is a statistically significant difference between the mean numbers of ulcers for rats with water diets and rats with solid diets.

Main Effect of Factor B The test statistic is $F_B = 0.28$. The critical value (Table A15) for $\alpha = 0.05$ is $F_{.05}(1, 28) = 4.20$. The observed value does not exceed the critical value; hence, the null hypothesis that the effects of no nutrition and a sucrose diet are the same *cannot* be rejected.

---

[10] The description of this study is modeled on K. -P. Ossenkopp, N. I. Wiener, and J. N. Nobrega, "Ventromedial Hypothalamic Lesions and Stomach Ulcers: Reduction by Non-Nutritive Bulk Ingested in the Post Lesion Period", *Physiology & Behavior* 24, (1980) 1125–1131. However, the data given here are fictitious.

TABLE 17.8 Data and Calculations for Example 1

(a) Data

		CONSISTENCY (FACTOR A)	
		WATER ($j = 1$)	SOLID ($j = 2$)
	NONE ($k = 1$)	1, 10, 4, 7, 8, 2, 13, 4	1, 8, 0, 0, 8, 0, 6, 2
NUTRITION (FACTOR B)	SUCROSE ($k = 2$)	12, 1, 1, 6, 11, 3, 2, 8	0, 1, 2, 11, 0, 0, 3, 1

(b) Totals

	$j = 1$	$j = 2$	
$k = 1$	$T_{11} = 49$	$T_{12} = 25$	$T_{1.} = 74$
$k = 2$	$T_{21} = 44$	$T_{22} = 18$	$T_{2.} = 62$
	$T_{.1} = 93$	$T_{.2} = 43$	$T = 136$

(c) Means

	$j = 1$	$j = 2$	
$k = 1$	$\bar{y}_{11} = 6.125$	$\bar{y}_{12} = 3.125$	$\bar{y}_{1.} = 4.625$
$k = 2$	$\bar{y}_{21} = 5.500$	$\bar{y}_{22} = 2.250$	$\bar{y}_{2.} = 3.875$
	$\bar{y}_{.1} = 5.812$	$\bar{y}_{.2} = 2.688$	$\bar{y} = 4.250$

(d) Terms in the SS

$$\Sigma y^2 = (1)^2 + (10)^2 + \ldots + (3)^2 + (1)^2 = 1,104$$

$$\sum_k \sum_j T_{kj}^2/n = [(49)^2 + (25)^2 + (44)^2 + (18)^2]/8 = 5,286/8 = 660.750$$

$$\sum_j T_{.j}^2/(nK) = [(93)^2 + (43)^2]/(8 \times 2) = 10,498/16 = 656.125$$

$$\sum_k T_{k.}^2/(nJ) = [(74)^2 + (62)^2]/(8 \times 2) = 9,320/16 = 582.500$$

$$T^2/N = (136)^2/32 = 18,496/32 = 578.000$$

(e) Sums of Squares

$$SS_{WG} = \Sigma y^2 - \sum_k \sum_j T_{kj}^2/n = 1,104 - 660.750 = 443.250$$

$$SS_A = \sum_j T_{.j}^2/(nK) - T^2/N = 656.125 - 578.000 = 78.125$$

$$SS_B = \sum_k T_{k.}^2/(nJ) - T^2/N = 582.500 - 578.000 = 4.500$$

$$SS_{AB} = \sum_k \sum_j T_{kj}^2/n - \sum_j T_{.j}^2/(nK) - \sum_k T_{k.}^2/(nJ) + T^2/N$$
$$= 660.750 - 656.125 - 582.500 + 578.000 = 0.125$$

$$SS_{Tot} = \Sigma y^2 - T^2/N = 1,104 - 578.00 = 526.000$$

Check: $526.000 = 443.250 + 78.125 + 4.500 + 0.125$

(f) Degrees of Freedom

$$df_{WG} = N - JK = 32 - (2 \times 2) = 32 - 4 = 28$$
$$df_A = J - 1 = 2 - 1 = 1$$
$$df_B = K - 1 = 2 - 1 = 1$$
$$df_{AB} = (J - 1)(K - 1) = (2 - 1) \times (2 - 1) = 1 \times 1 = 1$$
$$df_{Tot} = N - 1 = 32 - 1 = 31$$

Check: $31 = 28 + 1 + 1 + 1$

(g) Summary Table

SOURCE OF VARIABILITY	SS	df	MS	F
A (Consistency)	78.125	1	78.125	4.94[a]
B (Nutrition)	4.500	1	4.500	0.28 (ns)
$A \times B$	0.125	1	0.125	0.01 (ns)
Within groups	443.250	28	15.83	
Total	526.000	31		

[a] p-value < 0.05.

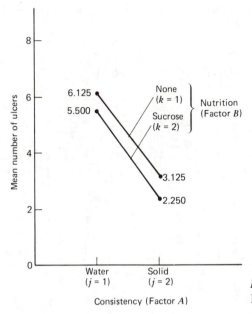

Figure 17.9 Regression functions in Example 1.

Interaction Effect The test statistic is $F_{AB} = 0.01$. The critical value (Table A15) for $\alpha = 0.05$ is $F_{.05}(1, 28) = 4.20$. The observed value does not exceed the critical value; hence, the null hypothesis that there is no interaction between nutrition and consistency *cannot* be rejected.

Summary The only significant factor in this study is the main effect of Factor A (consistency). This result was already suggested by Figure 17.9.

Example 2 Certain behaviors of mental patients can be affected by drugs. A study is carried out to determine the effect of four different types of drug (Factor B) on a behavior rating. Since the effect of the drugs differ for different types of patients, a second variable (Factor A — type of patient) is included in the study. One hundred and twenty patients are studied— 40 of each of three types. The patients are assigned at random, in equal numbers, to each of the four drugs. There are, therefore, 10 patients in each of the $3 \times 4 = 12$ conditions of the study.

The data and calculations are shown in Table 17.9. The original scores of each of the 120 patients are not given; however, the cell totals are given in Table 17.9*b*, and the sum of squares (Σy^2) of all the scores is given in Table 17.9*a*. The cell means (Table 17.9*c*) are plotted in Figure 17.10. The three regression functions, one for each patient type, appear not to be parallel, so we would expect a significant interaction between patient type and drug type in this study.

The summary table (Table 17.9*g*) shows that both the interaction ($A \times B$) and the main effect of Factor A (patient type) are significant. Note that the critical values of F for the three tests are different, since the degrees of freedom for the numerators of the three test statistics are different. The critical values (Table A15) are, for $\alpha = 0.01$,

$$A: F_{.01}(2, 108) = 4.8 \text{ (approximately)}$$
$$B: F_{.01}(3, 108) = 4.0 \text{ (approximately)}$$
$$C: F_{.01}(6, 108) = 3.0 \text{ (approximately)}$$

The significant effect of Factor A means that the mean behavior rating of patients of different type is different. This result is overshadowed, however, by the fact that the

TABLE 17.9 Data and Calculations for Example 2

(a) Data

Cell totals are given below; the sum of squares of all 120 scores (n = 10 scores in each cell) is $\Sigma y^2 = 6486$.

(b) Totals

PATIENT TYPE (FACTOR A)

	$j = 1$	$j = 2$	$j = 3$	
$k = 1$	$T_{11} = 83$	$T_{12} = 52$	$T_{13} = 60$	$T_{1.} = 195$
$k = 2$	$T_{21} = 76$	$T_{22} = 49$	$T_{23} = 64$	$T_{2.} = 189$
$k = 3$	$T_{31} = 36$	$T_{32} = 48$	$T_{33} = 102$	$T_{3.} = 186$
$k = 4$	$T_{41} = 41$	$T_{42} = 51$	$T_{43} = 94$	$T_{4.} = 186$
	$T_{.1} = 236$	$T_{.2} = 200$	$T_{.3} = 320$	$T = 756$

Drug Type (Factor B) — rows $k = 1 \ldots 4$.

(c) Means

	$j = 1$	$j = 2$	$j = 3$	
$k = 1$	$\bar{y}_{11} = 8.3$	$\bar{y}_{12} = 5.2$	$\bar{y}_{13} = 6.0$	$\bar{y}_{1.} = 6.5$
$k = 2$	$\bar{y}_{21} = 7.6$	$\bar{y}_{22} = 4.9$	$\bar{y}_{23} = 6.4$	$\bar{y}_{2.} = 6.3$
$k = 3$	$\bar{y}_{31} = 3.6$	$\bar{y}_{32} = 4.8$	$\bar{y}_{33} = 10.2$	$\bar{y}_{3.} = 6.2$
$k = 4$	$\bar{y}_{41} = 4.1$	$\bar{y}_{42} = 5.1$	$\bar{y}_{43} = 9.4$	$\bar{y}_{4.} = 6.2$
	$\bar{y}_{.1} = 5.9$	$\bar{y}_{.2} = 5.0$	$\bar{y}_{.3} = 8.0$	$\bar{y} = 6.3$

(d) Terms in the SS

$$\Sigma y^2 = 6{,}486 \text{ (given)}$$
$$\sum_k \sum_j T^2_{kj}/n = [(83)^2 + (52)^2 + \ldots + (94)^2\,]/10 = 52{,}588/10 = 5{,}258.8$$
$$\sum_j T^2_{.j}/(nK) = [(236)^2 + (200)^2 + (320)^2]/(10 \times 4) = 198{,}096/40 = 4{,}952.4$$
$$\sum_k T^2_{k.}/(nJ) = [(195)^2 + (189)^2 + (186)^2 + (186)^2]/(10 \times 3) = 142{,}938/30 = 4{,}764.6$$
$$T^2/N = (756)^2/120 = 571{,}536/120 = 4762.8$$

(e) Sums of Squares

$$SS_{WG} = \Sigma y^2 - \sum_k \sum_j T^2_{kj}/n = 6{,}486 - 5{,}258.8 = 1{,}227.2$$
$$SS_A = \sum_j T^2_{.j}/(nK) - T^2/N = 4{,}952.4 - 4{,}762.8 = 189.6$$
$$SS_B = \sum_k T^2_{k.}/(nJ) - T^2/N = 4{,}764.6 - 4{,}762.8 = 1.8$$
$$SS_{AB} = \sum_k \sum_j T^2_{kj}/n - \sum_j T^2_{.j}/(nK) - \sum_k T^2_{k.}/(nJ) + T^2/N$$
$$= 5{,}258.8 - 4{,}952.4 - 4{,}764.6 + 4{,}762.8 = 304.6$$
$$SS_{Tot} = \Sigma y^2 - T^2/N = 6{,}486 - 4{,}762.8 = 1{,}723.2$$

Check: $1723.2 = 1227.2 + 189.6 + 1.8 + 304.6$

(f) Degrees of Freedom

$$df_{WG} = N - JK = 120 - (3 \times 4) = 120 - 12 = 108$$
$$df_A = J - 1 = 3 - 1 = 2$$
$$df_B = K - 1 = 4 - 1 = 3$$
$$df_{AB} = (J - 1)(K - 1) = (3 - 1) \times (4 - 1) = 2 \times 3 = 6$$
$$df_{Tot} = N - 1 = 120 - 1 = 119$$

Check: $119 = 108 + 2 + 3 + 6$

(g) Summary Table

SOURCE OF VARIABILITY	SS	df	MS	F
A (patient type)	189.6	2	94.80	8.34[a]
B (drug type)	1.8	3	0.60	0.05 (ns)
$A \times B$	304.6	6	50.77	4.47[a]
Within groups	1,227.2	108	11.36	
Total	1,723.2	119		

[a] p-value < 0.01

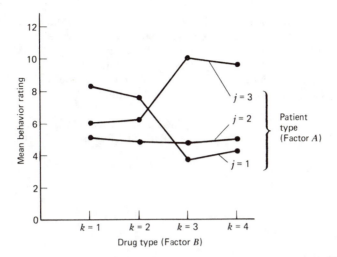

Figure 17.10 Regression functions in Example 2.

interaction ($A \times B$) is significant. Figure 17.10 shows clearly that for patient type 3, the greatest mean ratings are obtained for drug types 3 and 4, whereas for patient type 1, the greatest mean ratings are obtained for drug types 1 and 2. (For patient type 2, the mean behavior ratings are about the same for all drugs.) This nonparallelism of the regression functions shows that we must consider patient type as well as drug type when attempting to maximize the behavior rating. Hence, the fact that there is an overall difference between the mean ratings for the three patient types is less important than the significant interaction. It is customary in a two-factor analysis of variance to test the interaction first and to consider the main effects only if there is not a significant interaction.

17.10 Summary

The analysis of variance is used to study the relationship between an independent variable (or *factor*), which defines two or more groups or conditions; and a dependent variable, which is the score obtained by each subject in each group. The relationship can be studied by a regression function or a measure of the size of the relationship. The regression function is the graph of the means in each condition or *level*. The F-test is used to test whether the regression function is horizontal in the population: that is, whether the population means are equal. Two measures of the size of the relationship were described: the *coefficient of determination* (R^2) and $\hat{\omega}^2$.

The test statistic in the F-test is F_{obs}, which is the ratio of the *mean square between groups* and the *mean square within groups*. The mean square between groups is computed from the *sum of squares between groups* and the *degrees of freedom between groups*. Similarly, the mean square within groups is computed from the *sum of squares within groups* and the *degrees of freedom within groups*.

The *one-factor analysis of variance* summarized above can be extended to two independent variables; it is then called a *two-factor analysis of variance*. There are three ways in which the independent variables can affect the dependent variable: the two *main effects*, one for each factor; and the *interaction effect*, defined as the joint effect of the two factors that cannot be accounted for by the main effects of the two factors separately. Tests of these three effects were shown in detail; each test is an F-test.

Exercises

1 (Section 17.2) In a study of weight gain, 18 rats are assigned to one of three conditions: no sucrose, liquid sucrose, or solid sucrose. There are six rats in each condition. The weight gains, over 60 days, are given below.
(a) Plot the data and the regression function.
(b) Calculate the variance of the predicted scores and the total scores.
(c) Compute the coefficient of determination.

WEIGHT GAIN (GRAMS)

NO SUCROSE	LIQUID SUCROSE	SOLID SUCROSE
27	42	50
52	81	37
43	84	64
36	57	53
49	49	57
33	56	46

2 (Section 17.3)
(a) Calculate SS_{Tot}, SS_{BG}, SS_{WG} for the data of Question 1. Use the computational formulas.
(b) Calculate $R^2 = SS_{BG}/SS_{Tot}$ and compare its value to the coefficient of determination obtained in Question 1c.

3 (Section 17.4)
(a) What are the 95th and 99th percentiles of the F-distribution with 7 and 20 degrees of freedom?
(b) Find $F_{.05}(20, 7)$ and $F_{.01}(20, 7)$.

4 (Section 17.5) Carry out the F-test for the data of Question 1 as follows.
(a) Compute MS_{BG}, MS_{WG}, and F_{obs}.
(b) Make a summary table of the results (as in Table 17.4).
(c) Test F_{obs} for significance at the 0.05 level.
(d) State the conclusion of the study in words.

5 (Section 17.6) Compute two measures of relationship for the data in Question 1 and interpret these measures. Be sure to state what two variables these measures relate.

6 (Sections 17.3 and 17.5) It is not necessary to have the original data (such as given in Question 1) in order to carry out an analysis of variance, provided that the means, standard deviations, and sizes of each group are known. Consider a four-group study with the following statistics.

GROUP (j)	SIZE (n_j)	MEAN ($\bar{y}_j$)	STANDARD DEVIATION (s_j')
1	10	8.32	6.37
2	8	9.76	7.20
3	12	6.42	4.18
4	10	11.52	6.81

(a) Look at the *computational* formula for SS_{BG} in Section 17.3. What quantities are required in order to compute SS_{BG}? Calculate these quantities from the above statistics. Then calculate SS_{BG}.
(b) Look at the *definitional* (not computational) formula for SS_{WG}. The SS_{WG} is a sum, over groups, of the sum (in each group) of the squared deviations of the scores from the group mean. Each sum of squared deviations is simply the numerator of the variance in that group. Therefore, you should be able to calculate SS_{WG} from the given standard deviations.
(c) Complete the analysis of variance of these data, present a summary table, and make the significance test.

7 (Section 17.8) Make up 3×2 tables (three rows and two columns) of cell means to illustrate each of the following situations. These tables may be considered to be of population means, so that any differences between, say, the row means indicate a main effect for the row factor—and similarly for the main effect of the column factor and for the interaction effect.

(a) No main effects for either the row or the column factor and no interaction effect.

(b) Main effect for the row factor but no main effect for the column factor and no interaction effect.

(c) No main effect for the row factor but a main effect for the column factor and no interaction effect.

(d) No main effects for either the row factor or the column factor but an interaction effect.

(e) Main effect for the row factor, no main effect for the column factor, and an interaction effect.

For each of these tables, make a graph of the regression functions.

8 (Section 17.9) A two-factor study was carried out to determine the relationship between number of errors made in a learning task and two independent variables: induced anxiety of the subjects (high, medium, and low) and type of practice (massed and distributed). Four subjects were assigned to each of the six conditions. Their scores (number of errors) are recorded below. Test the two main effects and the interaction effect, reporting the results in a summary table and in words.

| | | INDUCED ANXIETY | | |
		HIGH	MEDIUM	LOW
PRACTICE	MASSED	9, 7, 8, 8	5, 7, 6, 6	3, 5, 5, 3
	DISTRIBUTED	12, 10, 11, 11	8, 8, 7, 6	7, 5, 7, 5

9 (Section 17.9) A study similar to the one presented in Question 8 was carried out with eight subjects in each condition. The sum of squares (Σy^2, where the sum is just over the eight subjects in the condition) and the mean in each condition are given below. Complete the analysis of variance.

PRACTICE	INDUCED ANXIETY	SUM OF SQUARES	MEAN
Massed	High	814	9.500
Massed	Medium	641	8.625
Massed	Low	520	7.125
Distributed	High	1146	11.375
Distributed	Medium	926	10.000
Distributed	Low	745	8.750

REVIEW CHAPTER D

In this chapter, the different types of variables are reviewed and used as the basis for a classification of most of the statistical procedures described thus far. The classification is illustrated by considering how four typical research studies could be analyzed.

Although previously stated, the requirements for the validity of statistical procedures are listed in this chapter, for convenience. If a requirement is severely violated, it sometimes is possible to substitute another procedure to which the requirement does not apply. These alternative procedures are listed. Many of them are called *nonparametric* or *distribution-free* procedures. The chapter concludes with a discussion of these procedures.

D.1 Types of Variables

The basis of any statistical analysis is a set of data, which consists of the values of one or more variables. These variables can be of many types. In this text, four principal types of variables are distinguished: *dichotomous, categorical, numerical*, and *rank*. The relationships among these four types and their relationships to other types of variables are summarized in Figure D.1.

The first distinction among variables is between numerical and nonnumerical variables. Numerical variables have values that are numbers. Numerical variables sometimes are called *quantitative* variables. Examples are distance jumped (1.36 m, 1.52 m, etc.) and score on a test (43, 76, etc.). Nonnumerical

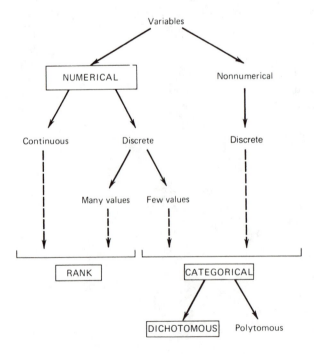

Figure D.1 Types of variables

variables have values that are words. Such variables sometimes are called *qualitative* variables. Examples are blood group (A, B, AB, etc.) and sex (male, female).

Numerical Variables

Numerical variables are either continuous or discrete. All numbers, in a certain range, are possible values of a continuous variable. Examples are distance jumped and time elapsed. On the other hand, there are gaps between the possible values of a discrete variable. The score on a test is usually an integer, and values between consecutive integers (e.g., 6.2) are not possible. Whereas the distinction between continuous and discrete variables is mathematically important, in practice all data is discrete, and hence all numerical variables usually may be analyzed by the same techniques, whether the variables are, theoretically, discrete or continuous.

Categorical and Dichotomous Variables

A numerical variable that is discrete may have many possible values or just a few possible values. There are many possible values for most test scores (e.g., all integers from 0 to 100). On the other hand, a rating of felt anxiety on an integer scale from 1 (low) to 5 (high) has only five possible values. Discrete variables with only a few values often are treated in the same way as nonnumerical variables (which are, by their nature, discrete). For example, an anxiety rating scale might be defined to have the following, nonnumerical, values: low, moderately low, average, moderately high, high. There is little practical difference between such a nonnumerical scale and a discrete scale extending from 1 to 5. Discrete variables with few values and nonnumerical variables may both be considered categorical variables.

A special case of a categorical variable is a variable with only two values, which is called a dichotomous variable. A polytomous variable is a categorical variable that is not dichotomous. Many techniques for polytomous variables also apply to dichotomous variables, and therefore, the techniques are said to apply to categorical variables. The principal types of variables are considered to be categorical and dichotomous variables, even though dichotomous variables are a subset of categorical ones.

Rank Variables

In some statistical techniques the values of a numerical variable are reduced to ranks, to make a rank variable. Ranking can be done if the variable is continuous or if it is discrete with many values. (See Figure D.1.)

D.2 Classification of Procedures

The four types of variables described in Section D.1 provide a convenient base from which to develop a classification of many of the statistical procedures described in this text. Most of the procedures are analyses of two variables; these are described first. A classification of one-variable procedures follows.

Two-Variable Procedures

The typical research study has two variables. One of these variables usually can be identified as the independent variable and the other as the dependent variable. Each of these variables can be dichotomous, categorical, numerical, or rank. There are, therefore, $4 \times 4 = 16$ combinations of type of independent variable with type of dependent variable. These 16 possibilities are the cells of

TABLE D.1 Two-variable Procedures

DEPENDENT VARIABLE	INDEPENDENT VARIABLE DICHOTOMOUS PROCEDURE	TEST	ESTIMATE	CATEGORICAL PROCEDURE	TEST	ESTIMATE
DICHOTOMOUS	*A* **1** Two independent proportions **2** Sign test	14.3, 14.4[a] 14.2	12.5, 16.5 —[b]	*E* **1** More than two independent proportions	14.5	16.5
CATEGORICAL	*B* (See cell *F*)			*F* **1** Tests in an r × c table	15.2	16.5
NUMERICAL	*C* **1** Two independent means **2** Two dependent means **3** Two independent medians	11.4 11.5 14.6	8.5, 12.4 12.4 —	*G* **1** One-factor analysis of variance **2** More than two independent medians	17.5 14.6	17.6 —
RANK	*D* **1** Rank-sum test for two independent groups **2** Rank-sum test for two dependent groups	15.5 15.6	— —	* * * * *[c]		

[a] The number associated with each test or estimate indicates the section in which the test or estimate is described.
[b] A dash indicates that the estimate is not described in this text.
[c] Cells for which no applicable procedure is described in this text are marked with asterisks.

Table D.1. Procedures for nine of these cells have been described in this text; they are listed in the appropriate places in the table.

Each procedure usually exists in two forms: a *test* of a hypothesis or an *estimate* of a parameter. This distinction can also be thought of as a distinction between (a) determining whether or not there is an effect (a *test*) and (b) measuring how large the effect is (an *estimate*). (See Section 16.1 for a detailed discussion of this distinction.) References are given in Table D.1 to the sections in which the particular test or estimate is described.

Table D.1 can be used in two ways. It provides a concise summary of a number of techniques, showing various relationships between them. It also can be used to select the appropriate procedure to apply to a given study of two variables. To do so, carry out the following steps.

1 Identify the independent variable in the study. What type of variable is it (dichotomous, categorical, numerical, or rank)? The type of the independent variable determines the *column* of Table D.1 to use.

2 Identify the dependent variable in the study. What type of variable is it (dichotomous, categorical, numerical, or rank)? The type of the dependent variable determines which *row* of Table D.1 to use.

3 Use the procedure in the cell at the intersection of the row and column. You may have to choose among several procedures. If no procedures are listed, you may be able to consider one of the variables as having a different type, so that a cell that does have a procedure listed can be obtained. Alternatively, you can consult more-advanced statistical texts.

TABLE D.1 Two-variable Procedures *(continued)*

	INDEPENDENT VARIABLE					
	NUMERICAL			RANK		
	PROCEDURE	TEST	ESTIMATE	PROCEDURE	TEST	ESTIMATE
D DICHOTOMOUS	* * * * *			* * * * *		
E						
P						
E CATEGORICAL	* * * * *			* * * * *		
N						
D						
E						
N						
T	*H*					
V NUMERICAL	**1** One Pearson correlation	16.2	8.6, 16.2, 16.3	* * * * *		
A	**2** Two independent Pearson correlations	16.2	—			
R						
I						
A				*I*		
B RANK	* * * * *			**1** Spearman correlation	16.4	16.4
L						
E						

Several examples of research studies and the selection of the appropriate procedure are described in Section D.3.

One-variable Procedures

Studies of one variable occur only infrequently. Such studies can be classified on the basis of the type of the variable. Table D.2 lists procedures for dichotomous, categorical, and numerical variables described in this text. This table can be used as a summary of a number of techniques, or as a way to select the appropriate procedure of a study—just as was explained above for two-variable procedures.

TABLE D.2 One-variable Procedures

		PROCEDURE	TEST	ESTIMATE
	DICHOTOMOUS	*J* **1** One proportion	9.1, 10.1	12.5
V				
A		*K*		
R CATEGORICAL		**1** Goodness-of-fit	15.3	—
I				
A		*L*		
B NUMERICAL		**1** One mean	11.2	12.3
L		**2** One median	14.6	—
E				
RANK		* * * * *		

Note. See footnotes for Table D.1.

The Sign Test

The sign test (Section 14.2) has been placed in cell A of Table D.1. However, the sign test does not fit into this classification as readily as the other procedures do. As described in Section 14.2, the sign test applies to two dependent, or matched, groups; hence, its location in the first column (dichotomous *independent* variable) of Table D.1 is appropriate. However, a dichotomous *dependent* variable is not measured separately for each subject. Rather, each matched pair is judged directly to be either plus or minus (Example 1 of Section 14.2) or each subject receives a numerical score (dependent variable) and each pair is rated as either plus or minus, depending on which member of the pair has a higher score (Example 2 of Section 14.2). Therefore, be cautious in using the classification of Table D.1 for the sign test.

**D.3
Four Typical
Studies**

This section briefly describes four typical research studies. (Similar studies are described in Section 1.1.) Attention is focused on the nature of the variables in the study and on what procedure, or procedures, might be used to analyze the data of the study. The discussion is based on the classification of statistical procedures described in Section D.2.

**The Visual Cliff
and Presence
of Mother**

A classic study of the development of an infant's perceptual abilities is the visual cliff experiment. An infant is placed near an edge, beyond which appears to be a drop of several feet. In fact, a horizontal transparent piece of plastic extends beyond the edge, to prevent the child from falling. Some infants refuse to crawl beyond the edge. Others crawl beyond the edge (onto the plastic), apparently oblivious of the cliff.

In the study considered here, the behavior of the infants at the visual cliff is studied when their mothers are coaxing them to cross the cliff and when their mothers are absent. Does the presence of the mother influence the willingness of the infants to cross the cliff?

The independent variable in this study is the presence or absence of the mother (a dichotomous variable). The dependent variable is the behavior of the infant. This variable could be simply whether or not the infant crosses the cliff (another dichotomous variable), or it could be the time the infant takes to cross the cliff (a numerical variable). The latter variable might be used if, in fact, all (or most) of the infants eventually cross the cliff.

Several methods of analysis are suggested by Table D.1. To begin, we must consider whether the infants in the mother-present condition are the same infants as those in the mother-absent condition. The groups are independent if the infants are different in the two conditions, and the groups are dependent if each infant is observed under both conditions.

Independent-Groups If the dependent variable is dichotomous, the study is a study of two proportions: the proportion of infants crossing the cliff when the mother is present and the proportion crossing when the mother is absent (A1—two independent proportions). If the dependent variable is numerical (time taken to cross the cliff), the means or medians could be studied (C1—two independent means—or C3—two independent medians).

Dependent-Groups Consider a numerical dependent variable (time). The analysis of means (C2—two dependent means) is based on the difference in times for each infant between the two conditions. If only the *sign* of this difference is analyzed, we have the sign test (A2).

Religiosity and Church Attendance

Consider a study of the relation between degree of religious belief (religiosity) and attendance at church. Let us take religiosity as the independent variable and attendance as the dependent variable. Suppose that religiosity is measured on a four-point scale ranging from 1 (no religious belief) to 4 (very strong religious belief). Religiosity is, therefore, a categorical variable. Church attendance could be recorded in a number of ways. Let us first define attendance as a categorical variable with three values: never attend, sometimes attend, and attend once or more a week.

The analysis of the relationship between the two categorical variables falls in cell *F* of Table D.1. A test of the hypothesis that there is no relationship (i.e., that the two variables are independent) is described in Section 15.2. An estimate of the size of the relationship can be made by statistics described in Section 16.5.

Alternatively, we might choose to measure attendance in a numerical way—say, as the number of times in one year that each person went to church. If we keep religiosity as a categorical variable, the study now falls in cell *G* of Table D.1. This cell indicates that the means or the medians of the dependent variable could be analyzed. Since the distributions of attendance in each of the groups defined by the religiosity variable probably are extremely skewed, it may be wise to analyze medians instead of means, since the medians are less affected by extreme scores than are means (Section 3.5).

Finally, suppose that each subject was given 20 statements about belief in religion and that the subject's score was the number of statements that he or she endorsed. Accordingly, religiosity would be a numerical variable, ranging from 0 to 20. We could then consider relating the two numerical variables, religiosity and attendance, by a Pearson correlation (cell *H* of Table D.1) or a Spearman correlation (cell *I* of Table D.1).

A Learning Study

In many learning studies, two conditions are defined and their effect on a dependent variable studied. Suppose that the two conditions are a 100% reinforcement condition and a 50% (partial) reinforcement condition. Suppose that the dependent variable is the number of trials to learn the task.

The independent variable is dichotomous and the dependent variable is numerical. The possible procedures are listed in cell *C* of Table D.1. The choice among the three procedures listed in that cell (two independent or dependent means or two independent medians) would be based on considerations similar to those described above for the visual cliff experiment. If we decided to use independent means, then measures of the size of the relationship between amount of reinforcement and learning would be the difference of the means (effect size) or the ratio of the difference of the means to the pooled standard deviation (standardized effect size)—see Section 8.5. Confidence intervals for the difference of the means are described in Section 12.4.

An alternative analysis that is sometimes used is to treat the dependent variable as a rank variable (cell *D* of Table D.1). The choice between the two rank-sum tests would be based on whether the two groups are independent or dependent. In this analysis, it would not be possible to make estimates of the size of the relationship between the independent and dependent variables.

Blood Groups   The four major blood groups have the following relative frequencies in the general U.S. population.

O: 45%; A: 41%; B: 10%; AB: 4%

Suppose that a blood distribution clinic is located in an area in which there is a large immigrant population. It is decided to determine whether the proportions in that special area are different from those in the general population. A random sample of 500 persons is selected and each person's blood type determined.

This study, unlike all the previous studies in this section, deals with only one variable (blood type). This variable is categorical, and the study falls in cell K of Table D.2. A goodness-of-fit test would be carried out to determine if the proportions in the sample differ from the proportions in the general population.

It would be useful to have a measure of the difference between the proportions in the area from which the sample was drawn and the proportions in the general U.S. population. However, such a measure is not described in this text. (There is no section number given under "Estimate" in cell K.)

It might be thought that there are *two* variables in this study—namely, location (the entire United States versus special area) and blood group. Although this way of looking at the study has some advantages, it should be noted that data are not collected from one of the locations (the entire United States). Rather, known proportions for that location are treated as population values and the sample values from the other location (special area) are tested against those population values. From this point of view, there is only one sample and only one variable in the study.

**D.4
Requirements
for the Validity
of Statistical
Procedures**

This section repeats much of the information presented in Sections C.3 and 15.4. One additional requirement, briefly mentioned in Section 16.2, is also included. As noted in Section C.3, in many studies it is difficult to determine whether a particular requirement is satisfied. Hence, these requirements are often called *assumptions*. For each requirement, it is stated whether the requirement can be relaxed and what alternative procedures may be used if the requirement is severely violated. These alternative procedures, which often are called "nonparametric" procedures, are discussed further in Section D.5.

REQUIREMENT 1: RANDOM SAMPLING

The subjects in the sample must be selected at random from the population of interest.

Alternative Requirement

The requirement of random sampling from a population may be replaced in two-group studies by the requirement of random assignment of subjects to the two conditions (Section 11.3).

Application

All cells of Tables D.1 and D.2.

Comment

Statistical inference requires randomness; the randomness may originate in random sampling or in randomization. If a study has neither feature, no inference made in the study is a valid statistical inference.

REQUIREMENT 2: INDEPENDENCE OF OBSERVATIONS

Each observation must be independent of every other observation in the study.

Alternative Requirement

None.

Application

All cells of Tables D.1 and D.2 except those designs in which dependence of the observations is built in by having two observations on each subject (correlational studies: that is, H1, H2, I1) or by having matched-pairs (dependent-groups designs: that is, A2, C2, D2).

Comment

See the discussions of this requirement in Sections C.3 and 15.4.

REQUIREMENT 3: NORMALITY OF POPULATION DISTRIBUTION

The distribution of scores in the population from which the sample is drawn must be normal. In the case of two or more populations, each population must be normally distributed.

Application

This requirement applies to all tests of hypotheses about means and to confidence-interval estimates of means (C1, C2, G1, L1). Normality is not required in order that point estimates and standard errors be valid, however.

Relaxation of Requirement

The requirement of normality can be relaxed considerably. If the sample size exceeds 10 the population can have almost any shape and the distribution of the statistic will still be close to its theoretical shape (normal, t, or F, as the case may be). If the sample size is 10 or less, the population distribution must be unimodal if the distribution of the statistic is to be close to its theoretical shape.

Alternative Procedures if Requirement Violated

Alternative procedures need be considered only if the requirement of normality is severely violated and the sample size is small. *If the original procedure is C1 (two independent means)*, we can analyze medians instead of means, changing the procedure to C3 (two independent medians). Or we can convert the dependent variable to ranks, changing the procedure to D1 (rank-sum test for two independent groups). *If the original procedure is C2 (two dependent means)*, we can again convert to ranks, changing the procedure to D2 (rank-sum test for two dependent groups). *If the original procedure is G1 (one-factor analysis of variance)*, we can analyze medians instead of means, changing the procedure to G2 (more than two independent medians). Rank-sum tests for more than two groups are described in more advanced texts. It is also possible, in all three procedures (C1, C2, and G1), to convert the dependent variable to a categorical variable and to apply an appropriate procedure from cells A, B, E, or F. If the original procedure is L1 (one mean), we can analyze the median, changing the procedure to L2 (one median).

Comment

The normality requirement can be relaxed because of the central limit theorem (Section 7.5). For further discussion, see Section C.3.

REQUIREMENT 4: EQUALITY OF VARIANCE

The variances of the population distributions must be equal.

Application

This requirement applies to tests and estimates of two or more independent means (C1, G1).

Relaxation of Requirement

The tests (t-test and F-test) and estimates (difference of means and coefficient of determination) can still be used if the population variances are different, provided that they are not *too* different. It is possible to evaluate, roughly, whether this requirement is met by comparing the *sample* variances: if the ratio of the largest sample variance to the smallest does not exceed, say, 4, it is safe to use the procedures. (For example, sample variances of 45.3 and 17.2 are alright, but variances of 100.3 and 7.6 are too different to allow the use of the tests and estimates.)

Alternative Procedures if Requirement Violated

If the variances are so different that even the relaxed requirement is violated, alternative procedures may be considered. The suggestions given above, for use when the normality requirement is violated, can be followed.

Comment

This requirement often is stated as a requirement of *homogeneity* of variance.

REQUIREMENT 5: LARGE EXPECTED FREQUENCIES

Each expected frequency must be 5 or more in order for the χ^2-distribution to be a satisfactory approximation to the distribution of χ^2_{obs}.

Application

All tests and estimates in cells A (except sign test), B, E, F, and K: that is, all procedures related to the χ^2-tests.

Relaxation of Requirement

The stated requirement is a rough "rule of thumb". It is sometimes relaxed to a requirement that no more than 10% of the cells may have expected frequencies between 1 and 5.

Alternative Procedures if Requirement Violated

It sometimes is possible to combine two or more categories so that the expected frequencies in the reduced, or collapsed, table satisfy the requirement. (See Example 2 in Section 15.2.) Such combinations should be made on a rational basis, such as combining adjacent categories.

Comment

Note that the requirement applies to the *expected*, not the *observed*, frequencies.

REQUIREMENT 6: BIVARIATE NORMALITY

The joint distribution of the two variables must be bivariate normal.

Definition

Bivariate normality will not be defined precisely here.[1] Roughly speaking, the requirement combines the requirements of normality and equality of variance with the requirement that the regression function be linear.

Application

This requirement applies to tests and estimates of the Pearson correlation (cell H).

Relaxation of Requirement

The data from any correlational study should be plotted to see if the regression function is at least approximately linear. The Pearson correlation does not have meaning if the regression is not linear. (See Section 16.1 for a summary of linear and nonlinear associations.) The scatterplot will also show if the variability about the regression line varies greatly across the range of the independent variable. As in the case of Requirement 4 (equality of variance), the sample variances do not have to be precisely equal, but they should not vary by more than, say, a factor of 4. The normality requirement is the hardest to judge, but it is probably the one that can be violated most safely, provided that the departure from normality is not too great.

Alternative Procedures if Requirement Violated

If the regression function appears to be monotonic but not linear, the Spearman correlation (cell I) may be the procedure of choice. If the regression function exhibits a more-complex nonlinear form, an advanced text should be consulted for the appropriate analysis.

D.5 Nonparametric and Distribution-free Procedures

The six requirements described in the preceding section can be divided into three groups.

- *Requirements 1 and 2* (random sampling and independence of observations) apply to all procedures except those in which dependence is built into the design (A2, C2, D2) and those in which there are two observations on each

---

[1] For a precise definition, see, for example, J. Neter and W. Wasserman, *Applied Linear Statistical Models* (Homewood, Ill.: Irwin, 1974).

subject (H1, H2, I1). These requirements are very strict and cannot be relaxed; no alternative procedures are available if the requirements are not met.

- *Requirement 5* (large expected frequencies) applies only to procedures in which a statistic, χ^2_{obs}, is approximated by the χ^2-distribution. The requirement serves simply to make the approximation sufficiently accurate. If the requirement is relaxed the approximation is not as good, but alternative procedures generally are not available.
- *Requirements 3, 4, and 6* (normality, equality of variance, and bivariate normality) have a different status from the others. They apply to tests of means or Pearson correlations. The requirements *can* be relaxed considerably without seriously affecting the validity of the procedures. Furthermore, if the requirements are severely violated, alternative procedures are available to which these requirements do not apply at all.

The alternative procedures that can be used instead of the tests and estimates of means and Pearson correlations involve medians, ranks, or categorical variables. These procedures are often described as *nonparametric* and *distribution-free* procedures. The two terms apply to both tests and estimates, but are most commonly applied just to tests of hypotheses, and are so defined here.

A *nonparametric test* is a test whose hypotheses are expressed in terms of the general locations or shapes of the population distributions rather than in terms of values of parameters of the population distributions.

A *distribution-free* test is a test in which the distribution of the test statistic is unaffected by the shapes of the population distributions.

Nonparametric and distribution-free procedures are used either (a) when the basic data are already in categories or ranks or (b) when the basic data are numerical but the requirements of normality, equality of variance, or bivariate normality are not met. Nonparametric and distribution-free procedures include

Chi-square tests. These tests essentially determine whether two or more population distributions are identical. They are nonparametric tests, since they do not involve specific parameters (e.g., the mean) of the distributions. The tests are distribution-free, since the distribution of χ^2_{obs} is, essentially, unaffected by the shapes of the population distributions. (Chapters 14 and 15).

Median tests. The median is a parameter, so these tests are parametric. However, the medians are tested by a χ^2-test that is unaffected by the particular shapes of the population distributions (Section 14.6).

Rank-sum tests. The hypotheses here refer to the identity of two population distributions, rather than to values of parameters. Hence, the tests are nonparametric. The test statistics (sums of ranks) are also unaffected by the population shapes; the tests are, therefore, distribution-free.

Spearman correlation. The Spearman correlation is a measure of monotonic association. It is a parameter of the joint distribution of the two variables, but its distribution is unaffected by the population distribution. Hence, the Spearman correlation is a distribution-free statistic.

Note from this summary that most nonparametric procedures are distribution-free, and vice versa. The two terms commonly are used interchangeably.[2]

D.6 Summary

Each variable in a study can be classified as belonging to one of four types: dichotomous, categorical, numerical, or rank. Most of the statistical procedures described in this text can be classified on the basis of the type of each variable in the analysis. The procedure to use in a particular study can then be determined from the types of the variables in the study.

The requirements for the validity of all the procedures described in this book were reviewed. These requirements include random sampling, independence of observations, normality of population distribution, equality of variance, large expected frequencies, and bivariate normality. Some of the requirements apply to all procedures, others to only a few. Some of the requirements cannot be relaxed, whereas others can be relaxed considerably. When some of the requirements are violated severely, the procedure can be replaced by another procedure that does not have the same requirement. Such procedures are called *nonparametric* or *distribution-free* procedures.

Exercises

1 (All sections) Find a published study that uses statistical techniques.
(a) Select one test or estimate that was reported in the study. Describe this test or estimate briefly, indicating the variables studied, how many groups were used, etc.
(b) Identify each variable, following the classification of types of variables in Section D.1.
(c) Identify the procedure used, following the classification of Section D.2.
(d) Which of the requirements, stated in Section D.4, are relevant to the procedure used in the study? Are any of these requirements discussed in the paper? Is it reasonable to believe that the requirements were met?

2 (Section D.2) Explain why the two-factor analysis of variance (Sections 17.8 and 17.9) cannot be included in Table D.1. How many variables are involved in such an analysis? What are the types of these variables?

---

[2] The distinction between nonparametric and distribution-free procedures is carefully made in L. A. Marascuilo and M. McSweeney, *Nonparametric and Distribution-Free Methods for the Social Sciences* (Monterey, Calif.: Wadsworth, 1977).

18 ADVANCED PROBABILITY

Statistics is based on probability theory. The results derived in this chapter provide the basis for some of the techniques and formulas that appear elsewhere in the text. Chapter 18 can be read after Chapters 1 to 6.

18.1 Ordered Samples or Permutations

The sample space is the set of all possible samples of a given size drawn from a given population. A number of sample spaces were described and illustrated by tree diagrams in Chapter 5. In this section (and the next), other kinds of sample spaces are introduced. Sample spaces may differ from each other in a number of ways. The method of sampling may be different (sampling with replacement or sampling without replacement). Or the definition of what are considered to be different samples can be varied: we may consider each different order of the elements in a sample to constitute a different sample (*ordered samples*) or we may consider samples to be different only if they contain a different selection of elements (*unordered samples*).

It is important to be able to count the number of samples in a given sample space. The sample space usually is too large to permit an actual count of the number of samples. This section illustrates general techniques and formulas for obtaining the number of samples in sample spaces of *ordered* samples. Techniques and formulas for sample spaces of *unordered samples* are the topic of Section 18.2. Some related counting formulas are described in Section 18.3.

Populations are sets of persons, or stimuli, or dosages, etc. The term *element* describes an individual member of a population. A population is, therefore, a set of elements. A sample space is also a set—a set of samples. It is conventional to use the term *sample point* for one sample in a sample space. A sample space is a set of sample points. ("Sample point" is synonymous with "sample", but is used to emphasize that the sample points form the set called the sample space.)

The number of elements in the population is denoted by N and the size of the samples drawn from the population by n. We want to determine the number of samples of size n that can be drawn from a population of size N. Suppose that the population consists of the elements A, B, C, D, E, ... Samples of three elements from this population include ABC, ADE, CAB, CDA, ADC, DAB, ACB, etc. Note that samples such as ABC and ACB, which contain the same elements in different orders, are considered to be different samples. Therefore, the samples are ordered samples. (Another term for an ordered sample is *permutation*.) Sample spaces for ordered samples can be represented by tree diagrams.

Ordered Samples with Replacement

In sampling with replacement, the sample is selected one element at a time, and each element is returned to the population before the next element is drawn. Samples of size n can be represented in an n-stage tree. The first stage represents the first element drawn, the second stage the second element, and so on. There are N branches at the first stage, since the population has N elements from which to choose. Each branch at the first stage is connected to N branches at the second stage, giving $N \times N$ paths through the first two stages of the tree. Similarly, there are $N \times N \times N = N^3$ paths through the first three stages of the tree, and N^n paths in the complete n-stage tree.

$$\boxed{\text{Number of ordered samples with replacement} = N^n}$$

Example 1

Each subject in an experiment is presented with two stimuli in succession. These two stimuli are to be selected from a set (population) of four stimuli. How many different sets of stimuli are possible? The stimuli are labeled 1, 2, 3, and 4. The possible ordered pairs of stimuli are shown as paths in Figure 18.1a. There are $4 \times 4 = 16$ pairs. Note that, for example, the pair 23 and the pair 32 are different sample points in this sample space.

Example 2

If there are 10 stimuli from which three are to be selected, with replacement, the number of ordered samples of three stimuli is $10 \times 10 \times 10 = 10^3 = 1,000$.

Ordered Samples without Replacement from a Large Population

If the population is very large, it matters little whether or not an element is returned to the population before the next element is selected. If the population is large, sampling without replacement is, to all intents and purposes, equivalent to sampling with replacement.

$$\boxed{\begin{array}{c}\text{Number of ordered samples without replacement}\\ \text{from a large population} = N^n\end{array}}$$

Example 3

Three persons are to be selected from a population of 1 million (10^6) persons. If the order in which the persons are selected is important (hence ordered samples), the number of possible samples is $10^6 \times 10^6 \times 10^6 = 10^{18}$.

Ordered Samples without Replacement from a Small Population

The difference between sampling with and sampling without replacement is important when the population is small. The n-stage tree for samples without replacement is similar to the tree for samples with replacement. The first stage represents the first element drawn from N elements (N branches). The second stage represents the second element drawn from the remaining $(N - 1)$ elements ($N - 1$ branches for each branch of the first stage). There are, therefore, $N \times (N - 1)$ paths through the first two stages of the tree. Similarly, there are $N \times (N - 1) \times (N - 2)$ paths through the first three stages of the tree. For an n-stage tree,

$$\boxed{\begin{array}{c}\text{Number of ordered samples without replacement from a small}\\ \text{population} = N \times (N - 1) \times (N - 2) \times \ldots \times (N - n + 1)\\ = \dfrac{N!}{(N - n)!}\end{array}}$$

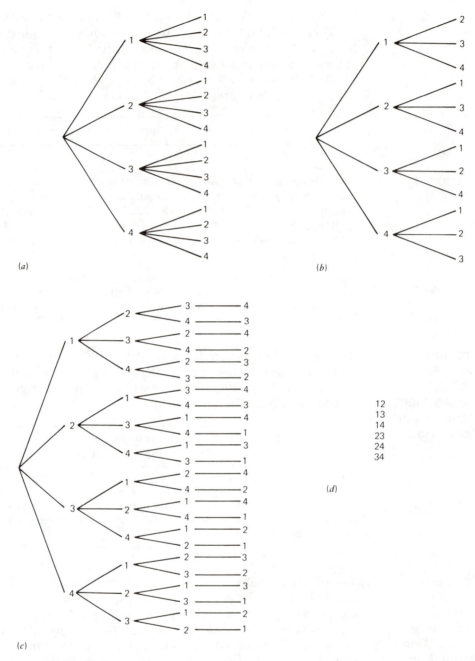

Figure 18.1 Some sample spaces. (a) Ordered samples with replacement. (b) Ordered samples without replacement. (c) Arrangements of a set. (d) Unordered samples without replacement.

Example 4   In Example 1, two stimuli were selected from four, with replacement. The sample space for selecting two stimuli from four, without replacement, is shown in Figure 18.1b. The number of samples is $4 \times 3 = 12$. Note that the stimulus pair 22 appears in Figure 18.1a but not in Figure 18.1b. There are four pairs (11, 22, 33, 44) in which both stimuli are the same—which accounts for the difference between the 16 samples in the former figure and the 12 samples in the latter.

Example 5   In Example 2, three stimuli were selected from 10, with replacement. When this selection is done without replacement, the number of samples is $10 \times 9 \times 8 = 720$. This answer can also be found from $10!/7!$.

Example 6   In Example 3, three persons were selected from 10^6 persons, without replacement. Strictly speaking, using the formula for samples without replacement, the number of samples is $10^6 \times (10^6 - 1) \times (10^6 - 2)$, which is slightly less than 10^{18}. The difference is insignificant, however, and the formula N^n can be used when the population is large.

Arrangements of a Set

The preceding formula for the number of ordered samples without replacement from a small population takes a simple form if the samples are the same size as the population ($n = N$). The number of samples is, then, exactly the number of ways in which a set (the population) can be arranged.

$$
\begin{aligned}
&\text{Number of arrangements of a set} \\
&= N \times (N - 1) \times (N - 2) \times \ldots \times 1 \\
&= N!
\end{aligned}
$$

The tree diagram has N stages. At the first stage, there are N branches; at the second stage, there are $(N - 1)$ branches for each branch of the first stage; at the third stage, there are $(N - 2)$ branches for each path through the first two stages of the tree; and so on until at the Nth stage there is only one branch for each path through the preceding $(N - 1)$ stages of the tree.

Example 7   Consider again the four stimuli. The ways in which all the stimuli can be arranged are shown in Figure 18.1c. There are $4 \times 3 \times 2 \times 1 = 4! = 24$ ways. (These 24 arrangements can be thought of as ordered samples of size 4, from a population of size 4. Figure 18.1c is simply an extension of Figure 18.1b to four stages.)

18.2 Unordered Samples or Combinations

The order in which the elements of a sample appear usually is not significant. A sample of three persons (A, B, and C) is the same sample whether we write it as ABC, ACB, or BAC. We are counting *unordered* samples when we consider all samples containing the same elements as just one sample. Such samples are called *combinations*. (Here we are, for simplicity, considering only sampling without replacement.) We simply list the different samples because it is difficult to represent unordered samples in a tree diagram.

The number of unordered samples is less than the number of ordered samples, since ordered samples with the same elements are counted individually in the total of ordered samples but as only one sample in the total of unordered samples. If, for example, the sample size is $n = 3$, there are $N!/(N - 3)! = N \times (N - 1) \times (N - 2)$ ordered samples. Consider ordered samples that contain the three elements A, B, and C. There are six such samples: ABC, ACB, BAC, BCA, CAB, CBA. The sample space of ordered samples can be arranged in sets of six samples, the six samples of each set being the permutations of the same three elements. In the sample space of *unordered* samples, each of these sets of six samples appears only once. This means that the number of unordered samples equals the number of ordered samples, divided by 6.

Before the general formula for the number of ordered samples is given, it must be noted that the number 6 is the number of permutations, or arrangements, of a set of three elements. In general, the sample size is n and we require the number of permutations of a set of n elements. This number is just $n!$. (Refer to the last formula of the preceding section—here the size of the permuted set is n, not N.)

The number of unordered samples of size n equals the number of ordered samples, divided by $n!$.

$$\text{Number of unordered samples without replacement}$$
$$= \frac{N!}{(N - n)!} \div n!$$
$$= \frac{N!}{n! (N - n)!}$$

This formula is given a special symbol: $\binom{N}{n}$, which is read "N on n". The symbol $\binom{N}{n}$ means "the number of unordered samples of size n that can be drawn from a population of size N". For example, $\binom{6}{3}$, which has the value 20, means the number of unordered samples of size 3 that can be drawn from a population of size 6. (Note that the symbol does *not* mean 6 divided by 3.)

Example 1

The unordered samples of two stimuli selected from a population of four stimuli are listed in Figure 18.1d. In this listing the order of the stimuli in each sample is arbitrary—for convenience, the numbers in the sample are in numeric order. The sample 24, for example, represents the sample in which the stimuli 2 and 4 appear, in either order. By comparing the list in Figure 18.1d with the ordered samples in Figure 18.1b, you can see that for each unordered sample there are 2 (=2!) ordered samples. Hence, the number of unordered samples = number of ordered samples ÷ 2, in this case.

Example 2

Suppose that two subjects are to be selected from a pool of five subjects. In how many ways can this be done? The answer is $\binom{5}{2} = 10$ ways. If the subjects are labeled A, B, C, D, E, the list of the 10 samples is AB, AC, AD, AE, BC, BD, BE, CD, CE, DE.

Subsets of a Set

Unordered samples without replacement are subsets of the original population or set of elements. Therefore, the number of subsets of size n that can be formed from a set of size N is

$$\text{Number of subsets} = \frac{N!}{n! (N - n)!} = \binom{N}{n}$$

Example 3

The list of samples Figure 18.1d can be considered to be the list of subsets of size 2 that can be formed from a set of size 4. There are $4!/(2!2!) = 6$ subsets. Similarly, the list of 10 samples in Example 2 are also the 10 subsets of size 2 that can be formed from a set of size 5.

18.3 The Product Rule

Many studies can be thought of as sequences of trials. Each trial is represented as a stage in the tree diagram. For example, if a sample is drawn with replacement, each trial or stage represents the selection of one member of the population. If 10 coins are tossed, each coin can be considered as one trial; the sample space consists of 10 stages, one for each trial. If a coin and a die are tossed, there are two trials—one for the coin and one for the die.

It is easy to count the number of different sequences in these sample spaces. Suppose there are N_1 possible results on the first trial, N_2 possible results on the second trial, and so on to N_n possible results on the nth trial.

$$\text{Number of different sequences} = N_1 \times N_2 \times \ldots \times N_n$$

All the formulas in Section 18.1 are special cases of this *product rule*. For example, the formula for the number of ordered samples with replacement, N^n, is a special case of the product rule applicable when $N_1 = N_2 = \ldots = N_n = N$. The examples below show other ways in which the product rule can be used.

Example 1

A die and a coin are tossed. How many possible results are there? The die has six possible results; the coin has two. The total number of possible results is $6 \times 2 = 12$. (For each of the six possible results of the die, there are two possible results of the coin, giving $6 \times 2 = 12$ results in all.)

Example 2

How many paths are there in a binomial tree with n stages? Each stage represents a trial with two possible results. Hence, the number of paths $= 2 \times 2 \times \ldots \times 2 = 2^n$.

Example 3

A sample of size 7 is selected, without replacement, from a population of 22 persons, 10 of whom are male and 12 of whom are female. How many samples are possible and, of these, how many have exactly three males and four females?

There are $\binom{22}{7} = 170{,}544$ possible samples. There are $\binom{10}{3} = 120$ ways to select three males and $\binom{12}{4} = 495$ ways to select four females. For *each* of the 120 ways in which the males can be selected, there are 495 ways to select the females. Hence, the number of possible samples with three males and four females is $120 \times 495 = 59{,}400$.

18.4 The Binomial Formula

Several examples of studies that lead to binomial trees and to the binomial distribution were given in Chapter 5. This section provides a proof of the binomial formula: that is, the formula for the binomial probabilities.

The binomial sample space for $n = 5$ is shown in Figure 18.2. In the tree, success is indicated by S and failure by F. The number of paths with, say, three successes is $5!/(3!2!) = 10$. This formula is a special case of the formula for the number of subsets of a given size that can be formed from a set (Section 18.2). Think of a binomial study with $n = 5$ as consisting of five trials. The number of

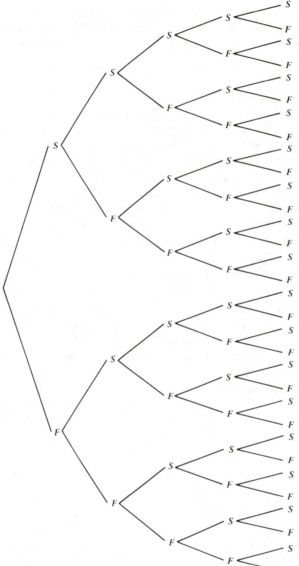

Figure 18.2 Binomial sample space.

samples with three successes is the number of ways of choosing three trials of the five to be S (and the remaining two trials to be F). The three trials that are S are subsets of size 3 formed from the set of five trials. Hence, the number of samples with three successes is the number of subsets of size 3 that can be formed from a set of size 5. The formula given in Section 18.2 states that this number is $\binom{5}{3}$. In general, the number of paths with y successes in n trials is $\binom{n}{y}$. This symbol often is called the *binomial coefficient*.

If the probabilities of S and F are p and q, respectively, the probability of a particular path with y successes and $(n - y)$ failures can be found by multiplying probabilities along the path in the tree. This probability is $p^y q^{n-y}$. Since there are $\binom{n}{y}$ paths with y successes, the total probability of y successes is given by the binomial formula

$$P(y) = \binom{n}{y}p^y q^{n-y} = \frac{n!}{y!\,(n-y)!}\,p^y q^{n-y}$$

This formula was given, without proof, in Section 5.6.

18.5 Events and Their Probabilities

The sample space provides a full description of the possible results in an experiment or survey. This description is too detailed for most purposes. Recall that in a binomial study, we usually group together the sample points having the same number of successes and consider the probability of that group (e.g., the probability of seven successes). Such a group of one or more sample points is given a special name: an *event*.

An event is a group of sample points that are considered together. If the event consists of only one sample point, it sometimes is called an *elementary event* for emphasis.

Example 1

Consider an experiment in which two dice are tossed. The 36 sample points in this experiment may be represented in a tree diagram (Figure 18.3*a*) or in a matrix (Figure 18.3*b*). The latter representation is convenient for representing certain events. Consider the event A, in which both dice are the same: event A occurs if the sample points are 11, 22, 33, 44, 55, or 66. These six sample points are labeled "A" in the figure. The event B is the event that the sum of the two dice is nine or greater. The sample points in B are 36, 45, 46, 54, 55, 56, 63, 64, 65, 66. The event C is the event that the first die is 3, and the second die is any value. This event consists of the sample points 31, 32, 33, 34, 35, 36.

In many sample spaces the sample points are equally likely: each sample point has the same probability. Random sampling was defined, in Section 5.3, as sampling in which every member of the population has an equal chance of being selected. The probabilities on the branches at any one stage of the tree diagram are equal; therefore, when the probabilities are multiplied along each path to obtain the probabilities of each sample point, all the sample points have the same probability.

It is easy to calculate the probability of any event when the sample points are equally likely. The probability is simply the ratio of the number of sample points in the event to the number of sample points in the sample space.

$$P(A) = \frac{\text{Number of sample points in } A}{\text{Number of sample points in sample space}}$$
$$\text{(all sample points equally likely)}$$

Example 2

When two dice are tossed, the 36 sample points are equally likely. The three events A, B, C, described in the preceding example, have 6, 10, and 6 sample points, respectively. Therefore, $P(A) = 6/36 = 1/6$; $P(B) = 10/36 = 5/18$; and, $P(C) = 6/36 = 1/6$.

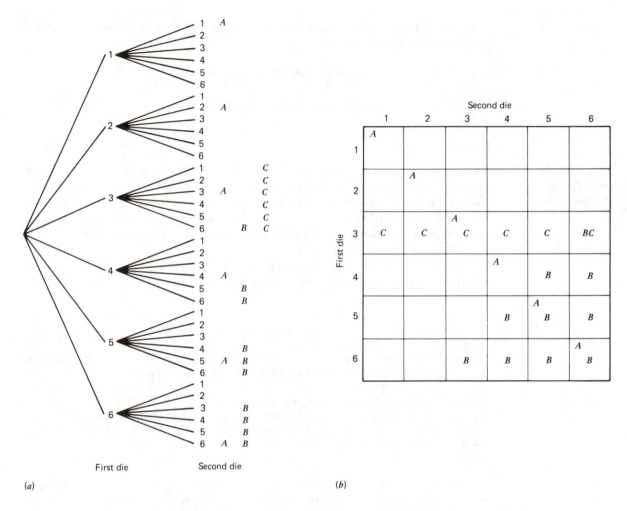

Figure 18.3 Events when two dice are tossed. (*a*) Tree diagram. (*b*) Alternative representation of sample space.

This simple technique cannot be used when the sample points have different probabilities. The probability of any event is obtained by adding the probabilities of the sample points that are included in that event.

$$P(A) = \text{sum of probabilities of sample points in } A$$

Example 3

Consider a binomial study with $n = 5$ and probability of success = 0.2; the sample points are *not* equally likely. What is the probability that four or five successes are obtained? The sample points in the specified event are *SSSSF, SSSFS, SSFSS, SFSSS, FSSSS*, and *SSSSS*. As you can see by drawing a tree diagram, each of the first five sample points has probability $(0.2)^4 \times 0.8 = 0.00128$, and the last sample point has probability $(0.2)^5 = 0.00032$. The probability of the specified event is calculated from the above rule as

$$\begin{aligned}
\text{Probability} &= 0.00128 + 0.00128 + 0.00128 + 0.00128 + 0.00128 \\
&\quad + 0.00032 \\
&= 0.00672
\end{aligned}$$

Example 4   The formula for $P(A)$ as the sum of the probabilities of the sample points in A is quite general and can also be used when the sample points are equally likely. If the number of sample points in the sample space is denoted by $n(S)$, the probability of each is $1/n(S)$. If event A has $n(A)$ sample points, its probability is the sum of $1/n(S)$ taken $n(A)$ times, or, simply, $n(A)/n(S)$, which is the formula for $P(A)$ when the sample points are equally likely.

Union, Intersection, and Complement of Events

Certain events can be defined in terms of other events. The *union* of two events, A and B, is the event that either A or B or both occur. The union of A and B is denoted by $A \cup B$. The *intersection* of two events, A and B, is the event that both A and B occur. The intersection of A and B is denoted by AB. (In some texts the notation $A \cap B$ is used.) The *complement* of an event A is the event that A does not occur. The complement of A is denoted by $\bar{A}$.

It sometimes is useful to represent the union, intersection, and complement of events by a *Venn diagram* (Figure 18.4a). Here the sample space is represented by a large rectangle; each of the elementary events is considered to be located inside the rectangle. Events (groups of elementary events) are repre-

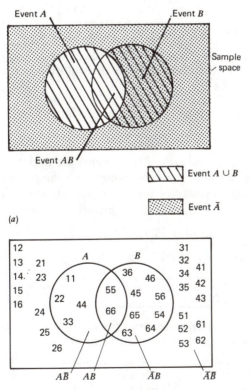

(a)

(b)

Figure 18.4   Venn diagrams. (a) General sample space. (b) Sample space of two dice.

sented by areas, often circles, inside the rectangle. The events A and B are shown as overlapping circles in Figure 18.4a. They overlap in order to indicate that some elementary events are in both A and B: that is, A and B can occur simultaneously. The set of elementary events in both A and B is the intersection of these events, AB, represented by the area of overlap.

The union of the two events is represented by the areas A and B taken together. The complement of the event A is all the area in the sample space except the area representing A.

Example 5

The sample space of two dice and the two events A and B, previously defined in Example 1, are represented in Figure 18.4b. The elementary events in A (11, 22, 33, 44, 55, 66) are shown. Two of these elementary events (55, 66) are also in B; hence, these two elementary events are placed in the intersection, AB. If, for example, both dice are 5, then both A (two dice the same) and B (sum of dice is nine or more) have occurred: that is, the intersection AB has occurred. The other eight elementary events in B that are not in A are placed in the part of the area representing B that is not part of the area for A. The many elementary events that are in neither A nor B are placed outside the areas of A and B but inside the rectangle representing the full sample space.

The probability of events and of their union, intersection, and complement are related by two formulas. First, the probability that an event occurs and the probability that it does not occur must sum to 1.0. Hence, the probability of the complement of an event is 1.0 minus the probability of the event.

$$\boxed{P(\bar{A}) = 1 - P(A)}$$

Example 6

In Example 2 (tossing two dice), the probability that A occurs is $P(A) = 1/6$. The probability that A does not occur (i.e., the probability that the two dice are not the same) is

$$P(\bar{A}) = 1 - P(A) = 1 - \frac{1}{6} = \frac{5}{6}.$$

Second, if the probabilities of two events are added, we obtain the probability of the union of the two events, except that the probability of the intersection is counted twice instead of just once (see Figure 18.4a). Hence, $P(AB)$ is the difference between the sum of the probabilities and $P(A \cup B)$.

$$P(AB) = P(A) + P(B) - P(A \cup B).$$

This formula usually is written in the following, equivalent, form.

$$\boxed{P(A \cup B) = P(A) + P(B) - P(AB)}$$

The above two formulas are true for all sample spaces. They hold if the sample points are equally likely or not equally likely.

Example 7

The events A and B in the example of tossing two dice have probabilities $P(A) = 1/6$ and $P(B) = 10/36$ (see Example 2). Suppose we want to obtain $P(A \cup B)$. We must first obtain $P(AB)$. Both A and B occur if and only if both dice are 5 or both are 6. Hence $P(AB) =$

2/36. The probability that either A or B occurs (i.e., the probability that the two dice are the same or that the dice sum to nine or more, or both) is

$$P(A \cup B) = P(A) + P(B) - P(AB)$$

$$= \frac{1}{6} + \frac{10}{36} - \frac{2}{36}$$

$$= \frac{6}{36} + \frac{10}{36} - \frac{2}{36}$$

$$= \frac{14}{36} = \frac{7}{18}$$

This answer can be checked, of course, by direct counting in the sample spaces (either in Figure 18.3b or Figure 18.4b).

Two events that cannot occur simultaneously are said to be *mutually exclusive*. There are no sample points in the intersection of two mutually exclusive events A and B; hence, $P(AB) = 0$. The formula, given above, for the union of two events takes a simple form if the events are mutually exclusive.

$$\boxed{P(A \cup B) = P(A) + P(B) \quad \text{(Mutually exclusive events)}}$$

Example 8

In a binomial study with $n = 5$ and probability of success $= 0.2$ (Example 3), the events A: 4 successes and B: 5 successes are mutually exclusive, since if one occurs the other cannot occur. Given that $P(A) = 0.00640$ and $P(B) = 0.00032$, the probability of getting either four or five successes is

$$P(A \cup B) = P(A) + P(B)$$
$$= 0.00640 + 0.00032 = 0.00672$$

**18.6
Independent
and Dependent
Trials or Stages**

This section defines what can be represented as a stage in a tree diagram, as well as what it means to say that a sequence of stages (or trials) is independent. If the stages are not independent, they are called dependent.

We are familiar with tree diagrams in which each stage represents a trial in an experiment. For example, Figure 18.5a shows a sequence of three trials: at each trial a coin is tossed. There are eight sample points in this tree: $H_1H_2H_3$, $H_1H_2T_3, \ldots, T_1T_2T_3$. Two events can be defined in terms of the result of the first coin: H_1 (first coin is heads) consisting of the four sample points $H_1H_2H_3$, $H_1H_2T_3, H_1T_2H_3, H_1T_2T_3$, and T_1 (first coin is tails) consisting of the four sample points $T_1H_2H_3, T_1H_2T_3, T_1T_2H_3, T_1T_2T_3$. Note that these two events, H_1 and T_1, are mutually exclusive. Similarly, two events H_2 (consisting of the sample points $H_1 H_2 H_3, H_1 H_2T_3, T_1 H_2 H_3, T_1 H_2 T_3$) and T_2 (consisting of the sample points $H_1T_2 H_3, H_1T_2T_3, T_1T_2 H_3, T_1T_2T_3$), defined in terms of the result of the second coin, are mutually exclusive. *A stage in a tree diagram always consists of two or more mutually exclusive events.* The first stage in Figure 18.5a consists of the two mutually exclusive events H_1 and T_1. The second stage consists of the two mutually exclusive events H_2 and T_2. Similarly, the third stage consists of two mutually exclusive events H_3 and T_3. For convenience, the events that define a stage can be called *stage events*.

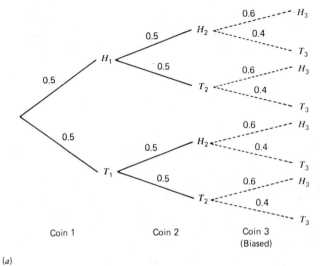

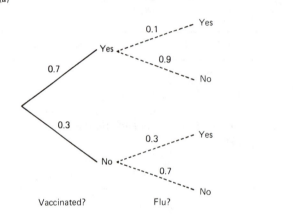

Figure 18.5 Stages in tree diagrams. (a) Three coins (one is biased). (b) Vaccination for the flu.

The stages represented in a tree diagram need not be trials in an experiment. The only requirement is that the branches at the stage represent mutually exclusive events. Consider Figure 18.5b, which represents some events in a study of the effectiveness of a flu vaccine. The first stage represents the distinction between those receiving the vaccine and those not receiving it. These two events are mutually exclusive. The second stage represents the distinction between those getting the flu (in a certain time interval) and those not getting it. Again, these are mutually exclusive events that may be represented as a stage. In the figure, 10% of those vaccinated get the flu, whereas 30% of those who are not vaccinated get the flu. Note that the sum of the probabilities of those getting and not getting the flu is 1.0 for the vaccinated persons (0.1 + 0.9), and that this sum is also 1.0 for the nonvaccinated persons (0.3 + 0.7). These results are consistent with the two events (getting and not getting the flu) being mutually exclusive.

We now want to define what it means to say that a sequence of trials (or stages) is independent or dependent. The definition is based on the tree diagram of the sample space. A sequence of trials is a sequence of *independent trials* if for each trial the subtrees are all identical. A sequence of trials is a sequence of *dependent trials* if for any trial the subtrees are not all identical.

Example 1

In Figure 18.5a the subtrees for the third trial are dashed. The four subtrees are all identical: each has two branches, and in each the probability of H is 0.6 and the probability of T is 0.4. Similarly, the two subtrees for the second trial are identical: each subtree has two branches and the two probabilities are 0.5 and 0.5 in each. Hence, the three trials are independent. Note carefully that the independence of the trials is a different characteristic from the fairness or bias of the coins. The third coin is assumed to be biased; nevertheless, the three coins are independent, since the bias of the third coin is the same for all possible tosses of the first two coins. (If the third coin was biased only when the first two coins were heads, the three coins would be dependent, not independent.)

Example 2

The subtrees for the second stage in Figure 18.5b are dashed. Note that these subtrees are different, and therefore that the two stages are dependent. We usually say that getting the flu is dependent on getting the vaccine. If you are vaccinated, there is a 10% chance of getting the flu; if you are not vaccinated, this chance is changed to 30%. Since these percentages (probabilities) are different, the two stages are dependent.

18.7 Conditional Probability

We have been referring to each probability in a tree diagram simply as a "probability". This section points out that there actually are three different types of probabilities in a tree diagram.

Three Types of Probabilities

The tree diagram in Figure 18.6 illustrates two aspects of the result of tossing two dice. (Refer back to Figure 18.4b to check the probabilities.) The first stage in the tree represents whether or not the dice are different: one branch represents "same" (event A) and the other branch represents "different" (event $\bar{A}$). The second stage in the tree represents the sum of the dice: one branch represents "nine or more" (event B) and the other branch represents "less than nine" (event $\bar{B}$).

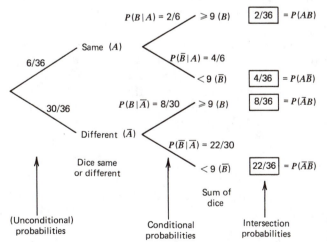

Figure 18.6 Conditional probability.

The four paths through the tree represent *intersections*. The top path represents the result that the dice are the same (*A*) and that the dice sum to nine or more (*B*). Hence, this path represents the event *AB*. The probability of this event is 2/36 (from Figure 18.4*b* and Example 7 of Section 18.5). The probabilities of paths in a tree are *intersection probabilities*. The other paths in the tree represent the intersections $A\bar{B}$ (*A* occurs and *B* does not occur), $\bar{A}B$ (*A* does not occur and *B* occurs), and $\bar{A}\bar{B}$ (both *A* and *B* do not occur). Their probabilities, obtained by counting elementary events in the sample space, are shown in Figure 18.6.

Next, consider the probabilities of the branches at the first stage of the tree, which are $P(A) = 6/36$ and $P(\bar{A}) = 30/36$. These are probabilities of basic events *A* and $\bar{A}$, and therefore are "regular" probabilities. They are sometimes called *unconditional probabilities*, for a reason given below. [The adjective "unconditional" actually is redundant; we usually will call $P(A)$ and similar probabilities simply "probabilities".]

Finally, consider the probabilities on the branches of the second stage. These probabilities are known as *conditional probabilities*. The events at the ends of the branches of the second stage are *conditional* on the events of the first stage. For example, the top branch at the second stage represents the event that the sum of the dice is nine or more (*B*), given that the dice are the same (*A*). This *conditional event* is denoted by $B|A$, which is read "*B* conditional on *A*" or "*B* given *A*". The probability of this conditional event is denoted by $P(B|A)$, which is read "the probability of *B* conditional on *A*" or "the probability of *B* given *A*". Conditional probabilities can be calculated from the rule that states that the probability of a path (intersection probability) is the product of the probabilities on the branches of the path. This means that

$$P(A) \cdot P(B|A) = P(AB)$$

Hence,

$$P(B|A) = \frac{P(AB)}{P(A)}$$

In Figure 18.6, we find $P(B|A) = P(AB)/P(A) = 2/36 \div 6/36 = 2/6$. The other conditional probabilities in the figure are calculated in the same way: that is, as the ratio of the intersection probability to the probability of the event in the first stage.

When there are more than two stages in a tree, the probabilities of, say, the fourth stage are conditional on the events in the branches of the first three stages. Such a probability is written, for example, $P(D|A\bar{B}C)$, denoting the probability that *D* occurs given that *A* and *C* have occured and *B* has not occurred. The probabilities in the first stage are not conditional on any events having occurred and therefore can be called unconditional probabilities.

Independence and Dependence of Events

Conditional and intersection probabilities are simple to calculate when the stages of the tree diagram are independent, since in that case all the subtrees at a given stage are identical. (Recall Section 18.6.) Consider, for example, the independent tossing of a coin and die (Figure 18.7).

Conditional Probabilities The independence of the two stages (trials) implies that each conditional probability in the top subtree at the second stage is

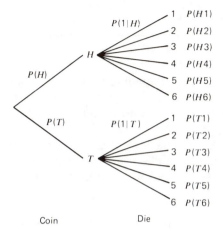

Figure 18.7 Independent events.

equal to the corresponding conditional probability in the bottom subtree. For example,

$$P(1|H) = P(1|T)$$

Let us calculate the probability that the die is 1.

$$P(1) = P(H1) + P(T1)$$
$$= P(1|H) \cdot P(H) + P(1|T) \cdot P(T)$$

Since $P(1|T) = P(1|H)$, this can be written as

$$P(1) = P(1|H) \cdot P(H) + P(1|H) \cdot P(T)$$
$$= P(1|H) [P(H) + P(T)]$$
$$= P(1|H) \cdot 1$$
$$= P(1|H)$$

Turning this formula around, we see that the *conditional* probability that the die is 1, given that the coin is heads, is equal to the *unconditional* probability that the die is 1.

This relationship holds for all events defined at the second stage. For example, $P(5|T) = P(5)$. By reversing the stages in the tree, we also can show that conditional probabilities of the coin, given a result on the die, are equal to unconditional probabilities of the coin—for example, $P(T|2) = P(T)$.

In general, if two stages are independent, and if A is an event defined at one stage and B is an event defined at the other stage, the conditional probabilities are equal to the unconditional probabilities.

$$\boxed{P(A|B) = P(A) \text{ and } P(B|A) = P(B)}$$

Intersection Probabilities The intersection probability of AB, obtained by multiplying probabilities along a path, is

$$P(AB) = P(A) \cdot P(B|A)$$

But $P(B|A) = P(B)$; therefore,

$$\boxed{P(AB) = P(A) \cdot P(B)}$$

In general, if two stages are independent, and if A is an event defined at one stage and B is an event defined at the other stage, the intersection probability of A and B is the product of the unconditional probabilities.

Independence of Events The formulas derived above on the basis of the independence of two *stages* can be used as the definition of the independence of two *events*. Two events A and B are independent if

$$P(A|B) = P(A)$$

As this formula is equivalent to $P(B|A) = P(B)$ and to $P(AB) = P(A) \cdot P(B)$, the independence of two events also can be defined in terms of either of these two formulas. Two events, A and B, are dependent if $P(A|B) \neq P(A)$.

[It should be noted that independence of stages implies the independence of every pair of events, one event defined at each stage; but that the independence of two stage events, A and B, one at each stage, does not necessarily imply that the two stages are independent. There are two situations in which we can infer that two stages are independent from the independence of two stage events: (1) if *every* event at one stage is independent of *every* event at the other stage, then the stages are independent; (2) if A is independent of B and one stage consists of A and its complement, $\bar{A}$, and the other stage consists of B and $\bar{B}$, then the two stages are independent.]

Example

A fair die is tossed three times. What is the probability that it is even on each of the first two tosses and 5 or 6 on the last toss?

Since the tosses are independent, the events on the different tosses are independent. Hence, the intersection probability can be calculated as the product of the unconditional probabilities:

$$P(\text{even, even, 5 or 6}) = P(\text{even}) \cdot P(\text{even}) \cdot P(\text{5 or 6})$$
$$= \frac{3}{6} \times \frac{3}{6} \times \frac{2}{6} = \frac{1}{2} \times \frac{1}{2} \times \frac{1}{3} = \frac{1}{12}$$

18.8 Summary

Several ways of counting the number of samples in a sample space were described. The number of *ordered samples* is called the number of *permutations*. The number of *unordered samples* is called the number of *combinations*. The *binomial formula* was derived.

An *event* is a group, or set, of samples (*sample points*). The probability of an event can be calculated from the number of sample points in the event, if the sample points are equally likely; or by adding the probabilities of the sample points in the event, if they are not equally likely. New events can be defined by the *union*, *intersection*, or *complement* of other events.

The terms *independence* and *dependence* were defined for *trials* (or *stages*) and for *events*. These definitions depend on the distinction between three types of probability: *intersection (joint)*, *conditional*, and *unconditional* (*marginal*) probabilities.

Exercises

1 (Section 18.1) A type of combination lock has 36 positions on its dial (1 through 36). How many different combinations of three numbers can be manufactured?

2 (Section 18.1) Extension-telephone numbers are four digits long. Any digit from 1 to 9 may appear in any position in the extension number; zero is not used in extension numbers.
(a) How many different extension numbers are possible?
(b) How many extension numbers have all four digits different?

3 (Section 18.1) Twelve subjects are selected to take part in an experiment. Each subject is to read a short story. There are 12 different stories, and each subject will read one of them. In how many ways can the 12 stories be assigned to the 12 subjects so that every subject reads a different one? (Leave your answer as an expression).

4 (Section 18.2) There are 10 persons in a department and a committee of seven persons is to be formed.
(a) In how many ways can the committee be selected?
(b) In how many ways would committees of three persons be selected from the 10 persons in the department?
(c) Can you explain the relationship between the answers to **(a)** and **(b)**?

5 (Section 18.2) Samples of size 10 are drawn from a population of 50 persons, 25 of whom are male and 25 of whom are female. (Leave your answers as expressions).
(a) How many different samples can be drawn?
(b) How many of these samples are all female?

6 (Section 18.3) Suppose that people are classified on the basis of a survey by sex, age (four categories), and income (six categories). Into how many different groups are the people classified?

7 (Section 18.3) This question is a continuation of Question 5. Use the product rule to calculate how many samples have five females and five males. (Leave your answer as an expression).

8 (Section 18.4) The chance that a given person will have a car accident in one year is 0.01. If the probabilities of persons having car accidents are independent, what is the probability, in a group of four persons, that at least one of them had an accident in the last year?

9 (Section 18.5) This question is a continuation of Question 2. Assume that all extension numbers are equally likely.
(a) Let A be the event that all the digits in the extension number are different. What is $P(A)$?
(b) Let B be the event that all the digits are the same. What is $P(B)$?
(c) Let C be the event that all the four digits in the number are odd (1, 3, 5, 7, 9). What is $P(C)$?
(d) Let D be the event that all the four digits are even (2, 4, 6, 8). What is $P(D)$?
(e) What are the probabilities of the following events.
  (i) Not all the digits are different (i.e., there are some duplicate digits).
  (ii) All the digits are even or all the digits are odd.
  (iii) All the digits are the same or all the digits are even (or both).
  (iv) All the digits are even and they are all different.
  (v) All the digits are even *or* they are all different (or both).

10 (Section 18.5) This question is a continuation of Questions 5 and 7. (Leave your answers as expressions).
(a) What is the probability that a randomly drawn sample is all female?
(b) What is the probability that a randomly drawn sample has five females and five males?

11 (Section 18.5) There are four aces in a deck of 52 playing cards. What is the probability that a randomly dealt hand of 13 cards contains exactly three aces? (Leave your answer as an expression).

12 (Section 18.6) You are going to attend a dinner party. If it is a hot day, the probability that hot soup will be served is 0.3 and the probability that cold soup will be served is

0.7. If it is a cool day, the probabilities are 0.6 and 0.4, respectively. Draw a tree diagram to represent this situation and state whether the temperature of the soup that will be served is dependent on or independent of the weather.

13 (Section 18.6) A certain gene G is linked to a disease of the spinal cord S. The gene occurs in 10% of the population. Of those persons having the gene, 40% show the symptoms of the disease. Of those persons not having the gene, only 5% show the symptoms.

(a) Is the disease independent of having the gene? Prove your answer.

(b) Change one or more numbers in the question so that having the gene is independent of having the disease.

14 (Section 18.7) This question is a continuation of Questions 5 and 10a. Recalculate the probability that the sample is all female by using conditional probabilities. Consider the probability that the first person selected is female, the probability that the second person is female given that the first person is female, and so on. (Leave your answer as an expression). This question illustrates that many probability questions, including others in this set, can be solved in several ways.

15 (Section 18.7) The 10 questions on a final exam are chosen at random from a pool of 50 questions. The students in the course have had the list of the 50 questions before the exam so that they may study them. Suppose a student has time to study only 20 of the questions. What is the probability that all 10 questions on the exam are questions that he has studied? (Leave your answer as an expression).

Answers To Exercises

Chapter 3

1 (a) 44.0

2 (a) 26.5

(b) 27 or 25–29

3 (a) 24.6

5 (a) 32.067347

(b) 32.07

6 46.99, 46.98, 47.00, 46.0, 46.0

Chapter 4

1 (a) 12, 79

(b) 86

2 (a) 88

(b) 53

3 (a) 5.7, 6.2, 7.2

(b) 12

(c) 6.2

4 (a) 7, 2, 3, 1.5

(b) 5.1, 1.5, 2.7, 1.4

5 (a) 8.3, 2.9

(b) 275.2, 16.6

Chapter 5

1 (b) 36

2 (b) 4

(d) 32

3 (a) 1/3, 2/3

(b) 10

(c) .0329

(d) .329

4 (a) 1/3, 2/3

6 (c) .9546, .0447, .0007, .0000

7 (a) .0162

(b) .3770

Chapter 6

3 (a) .029, .970

(b) .006

(c) .205, .377

(d) .512, .575

4 (a) Yes, $P(y \geqq 18) = 0.021$

(b) .72

(c) No, $P(y \geqq 16) = .114$; .64

5 (b) 6.0, 1.4, 1.18

(c) Bernoulli: 0.2, 0.4; binomial: 1.0, 0.89

Chapter 7

1 (a)   34, 1.414
   (b) $z = -0.71$ for $x = 33$, etc.
   (c) $z = -0.71$ for $y = 11$, etc.
   (d)   22nd percentile for $x = 33$ and $y = 11$, etc.
   (e)   Standard scores
   (f)   Uniform distribution
2 (a)   .6826, .9974, .4772, .8400
   (b)   0, 1, normal; $(-1, 1)$, $(-3, 3)$, $(-2, 0)$, $(-1, 3)$
3 (a)   .4686
   (b)   .8873
   (c) -0.674
   (d) $(-1.645, 1.645)$
   (e)   .253
   (f)   .1156
   (g)   56.74
   (h)   (38.5, 61.5)
   (i)   .6915
   (j)   0.0
4 (a)   .0885
   (b)   .0386
   (c) $-.674$, .000, $+.674$
   (d) $(-1.282, 1.282)$
   (e)   .95
   (f) $-.95$
   (g)   .18
   (h)   (89.89, 110.11)
   (i)   .8413
   (j)   .25
5   0.8413
6   53.42 minutes
7 (a)   Normal, 65, 0.96
   (b)   1.000
   (c)   .6826, .9544
8   .24
9 (a) $P(\bar{y} < 98.23) = 0.10$; accept hypothesis that mean $= 100$
   (b) 98.23 ± 2.78

Chapter 8

2 (a)   Relationship
   (b)   No relationship
4 (a)   4.3, 0.0
   (b)   1.82
5   0.5, approximately
6 (a)   0.74
   (b) -0.98

Chapter B

2 (a)   102.4
   (b)   No, $P(y \geqq 18) = 0.005$

(c) Yes, $P(\bar{y} \geq 101) = 0.17$

(d) 0.59

(e) Population means are different, $P\{(\bar{y}_1 - \bar{y}_2) < -3.28\} = P\{z < -2.05\} = 0.02$

Chapter 9

1 0.127, 0.048

2 (a) 16.5

(b) $(y > 16.5)$, $(y < 16.5)$

(c) 0.273

3 Accept H_0

4 (a) 0.586

(b) 0.713

(c) 0.021

5 (a) 0.128, 0.048

(b) 16.5, 0.274

6 (a) 0.0228, $\beta < 0.0001$

(b) Accept H_1

7 (a) 52.06; accept H_0 for $(\bar{y} < 52.06)$; accept H_1 for $(\bar{y} > 52.06)$

(b) 1.645; accept H_0 for $(z_{obs} < 1.645)$; accept H_1 for $(z_{obs} > 1.645)$

(c) Accept H_0

8 (a) $\bar{y}_{crit} = 1.97$; accept H_1

(b) $z_{crit} = 2.326$; accept H_1

(c) 0.11

9 0.13, 0.04, 0.58, 0.72, 0.02

Chapter 10

1 (a) 50, 2

(b) 53, 2

(c) 53.28

(d) 0.56

(e) Less

2 Reject H_0

3 $z_{obs} = -5.54$, reject H_0

5 (b) Fail to reject H_0

6 $z_{obs} = 2.69$, reject H_0

7 (a) .022, $<.05$; .004, $< .005$; .172, $> .10$; .0668, $> .05$; .0062, $< .01$

(b) .044, $< .05$; .008, $< .01$; .344, $> .10$; .1336, $> .10$; .0124, $< .05$

8 (a) 0.032, 0.0072

(b) p-value $< .05$, p-value $< .01$

(d) p-value $= 0.2186$, p-value $> .20$

9 $z_{obs} = -1.735$, fail to reject H_0

Chapter 11

1 3.707, .05, .05, .10, 1.341, .025, .01, .01, .01

2 $t_{obs} = 4.42$, reject H_0

4 $t_{obs} = 1.06$, fail to reject H_0 (p-value $> .20$, two-tailed)

5 $t_{obs} = -2.57$, fail to reject H_0

6 $t_{obs} = 1.625$, fail to reject H_0

Chapter 12

3 (a) 51.3, 3.1

  (b) (46.2, 56.4)

  (d) (47.0, 55.7)

4 (a) $(-7.5, +0.3)$

  (b) $(-2.4, +6.1)$

5 (a) 0.810, 0.012

  (b) (0.79, 0.83)

6 (a) 0.16, 0.06

  (b) (0.04, 0.28)

7 69

8 (a) 92

  (b) 96

Chapter 13

1 (a) 0.238

  (b) 0.957

  (d) 0.341, 0.991

2 (a) 236.98, 0.63, 0.37

  (c) 233.50, 0.05, 0.95

  (d) 237.73, 0.72, 0.28

3 0.37, 0.95, 0.28

4 0.46

5 375

6 (a) 0.50

  (b) 65 per condition

  (c) 0.84σ

Chapter C

1 (a) (94.8, 115.8)

  (b) Fail to reject H_0

  (c) $t_{obs} = 1.48$

2 (a) $t_{obs} = -3.77$

  (b) 90% C. I. will not include zero

  (c) $(-38.4, -14.2)$

5 (a) $y = 8$, fail to reject H_0 ($\alpha = 0.05$)

  (b) $t_{obs} = 2.46$, reject H_0 ($\alpha = 0.05$)

8 (a) $\bar{y} < 116.80$ or $z_{obs} < -1.282$

  (b) Fail to reject H_0

  (c) 0.37

Chapter 14

1 $y = 3$, not significant

2 (a) 3.841, 9.210, 4.605, 0.103

  (b) $0.005 < \text{prob} < 0.01$

  (c) $0.025 < \text{prob} < 0.05$

3 $\chi^2_{obs} = 12.97$, significant at 0.005 level

4 $z_{obs} = -3.56$, p-value < 0.0005

5 $\chi^2_{obs} = 16.77$, significant at 0.005 level

6 $\chi^2_{obs} = 4.16$, not significant

7 $y = 17$, not significant

8 $\chi^2_{obs} = 18.40$, significant at 0.005 level

9 (a) Test of two independent proportions

    (b) Test of five independent medians

    (c) Sign test

    (d) Test of independence of two dichotomous variables (= test of two independent proportions)

Chapter 15

1 $\chi^2_{obs} = 16.62$, significant

2 (a) $\chi^2_{obs} = 25.66$, significant at 0.005 level

    (b) χ^2_{obs} is not well approximated by χ^2

    (c) $\chi^2_{obs} = 17.48$, significant at 0.005 level; A, B and C, D are adjacent categories

3 $\chi^2_{obs} = 4.20$, not significant

4 $W_S = 60$ (or 111), significant at 0.05 level, two-tailed

5 $T_+ = 102.5$ (or 17.5), significant at 0.01 level, one-tailed

6 (a) Test of homogeneity of four independent distributions

    (b) Rank-sum test for two independent groups

    (c) Test of two independent proportions

    (d) Rank-sum test for two dependent groups (or sign test)

    (e) Test of independence of two categorical variables (= test of homogeneity of five independent distributions)

Chapter 16

2 $r = 0.41$, not significant, p-value > 0.05

3 (a) $z_{obs} = 1.487$, not significant

    (b) (0.20, 0.85)

    (c) Yes

4 $z_{obs} = 0.810$, not significant, p-value > 0.40

5 (a) $r = -0.503$

    (b) $b = -0.291$, $a = 4.98$

6 (b) $s'^2_{Tot} = 2.25$, $s'^2_{Pred} = 0.57$

    (c) ratio $= 0.25$

7 (a) 0.40, 0.60

    (b) 0.40, 0.60

8 (a) $r_S = -0.782$

9 (a) Significant

    (b) (0.17, 0.57)

10 Q1: $\hat{\phi}'^2 = 0.028$, 95% C. I. $= (0.01, 0.06)$

    Q2: $\hat{\phi}'^2 = 0.214$, 95% C. I. $= (0.08, 0.41)$

11 (a) $y_{Pred} = 48.66 - 0.494x$

    (b) $x_{Pred} = 67.24 - 0.953y$

    (e) 0.47, 0.53

Chapter 17

1 (b) $s'^2_{Pred} = 77.08$; $s'^2_{Tot} = 208.87$

    (c) $R^2 = 0.37$

2 (a) $SS_{Tot} = 3759.778$; $SS_{BG} = 1387.444$; $SS_{WG} = 2372.334$

    (b) $R^2 = 0.37$

3 (a) 2.51, 3.70

(b) 3.44, 6.16

4 (a) $MS_{BG} = 693.72$; $MS_{WG} = 158.16$; $F_{obs} = 4.39$

(c) Significant

5 $R^2 = 0.37$; $\hat{\omega}^2 = 0.27$

6 (a) $SS_{BG} = 151.58$

(b) $SS_{WG} = 1493.92$

(c) $F_{obs} = 1.22$, not significant (p-value > 0.05)

8 $F_A = 44.67$ (p-value < 0.01); $F_B = 28.00$ (p-value < 0.01); $F_{AB} = 1.66$ (ns)

9 $F_A = 1.70$ (ns); $F_B = 2.14$ (ns); $F_{AB} = 0.02$ (ns)

Chapter 18

1 46656

2 (a) 6561

(b) 3024

3 12!

4 (a) 120

(b) 120

5 (a) 50!/(10!40!)

(b) 25!/(10!15!)

6 48

7 $[25!/(5!20!)]^2$

8 0.0394, to four decimals; Table A1 gives 0.040

9 (a) 0.46

(b) 0.001

(c) 0.095

(d) 0.039

(e) 0.54, 0.134, 0.040, 0.0037, 0.496

10 (a) 25!40!/(15!50!)

(b) $10!40!(25!)^2/[50!(5!20!)^2]$

11 4!48!13!39!/[3!1!10!38!52!]

12 Dependent

13 (a) No

14 25!40!/(15!50!)

15 20!40!/(10!50!)

Appendix

TABLE A1 Individual Binomial Probabilities

n	y	.01	.05	.10	.20	.30	.40	.50	.60	.70	.80	.90	.95	.99	y
2	0	980	902	810	640	490	360	250	160	090	040	010	002	0+	0
	1	020	095	180	320	420	480	500	480	420	320	180	095	020	1
	2	0+	002	010	040	090	160	250	360	490	640	810	902	980	2
3	0	970	857	729	512	343	216	125	064	027	008	001	0+	0+	0
	1	029	135	243	384	441	432	375	288	189	096	027	007	0+	1
	2	0+	007	027	096	189	288	375	432	441	384	243	135	029	2
	3	0+	0+	001	008	027	064	125	216	343	512	729	857	970	3
4	0	961	815	656	410	240	130	062	026	008	002	0+	0+	0+	0
	1	039	171	292	410	412	346	250	154	076	026	004	0+	0+	1
	2	001	014	049	154	265	346	375	346	265	154	049	014	001	2
	3	0+	0+	004	026	076	154	250	346	412	410	292	171	039	3
	4	0+	0+	0+	002	008	026	062	130	240	410	656	815	961	4
5	0	951	774	590	328	168	078	031	010	002	0+	0+	0+	0+	0
	1	048	204	328	410	360	259	156	077	028	006	0+	0+	0+	1
	2	001	021	073	205	309	346	312	230	132	051	008	001	0+	2
	3	0+	001	008	051	132	230	312	346	309	205	073	021	001	3
	4	0+	0+	0+	006	028	077	156	259	360	410	328	204	048	4
	5	0+	0+	0+	0+	002	010	031	078	168	328	590	774	951	5
6	0	941	735	531	262	118	047	016	004	001	0+	0+	0+	0+	0
	1	057	232	354	393	303	187	094	037	010	002	0+	0+	0+	1
	2	001	031	098	246	324	311	234	138	060	015	001	0+	0+	2
	3	0+	002	015	082	185	276	312	276	185	082	015	002	0+	3
	4	0+	0+	001	015	060	138	234	311	324	246	098	031	001	4
	5	0+	0+	0+	002	010	037	094	187	303	393	354	232	057	5
	6	0+	0+	0+	0+	001	004	016	047	118	262	531	735	941	6
7	0	932	698	478	210	082	028	008	002	0+	0+	0+	0+	0+	0
	1	066	257	372	367	247	131	055	017	004	0+	0+	0+	0+	1
	2	002	041	124	275	318	261	164	077	025	004	0+	0+	0+	2
	3	0+	004	023	115	227	290	273	194	097	029	003	0+	0+	3
	4	0+	0+	003	029	097	194	273	290	227	115	023	004	0+	4
	5	0+	0+	0+	004	025	077	164	261	318	275	124	041	002	5
	6	0+	0+	0+	0+	004	017	055	131	247	367	372	257	066	6
	7	0+	0+	0+	0+	0+	002	008	028	082	210	478	698	932	7
8	0	923	663	430	168	058	017	004	001	0+	0+	0+	0+	0+	0
	1	075	279	383	336	198	090	031	008	001	0+	0+	0+	0+	1
	2	003	051	149	294	296	209	109	041	010	001	0+	0+	0+	2
	3	0+	005	033	147	254	279	219	124	047	009	0+	0+	0+	3
	4	0+	0+	005	046	136	232	273	232	136	046	005	0+	0+	4
	5	0+	0+	0+	009	047	124	219	279	254	147	033	005	0+	5
	6	0+	0+	0+	001	010	041	109	209	296	294	149	051	003	6
	7	0+	0+	0+	0+	001	008	031	090	198	336	383	279	075	7
	8	0+	0+	0+	0+	0+	001	004	017	058	168	430	663	923	8

TABLE A1 *(continued)*

n	y	.01	.05	.10	.20	.30	.40	p .50	.60	.70	.80	.90	.95	.99	y
9	0	914	630	387	134	040	010	002	0+	0+	0+	0+	0+	0+	0
	1	083	299	387	302	156	060	018	004	0+	0+	0+	0+	0+	1
	2	003	063	172	302	267	161	070	021	004	0+	0+	0+	0+	2
	3	0+	008	045	176	267	251	164	074	021	003	0+	0+	0+	3
	4	0+	001	007	066	172	251	246	167	074	017	001	0+	0+	4
	5	0+	0+	001	017	074	167	246	251	172	066	007	001	0+	5
	6	0+	0+	0+	003	021	074	164	251	267	176	045	008	0+	6
	7	0+	0+	0+	0+	004	021	070	161	267	302	172	063	003	7
	8	0+	0+	0+	0+	0+	004	018	060	156	302	387	299	083	8
	9	0+	0+	0+	0+	0+	0+	002	010	040	134	387	630	914	9
10	0	904	599	349	107	028	006	001	0+	0+	0+	0+	0+	0+	0
	1	091	315	387	268	121	040	010	002	0+	0+	0+	0+	0+	1
	2	004	075	194	302	233	121	044	011	001	0+	0+	0+	0+	2
	3	0+	010	057	201	267	215	117	042	009	001	0+	0+	0+	3
	4	0+	001	011	088	200	251	205	111	037	006	0+	0+	0+	4
	5	0+	0+	001	026	103	201	246	201	103	026	001	0+	0+	5
	6	0+	0+	0+	006	037	111	205	251	200	088	011	001	0+	6
	7	0+	0+	0+	001	009	042	117	215	267	201	057	010	0+	7
	8	0+	0+	0+	0+	001	011	044	121	233	302	194	075	004	8
	9	0+	0+	0+	0+	0+	002	010	040	121	268	387	315	091	9
	10	0+	0+	0+	0+	0+	0+	001	006	028	107	349	599	904	10
11	0	895	569	314	086	020	004	0+	0+	0+	0+	0+	0+	0+	0
	1	099	329	384	236	093	027	005	001	0+	0+	0+	0+	0+	1
	2	005	087	213	295	200	089	027	005	001	0+	0+	0+	0+	2
	3	0+	014	071	221	257	177	081	023	004	0+	0+	0+	0+	3
	4	0+	001	016	111	220	236	161	070	017	002	0+	0+	0+	4
	5	0+	0+	002	039	132	221	226	147	057	010	0+	0+	0+	5
	6	0+	0+	0+	010	057	147	226	221	132	039	002	0+	0+	6
	7	0+	0+	0+	002	017	070	161	236	220	111	016	001	0+	7
	8	0+	0+	0+	0+	004	023	081	177	257	221	071	014	0+	8
	9	0+	0+	0+	0+	001	005	027	089	200	295	213	087	005	9
	10	0+	0+	0+	0+	0+	001	005	027	093	236	384	329	099	10
	11	0+	0+	0+	0+	0+	0+	0+	004	020	086	314	569	895	11
12	0	886	540	282	069	014	002	0+	0+	0+	0+	0+	0+	0+	0
	1	107	341	377	206	071	017	003	0+	0+	0+	0+	0+	0+	1
	2	006	099	230	283	168	064	016	002	0+	0+	0+	0+	0+	2
	3	0+	017	085	236	240	142	054	012	001	0+	0+	0+	0+	3
	4	0+	002	021	133	231	213	121	042	008	001	0+	0+	0+	4
	5	0+	0+	004	053	158	227	193	101	029	003	0+	0+	0+	5
	6	0+	0+	0+	016	079	177	226	177	079	016	0+	0+	0+	6
	7	0+	0+	0+	003	029	101	193	227	158	053	004	0+	0+	7
	8	0+	0+	0+	001	008	042	121	213	231	133	021	002	0+	8
	9	0+	0+	0+	0+	001	012	054	142	240	236	085	017	0+	9

TABLE A1 *(continued)*

n	y	.01	.05	.10	.20	.30	.40	.50	.60	.70	.80	.90	.95	.99	y
12	10	0+	0+	0+	0+	0+	002	016	064	168	283	230	099	006	10
	11	0+	0+	0+	0+	0+	0+	003	017	071	206	377	341	107	11
	12	0+	0+	0+	0+	0+	0+	0+	002	014	069	282	540	886	12
13	0	878	513	254	055	010	001	0+	0+	0+	0+	0+	0+	0+	0
	1	115	351	367	179	054	011	002	0+	0+	0+	0+	0+	0+	1
	2	007	111	245	268	139	045	010	001	0+	0+	0+	0+	0+	2
	3	0+	021	100	246	218	111	035	006	001	0+	0+	0+	0+	3
	4	0+	003	028	154	234	184	087	024	003	0+	0+	0+	0+	4
	5	0+	0+	006	069	180	221	157	066	014	001	0+	0+	0+	5
	6	0+	0+	001	023	103	197	209	131	044	006	0+	0+	0+	6
	7	0+	0+	0+	006	044	131	209	197	103	023	001	0+	0+	7
	8	0+	0+	0+	001	014	066	157	221	180	069	006	0+	0+	8
	9	0+	0+	0+	0+	003	024	087	184	234	154	028	003	0+	9
	10	0+	0+	0+	0+	001	006	035	111	218	246	100	021	0+	10
	11	0+	0+	0+	0+	0+	001	010	045	139	268	245	111	007	11
	12	0+	0+	0+	0+	0+	0+	002	011	054	179	367	351	115	12
	13	0+	0+	0+	0+	0+	0+	0+	001	010	055	254	513	878	13
14	0	869	488	229	044	007	001	0+	0+	0+	0+	0+	0+	0+	0
	1	123	359	356	154	041	007	001	0+	0+	0+	0+	0+	0+	1
	2	008	123	257	250	113	032	006	001	0+	0+	0+	0+	0+	2
	3	0+	026	114	250	194	085	022	003	0+	0+	0+	0+	0+	3
	4	0+	004	035	172	229	155	061	014	001	0+	0+	0+	0+	4
	5	0+	0+	008	086	196	207	122	041	007	0+	0+	0+	0+	5
	6	0+	0+	001	032	126	207	183	092	023	002	0+	0+	0+	6
	7	0+	0+	0+	009	062	157	209	157	062	009	0+	0+	0+	7
	8	0+	0+	0+	002	023	092	183	207	126	032	001	0+	0+	8
	9	0+	0+	0+	0+	007	041	122	207	196	086	008	0+	0+	9
	10	0+	0+	0+	0+	001	014	061	155	229	172	035	004	0+	10
	11	0+	0+	0+	0+	0+	003	022	085	194	250	114	026	0+	11
	12	0+	0+	0+	0+	0+	001	006	032	113	250	257	123	008	12
	13	0+	0+	0+	0+	0+	0+	001	007	041	154	356	359	123	13
	14	0+	0+	0+	0+	0+	0+	0+	001	007	044	229	488	869	14
15	0	860	463	206	035	005	0+	0+	0+	0+	0+	0+	0+	0+	0
	1	130	366	343	132	031	005	0+	0+	0+	0+	0+	0+	0+	1
	2	009	135	267	231	092	022	003	0+	0+	0+	0+	0+	0+	2
	3	0+	031	129	250	170	063	014	002	0+	0+	0+	0+	0+	3
	4	0+	005	043	188	219	127	042	007	001	0+	0+	0+	0+	4
	5	0+	001	010	103	206	186	092	024	003	0+	0+	0+	0+	5
	6	0+	0+	002	043	147	207	153	061	012	001	0+	0+	0+	6
	7	0+	0+	0+	014	081	177	196	118	035	003	0+	0+	0+	7
	8	0+	0+	0+	003	035	118	196	177	081	014	0+	0+	0+	8
	9	0+	0+	0+	001	012	061	153	207	147	043	002	0+	0+	9

TABLE A1 *(continued)*

n	y	.01	.05	.10	.20	.30	.40	*p* .50	.60	.70	.80	.90	.95	.99	y
15	10	0+	0+	0+	0+	003	024	092	186	206	103	010	001	0+	10
	11	0+	0+	0+	0+	001	007	042	127	219	188	043	005	0+	11
	12	0+	0+	0+	0+	0+	002	014	063	170	250	129	031	0+	12
	13	0+	0+	0+	0+	0+	0+	003	022	092	231	267	135	009	13
	14	0+	0+	0+	0+	0+	0+	0+	005	031	132	343	366	130	14
	15	0+	0+	0+	0+	0+	0+	0+	0+	005	035	206	463	860	15
16	0	851	440	185	028	003	0+	0+	0+	0+	0+	0+	0+	0+	0
	1	138	371	329	113	023	003	0+	0+	0+	0+	0+	0+	0+	1
	2	010	146	275	211	073	015	002	0+	0+	0+	0+	0+	0+	2
	3	0+	036	142	246	146	047	009	001	0+	0+	0+	0+	0+	3
	4	0+	006	051	200	204	101	028	004	0+	0+	0+	0+	0+	4
	5	0+	001	014	120	210	162	067	014	001	0+	0+	0+	0+	5
	6	0+	0+	003	055	165	198	122	039	006	0+	0+	0+	0+	6
	7	0+	0+	0+	020	101	189	175	084	019	001	0+	0+	0+	7
	8	0+	0+	0+	006	049	142	196	142	049	006	0+	0+	0+	8
	9	0+	0+	0+	001	019	084	175	189	101	020	0+	0+	0+	9
	10	0+	0+	0+	0+	006	039	122	198	165	055	003	0+	0+	10
	11	0+	0+	0+	0+	001	014	067	162	210	120	014	001	0+	11
	12	0+	0+	0+	0+	0+	004	028	101	204	200	051	006	0+	12
	13	0+	0+	0+	0+	0+	001	009	047	146	246	142	036	0+	13
	14	0+	0+	0+	0+	0+	0+	002	015	073	211	275	146	010	14
	15	0+	0+	0+	0+	0+	0+	0+	003	023	113	329	371	138	15
	16	0+	0+	0+	0+	0+	0+	0+	0+	003	028	185	440	851	16
17	0	843	418	167	023	002	0+	0+	0+	0+	0+	0+	0+	0+	0
	1	145	374	315	096	017	002	0+	0+	0+	0+	0+	0+	0+	1
	2	012	158	280	191	058	010	001	0+	0+	0+	0+	0+	0+	2
	3	001	041	156	239	125	034	005	0+	0+	0+	0+	0+	0+	3
	4	0+	008	060	209	187	080	018	002	0+	0+	0+	0+	0+	4
	5	0+	001	017	136	208	138	047	008	001	0+	0+	0+	0+	5
	6	0+	0+	004	068	178	184	094	024	003	0+	0+	0+	0+	6
	7	0+	0+	001	027	120	193	148	057	009	0+	0+	0+	0+	7
	8	0+	0+	0+	008	064	161	185	107	028	002	0+	0+	0+	8
	9	0+	0+	0+	002	028	107	185	161	064	008	0+	0+	0+	9
	10	0+	0+	0+	0+	009	057	148	193	120	027	001	0+	0+	10
	11	0+	0+	0+	0+	003	024	094	184	178	068	004	0+	0+	11
	12	0+	0+	0+	0+	001	008	047	138	208	136	017	001	0+	12
	13	0+	0+	0+	0+	0+	002	018	080	187	209	060	008	0+	13
	14	0+	0+	0+	0+	0+	0+	005	034	125	239	156	041	001	14
	15	0+	0+	0+	0+	0+	0+	001	010	058	191	280	158	012	15
	16	0+	0+	0+	0+	0+	0+	0+	002	017	096	315	374	145	16
	17	0+	0+	0+	0+	0+	0+	0+	0+	002	023	167	418	843	17

TABLE A1 *(continued)*

n	y	.01	.05	.10	.20	.30	.40	p .50	.60	.70	.80	.90	.95	.99	y
18	0	835	397	150	018	002	0+	0+	0+	0+	0+	0+	0+	0+	0
	1	152	376	300	081	013	001	0+	0+	0+	0+	0+	0+	0+	1
	2	013	168	284	172	046	007	001	0+	0+	0+	0+	0+	0+	2
	3	001	047	168	230	105	025	003	0+	0+	0+	0+	0+	0+	3
	4	0+	009	070	215	168	061	012	001	0+	0+	0+	0+	0+	4
	5	0+	001	022	151	202	115	033	004	0+	0+	0+	0+	0+	5
	6	0+	0+	005	082	187	166	071	015	001	0+	0+	0+	0+	6
	7	0+	0+	001	035	138	189	121	037	005	0+	0+	0+	0+	7
	8	0+	0+	0+	012	081	173	167	077	015	001	0+	0+	0+	8
	9	0+	0+	0+	003	039	128	185	128	039	003	0+	0+	0+	9
	10	0+	0+	0+	001	015	077	167	173	081	012	0+	0+	0+	10
	11	0+	0+	0+	0+	005	037	121	189	138	035	001	0+	0+	11
	12	0+	0+	0+	0+	001	015	071	166	187	082	005	0+	0+	12
	13	0+	0+	0+	0+	0+	004	033	115	202	151	022	001	0+	13
	14	0+	0+	0+	0+	0+	001	012	061	168	215	070	009	0+	14
	15	0+	0+	0+	0+	0+	0+	003	025	105	230	168	047	001	15
	16	0+	0+	0+	0+	0+	0+	001	007	046	172	284	168	013	16
	17	0+	0+	0+	0+	0+	0+	0+	001	013	081	300	376	152	17
	18	0+	0+	0+	0+	0+	0+	0+	0+	002	018	150	397	835	18
19	0	826	377	135	014	001	0+	0+	0+	0+	0+	0+	0+	0+	0
	1	159	377	285	068	009	001	0+	0+	0+	0+	0+	0+	0+	1
	2	014	179	285	154	036	005	0+	0+	0+	0+	0+	0+	0+	2
	3	001	053	180	218	087	017	002	0+	0+	0+	0+	0+	0+	3
	4	0+	011	080	218	149	047	007	001	0+	0+	0+	0+	0+	4
	5	0+	002	027	164	192	093	022	002	0+	0+	0+	0+	0+	5
	6	0+	0+	007	095	192	145	052	008	001	0+	0+	0+	0+	6
	7	0+	0+	001	044	153	180	096	024	002	0+	0+	0+	0+	7
	8	0+	0+	0+	017	098	180	144	053	008	0+	0+	0+	0+	8
	9	0+	0+	0+	005	051	146	176	098	022	001	0+	0+	0+	9
	10	0+	0+	0+	001	022	098	176	146	051	005	0+	0+	0+	10
	11	0+	0+	0+	0+	008	053	144	180	098	017	0+	0+	0+	11
	12	0+	0+	0+	0+	002	024	096	180	153	044	001	0+	0+	12
	13	0+	0+	0+	0+	001	008	052	145	192	095	007	0+	0+	13
	14	0+	0+	0+	0+	0+	002	022	093	192	164	027	002	0+	14
	15	0+	0+	0+	0+	0+	001	007	047	149	218	080	011	0+	15
	16	0+	0+	0+	0+	0+	0+	002	017	087	218	180	053	001	16
	17	0+	0+	0+	0+	0+	0+	0+	005	036	154	285	179	014	17
	18	0+	0+	0+	0+	0+	0+	0+	001	009	068	285	377	159	18
	19	0+	0+	0+	0+	0+	0+	0+	0+	001	014	135	377	826	19
20	0	818	358	122	012	001	0+	0+	0+	0+	0+	0+	0+	0+	0
	1	165	377	270	058	007	0+	0+	0+	0+	0+	0+	0+	0+	1
	2	016	189	285	137	028	003	0+	0+	0+	0+	0+	0+	0+	2
	3	001	060	190	205	072	012	001	0+	0+	0+	0+	0+	0+	3
	4	0+	013	090	218	130	035	005	0+	0+	0+	0+	0+	0+	4

TABLE A1 *(continued)*

n	y	.01	.05	.10	.20	.30	.40	p .50	.60	.70	.80	.90	.95	.99	y
20	5	0+	002	032	175	179	075	015	001	0+	0+	0+	0+	0+	5
	6	0+	0+	009	109	192	124	037	005	0+	0+	0+	0+	0+	6
	7	0+	0+	002	055	164	166	074	015	001	0+	0+	0+	0+	7
	8	0+	0+	0+	022	114	180	120	035	004	0+	0+	0+	0+	8
	9	0+	0+	0+	007	065	160	160	071	012	0+	0+	0+	0+	9
	10	0+	0+	0+	002	031	117	176	117	031	002	0+	0+	0+	10
	11	0+	0+	0+	0+	012	071	160	160	065	007	0+	0+	0+	11
	12	0+	0+	0+	0+	004	035	120	180	114	022	0+	0+	0+	12
	13	0+	0+	0+	0+	001	015	074	166	164	055	002	0+	0+	13
	14	0+	0+	0+	0+	0+	005	037	124	192	109	009	0+	0+	14
	15	0+	0+	0+	0+	0+	001	015	075	179	175	032	002	0+	15
	16	0+	0+	0+	0+	0+	0+	005	035	130	218	090	013	0+	16
	17	0+	0+	0+	0+	0+	0+	001	012	072	205	190	060	001	17
	18	0+	0+	0+	0+	0+	0+	0+	003	028	137	285	189	016	18
	19	0+	0+	0+	0+	0+	0+	0+	0+	007	058	270	377	165	19
	20	0+	0+	0+	0+	0+	0+	0+	0+	001	012	122	358	818	20
21	0	810	341	109	009	001	0+	0+	0+	0+	0+	0+	0+	0+	0
	1	172	376	255	048	005	0+	0+	0+	0+	0+	0+	0+	0+	1
	2	017	198	284	121	022	002	0+	0+	0+	0+	0+	0+	0+	2
	3	001	066	200	192	058	009	001	0+	0+	0+	0+	0+	0+	3
	4	0+	016	100	216	113	026	003	0+	0+	0+	0+	0+	0+	4
	5	0+	003	038	183	164	059	010	001	0+	0+	0+	0+	0+	5
	6	0+	0+	011	122	188	105	026	003	0+	0+	0+	0+	0+	6
	7	0+	0+	003	065	172	149	055	009	0+	0+	0+	0+	0+	7
	8	0+	0+	001	029	129	174	097	023	002	0+	0+	0+	0+	8
	9	0+	0+	0+	010	080	168	140	050	006	0+	0+	0+	0+	9
	10	0+	0+	0+	003	041	134	168	089	018	001	0+	0+	0+	10
	11	0+	0+	0+	001	018	089	168	134	041	003	0+	0+	0+	11
	12	0+	0+	0+	0+	006	050	140	168	080	010	0+	0+	0+	12
	13	0+	0+	0+	0+	002	023	097	174	129	029	001	0+	0+	13
	14	0+	0+	0+	0+	0+	009	055	149	172	065	003	0+	0+	14
	15	0+	0+	0+	0+	0+	003	026	105	188	122	011	0+	0+	15
	16	0+	0+	0+	0+	0+	001	010	059	164	183	038	003	0+	16
	17	0+	0+	0+	0+	0+	0+	003	026	113	216	100	016	0+	17
	18	0+	0+	0+	0+	0+	0+	001	009	058	192	200	066	001	18
	19	0+	0+	0+	0+	0+	0+	0+	002	022	121	284	198	017	19
	20	0+	0+	0+	0+	0+	0+	0+	0+	005	048	255	376	172	20
	21	0+	0+	0+	0+	0+	0+	0+	0+	001	009	109	341	810	21
22	0	802	324	098	007	0+	0+	0+	0+	0+	0+	0+	0+	0+	0
	1	178	375	241	041	004	0+	0+	0+	0+	0+	0+	0+	0+	1
	2	019	207	281	107	017	001	0+	0+	0+	0+	0+	0+	0+	2
	3	001	073	208	178	047	006	0+	0+	0+	0+	0+	0+	0+	3
	4	0+	018	110	211	096	019	002	0+	0+	0+	0+	0+	0+	4

TABLE A1 (continued)

n	y	.01	.05	.10	.20	.30	.40	p .50	.60	.70	.80	.90	.95	.99	y
22	5	0+	003	044	190	149	046	006	0+	0+	0+	0+	0+	0+	5
	6	0+	001	014	134	181	086	018	001	0+	0+	0+	0+	0+	6
	7	0+	0+	004	077	177	131	041	005	0+	0+	0+	0+	0+	7
	8	0+	0+	001	036	142	164	076	014	001	0+	0+	0+	0+	8
	9	0+	0+	0+	014	095	170	119	034	003	0+	0+	0+	0+	9
	10	0+	0+	0+	005	053	148	154	066	010	0+	0+	0+	0+	10
	11	0+	0+	0+	001	025	107	168	107	025	001	0+	0+	0+	11
	12	0+	0+	0+	0+	010	066	154	148	053	005	0+	0+	0+	12
	13	0+	0+	0+	0+	003	034	119	170	095	014	0+	0+	0+	13
	14	0+	0+	0+	0+	001	014	076	164	142	036	001	0+	0+	14
	15	0+	0+	0+	0+	0+	005	041	131	177	077	004	0+	0+	15
	16	0+	0+	0+	0+	0+	001	018	086	181	134	014	001	0+	16
	17	0+	0+	0+	0+	0+	0+	006	046	149	190	044	003	0+	17
	18	0+	0+	0+	0+	0+	0+	002	019	096	211	110	018	0+	18
	19	0+	0+	0+	0+	0+	0+	0+	006	047	178	208	073	001	19
	20	0+	0+	0+	0+	0+	0+	0+	001	017	107	281	207	019	20
	21	0+	0+	0+	0+	0+	0+	0+	0+	004	041	241	375	178	21
	22	0+	0+	0+	0+	0+	0+	0+	0+	0+	007	098	324	802	22
23	0	794	307	089	006	0+	0+	0+	0+	0+	0+	0+	0+	0+	0
	1	184	372	226	034	003	0+	0+	0+	0+	0+	0+	0+	0+	1
	2	020	215	277	093	013	001	0+	0+	0+	0+	0+	0+	0+	2
	3	001	079	215	163	038	004	0+	0+	0+	0+	0+	0+	0+	3
	4	0+	021	120	204	082	014	001	0+	0+	0+	0+	0+	0+	4
	5	0+	004	051	194	133	035	004	0+	0+	0+	0+	0+	0+	5
	6	0+	001	017	145	171	070	012	001	0+	0+	0+	0+	0+	6
	7	0+	0+	005	088	178	113	029	003	0+	0+	0+	0+	0+	7
	8	0+	0+	001	044	153	151	058	009	0+	0+	0+	0+	0+	8
	9	0+	0+	0+	018	109	168	097	022	002	0+	0+	0+	0+	9
	10	0+	0+	0+	006	065	157	136	046	005	0+	0+	0+	0+	10
	11	0+	0+	0+	002	033	123	161	082	014	0+	0+	0+	0+	11
	12	0+	0+	0+	0+	014	082	161	123	033	002	0+	0+	0+	12
	13	0+	0+	0+	0+	005	046	136	157	065	006	0+	0+	0+	13
	14	0+	0+	0+	0+	002	022	097	168	109	018	0+	0+	0+	14
	15	0+	0+	0+	0+	0+	009	058	151	153	044	001	0+	0+	15
	16	0+	0+	0+	0+	0+	003	029	113	178	088	005	0+	0+	16
	17	0+	0+	0+	0+	0+	001	012	070	171	145	017	001	0+	17
	18	0+	0+	0+	0+	0+	0+	004	035	133	194	051	004	0+	18
	19	0+	0+	0+	0+	0+	0+	001	014	082	204	120	021	0+	19
	20	0+	0+	0+	0+	0+	0+	0+	004	038	163	215	079	001	20
	21	0+	0+	0+	0+	0+	0+	0+	001	013	093	277	215	020	21
	22	0+	0+	0+	0+	0+	0+	0+	0+	003	034	226	372	184	22
	23	0+	0+	0+	0+	0+	0+	0+	0+	0+	006	089	307	794	23

TABLE A1 *(continued)*

n	y	.01	.05	.10	.20	.30	.40	p .50	.60	.70	.80	.90	.95	.99	y
24	0	786	292	080	005	0+	0+	0+	0+	0+	0+	0+	0+	0+	0
	1	190	369	213	028	002	0+	0+	0+	0+	0+	0+	0+	0+	1
	2	022	223	272	081	010	001	0+	0+	0+	0+	0+	0+	0+	2
	3	002	086	221	149	031	003	0+	0+	0+	0+	0+	0+	0+	3
	4	0+	024	129	196	069	010	001	0+	0+	0+	0+	0+	0+	4
	5	0+	005	057	196	118	027	003	0+	0+	0+	0+	0+	0+	5
	6	0+	001	020	155	160	056	008	0+	0+	0+	0+	0+	0+	6
	7	0+	0+	006	100	176	096	021	002	0+	0+	0+	0+	0+	7
	8	0+	0+	001	053	160	136	044	005	0+	0+	0+	0+	0+	8
	9	0+	0+	0+	024	122	161	078	014	001	0+	0+	0+	0+	9
	10	0+	0+	0+	009	079	161	117	032	003	0+	0+	0+	0+	10
	11	0+	0+	0+	003	043	137	149	061	008	0+	0+	0+	0+	11
	12	0+	0+	0+	001	020	099	161	099	020	001	0+	0+	0+	12
	13	0+	0+	0+	0+	008	061	149	137	043	003	0+	0+	0+	13
	14	0+	0+	0+	0+	003	032	117	161	079	009	0+	0+	0+	14
	15	0+	0+	0+	0+	001	014	078	161	122	024	0+	0+	0+	15
	16	0+	0+	0+	0+	0+	005	044	136	160	053	001	0+	0+	16
	17	0+	0+	0+	0+	0+	002	021	096	176	100	006	0+	0+	17
	18	0+	0+	0+	0+	0+	0+	008	056	160	155	020	001	0+	18
	19	0+	0+	0+	0+	0+	0+	003	027	118	196	057	005	0+	19
	20	0+	0+	0+	0+	0+	0+	001	010	069	196	129	024	0+	20
	21	0+	0+	0+	0+	0+	0+	0+	003	031	149	221	086	002	21
	22	0+	0+	0+	0+	0+	0+	0+	001	010	081	272	223	022	22
	23	0+	0+	0+	0+	0+	0+	0+	0+	002	028	213	369	190	23
	24	0+	0+	0+	0+	0+	0+	0+	0+	0+	005	080	292	786	24
25	0	778	277	072	004	0+	0+	0+	0+	0+	0+	0+	0+	0+	0
	1	196	365	199	024	001	0+	0+	0+	0+	0+	0+	0+	0+	1
	2	024	231	266	071	007	0+	0+	0+	0+	0+	0+	0+	0+	2
	3	002	093	226	136	024	002	0+	0+	0+	0+	0+	0+	0+	3
	4	0+	027	138	187	057	007	0+	0+	0+	0+	0+	0+	0+	4
	5	0+	006	065	196	103	020	002	0+	0+	0+	0+	0+	0+	5
	6	0+	001	024	163	147	044	005	0+	0+	0+	0+	0+	0+	6
	7	0+	0+	007	111	171	080	014	001	0+	0+	0+	0+	0+	7
	8	0+	0+	002	062	165	120	032	003	0+	0+	0+	0+	0+	8
	9	0+	0+	0+	029	134	151	061	009	0+	0+	0+	0+	0+	9
	10	0+	0+	0+	012	092	161	097	021	001	0+	0+	0+	0+	10
	11	0+	0+	0+	004	054	147	133	043	004	0+	0+	0+	0+	11
	12	0+	0+	0+	001	027	114	155	076	011	0+	0+	0+	0+	12
	13	0+	0+	0+	0+	011	076	155	114	027	001	0+	0+	0+	13
	14	0+	0+	0+	0+	004	043	133	147	054	004	0+	0+	0+	14
	15	0+	0+	0+	0+	001	021	097	161	092	012	0+	0+	0+	15
	16	0+	0+	0+	0+	0+	009	061	151	134	029	0+	0+	0+	16
	17	0+	0+	0+	0+	0+	003	032	120	165	062	002	0+	0+	17
	18	0+	0+	0+	0+	0+	001	014	080	171	111	007	0+	0+	18
	19	0+	0+	0+	0+	0+	0+	005	044	147	163	024	001	0+	19

TABLE A1 *(continued)*

n	y	.01	.05	.10	.20	.30	.40	p .50	.60	.70	.80	.90	.95	.99	y
25	20	0+	0+	0+	0+	0+	0+	002	020	103	196	065	006	0+	20
	21	0+	0+	0+	0+	0+	0+	0+	007	057	187	138	027	0+	21
	22	0+	0+	0+	0+	0+	0+	0+	002	024	136	226	093	002	22
	23	0+	0+	0+	0+	0+	0+	0+	0+	007	071	266	231	024	23
	24	0+	0+	0+	0+	0+	0+	0+	0+	001	024	199	365	196	24
	25	0+	0+	0+	0+	0+	0+	0+	0+	0+	004	072	277	778	25

Source: From F. Mosteller, R. E. K. Rourke, and G. B. Thomas, Jr., *Probability and Statistics* © 1961. Addison-Wesley, Reading, MA. Table IV, Part A. Reprinted with permission.

TABLE A2 Cumulative Binomial Probabilities

Lower cumulative →

n	y	.01	.05	.10	.20	.30	.40	p .50	.60	.70	.80	.90	.95	.99		
5	0.5	951	774	590	328	168	078	031	010	002	000	000	000	000	4.5	
	1.5	999	977	919	737	528	337	188	087	031	007	000	000	000	3.5	
	2.5	1000	999	991	942	837	683	500	317	163	058	009	001	000	2.5	
	3.5	1000	1000	1000	993	969	913	812	663	472	263	081	023	001	1.5	
	4.5	1000	1000	1000	1000	998	990	969	922	832	672	410	226	049	0.5	5
10	0.5	904	599	349	107	028	006	001	000	000	000	000	000	000	9.5	
	1.5	996	914	736	376	149	046	011	002	000	000	000	000	000	8.5	
	2.5	1000	988	930	678	383	167	055	012	002	000	000	000	000	7.5	
	3.5	1000	999	987	879	650	382	172	055	011	001	000	000	000	6.5	
	4.5	1000	1000	998	967	850	633	377	166	047	006	000	000	000	5.5	
	5.5	1000	1000	1000	994	953	834	623	367	150	033	002	000	000	4.5	
	6.5	1000	1000	1000	999	989	945	828	618	350	121	013	001	000	3.5	
	7.5	1000	1000	1000	1000	998	988	945	833	617	322	070	012	000	2.5	
	8.5	1000	1000	1000	1000	1000	998	989	954	851	624	264	086	004	1.5	
	9.5	1000	1000	1000	1000	1000	1000	999	994	972	893	651	401	096	0.5	10
		.99	.95	.90	.80	.70	.60	.50	.40	.30	.20	.10	.05	.01	y	n

p

Upper cumulative ←

TABLE A2 *(continued)*

n	y	.01	.05	.10	.20	.30	.40	.50	.60	.70	.80	.90	.95	.99	
15	0.5	860	463	206	035	005	000	000	000	000	000	000	000	000	14.5
	1.5	990	829	549	167	035	005	000	000	000	000	000	000	000	13.5
	2.5	1000	964	816	398	127	027	004	000	000	000	000	000	000	12.5
	3.5	1000	995	944	648	297	091	018	002	000	000	000	000	000	11.5
	4.5	1000	999	987	836	515	217	059	009	001	000	000	000	000	10.5
	5.5	1000	1000	998	939	722	403	151	034	004	000	000	000	000	9.5
	6.5	1000	1000	1000	982	869	610	304	095	015	001	000	000	000	8.5
	7.5	1000	1000	1000	996	950	787	500	213	050	004	000	000	000	7.5
	8.5	1000	1000	1000	999	985	905	696	390	131	018	000	000	000	6.5
	9.5	1000	1000	1000	1000	996	966	849	597	278	061	002	000	000	5.5
	10.5	1000	1000	1000	1000	999	991	941	783	485	164	013	001	000	4.5
	11.5	1000	1000	1000	1000	1000	998	982	909	703	352	056	005	000	3.5
	12.5	1000	1000	1000	1000	1000	1000	996	973	873	602	184	036	000	2.5
	13.5	1000	1000	1000	1000	1000	1000	1000	995	965	833	451	171	010	1.5
	14.5	1000	1000	1000	1000	1000	1000	1000	1000	995	965	794	537	140	0.5 · 15
20	0.5	818	358	122	012	001	000	000	000	000	000	000	000	000	19.5
	1.5	983	736	392	069	008	001	000	000	000	000	000	000	000	18.5
	2.5	999	925	677	206	035	004	000	000	000	000	000	000	000	17.5
	3.5	1000	984	867	411	107	016	001	000	000	000	000	000	000	16.5
	4.5	1000	997	957	630	238	051	006	000	000	000	000	000	000	15.5
	5.5	1000	1000	989	804	416	126	021	002	000	000	000	000	000	14.5
	6.5	1000	1000	998	913	608	250	058	006	000	000	000	000	000	13.5
	7.5	1000	1000	1000	968	772	416	132	021	001	000	000	000	000	12.5
	8.5	1000	1000	1000	990	887	596	252	057	005	000	000	000	000	11.5
	9.5	1000	1000	1000	997	952	755	412	128	017	001	000	000	000	10.5
	10.5	1000	1000	1000	999	983	872	588	245	048	003	000	000	000	9.5
	11.5	1000	1000	1000	1000	995	943	748	404	113	010	000	000	000	8.5
	12.5	1000	1000	1000	1000	999	979	868	584	228	032	000	000	000	7.5
	13.5	1000	1000	1000	1000	1000	994	942	750	392	087	002	000	000	6.5
	14.5	1000	1000	1000	1000	1000	998	979	874	584	196	011	000	000	5.5
	15.5	1000	1000	1000	1000	1000	1000	994	949	762	370	043	003	000	4.5
	16.5	1000	1000	1000	1000	1000	1000	999	984	893	589	133	016	000	3.5
	17.5	1000	1000	1000	1000	1000	1000	1000	996	965	794	323	075	001	2.5
	18.5	1000	1000	1000	1000	1000	1000	1000	999	992	931	608	264	017	1.5
	19.5	1000	1000	1000	1000	1000	1000	1000	1000	999	988	878	642	182	0.5 20
		.99	.95	.90	.80	.70	.60	.50	.40	.30	.20	.10	.05	.01	y n

Lower cumulative →

p

p

Upper cumulative ←

TABLE A2 *(continued)*

n	y	.01	.05	.10	.20	.30	.40	.50	.60	.70	.80	.90	.95	.99		
25	0.5	778	277	072	004	000	000	000	000	000	000	000	000	000	24.5	
	1.5	974	642	271	027	002	000	000	000	000	000	000	000	000	23.5	
	2.5	998	873	537	098	009	000	000	000	000	000	000	000	000	22.5	
	3.5	1000	966	764	234	033	002	000	000	000	000	000	000	000	21.5	
	4.5	1000	993	902	421	090	009	000	000	000	000	000	000	000	20.5	
	5.5	1000	999	967	617	193	029	002	000	000	000	000	000	000	19.5	
	6.5	1000	1000	991	780	341	074	007	000	000	000	000	000	000	18.5	
	7.5	1000	1000	998	891	512	154	022	001	000	000	000	000	000	17.5	
	8.5	1000	1000	1000	953	677	274	054	004	000	000	000	000	000	16.5	
	9.5	1000	1000	1000	983	811	425	115	013	000	000	000	000	000	15.5	
	10.5	1000	1000	1000	994	902	586	212	034	002	000	000	000	000	14.5	
	11.5	1000	1000	1000	998	956	732	345	078	006	000	000	000	000	13.5	
	12.5	1000	1000	1000	1000	983	846	500	154	017	000	000	000	000	12.5	
	13.5	1000	1000	1000	1000	994	922	655	268	044	002	000	000	000	11.5	
	14.5	1000	1000	1000	1000	998	966	788	414	098	006	000	000	000	10.5	
	15.5	1000	1000	1000	1000	1000	987	885	575	189	017	000	000	000	9.5	
	16.5	1000	1000	1000	1000	1000	996	946	726	323	047	000	000	000	8.5	
	17.5	1000	1000	1000	1000	1000	999	978	846	488	109	002	000	000	7.5	
	18.5	1000	1000	1000	1000	1000	1000	993	926	659	220	009	000	000	6.5	
	19.5	1000	1000	1000	1000	1000	1000	998	971	807	383	033	001	000	5.5	
	20.5	1000	1000	1000	1000	1000	1000	1000	991	910	579	098	007	000	4.5	
	21.5	1000	1000	1000	1000	1000	1000	1000	998	967	766	236	034	000	3.5	
	22.5	1000	1000	1000	1000	1000	1000	1000	1000	991	902	463	127	002	2.5	
	23.5	1000	1000	1000	1000	1000	1000	1000	1000	998	973	729	358	026	1.5	
	24.5	1000	1000	1000	1000	1000	1000	1000	1000	1000	996	928	723	222	0.5	25
		.99	.95	.90	.80	.70	.60	.50	.40	.30	.20	.10	.05	.01	y	n

Source: From William Mendenhall & Madelaine Ramey, *Statistics for Psychology* (1973). Copyright © 1973 by Wadsworth Publishing Company, Inc. Reprinted by permission of the publisher.

TABLE A3 The Normal Probability Distribution

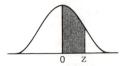

z	.00	.01	.02	.03	.04	.05	.06	.07	.08	.09
0.0	.0000	.0040	.0080	.0120	.0160	.0199	.0239	.0279	.0319	.0359
0.1	.0398	.0438	.0478	.0517	.0557	.0596	.0636	.0675	.0714	.0753
0.2	.0793	.0832	.0871	.0910	.0948	.0987	.1026	.1064	.1103	.1141
0.3	.1179	.1217	.1255	.1293	.1331	.1368	.1406	.1443	.1480	.1517
0.4	.1554	.1591	.1628	.1664	.1700	.1736	.1772	.1808	.1844	.1879
0.5	.1915	.1950	.1985	.2019	.2054	.2088	.2123	.2157	.2190	.2224
0.6	.2257	.2291	.2324	.2357	.2389	.2422	.2454	.2486	.2517	.2549
0.7	.2580	.2611	.2642	.2673	.2704	.2734	.2764	.2794	.2823	.2852
0.8	.2881	.2910	.2939	.2967	.2995	.3023	.3051	.3078	.3106	.3133
0.9	.3159	.3186	.3212	.3238	.3264	.3289	.3315	.3340	.3365	.3389
1.0	.3413	.3438	.3461	.3485	.3508	.3531	.3554	.3577	.3599	.3621
1.1	.3643	.3665	.3686	.3708	.3729	.3749	.3770	.3790	.3810	.3830
1.2	.3849	.3869	.3888	.3907	.3925	.3944	.3962	.3980	.3997	.4015
1.3	.4032	.4049	.4066	.4082	.4099	.4115	.4131	.4147	.4162	.4177
1.4	.4192	.4207	.4222	.4236	.4251	.4265	.4279	.4292	.4306	.4319
1.5	.4332	.4345	.4357	.4370	.4382	.4394	.4406	.4418	.4429	.4441
1.6	.4452	.4463	.4474	.4484	.4495	.4505	.4515	.4525	.4535	.4545
1.7	.4554	.4564	.4573	.4582	.4591	.4599	.4608	.4616	.4625	.4633
1.8	.4641	.4649	.4656	.4664	.4671	.4678	.4686	.4693	.4699	.4706
1.9	.4713	.4719	.4726	.4732	.4738	.4744	.4750	.4756	.4761	.4767
2.0	.4772	.4778	.4783	.4788	.4793	.4798	.4803	.4808	.4812	.4817
2.1	.4821	.4826	.4830	.4834	.4838	.4842	.4846	.4850	.4854	.4857
2.2	.4861	.4864	.4868	.4871	.4875	.4878	.4881	.4884	.4887	.4890
2.3	.4893	.4896	.4898	.4901	.4904	.4906	.4909	.4911	.4913	.4916
2.4	.4918	.4920	.4922	.4925	.4927	.4929	.4931	.4932	.4934	.4936
2.5	.4938	.4940	.4941	.4943	.4945	.4946	.4948	.4949	.4951	.4952
2.6	.4953	.4955	.4956	.4957	.4959	.4960	.4961	.4962	.4963	.4964
2.7	.4965	.4966	.4967	.4968	.4969	.4970	.4971	.4972	.4973	.4974
2.8	.4974	.4975	.4976	.4977	.4977	.4978	.4979	.4979	.4980	.4981
2.9	.4981	.4982	.4982	.4983	.4984	.4984	.4985	.4985	.4986	.4986
3.0	.4987	.4987	.4987	.4988	.4988	.4989	.4989	.4989	.4990	.4990

Source: From E. S. Pearson & H. O. Hartley (Editors), *Biometrika Tables for Statisticians*, Volume 1, Third Edition (1966). By permission of Biometrika Trustees.

TABLE A4 Backward Normal Table—
Critical Values of z

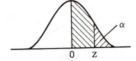

Area from 0 to Z	Area in tail (α)	z
.00	.50	.000
.05	.45	.126
.10	.40	.253
.15	.35	.385
.20	.30	.524
.25	.25	.674
.30	.20	.842
.35	.15	1.036
.40	.10	1.282
.45	.05	1.645
.475	.025	1.960
.49	.01	2.326
.495	.005	2.576
.4975	.0025	2.807
.499	.001	3.090
.4995	.0005	3.291
.49975	.00025	3.481
.4999	.0001	3.719

Source: From E. S. Pearson & H. O. Hartley (Editors), *Biometrika Tables for Statisticians*, Volume 1, Third Edition (1966). By permission of Biometrika Trustees.

TABLE A5 Random Digits

LINE/COL.	(1)	(2)	(3)	(4)	(5)	(6)	(7)	(8)	(9)	(10)	(11)	(12)	(13)	(14)
1	10480	15011	01536	02011	81647	91646	69179	14194	62590	36207	20969	99570	91291	90700
2	22368	46573	25595	85393	30995	89198	27982	53402	93965	34095	52666	19174	39615	99505
3	24130	48360	22527	97265	76393	64809	15179	24830	49340	32081	30680	19655	63348	58629
4	42167	93093	06243	61680	07856	16376	39440	53537	71341	57004	00849	74917	97758	16379
5	37570	39975	81837	16656	06121	91782	60468	81305	49684	60672	14110	06927	01263	54613
6	77921	06907	11008	42751	27756	53498	18602	70659	90655	15053	21916	81825	44394	42880
7	99562	72905	56420	69994	98872	31016	71194	18738	44013	48840	63213	21069	10634	12952
8	96301	91977	05463	07972	18876	20922	94595	56869	69014	60045	18425	84903	42508	32307
9	86579	14342	63661	10281	17453	18103	57740	84378	25331	12566	58678	44947	05585	56941
10	85475	36857	43342	53988	53060	59533	38867	62300	08158	17983	16439	11458	18593	64952
11	28918	69578	88231	33276	70997	79936	56865	05859	90106	31595	01547	85590	91610	78188
12	63553	40961	48235	03427	49626	69445	18663	72695	52180	20847	12234	90511	33703	90322
13	09429	93969	52636	92737	88974	33488	36320	17617	30015	08272	84115	27156	30613	74952
14	10365	61129	87529	85689	48237	52267	67689	93394	01511	26358	85104	20285	29975	89868
15	07119	97336	71048	08178	77233	13916	47564	81056	97735	85977	29372	74461	28551	90707
16	51085	12765	51821	51259	77452	16308	60756	92144	49442	53900	70960	63990	75601	40719
17	02368	21382	52404	60268	89368	19885	55322	44819	01188	65255	64835	44919	05944	55157
18	01011	54092	33362	94904	31273	04146	18594	29852	71585	85030	51132	01915	92747	64951
19	52162	53916	46369	58586	23216	14513	83149	98736	23495	64350	94738	17752	35156	35749
20	07056	97628	33787	09998	42698	06691	76988	13602	51851	46104	88916	19509	25625	58104
21	48663	91245	85828	14346	09172	30168	90229	04734	59193	22178	30421	61666	99904	32812
22	54164	58492	22421	74103	47070	25306	76468	26384	58151	06646	21524	15227	96909	44592
23	32639	32363	05597	24200	13363	38005	94342	28728	35806	06912	17012	64161	18296	22851
24	29334	27001	87637	87308	58731	00256	45834	15398	46557	41135	10367	07684	36188	18510
25	02488	33062	28834	07351	19731	92420	60952	61280	50001	67658	32586	86679	50720	94953
26	81525	72295	04839	96423	24878	82651	66566	14778	76797	14780	13300	87074	79666	95725
27	29676	20591	68086	26432	46901	20849	89768	81536	86645	12659	92259	57102	80428	25280
28	00742	57392	39064	66432	84673	40027	32832	61362	98947	96067	64760	64584	96096	98253
29	05366	04213	25669	26422	44407	44048	37937	63904	45766	66134	75470	66520	34693	90449
30	91921	26418	64117	94305	26766	25940	39972	22209	71500	64568	91402	42416	07844	69618
31	00582	04711	87917	77341	42206	35126	74087	99547	81817	42607	43808	76655	62028	76630
32	00725	69884	62797	56170	86324	88072	76222	36086	84637	93161	76038	65855	77919	88006
33	69011	65797	95876	55293	18988	27354	26575	08625	40801	59920	29841	80150	12777	48501
34	25976	57948	29888	88604	67917	48708	18912	82271	65424	69774	33611	54262	85963	03547
35	09763	83473	73577	12908	30883	18317	28290	35797	05998	41688	34952	37888	38917	88050
36	91567	42595	27958	30134	04024	86385	29880	99730	55536	84855	29080	09250	79656	73211
37	17955	56349	90999	49127	20044	59931	06115	20542	18059	02008	73708	83517	36103	42791
38	46503	18584	18845	49618	02304	51038	20655	58727	28168	15475	56942	53389	20562	87338
39	92157	89634	94824	78171	84610	82834	09922	25417	44137	48413	25555	21246	35509	20468
40	14577	62765	35605	81263	39667	47358	56873	56307	61607	49518	89656	20103	77490	18062
41	98427	07523	33362	64270	01638	92477	66969	98420	04880	45585	46565	04102	46880	45709
42	34914	63976	88720	82765	34476	17032	87589	40836	32427	70002	70663	88863	77775	69348
43	70060	28277	39475	46473	23219	53416	94970	25832	69975	94884	19661	72828	00102	66794
44	53976	54914	06990	67245	68350	82948	11398	42878	80287	88267	47363	46634	06541	97809
45	76072	29515	40980	07391	58745	25774	22987	80059	39911	96189	41151	14222	60697	59583
46	90725	52210	83974	29992	65831	38857	50490	83765	55657	14361	31720	57375	56228	41546
47	64364	67412	33339	31926	14883	24413	59744	92351	97473	89286	35931	04110	23726	51900
48	08962	00358	31662	25388	61642	34072	81249	35648	56891	69352	48373	45578	78547	81788
49	95012	68379	93526	70765	10593	04542	76463	54328	02349	17247	28865	14777	62730	92277
50	15664	10493	20492	38391	91132	21999	59516	81652	27195	48223	46751	22923	32261	85653

TABLE A5 *(continued)*

LINE/COL.	(1)	(2)	(3)	(4)	(5)	(6)	(7)	(8)	(9)	(10)	(11)	(12)	(13)	(14)
51	16408	81899	04153	53381	79401	21438	83035	93250	36693	31238	59649	91754	72772	02338
52	18629	81953	05520	91962	04739	13092	94662	24822	94730	06496	35090	04822	86772	98289
53	73115	35101	47498	87637	99016	71060	88824	71013	18735	20286	23153	72924	35165	43040
54	57491	16703	23167	49323	45021	33132	12544	41035	80780	45393	44812	12515	98931	91202
55	30405	83946	23792	14422	15059	45799	22716	19792	09983	74353	68668	30429	70735	25499
56	16631	35006	85900	98275	32388	52390	16815	69298	8273?	38480	73817	32523	41961	44437
57	96773	20206	42559	78985	05300	22164	24369	54224	35083	19687	11052	91491	60383	19746
58	38935	64202	14349	82674	66523	44133	00697	35552	35970	19124	63318	29686	03387	59846
59	31624	76384	17403	53363	44167	64486	64758	75366	76554	31601	12614	33072	60332	92325
60	78919	19474	23632	27889	47914	02584	37680	20801	72152	39339	34806	08930	85001	87820
61	03931	33309	57047	74211	63445	17361	62825	39908	05607	91284	68833	25570	38818	46920
62	74426	33278	43972	10119	89917	15665	52872	73823	73144	88662	88970	74492	51805	99378
63	09066	00903	20795	95452	92648	45454	09552	88815	16553	51125	79375	97596	16296	66092
64	42238	12426	87025	14267	20979	04508	64535	31355	86064	29472	47689	05974	52468	16834
65	16153	08002	26504	41744	81959	65642	74240	56302	00033	67107	77510	70625	28725	34191
66	21457	40742	29820	96783	29400	21840	15035	34537	33310	06116	95240	15957	16572	06004
67	21581	57802	02050	89728	17937	37621	47075	42080	97403	48626	68995	43805	33386	21597
68	55612	78095	83197	33732	05810	24813	86902	60397	16489	03264	88525	42786	05269	92532
69	44657	66999	99324	51281	84463	60563	79312	93454	68876	25471	93911	25650	12682	73572
70	91340	84979	46949	81973	37949	61023	43997	15263	80644	43942	89203	71795	99533	50501
71	91227	21199	31935	27022	84067	05462	35216	14486	29891	68607	41867	14951	91696	85065
72	50001	38140	66321	19924	72163	09538	12151	06878	91903	18749	34405	56087	82790	70925
73	65390	05224	72958	28609	81406	39147	25549	48542	42627	45233	57202	94617	23772	07896
74	27504	96131	83944	41575	10573	08619	64482	73923	36152	05184	94142	25299	84387	34925
75	37169	94851	39117	89632	00959	16487	65536	49071	39782	17095	02330	74301	00275	48280
76	11508	70225	51111	38351	19444	66499	71945	05422	13442	78675	84081	66938	93654	59894
77	37449	30362	06694	54690	04052	53115	62757	95348	78662	11163	81651	50245	34971	52924
78	46515	70331	85922	38329	57015	15765	97161	17869	45349	61796	66345	81073	49106	79860
79	30986	81223	42416	58353	21532	30502	32305	86482	05174	07901	54339	58861	74818	46942
80	63798	64995	46583	09765	44160	78128	83991	42865	92520	83531	80377	35909	81250	54238
81	82486	84846	99254	67632	43218	50076	21361	64816	51202	88124	41870	52689	51275	83556
82	21885	32906	92431	09060	64297	51674	64126	62570	26123	05155	59194	52799	28225	85762
83	60336	98782	07408	53458	13564	59089	26445	29789	85205	41001	12535	12133	14645	23541
84	43937	46891	24010	25560	86355	33941	25786	54990	71899	15475	95434	98227	21824	19585
85	97656	63175	89303	16275	07100	92063	21942	18611	47348	20203	18534	03862	78095	50136
86	03299	01221	05418	38982	55758	92237	26759	86367	21216	98442	08303	56613	91511	75928
87	79626	06486	03574	17668	07785	76020	79924	25651	83325	88428	85076	72811	22717	50585
88	85636	68335	47539	03129	65651	11977	02510	26113	99447	68645	34327	15152	55230	93448
89	18039	14367	61337	06177	12143	46609	32989	74014	64708	00533	35398	58408	13261	47908
90	08362	15656	60627	36478	65648	16764	53412	09013	07832	41574	17639	82163	60859	75567
91	79556	29068	04142	16268	15387	12856	66227	38358	22478	73373	88732	09443	82558	05250
92	92608	82674	27072	32534	17075	27698	98204	63863	11951	34648	88022	56148	34925	57031
93	23982	25835	40055	67006	12293	02753	14827	22235	35071	99704	37543	11601	35503	85171
94	09915	96306	05908	97901	28395	14186	00821	80703	70426	75647	76310	88717	37890	40129
95	50937	33300	26695	62247	69927	76123	50842	43834	86654	70959	79725	93872	28117	19233
96	42488	78077	69882	61657	34136	79180	97526	43092	04098	73571	80799	76536	71255	64239
97	46764	86273	63003	93017	31204	36692	40202	35275	57306	55543	53203	18098	47625	88684
98	03237	45430	55417	63282	90816	17349	88298	90183	36600	78406	06216	95787	42579	90730
99	86591	81482	52667	61583	14972	90053	89534	76036	49199	43716	97548	04379	46370	28672
100	38534	01715	94964	87288	65680	43772	39560	12918	86537	62738	19636	51132	25739	56947

Source: Reprinted with permission from William H. Beyer (Editor), *Handbook of Tables for Probability and Statistics,* Second Edition (1968). Copyright The Chemical Rubber Co., CRC Press, Inc.

TABLE A6 Critical Values of *t*

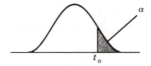

df	$t_{.100}$	$t_{.050}$	$t_{.025}$	$t_{.010}$	$t_{.005}$	$t_{.0005}$	df
1	3.078	6.314	12.706	31.821	63.657	636.619	1
2	1.886	2.920	4.303	6.965	9.925	31.598	2
3	1.638	2.353	3.182	4.541	5.841	12.924	3
4	1.533	2.132	2.776	3.747	4.604	8.610	4
5	1.476	2.015	2.571	3.365	4.032	6.869	5
6	1.440	1.943	2.447	3.143	3.707	5.959	6
7	1.415	1.895	2.365	2.998	3.499	5.408	7
8	1.397	1.860	2.306	2.896	3.355	5.041	8
9	1.383	1.833	2.262	2.821	3.250	4.781	9
10	1.372	1.812	2.228	2.764	3.169	4.587	10
11	1.363	1.796	2.201	2.718	3.106	4.437	11
12	1.356	1.782	2.179	2.681	3.055	4.318	12
13	1.350	1.771	2.160	2.650	3.012	4.221	13
14	1.345	1.761	2.145	2.624	2.977	4.140	14
15	1.341	1.753	2.131	2.602	2.947	4.073	15
16	1.337	1.746	2.120	2.583	2.921	4.015	16
17	1.333	1.740	2.110	2.567	2.898	3.965	17
18	1.330	1.734	2.101	2.552	2.878	3.922	18
19	1.328	1.729	2.093	2.539	2.861	3.883	19
20	1.325	1.725	2.086	2.528	2.845	3.850	20
21	1.323	1.721	2.080	2.518	2.831	3.819	21
22	1.321	1.717	2.074	2.508	2.819	3.792	22
23	1.319	1.714	2.069	2.500	2.807	3.767	23
24	1.318	1.711	2.064	2.492	2.797	3.745	24
25	1.316	1.708	2.060	2.485	2.787	3.725	25
26	1.315	1.706	2.056	2.479	2.779	3.707	26
27	1.314	1.703	2.052	2.473	2.771	3.690	27
28	1.313	1.701	2.048	2.467	2.763	3.674	28
29	1.311	1.699	2.045	2.462	2.756	3.659	29
30	1.310	1.697	2.042	2.457	2.750	3.646	30
40	1.303	1.684	2.021	2.423	2.704	3.551	40
60	1.296	1.671	2.000	2.390	2.660	3.460	60
120	1.289	1.658	1.980	2.358	2.617	3.373	120
∞	1.282	1.645	1.960	2.326	2.576	3.291	∞

Source: Reprinted with permission from William H. Beyer (Editor), *Handbook of Tables for Probability and Statistics,* Second Edition (1968). Copyright The Chemical Rubber Co., CRC Press, Inc.

TABLE A7 Power of a Statistical Test

δ	One-tailed α							
	.05		.025		.01		.005	
	Two-tailed α							
		.10		.05		.02		.01
0.0	.05	.10	.025	.05	.01	.02	.005	.01
0.1	.06	.10	.03	.05	.02		.01	
0.2	.07	.11	.05		.02		.01	
0.3	.09	.12	.06		.03		.01	
0.4	.11	.13	.07		.03		.01	
0.5	.14		.08		.03		.02	
0.6	.16		.09		.04		.02	
0.7	.18		.11		.05		.03	
0.8	.21		.13		.06		.04	
0.9	.23		.15		.08		.05	
1.0	.26		.17		.09		.06	
1.1	.30		.20		.11		.07	
1.2	.33		.22		.13		.08	
1.3	.37		.26		.15		.10	
1.4	.40		.29		.18		.12	
1.5	.44		.32		.20		.14	
1.6	.48		.36		.23		.16	
1.7	.52		.40		.27		.19	
1.8	.56		.44		.30		.22	
1.9	.60		.48		.33		.25	
2.0	.64		.52		.37		.28	
2.1	.68		.56		.41		.32	
2.2	.71		.59		.45		.35	
2.3	.74		.63		.49		.39	
2.4	.77		.67		.53		.43	
2.5	.80		.71		.57		.47	
2.6	.83		.74		.61		.51	
2.7	.85		.77		.65		.55	
2.8	.88		.80		.68		.59	
2.9	.90		.83		.72		.63	
3.0	.91		.85		.75		.66	

TABLE A7 *(continued)*

δ	One-tailed α			
	.05	.025	.01	.005
	Two-tailed α			
	.10	.05	.02	.01
3.1	.93	.87	.78	.70
3.2	.94	.89	.81	.73
3.3	.96	.91	.83	.77
3.4	.96	.93	.86	.80
3.5	.97	.94	.88	.82
3.6	.97	.95	.90	.85
3.7	.98	.96	.92	.87
3.8	.98	.97	.93	.89
3.9	.99	.97	.94	.91
4.0	.99	.98	.95	.92
4.1	.99	.98	.96	.94
4.2	.99	.99	.97	.95
4.3	*	.99	.98	.96
4.4		.99	.98	.97
4.5		.99	.99	.97
4.6		*	.99	.98
4.7			.99	.98
4.8			.99	.99
4.9			.99	.99
5.0			*	.99
5.1				.99
5.2				*

* The power at and below this point is greater than .995.
Source: Reprinted with permission from J. Welkowitz, R. B. Ewen, and J. Cohen, *Introductory Statistics for the Behavioral Sciences*, New York: Academic Press (1976).

TABLE A8 Critical Values of χ^2

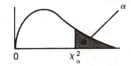

df	$\chi^2_{.995}$	$\chi^2_{.990}$	$\chi^2_{.975}$	$\chi^2_{.950}$	$\chi^2_{.900}$	$\chi^2_{.100}$	$\chi^2_{.050}$	$\chi^2_{.025}$	$\chi^2_{.010}$	$\chi^2_{.005}$	df
1	.000	.000	.001	.004	.016	2.706	3.841	5.024	6.635	7.879	1
2	.010	.020	.051	.103	.211	4.605	5.991	7.378	9.210	10.60	2
3	.072	.115	.216	.352	.584	6.251	7.815	9.348	11.34	12.84	3
4	.207	.297	.484	.711	1.064	7.779	9.488	11.14	13.28	14.86	4
5	.412	.554	.831	1.145	1.610	9.236	11.07	12.83	15.09	16.75	5
6	.676	.872	1.237	1.635	2.204	10.64	12.59	14.45	16.81	18.55	6
7	.989	1.239	1.690	2.167	2.833	12.02	14.07	16.01	18.48	20.28	7
8	1.344	1.646	2.180	2.733	3.490	13.36	15.51	17.53	20.09	21.95	8
9	1.735	2.088	2.700	3.325	4.168	14.68	16.92	19.02	21.67	23.59	9
10	2.156	2.558	3.247	3.940	4.865	15.99	18.31	20.48	23.21	25.19	10
11	2.603	3.053	3.816	4.575	5.578	17.28	19.68	21.92	24.72	26.76	11
12	3.074	3.571	4.404	5.226	6.304	18.55	21.03	23.34	26.22	28.30	12
13	3.565	4.107	5.009	5.892	7.042	19.81	22.36	24.74	27.69	29.82	13
14	4.075	4.660	5.629	6.571	7.790	21.06	23.68	26.12	29.14	31.32	14
15	4.601	5.229	6.262	7.261	8.547	22.31	25.00	27.49	30.58	32.80	15
16	5.142	5.812	6.908	7.962	9.312	23.54	26.30	28.85	32.00	34.27	16
17	5.697	6.408	7.564	8.672	10.09	24.77	27.59	30.19	33.41	35.72	17
18	6.265	7.015	8.231	9.390	10.86	25.99	28.87	31.53	34.81	37.16	18
19	6.844	7.633	8.907	10.12	11.65	27.20	30.14	32.85	36.19	38.58	19
20	7.434	8.260	9.591	10.85	12.44	28.41	31.41	34.17	37.57	40.00	20
21	8.034	8.897	10.28	11.59	13.24	29.62	32.67	35.48	38.93	41.40	21
22	8.643	9.542	10.98	12.34	14.04	30.81	33.92	36.78	40.29	42.80	22
23	9.260	10.20	11.69	13.09	14.85	32.01	35.17	38.08	41.64	44.18	23
24	9.886	10.86	12.40	13.85	15.66	33.20	36.42	39.36	42.98	45.56	24
25	10.52	11.52	13.12	14.61	16.47	34.38	37.65	40.65	44.31	46.93	25
26	11.16	12.20	13.84	15.38	17.29	35.56	38.89	41.92	45.64	48.29	26
27	11.81	12.88	14.57	16.15	18.11	36.74	40.11	43.19	46.96	49.64	27
28	12.46	13.56	15.31	16.93	18.94	37.92	41.34	44.46	48.28	50.99	28
29	13.12	14.26	16.05	17.71	19.77	39.09	42.56	45.72	49.59	52.34	29
30	13.79	14.95	16.79	18.49	20.60	40.26	43.77	46.98	50.89	53.67	30
40	20.71	22.16	24.43	26.51	29.05	51.81	55.76	59.34	63.69	66.77	40
50	27.99	29.71	32.36	34.76	37.69	63.17	67.50	71.42	76.15	79.49	50
60	35.53	37.48	40.48	43.19	46.46	74.40	79.08	83.30	88.38	91.95	60
70	43.28	45.44	48.76	51.74	55.33	85.53	90.53	95.02	100.4	104.2	70
80	51.17	53.54	57.15	60.39	64.28	96.58	101.9	106.6	112.3	116.3	80
90	59.20	61.75	65.65	69.13	73.29	107.6	113.1	118.1	124.1	128.3	90
100	67.33	70.06	74.22	77.93	82.36	118.5	124.3	129.6	135.8	140.2	100

Source: From E. S. Pearson & H. O. Hartley, (Editors), *Biometrika Tables for Statisticians*, Volume 1, Third Edition (1966). By permission of Biometrika Trustees.

TABLE A9 Rejection Regions for Rank-sum Test for Two Independent Groups

		\multicolumn{8}{c}{One-tailed α}							

		.05		.025		.01		.005	
		\multicolumn{8}{c}{Two-tailed α}							
n_S	n_L	.10		.05		.02		.01	
2	5	$W_S \leq 3, W_S \geq 13$							
	6	3	15						
	7	3	17						
	8	4	18	$W_S \leq 3, W_S \geq 19$					
	9	4	20	3	21				
	10	4	22	3	23				
3	3	6	15						
	4	6	18						
	5	7	20	6	21				
	6	8	22	7	23				
	7	8	25	7	26	$W_S \leq 6, W_S \geq 27$			
	8	9	27	8	28	6	30		
	9	10	29	8	31	7	32	$W_S \leq 6, W_S \geq 33$	
	10	10	32	9	33	7	35	6	36
4	4	11	25	10	26				
	5	12	28	11	29	10	30		
	6	13	31	12	32	11	33	10	34
	7	14	34	13	35	11	37	10	38
	8	15	37	14	38	12	40	11	41
	9	16	40	15	41	13	43	11	45
	10	17	43	15	45	13	47	12	48
5	5	19	36	17	38	16	39	15	40
	6	20	40	18	42	17	43	16	44
	7	21	44	20	45	18	47	17	48
	8	23	47	21	49	19	51	18	52
	9	24	51	22	53	20	55	18	57
	10	26	54	23	57	21	59	19	61
6	6	28	50	26	52	24	54	23	55
	7	29	55	27	57	25	59	24	60
	8	31	59	29	61	27	63	25	65
	9	33	63	31	65	28	68	26	70
	10	35	67	32	70	29	73	28	74
7	7	39	66	36	69	34	71	32	73
	8	41	71	38	74	36	76	34	78
	9	43	76	40	79	37	82	35	84
	10	45	81	42	84	39	87	37	89
8	8	51	85	49	87	46	90	44	92
	9	54	90	51	93	48	96	45	99
	10	56	96	53	99	50	102	47	105
9	9	66	105	63	108	59	112	57	114
	10	69	111	65	115	61	119	59	121
10	10	82	128	78	132	74	136	71	139

Source: From F. Wilcoxon, S. K. Katti, and Roberta Wilcox, Critical values and probability levels for the Wilcoxon rank sum test and the Wilcoxon signed rank test, 1968, published by the Lederle Laboratories Division, American Cynamid Company. Reproduced with the permission of American Cyanamid Company.

TABLE A10 Rejection Regions for Rank-sum Test for Two Dependent Groups

	One-tailed α							
	.05		.025		.01		.005	
	Two-tailed α							
n	.10		.05		.02		.01	
5	$T_+ \leq$ 0,	$T_+ \geq 15$						
6	2	19	$T_+ \leq$ 0,	$T_+ \geq 21$				
7	3	25	2	26	$T_+ \leq$ 0,	$T_+ \geq 28$		
8	5	31	3	33	1	35	$T_+ \leq$ 0,	$T_+ \geq 36$
9	8	37	5	40	3	42	1	44
10	10	45	8	47	5	50	3	52
11	13	53	10	56	7	59	5	61
12	17	61	13	65	10	68	7	71
13	21	70	17	74	12	79	10	81
14	25	80	21	84	16	89	13	92
15	30	90	25	95	19	101	16	104
16	35	101	30	106	23	113	20	116
17	41	112	35	118	28	125	24	129
18	47	124	40	131	33	138	28	143
19	53	137	46	144	38	152	33	157
20	60	150	52	158	43	167	38	172

Source: From F. Wilcoxon, S. K. Katti, and Roberta Wilcox, Critical values and probability levels for the Wilcoxon rank sum test and the Wilcoxon signed rank test, 1968, published by the Lederle Laboratories Division, American Cynamid Company. Reproduced with the permission of American Cyanamid Company.

TABLE A11 Critical Values of *r* (Pearson Correlation)

n	One-tailed α					
	.10	.05	.025	.010	.005	.001
3	.951	.988	.997	1.000	1.000	1.000
4	.800	.900	.950	.980	.990	.998
5	.687	.805	.878	.934	.959	.986
6	.608	.729	.811	.882	.917	.963
7	.551	.669	.754	.833	.875	.935
8	.507	.621	.707	.789	.834	.905
9	.472	.582	.666	.750	.798	.875
10	.443	.549	.632	.715	.765	.847
11	.419	.521	.602	.685	.735	.820
12	.398	.497	.576	.658	.708	.795
13	.380	.476	.553	.634	.684	.772
14	.365	.458	.532	.612	.661	.750
15	.351	.441	.514	.592	.641	.730
16	.338	.426	.497	.574	.623	.711
17	.327	.412	.482	.558	.606	.694
18	.317	.400	.468	.543	.590	.678
19	.308	.389	.456	.529	.575	.662
20	.299	.378	.444	.516	.561	.648
21	.291	.369	.433	.503	.549	.635
22	.284	.360	.423	.492	.537	.622
23	.277	.352	.413	.482	.526	.610
24	.271	.344	.404	.472	.515	.599
25	.265	.337	.396	.462	.505	.588
26	.260	.330	.388	.453	.496	.578
27	.255	.323	.381	.445	.487	.568
28	.250	.317	.374	.437	.479	.559
29	.245	.311	.367	.430	.471	.550
30	.241	.306	.361	.423	.463	.541

TABLE A11 *(continued)*

	One-tailed α					
n	.10	.05	.025	.010	.005	.001
32	.233	.296	.349	.409	.449	.526
34	.225	.287	.339	.397	.436	.511
36	.219	.279	.329	.386	.424	.498
38	.213	.271	.320	.376	.413	.486
40	.207	.264	.312	.367	.403	.474
42	.202	.257	.304	.358	.393	.463
44	.197	.251	.297	.350	.384	.453
46	.192	.246	.291	.342	.376	.444
48	.188	.240	.285	.335	.368	.435
50	.184	.235	.279	.328	.361	.427
55	.175	.224	.266	.313	.345	.408
60	.168	.214	.254	.300	.330	.391
65	.161	.206	.244	.288	.317	.376
70	.155	.198	.235	.278	.306	.363
75	.150	.191	.227	.268	.296	.351
80	.145	.185	.220	.260	.286	.340
85	.140	.180	.213	.252	.278	.331
90	.136	.174	.207	.245	.270	.322
95	.133	.170	.202	.238	.263	.313
100	.129	.165	.197	.232	.256	.305
110	.123	.158	.187	.222	.245	.292
120	.118	.151	.179	.212	.234	.279
130	.113	.145	.172	.204	.225	.269
140	.109	.140	.166	.196	.217	.259
150	.105	.135	.160	.190	.210	.250
160	.102	.131	.155	.184	.203	.243
180	.096	.123	.146	.173	.192	.229
200	.091	.117	.139	.164	.182	.217

Source: From *Statistics Tables* by H. R. Neave (George Allen & Unwin, 1978).

TABLE A12 Fisher's Z

r	Z	r	Z
.00	.000	.50	.549
.01	.010	.51	.563
.02	.020	.52	.577
.03	.030	.53	.590
.04	.040	.54	.604
.05	.050	.55	.618
.06	.060	.56	.633
.07	.070	.57	.648
.08	.080	.58	.663
.09	.090	.59	.678
.10	.100	.60	.693
.11	.110	.61	.709
.12	.121	.62	.725
.13	.131	.63	.741
.14	.141	.64	.758
.15	.151	.65	.775
.16	.161	.66	.793
.17	.172	.67	.811
.18	.182	.68	.829
.19	.192	.69	.848
.20	.203	.70	.867
.21	.214	.71	.887
.22	.224	.72	.908
.23	.234	.73	.929
.24	.245	.74	.950
.25	.256	.75	.973
.26	.266	.76	.996
.27	.277	.77	1.020
.28	.288	.78	1.045
.29	.299	.79	1.071
.30	.309	.80	1.099
.31	.321	.81	1.127
.32	.332	.82	1.157
.33	.343	.83	1.188
.34	.354	.84	1.221
.35	.366	.85	1.256
.36	.377	.86	1.293
.37	.389	.87	1.333
.38	.400	.88	1.376
.39	.412	.89	1.422
.40	.424	.90	1.472
.41	.436	.91	1.528
.42	.448	.92	1.589
.43	.460	.93	1.658
.44	.472	.94	1.738
.45	.485	.95	1.832
.46	.497	.96	1.946
.47	.510	.97	2.092
.48	.523	.98	2.298
.49	.536	.99	2.647

Note. For negative values of r, place a negative sign on the value of Z found in the table.

Source: Reprinted with permission from William H. Beyer (Editor), *Handbook of Tables for Probability and Statistics,* Second Edition (1968). Copyright The Chemical Rubber Co., CRC Press, Inc.

TABLE A13 Confidence Intervals for ρ

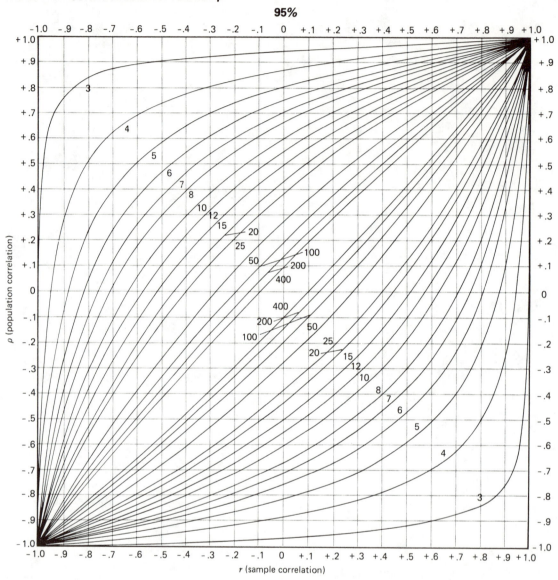

95%

Source: Reprinted with permission from William H. Beyer (Editor), *Handbook of Tables for Probability and Statistics*, Second Edition (1968). Copyright The Chemical Rubber Co., CRC Press, Inc.

TABLE A13 *(continued)*

99%

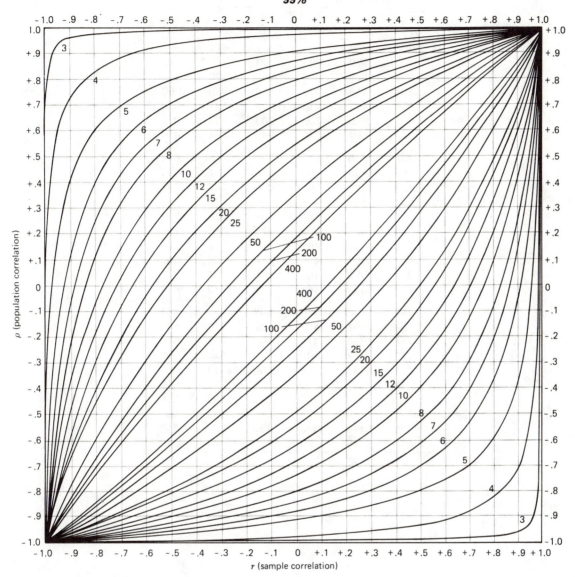

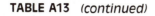

TABLE A14 Critical Values of r_S (Spearman Correlation)

			One-tailed α			
n	.10	.05	.025	.010	.005	.001
4	1.000	1.000	—	—	—	—
5	.800	.900	1.000	1.000	—	—
6	.657	.829	.886	.943	1.000	—
7	.571	.714	.786	.893	.929	1.000
8	.524	.643	.738	.833	.881	.952
9	.483	.600	.700	.783	.833	.917
10	.455	.564	.648	.745	.794	.879
11	.427	.536	.618	.709	.755	.845
12	.406	.503	.587	.678	.727	.818
13	.385	.484	.560	.648	.703	.791
14	.367	.464	.538	.626	.679	.771
15	.354	.446	.521	.604	.654	.750
16	.341	.429	.503	.582	.635	.729
17	.328	.414	.488	.566	.618	.711
18	.317	.401	.472	.550	.600	.692
19	.309	.391	.460	.535	.584	.675
20	.299	.380	.447	.522	.570	.662
21	.292	.370	.436	.509	.556	.647
22	.284	.361	.425	.497	.544	.633
23	.278	.353	.416	.486	.532	.621
24	.271	.344	.407	.476	.521	.609
25	.265	.337	.398	.466	.511	.597
26	.259	.331	.390	.457	.501	.586
27	.255	.324	.383	.449	.492	.576
28	.250	.318	.375	.441	.483	.567
29	.245	.312	.368	.433	.475	.558
30	.240	.306	.362	.425	.467	.549

TABLE A14 *(continued)*

n	One-tailed α					
	.10	.05	.025	.010	.005	.001
31	.236	.301	.356	.419	.459	.540
32	.232	.296	.350	.412	.452	.532
33	.229	.291	.345	.405	.446	.525
34	.225	.287	.340	.400	.439	.517
35	.222	.283	.335	.394	.433	.510
36	.219	.279	.330	.388	.427	.503
37	.215	.275	.325	.383	.421	.497
38	.212	.271	.321	.378	.415	.491
39	.210	.267	.317	.373	.410	.485
40	.207	.264	.313	.368	.405	.479
41	.204	.261	.309	.364	.400	.473
42	.202	.257	.305	.359	.396	.468
43	.199	.254	.301	.355	.391	.462
44	.197	.251	.298	.351	.386	.457
45	.194	.248	.294	.347	.382	.452
46	.192	.246	.291	.343	.378	.448
47	.190	.243	.288	.340	.374	.443
48	.188	.240	.285	.336	.370	.439
49	.186	.238	.282	.333	.366	.434
50	.184	.235	.279	.329	.363	.430
51	.182	.233	.276	.326	.359	.426
52	.180	.231	.274	.323	.356	.422
53	.179	.228	.271	.320	.352	.418
54	.177	.226	.268	.317	.349	.414
55	.175	.224	.266	.314	.346	.411
56	.174	.222	.264	.311	.343	.407
57	.172	.220	.261	.308	.340	.404
58	.171	.218	.259	.306	.337	.400
59	.169	.216	.257	.303	.334	.397
60	.168	.214	.255	.301	.331	.394

Note: For $n > 60$, the critical values in Table A11 (Pearson correlation) may be used.
Source: From *Statistics Tables* by H. R. Neave (George Allen & Unwin 1978).

TABLE A15 Critical Values of F

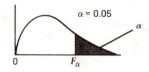

		Numerator Degrees of Freedom df_1							
	1	2	3	4	5	6	7	8	9
1	161.4	199.5	215.7	224.6	230.2	234.0	236.8	238.9	240.5
2	18.51	19.00	19.16	19.25	19.30	19.33	19.35	19.37	19.38
3	10.13	9.55	9.28	9.12	9.01	8.94	8.89	8.85	8.81
4	7.71	6.94	6.59	6.39	6.26	6.16	6.09	6.04	6.00
5	6.61	5.79	5.41	5.19	5.05	4.95	4.88	4.82	4.77
6	5.99	5.14	4.76	4.53	4.39	4.28	4.21	4.15	4.10
7	5.59	4.74	4.35	4.12	3.97	3.87	3.79	3.73	3.68
8	5.32	4.46	4.07	3.84	3.69	3.58	3.50	3.44	3.39
9	5.12	4.26	3.86	3.63	3.48	3.37	3.29	3.23	3.18
10	4.96	4.10	3.71	3.48	3.33	3.22	3.14	3.07	3.02
11	4.84	3.98	3.59	3.36	3.20	3.09	3.01	2.95	2.90
12	4.75	3.89	3.49	3.26	3.11	3.00	2.91	2.85	2.80
13	4.67	3.81	3.41	3.18	3.03	2.92	2.83	2.77	2.71
14	4.60	3.74	3.34	3.11	2.96	2.85	2.76	2.70	2.65
15	4.54	3.68	3.29	3.06	2.90	2.79	2.71	2.64	2.59
16	4.49	3.63	3.24	3.01	2.85	2.74	2.66	2.59	2.54
17	4.45	3.59	3.20	2.96	2.81	2.70	2.61	2.55	2.49
18	4.41	3.55	3.16	2.93	2.77	2.66	2.58	2.51	2.46
19	4.38	3.52	3.13	2.90	2.74	2.63	2.54	2.48	2.42
20	4.35	3.49	3.10	2.87	2.71	2.60	2.51	2.45	2.39
21	4.32	3.47	3.07	2.84	2.68	2.57	2.49	2.42	2.37
22	4.30	3.44	3.05	2.82	2.66	2.55	2.46	2.40	2.34
23	4.28	3.42	3.03	2.80	2.64	2.53	2.44	2.37	2.32
24	4.26	3.40	3.01	2.78	2.62	2.51	2.42	2.36	2.30
25	4.24	3.39	2.99	2.76	2.60	2.49	2.40	2.34	2.28
26	4.23	3.37	2.98	2.74	2.59	2.47	2.39	2.32	2.27
27	4.21	3.35	2.96	2.73	2.57	2.46	2.37	2.31	2.25
28	4.20	3.34	2.95	2.71	2.56	2.45	2.36	2.29	2.24
29	4.18	3.33	2.93	2.70	2.55	2.43	2.35	2.28	2.22
30	4.17	3.32	2.92	2.69	2.53	2.42	2.33	2.27	2.21
40	4.08	3.23	2.84	2.61	2.45	2.34	2.25	2.18	2.12
60	4.00	3.15	2.76	2.53	2.37	2.25	2.17	2.10	2.04
120	3.92	3.07	2.68	2.45	2.29	2.17	2.09	2.02	1.96
∞	3.84	3.00	2.60	2.37	2.21	2.10	2.01	1.94	1.88

Denominator Degrees of Freedom df_2

TABLE A15 *(continued)*

				Numerator Degrees of Freedom df_1						
10	12	15	20	24	30	40	60	120	∞	
241.9	243.9	245.9	248.0	249.1	250.1	251.1	252.2	253.3	254.3	1
19.40	19.41	19.43	19.45	19.45	19.46	19.47	19.48	19.49	19.50	2
8.79	8.74	8.70	8.66	8.64	8.62	8.59	8.57	8.55	8.53	3
5.96	5.91	5.86	5.80	5.77	5.75	5.72	5.69	5.66	5.63	4
4.74	4.68	4.62	4.56	4.53	4.50	4.46	4.43	4.40	4.36	5
4.06	4.00	3.94	3.87	3.84	3.81	3.77	3.74	3.70	3.67	6
3.64	3.57	3.51	3.44	3.41	3.38	3.34	3.30	3.27	3.23	7
3.35	3.28	3.22	3.15	3.12	3.08	3.04	3.01	2.97	2.93	8
3.14	3.07	3.01	2.94	2.90	2.86	2.83	2.79	2.75	2.71	9
2.98	2.91	2.85	2.77	2.74	2.70	2.66	2.62	2.58	2.54	10
2.85	2.79	2.72	2.65	2.61	2.57	2.53	2.49	2.45	2.40	11
2.75	2.69	2.62	2.54	2.51	2.47	2.43	2.38	2.34	2.30	12
2.67	2.60	2.53	2.46	2.42	2.38	2.34	2.30	2.25	2.21	13
2.60	2.53	2.46	2.39	2.35	2.31	2.27	2.22	2.18	2.13	14
2.54	2.48	2.40	2.33	2.29	2.25	2.20	2.16	2.11	2.07	15
2.49	2.42	2.35	2.28	2.24	2.19	2.15	2.11	2.06	2.01	16
2.45	2.38	2.31	2.23	2.19	2.15	2.10	2.06	2.01	1.96	17
2.41	2.34	2.27	2.19	2.15	2.11	2.06	2.02	1.97	1.92	18
2.38	2.31	2.23	2.16	2.11	2.07	2.03	1.98	1.93	1.88	19
2.35	2.28	2.20	2.12	2.08	2.04	1.99	1.95	1.90	1.84	20
2.32	2.25	2.18	2.10	2.05	2.01	1.96	1.92	1.87	1.81	21
2.30	2.23	2.15	2.07	2.03	1.98	1.94	1.89	1.84	1.78	22
2.27	2.20	2.13	2.05	2.01	1.96	1.91	1.86	1.81	1.76	23
2.25	2.18	2.11	2.03	1.98	1.94	1.89	1.84	1.79	1.73	24
2.24	2.16	2.09	2.01	1.96	1.92	1.87	1.82	1.77	1.71	25
2.22	2.15	2.07	1.99	1.95	1.90	1.85	1.80	1.75	1.69	26
2.20	2.13	2.06	1.97	1.93	1.88	1.84	1.79	1.73	1.67	27
2.19	2.12	2.04	1.96	1.91	1.87	1.82	1.77	1.71	1.65	28
2.18	2.10	2.03	1.94	1.90	1.85	1.81	1.75	1.70	1.64	29
2.16	2.09	2.01	1.93	1.89	1.84	1.79	1.74	1.68	1.62	30
2.08	2.00	1.92	1.84	1.79	1.74	1.69	1.64	1.58	1.51	40
1.99	1.92	1.84	1.75	1.70	1.65	1.59	1.53	1.47	1.39	60
1.91	1.83	1.75	1.66	1.61	1.55	1.50	1.43	1.35	1.25	120
1.83	1.75	1.67	1.57	1.52	1.46	1.39	1.32	1.22	1.00	∞

TABLE A15 *(continued)*

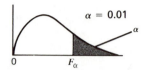

$\alpha = 0.01$

	Numerator Degrees of Freedom df_1								
	1	2	3	4	5	6	7	8	9
1	4052.	4999.5	5403.	5625.	5764.	5859.	5928.	5982.	6022.
2	98.50	99.00	99.17	99.25	99.30	99.33	99.36	99.37	99.39
3	34.12	30.82	29.46	28.71	28.24	27.91	27.67	27.49	27.35
4	21.20	18.00	16.69	15.98	15.52	15.21	14.98	14.80	14.66
5	16.26	13.27	12.06	11.39	10.97	10.67	10.46	10.29	10.16
6	13.75	10.92	9.78	9.15	8.75	8.47	8.26	8.10	7.98
7	12.25	9.55	8.45	7.85	7.46	7.19	6.99	6.84	6.72
8	11.26	8.65	7.59	7.01	6.63	6.37	6.18	6.03	5.91
9	10.56	8.02	6.99	6.42	6.06	5.80	5.61	5.47	5.35
10	10.04	7.56	6.55	5.99	5.64	5.39	5.20	5.06	4.94
11	9.65	7.21	6.22	5.67	5.32	5.07	4.89	4.74	4.63
12	9.33	6.93	5.95	5.41	5.06	4.82	4.64	4.50	4.39
13	9.07	6.70	5.74	5.21	4.86	4.62	4.44	4.30	4.19
14	8.86	6.51	5.56	5.04	4.69	4.46	4.28	4.14	4.03
15	8.68	6.36	5.42	4.89	4.56	4.32	4.14	4.00	3.89
16	8.53	6.23	5.29	4.77	4.44	4.20	4.03	3.89	3.78
17	8.40	6.11	5.18	4.67	4.34	4.10	3.93	3.79	3.68
18	8.29	6.01	5.09	4.58	4.25	4.01	3.84	3.71	3.60
19	8.18	5.93	5.01	4.50	4.17	3.94	3.77	3.63	3.52
20	8.10	5.85	4.94	4.43	4.10	3.87	3.70	3.56	3.46
21	8.02	5.78	4.87	4.37	4.04	3.81	3.64	3.51	3.40
22	7.95	5.72	4.82	4.31	3.99	3.76	3.59	3.45	3.35
23	7.88	5.66	4.76	4.26	3.94	3.71	3.54	3.41	3.30
24	7.82	5.61	4.72	4.22	3.90	3.67	3.50	3.36	3.26
25	7.77	5.57	4.68	4.18	3.85	3.63	3.46	3.32	3.22
26	7.72	5.53	4.64	4.14	3.82	3.59	3.42	3.29	3.18
27	7.68	5.49	4.60	4.11	3.78	3.56	3.39	3.26	3.15
28	7.64	5.45	4.57	4.07	3.75	3.53	3.36	3.23	3.12
29	7.60	5.42	4.54	4.04	3.73	3.50	3.33	3.20	3.09
30	7.56	5.39	4.51	4.02	3.70	3.47	3.30	3.17	3.07
40	7.31	5.18	4.31	3.83	3.51	3.29	3.12	2.99	2.89
60	7.08	4.98	4.13	3.65	3.34	3.12	2.95	2.82	2.72
120	6.85	4.79	3.95	3.48	3.17	2.96	2.79	2.66	2.56
∞	6.63	4.61	3.78	3.32	3.02	2.80	2.64	2.51	2.41

(Denominator Degrees of Freedom df_2)

TABLE A15 *(continued)*

				Numerator Degrees of Freedom df_1						
10	12	15	20	24	30	40	60	120	∞	
6056.	6106.	6157.	6209.	6235.	6261.	6287.	6313.	6339.	6366.	1
99.40	99.42	99.43	99.45	99.46	99.47	99.47	99.48	99.49	99.50	2
27.23	27.05	26.87	26.69	26.60	26.50	26.41	26.32	26.22	26.13	3
14.55	14.37	14.20	14.02	13.93	13.84	13.75	13.65	13.56	13.46	4
10.05	9.89	9.72	9.55	9.47	9.38	9.29	9.20	9.11	9.02	5
7.87	7.72	7.56	7.40	7.31	7.23	7.14	7.06	6.97	6.88	6
6.62	6.47	6.31	6.16	6.07	5.99	5.91	5.82	5.74	5.65	7
5.81	5.67	5.52	5.36	5.28	5.20	5.12	5.03	4.95	4.86	8
5.26	5.11	4.96	4.81	4.73	4.65	4.57	4.48	4.40	4.31	9
4.85	4.71	4.56	4.41	4.33	4.25	4.17	4.08	4.00	3.91	10
4.54	4.40	4.25	4.10	4.02	3.94	3.86	3.78	3.69	3.60	11
4.30	4.16	4.01	3.86	3.78	3.70	3.62	3.54	3.45	3.36	12
4.10	3.96	3.82	3.66	3.59	3.51	3.43	3.34	3.25	3.17	13
3.94	3.80	3.66	3.51	3.43	3.35	3.27	3.18	3.09	3.00	14
3.80	3.67	3.52	3.37	3.29	3.21	3.13	3.05	2.96	2.87	15
3.69	3.55	3.41	3.26	3.18	3.10	3.02	2.93	2.84	2.75	16
3.59	3.46	3.31	3.16	3.08	3.00	2.92	2.83	2.75	2.65	17
3.51	3.37	3.23	3.08	3.00	2.92	2.84	2.75	2.66	2.57	18
3.43	3.30	3.15	3.00	2.92	2.84	2.76	2.67	2.58	2.49	19
3.37	3.23	3.09	2.94	2.86	2.78	2.69	2.61	2.52	2.42	20
3.31	3.17	3.03	2.88	2.80	2.72	2.64	2.55	2.46	2.36	21
3.26	3.12	2.98	2.83	2.75	2.67	2.58	2.50	2.40	2.31	22
3.21	3.07	2.93	2.78	2.70	2.62	2.54	2.45	2.35	2.26	23
3.17	3.03	2.89	2.74	2.66	2.58	2.49	2.40	2.31	2.21	24
3.13	2.99	2.85	2.70	2.62	2.54	2.45	2.36	2.27	2.17	25
3.09	2.96	2.81	2.66	2.58	2.50	2.42	2.33	2.23	2.13	26
3.06	2.93	2.78	2.63	2.55	2.47	2.38	2.29	2.20	2.10	27
3.03	2.90	2.75	2.60	2.52	2.44	2.35	2.26	2.17	2.06	28
3.00	2.87	2.73	2.57	2.49	2.41	2.33	2.23	2.14	2.03	29
2.98	2.84	2.70	2.55	2.47	2.39	2.30	2.21	2.11	2.01	30
2.80	2.66	2.52	2.37	2.29	2.20	2.11	2.02	1.92	1.80	40
2.63	2.50	2.35	2.20	2.12	2.03	1.94	1.84	1.73	1.60	60
2.47	2.34	2.19	2.03	1.95	1.86	1.76	1.66	1.53	1.38	120
2.32	2.18	2.04	1.88	1.79	1.70	1.59	1.47	1.32	1.00	∞

Index

Prepared with the assistance of Grant Austin.

THE PRINCIPLES

P1: Summarization Principle.
Statistical summaries should
    (a) Be as useful as possible.
    (b) Represent the data fairly.
    (c) Be easy to interpret.
When the summary is a frequency distribution, the intervals should be "nice". (Section 2.3)

P2: The Principle of Reasonableness.
Check the results of all calculations for reasonableness. The way to do this depends on the problem. You may have to look at the distribution of scores, or at a graph, or you may have to apply some theoretical result. You may even have to invent a new way to check the reasonableness of your answer. (Section 3.6)

P3: The Principle of Usefulness.
Report a useful and appropriate number of digits in any calculated number. The useful number of digits depends principally on the number of digits in the original data and on the size of the data set. It also depends on the use to which the calculated result will be put. Digits that are essentially "noise" should not be reported. (Section 3.6)

P4: The Variability Principle.
The standard deviation is approximately one-half the 68% range. This approximation is quite accurate for unimodal distributions that are not extremely skewed. The principle applies to both empirical and idealized distributions. (Section 4.3)

P5: Characteristics of Tree Diagrams.
A series of choices or possibilities can be represented by a tree diagram. A set of choices, among which one choice is to be made, is called a *stage* of the diagram. The individual choices at any one stage are called *branches* of the tree. A complete series of choices, one choice at each stage, is called a *path* through the tree. The tree diagram as a whole (all the paths) represents all possible series of choices. (Section A.3)

P6: The Basic Principle of Statistical Inference.
In order to make an inference about a population parameter from a statistic computed from a random sample, we must consider the distribution of that statistic. The distribution of the statistic is derived from the probabilities of all samples in the sample space. The inference about the population parameter is made on the basis of (1) the distribution of the statistic and (2) the value of the statistic in the sample. (Section 6.4)

P7: The Central Limit Theorem.
The distribution of the sample mean, based on random samples of size n drawn from a population with mean μ and standard deviation σ, has the following characteristics: the mean, $\mu_{\bar{y}}$, is exactly equal to the population mean, μ; the standard deviation, $\sigma_{\bar{y}}$, is exactly equal to the population standard deviation divided by the square root of the sample size, $\sigma/\sqrt{n}$; the shape is approximately normal. The approximation of the shape to normality improves rapidly with increasing sample size such that for $n > 10$, the shape can be taken to be normal. Furthermore, if the population is normally distributed, the distribution of the sample mean is exactly normal, even for small sample sizes. (Section 7.5)

P8: Definition of Statistics.
Statistics is a set of techniques for making inferences about a single variable or about the relationship between two or more variables from observations of a sample drawn from a population. (Section 8.1)

P9: Statistical Significance and Scientific Significance.
The statistical significance of an observed result indicates which of two statistical hypotheses about a population is supported by the result, given a certain probability of a Type I error. The statistical hypotheses are expressed in terms of the variables measured in the study. The scientific significance of a result goes beyond a simple report of statistical significance. The justification of scientific significance requires much more than just statistical significance. The scientific significance is expressed in terms of more general or theoretical variables and may take into account the results of other studies. (Section 10.2)

P10: The Basic Principle of Experimental Design.
The relationship between an independent and a dependent variable can be investigated in an experiment involving two or more groups. Each group is defined by a condition, that is, by a value of the independent (manipulated) variable. The subjects are measured on the dependent (observed) variable. The effect of each irrelevant variable is minimized by (*a*) holding the variable constant for all subjects or (*b*) randomizing the effect of the variable or (*c*) matching the subjects on the variable. (Section 11.3)

P11: Estimation and Hypothesis Testing.
A result stated as a test of a hypothesis can be reexpressed as an estimate, and vice versa. For many purposes, an estimate with its standard error (or a confidence interval estimate) is more useful than a statement of the significance of the result. (Section C.1)

P12: Statistical Significance and Effect Size.
A statistically significant result shows that there is an effect, but the effect may be large or small. In evaluating a statistical study, it is important to consider both the effect size and whether the result is statistically significant. (Section C.4)

P13: The Principle of Prediction.
Consider the prediction of a dependent variable from one (or more than one) independent variable. The total variance of the dependent variable equals the sum of the following two variances: the variance of that part of the dependent variable which can be predicted from the independent variable(s) (the predicted variance) and the variance of that part of the dependent variable that cannot be predicted from the independent variable(s) (the unpredicted variance). In brief, $s'^2_{Tot} = s'^2_{Pred} + s'^2_{Unpred}$. (Section 16.3)

P14: Definition of Interaction.
The interaction effect of two factors (independent variables) on a third (dependent) variable is the joint effect of the two factors that cannot be accounted for by the main effects of the two factors separately. There is an interaction if the effect of one factor is not the same for all levels of the other factor. Interaction may be defined in terms of the regression functions of the dependent variable, *y*, on one factor—say, *A*—for different levels of the other factor—say, *B*. *There is no interaction* if all the regression functions of *y* on *A* are parallel: that is, they are the same except for the addition or subtraction of a constant. *There is an interaction* if one pair (or more) of the regression functions of *y* on *A* are not parallel: that is, if the differences between the pair of regression functions are not constant. (Section 17.8)